Millets and Millet Technology

Anil Kumar • Manoj Kumar Tripathi •
Dinesh Joshi • Vishnu Kumar
Editors

Millets and Millet Technology

Springer

Editors
Anil Kumar
Directorate of Education
Rani Lakshmi Bai Central Agricultural University
Jhansi, Uttar Pradesh, India

Manoj Kumar Tripathi
ICAR-Central Institute of Agricultural Engineering
Bhopal, Madhya Pradesh, India

Dinesh Joshi
Division of Crop Improvement
ICAR-Vivekananda Institute of Hill Agriculture
Almora, Uttarakhand, India

Vishnu Kumar
Department of Genetics & Plant Breeding
Rani Lakshmi Bai Central Agricultural University
Jhansi, Uttar Pradesh, India

ISBN 978-981-16-0675-5 ISBN 978-981-16-0676-2 (eBook)
https://doi.org/10.1007/978-981-16-0676-2

This Springer imprint is published by the registered company Springer Nature Singapore Pte Ltd.
The registered company address is: 152 Beach Road, #21-01/04 Gateway East, Singapore 189721, Singapore

Foreword

World is facing agrarian as well as nutritional challenges. Globally, more than 800 million people are undernourished while >2 billion people have one or more chronic macro- and micronutrient deficiencies (MNDs) notably the minerals calcium (Ca), iodine (I), iron (Fe), selenium (Se), zinc (Zn), and vitamins (e.g., vitamin A). Food-based solutions for dealing with micronutrient deficiencies, although extremely challenging, are potentially sustainable, affordable, effective, and feasible approaches for addressing macro- and micronutrient malnutrition. The whole grains of millets are rich in protein, minerals, and excellent source of other biologically important molecules. Millets rank sixth in the world cereal grain production. In Africa and Asia, these underutilized grains play a major role in the food security of millions of people. They can strongly resist the conditions of drought and even can grow in rain fed region. India is known to be the leading producer of both large and small millets. Considering their climate resilience, role in nutritional and health security, the Government of India has declared the year 2018 as "National year of Millets" and year 2023 as "International Year of Millets" by United Nations. These nutri-cereals harbor vitamins, minerals, essential fatty acids, phyto-chemicals, and antioxidants can help to eradicate the plethora of nutritional deficiency diseases. Millets cultivation can keep dry lands productive and ensure future food and nutritional security.

Millets are orphan crops with tremendous potential but underexplored source of nutraceutical properties as compared to other regularly consumed cereals. Regular consumption of millets can reduce the chance of various live-threatening diseases such as diabetes, obesity, cardiovascular diseases, osteoporosis, and even age-associated diseases. It is not common in our diets so the chance of incorporating it into various types of food products holds a vast scope to study and research for scientific rationalization of its health healing properties and moreover millets can also probably transform food products into magical food products, i.e., super foods using various agri-processing and other modern technologies integrating the fundamental knowledge of genomics, bioinformatics, biotechnology, and nanotechnology. Promotion of millets production and value addition contribute significantly to meet many of the Sustainable Development Goals (SDGs). As significant scientific advances in regulatory, technological, and policy changes, mainstreaming of millets for human welfare will be needed. This book *Millet and Millet Technology* will be proven useful for developing thoughts and solutions that will not be achieved unless enough resources are made available for research and implementation.

In order to aid to nutritional security, development of the novel, value-added food, and healthcare products are the need of time. Therefore, using traditional and modern or advance techniques to process millets for value addition or convenient food products can be a great idea to introduce it for better consumption. Due to its reasonably high grain mineral contents and nutritionally good quality, millets can be used for formulating diets for pregnant and lactating women as well as for growing children. Bioactive substances in foods can represent "extra-nutritional" constituents naturally present in small quantities in the food matrix, produced upon either in vivo or industrial enzymatic digestion, the latter being a result of food-processing activities.

It gives me great honor and pleasure to write this Foreword to the latest book of Dr. Anil Kumar and his editorial team known to me for many years as they are part of our own Central Agricultural University and Indian Council of Agricultural Research where I earlier served as Deputy Director General (Education). This, third book of Dr. Kumar's group from Springer Nature, indicates his keen interest in providing with his tradition of thoroughly reading and conceptualization of thought-provoking topics written by different authors of this book. This book having 21 chapters is a milestone and will evoke a great interest to researchers and students of the various academic institutions, scientists, processors, and consumers.

Rani Lakshmi Bai Central Agricultural University
Jhansi, Uttar Pradesh, India
Arvind Kumar

Preface

Globalization and industrialization of agriculture although benefitted with adequate food supplies, its negative side effects have also been realized particularly with respect to loss of agro-biodiversity and sustenance of minor crops. Narrowing of diversity in crop species contributing to the world's food supplies has been considered a potential threat to global food security. Only 12 crops provide 75% of the world's food supplies and three major crops, rice, wheat, and maize, provide 50% of global dietary requirements. While these crops are the primary suppliers of carbohydrates, they lack essential amino acids and minerals for a balanced nutrition. This lack of nutritional quality in human diet affects millions of people worldwide leading to hidden hunger, which results from insufficient vitamins and micronutrients, e.g., zinc (Zn), magnesium (Mg), and iron (Fe) intake. For instance, nearly half of the world's micronutrient deficient population live in developing countries. This is mainly because of overdependence on starchy food such as rice, wheat, and maize. Supplementation of major cereals with alternative crops possessing better nutritive value and nutraceutical properties would be an effective strategy for dietary diversification and reduction of hidden hunger.

Sustainable crop alternatives are needed to meet the world hunger and to increase the income of farmers. Role of millets cannot be ignored for achieving sustainable means for nutritional security. Millets are the crops with high potential as source of underexplored nutraceutical properties and have the potential to act as alternative food-grain in most parts of the world. The gluten-free nature of protein, low glycemic index, high concentration of micronutrients, bioactive flavonoids, and small peptides makes them promising crops for the future agricultural systems to combat hidden hunger. Additionally, the C4 photosynthetic pathway and the ability to withstand environmental stress and low input marginal agriculture make them golden crops of future. Millets are not only the staple food but also serve as an important source of high-quality fodder for livestock in semi-arid regions of the world. In order to promote millets for nutritional and health security due to their richness in nutritionally and biologically important ingredients, it is mandatory to bring them in mainstream of cultivation and value addition. The role of millets in formulating the modern foods like multigrain and gluten-free cereal products is well known. Besides agro-processing, various modern sciences like biotechnology and nanotechnology can also be used to convert agricultural produce into the form which is most acceptable by consumers. Usually, with the help of such technologies, essential nutrients can be added to farm produce or agricultural commodities can be produced in their organic state. Also, the agricultural produce can be modified so as to increase bioavailability of some essential nutrients. As a result of this transformation, the nutritional quality of food is increased and farmers can earn more from their agriculture by selling their produce at higher price to consumers. On the other hand, consumers are ready to pay more price for value-added products as they want to get high-quality nutritious and healthy products to avoid nutrition-related deficiencies or lifestyle-related diseases. Due to increased awareness regarding the health-promoting profile of millets, preference towards their consumption has been observed. Because the consumers have become more health conscious, they are willing to pay more price to buy quality products. Thus, the book *Millet and Millet Technology* is an attempt to put forward various technological approaches by which millets can be proposed as nutraceutical crops from its status of orphan crops. Such opportunities can create employment opportunities for weaker sections of society especially women and rural youths who can establish small industries for the development and marketing of value-added products.

In recent years, there has been a revived interest in cultivation of millets for their use as potential nutritional substitute in food formulations, pharmaceutical use, animal feed and for commercial starch production. Calling them as "climate resilient crops" and "powerhouse of nutrients" the Government of India has declared eight millet crops as nutri-cereals and the year 2018 as "National Year of Millets." Owing to their tremendous agricultural potential, the Food and Agriculture Organization (FAO) declared 2023 as the "International Year of Millets." Keeping in view the global scenario, this book is an attempt to bring together the historical perspective, present status, and future outlook of millets. This book is an attempt to increase awareness among the masses about diversifying and adapting our cropping systems to future consumer needs (qualitative food security) and a changing environment. Our suggestions in the form of this book are widely transferable to many minor crops of immense nutraceutical potential, which have been neglected in the last decades.

The edited book, *Millet and Millet Technology*, comprising 21 chapters will cover the current research and development in the field of millet nutrition, food processing, and novel processing technologies; value-added products, human health, omics approach for grain quality improvement and nutraceutical traits, quality management, and millets-based entrepreneurship development. Chapter 1 of the book is basically an introduction to millets to readers. It gives a cursory impression of the way millets have been adopted and utilized by different societies in the course of history. It further highlights the nutritional and medicinal importance of millets for a healthy lifestyle. Finally, it provides a brief overview of policies to design and execute millet-based climate resilient agricultural technologies. Chapter 2 is devoted to the global production scenario of minor millets and their economic importance. It provides an overview of major production constraints associated with millets and suggested future prospects that can accelerate millet production and productivity.

Minor millets have always been an integral part of tribal agriculture. They are the heritage grains adapted to ecological niches of tribal agriculture where crop substitution is very tough. Therefore, understanding the ethno-botany, traditional knowledge, and its protection in natural habitat is essential for ensuring future food and nutritional security. All these aspects are narrated in Chap. 3. Asia and Africa are the home of the world's largest malnourished population. It has been observed that proportion of malnutrition among the vulnerable section of the society is quite high in these countries. For instance, proportion of children with stunted growth and anemic pregnant women is quite high in these countries compared to the rest of the world. Millets being the native of Asia and Africa have the potential to solve the malnutrition problem of the world. Therefore, Chaps. 4 and 5 are an attempt to understand the unparallel macro- and micronutrient profile, flavonoids, and nutraceutical potential of minor millets for their effective utilization in combating hidden hunger and malnutrition. Millets are nutritionally similar or superior to some major cereal grains. The additional benefits of the millets like gluten-free proteins, high fiber content, low glycemic index, and richness in bioactive compounds made them a suitable healthy food. Similarly, millet starch has very high medicinal and commercial value. The slow digestion of millet starch results in low glycemic load on human belly; therefore, it is an ideal food for people suffering from diabetes and obesity. Chapter 6 covering this aspect is an attempt to compile the latest advances in understanding the structural and functional properties of millet starch. It has been estimated that 97% of the millets are consumed where they are produced. India is one of the largest producers and consumers of millets in the world. They are heavily consumed in the form of various traditional dishes from the southern states to North West and North Eastern Himalayan region. Chapter 7 has covered the consumption pattern of millets in the country.

Millet grains are the source of high-quality protein and nutritive substitute for cereal protein. Millet seed protein is especially preferred as a substitute of wheat in diets of patients with celiac disease due to its gluten-free properties. Therefore, Chap. 8 is the holistic compilation of biologically and nutritionally important proteins and bioactive peptides of millets. While the most widely known health benefit of millets is that the grains are effectively gluten-free, there are several reports of the presence of a variety of bioactive compound with potential nutraceutical properties in their grains. A detailed account of such compounds is dealt with in Chap. 9.

Chapter 10 explores the genomics approaches such as large-scale genotyping and next-generation sequencing for millets improvement and searching genes governing traits of economic importance. Processing is the most important technology for value addition of millets. Therefore, Chaps. 11–14 cover the various aspects associated with millet processing and value addition technology. In Africa, alcoholic and nonalcoholic beverages made from malted and sprouted finger millet, pearl millet, and sorghum are widely consumed. For instance, Nigeria is the largest producer of sorghum beer. Chapter 15 of the book provides a detailed account of millet-based traditional food beverages. Diabetes mellitus is a metabolic disorder characterized by hyperglycemia owing to insufficient or inefficient insulin secretion. Globally, half a billion people are living with diabetes, and in India it is increasing at an alarming rate. The nutraceutical potency of millets aids in lowering the glycemic index (GI) of food containing millets. Therefore, Chap. 16 of the book is specifically devoted to the various millet-based value-added food products for diabetics, their chemical composition, and glycemic index. Chapter 17 provides the current knowledge and emerging insights of genomics, transcriptomics, proteomics, and metabolomics for improving the grain quality like micronutrient accumulation, protein quality, and bioactive compounds.

The processing aspects and development of complete value chain of millets have a huge scope for entrepreneurship development of rural livelihoods promotion. All these developments and strategies are narrated in Chap. 18. In the last decades, several incidents have occurred in the agro-food sectors, such as an incidence of food-borne diseases and production of higher risk food products. Owing to their unparallel nutraceutical potential, millet grains are being used in the development of various food products. Therefore, quality management systems occur for millets from many legal forums, such quality management system utilizing multidisciplinary approach for millets is elaborated in Chap. 19. The last two chapters (Chaps. 20 and 21) of the book deal with the concept of value chain models, demand creation measures, generic incubation, and different business and incubation opportunities possible in the development of value chain on millets.

The chapters of the book have been authored by various experts keeping in view syllabi of different research institutions, researchers, students as well as requirement of the industry. This book will serve as an instructional material for researchers in food science, microbiology, process engineering, biochemistry, biotechnology, and reference material for those working in industry and R & D labs. Efforts have been made to avoid overlapping in contents; however, some overlapping in isolated spots is unavoidable.

We express our sincere thanks to all the contributors of the 21 chapters. We sincerely appreciate their continuous cooperation starting from first submission of drafts to revision of the chapters as suggested by editors. We also thank our family members for bearing us throughout the process of editing of this book.

Jhansi, Uttar Pradesh, India — Anil Kumar
Bhopal, Madhya Pradesh, India — Manoj Kumar Tripathi
Almora, Uttarakhand, India — Dinesh Joshi
Jhansi, Uttar Pradesh, India — Vishnu Kumar

Contents

Editors and Contributors

About the Editors

Anil Kumar is Director, Education, Rani Lakshmi Bai Central Agricultural University, Jhansi (UP). He has 30 years of research, teaching and administrative experience in various capacities. He had worked as Professor & Head, (2006–2018), Associate Professor (2003–2006) and Assistant Professor (1993–1999) Dept. of Molecular Biology & Genetic Engineering, G.B. Pant University of Agriculture & Technology, Pantnagar. The research interests focus on Plant Molecular & Cell Biology, Functional Genomics, Immunodiagnostic applications in Agriculture, Host-Pathogen Interaction and solving problems of agricultural importance employing the *OMICS* tools and techniques of molecular biology and immunology. He has to his credit more than 250 research papers, reviews, and also several chapters contributed to books, popular articles, 16 Books/Bulletins/Manuals. He has been recipient of several prestigious awards and honors of national importance and also a highly cited scientist who contributed for augmenting biotechnology education in the country.

Manoj Kumar Tripathi is currently working as Principal Scientist at ICAR-Central Institute of Agricultural Engineering, Bhopal, India. He has 17 years of research, teaching and administrative experience. His major areas of professional interest include medicinal compounds, functional Food, nutraceutical, probiotic fermented food, food safety and food quality. He has published more than 80 research papers and 85 abstracts in International and National journal of repute, more than 40 book chapters, 30 popular articles. He has authored/ edited 12 books and 5 technical bulletins. He has been recipient of several prestigious awards and honours. He is also working as an editor-In-chief, editor and member of editorial board of more than 10 reputed National and International journals.

Dinesh Joshi is currently working as Senior Scientist at ICAR-Vivekananda Institute of Hill Agriculture, Almora, Uttarakhand, India. He worked on large scale molecular characterization of millets (sorghum) and established unique molecular identification profiles of important sorghum landraces. Currently, his area of specialization includes breeding and genomics of small millets and pseudo-cereals and attained recognition as excellent millets' plant breeder. He has to his credit 30 referred research papers and 15 book chapters.

Vishnu Kumar is currently working as Associate Professor (Genetics & Plant Breeding) at Rani Lakshmi Bai Central Agricultural University, Jhansi. His area of specialization is genetic enhancement of coarse cereals and millets for grain physical and quality traits coupled with grain yield improvement and biotic stress resistance. He has published 70 research papers in referred journals.

Contributors

Aparna Agrawal Department of Molecular Biology & Genetic Engineering, G.B. Pant University of Agriculture and Technology, Pantnagar, Uttarakhand, India

Sanjeev Agrawal Department of Biochemistry, G.B. Pant University of Agriculture and Technology, Pantnagar, Uttarakhand, India

Chhavi Arya Department of Home Science, DSB Campus, Kumaun University, Nainital, Uttarakhand, India

M. I. Bhat Animal Biochemistry Division, ICAR-National Dairy Research Institute, Karnal, Haryana, India

Venkatesh Bhat ICAR-Indian Institute of Millets Research, Hyderabad, Andhra Pradesh, India

Anju Bisht Department of Home Science, MB Government PG College, Nainital, Uttarakhand, India

J. K. Bisht ICAR-Vivekananda Institute of Hill Agriculture, Almora, Uttarakhand, India

T. S. Bisht Department of Horticulture, School of Agriculture and Allied Sciences, HNB Garhwal University, Srinagar Garhwal (A Central University), Srinagar Garhwal, Uttarakhand, India

Subir Kumar Chakraborty Agro Produce Processing Division, ICAR-Central Institute of Agricultural Engineering, Bhopal, Madhya Pradesh, India

Shalini Chakraborty Fruit Research Station (Enthkhedi), Rajmata Vijayaraje Scindia Agricultural University, Gwalior, Madhya Pradesh, India

V. Ratnavathi Chamarthy ICAR-Indian Institute of Millets Research, Hyderabad, Andhra Pradesh, India

Ajay Chandra G.B. Pant University of Agriculture and Technology, Pantnagar, Uttarakhand, India

Rakesh Kumar Choudhary Rani Lakshmi Bai Central Agricultural University, Jhansi, Uttar Pradesh, India

B. Dayakar Rao ICAR-Indian Institute of Millets Research, Hyderabad, Andhra Pradesh, India

S. D. Deshpande ICAR-Central Institute of Agricultural Engineering, Bhopal, Madhya Pradesh, India

Sumedha Deshpande ICAR-Central Institute of Agricultural Engineering, Bhopal, Madhya Pradesh, India

Supriya Gupta Department of Molecular Biology and Genetic Engineering, College of Basic Sciences & Humanities, G.B. Pant University of Agriculture & Technology, Pantnagar, Uttarakhand, India

Kavita Gururani Department of Molecular Biology and Genetic Engineering, College of Basic Sciences & Humanities, G.B. Pant University of Agriculture & Technology, Pantnagar, Uttarakhand, India

Rajpal S. Jadam ICAR-Central Institute of Agricultural Engineering, Bhopal, Madhya Pradesh, India

Dinesh Joshi ICAR-Vivekananda Institute of Hill Agriculture, Almora, Uttarakhand, India

Mohammad Hassan Kamani Department of Grain Science and Technology, CSIR-Central Food Technological Research Institute, Mysore, Karnataka, India

Lakshmi Kant ICAR-Vivekananda Institute of Hill Agriculture, Almora, Uttarakhand, India

Rajeev Kapila Animal Biochemistry Division, ICAR-National Dairy Research Institute, Karnal, Haryana, India

Suman Kapila Animal Biochemistry Division, ICAR-National Dairy Research Institute, Karnal, Haryana, India

A. K. Karnatak V.C.S.G. Uttarakhand University of Horticulture and Forestry, Bharsar, Uttarakhand, India

Adinath Kate ICAR-Central Institute of Agricultural Engineering, Bhopal, Madhya Pradesh, India

Akshit Kukreti Plant Pathology Division, College of Forestry, Rancihauri, V.C.S. G. Uttarakhand University of Horticulture and Forestry, Bharsar, Uttarakhand, India

Anil Kumar Directorate of Education, Rani Lakshmi Bai Central Agricultural University, Jhansi, Uttar Pradesh, India

Vishnu Kumar Department of Genetics & Plant Breeding, Rani Central Lakshmi Bai Agricultural University, Jhansi, Uttar Pradesh, India

Rajendra Prasad Meena ICAR-Vivekananda Institute of Hill Agriculture, Almora, Uttarakhand, India

M. S. Meera Department of Grain Science and Technology, CSIR-Central Food Technological Research Institute, Mysore, Karnataka, India

K. K. Mishra ICAR-Vivekananda Institute of Hill Agriculture, Almora, Uttarakhand, India

Debabandya Mohapatra ICAR-Central Institute of Agricultural Engineering, Bhopal, Madhya Pradesh, India

T. Niranjan ICAR-Indian Institute of Millets Research, Hyderabad, Andhra Pradesh, India

P. K. Nishad ICAR-Central Institute of Agricultural Engineering, Bhopal, Madhya Pradesh, India

Sri Devi Nune Nutrihub, ICAR-Indian Institute of Millets Research, Hyderabad, Andhra Pradesh, India

Dinesh Pandey Department of Molecular Biology and Genetic Engineering, College of Basic Sciences & Humanities, G.B. Pant University of Agriculture & Technology, Pantnagar, Uttarakhand, India

Sharad Pandey Himgiri Zee University, Dehradun, Uttarakhand, India

Amit U. Paschapur ICAR-Vivekananda Institute of Hill Agriculture, Almora, Uttarakhand, India

Rajesh Kumar Pathak Department of Biotechnology, GovindBallabh Pant Institute of Engineering & Technology, Pauri Garhwal, Uttarakhand, India

Laxmi Rawat Plant Pathology Division, College of Forestry, Rancihauri, V.C.S.G. Uttarakhand University of Horticulture and Forestry, Bharsar, Uttarakhand, India

Jyoti Semwal Department of Grain Science and Technology, CSIR-Central Food Technological Research Institute, Mysore, Karnataka, India

Pallavi Shah Department of Molecular Biology and Genetic Engineering, G.B. Pant University of Agriculture and Technology, Pantnagar, Uttarakhand, India

Divya Sharma Department of Molecular Biology & Genetic Engineering, G.B. Pant University of Agriculture and Technology, Pantnagar, Uttarakhand, India

Anupama Singh National Institute of Food Technology and Entrepreneurship Management, Sonipat, Haryana, India

Prabha Singh ICAR-Indian Grassland and Fodder Research Institute, Jhansi, Uttar Pradesh, India

Vaishali Singh SHAUTS, Prayagraj, Uttar Pradesh, India

Sarita Srivastava Department of Food and Nutrition, College of Home Science, G. B. Pant University of Agriculture and Technology, Pantnagar, Uttarakhand, India

M. Sujatha ICAR-Indian Institute of Millets Research, Hyderabad, Andhra Pradesh, India

Sushma Tiwari Rajmata Vijayaraje Scindia Krsihi Vishva Vidyalaya, Gwalior, Madhya Pradesh, India

Apoorv Tiwari Department of Computational Biology and Bioinformatics, Jacob Institute of Biotechnology and Bio-Engineering, Sam Higginbottom University of Agriculture, Technology and Sciences, Allahabad, Uttar Pradesh, India

Ram Sewak Singh Tomar Department of Biotechnology & Crop Improvement, Rani Lakshmi Bai Central Agricultural University, Jhansi, Uttar Pradesh, India

Vilas A. Tonapi ICAR-Indian Institute of Millets Research, Hyderabad, Andhra Pradesh, India

Manoj Kumar Tripathi ICAR-Central Institute of Agricultural Engineering, Bhopal, Madhya Pradesh, India

Vivek Chandra Verma DIHAR-DRDO, Chandigarh, India

Millets for Life: A Brief Introduction

1

Amit U. Paschapur, Dinesh Joshi, K. K. Mishra, Lakshmi Kant, Vishnu Kumar, and Anil Kumar

Abstract

Millets or nutri-cereals are high-energy foods; that were domesticated and cultivated as early as 10,000 years ago. The millets cultivation is taken up usually in degraded and marginal lands that receive very less rainfall and are poor in soil nutrient content. Seven important millets cultivated globally are finger millet, pearl millet, foxtail millet, barnyard millet, proso millet, kodo millet, and little millet. Overdependence on cereals after the green revolution and the present-day sedentary lifestyle of people has proliferated health-related disorders like obesity, diabetes, coronary diseases, gastrointestinal disorders and risk of colon, breast, and oesophageal cancer. The only way to fight back is through the introduction of nutritionally rich millets in our daily diets. Millets are unique for their richness in dietary fibers, antioxidants, minerals, phytochemicals, polyphenols, and proteins; that act as elixir to fight against health-related disorders. Recent global phenomenon of climate change has lead to a decrease in the yield of major staple cereals and has paved path for introduction of millets into agriculture production system to formulate climate resilient cropping systems because millets are C4 plants with very superior photosynthetic efficiency, short duration, higher dry matter production capacity, and a high degree of tolerance to heat and drought. Keeping the

A. U. Paschapur · D. Joshi (✉) · K. K. Mishra · L. Kant
ICAR-Vivekananda Parvatiya Krishi Anusandhana Sansthan, Almora, Uttarakhand, India
e-mail: amit.paschapur@icar.gov.in; Dinesh.Joshi@icar.gov.in; Lakshmi.Kant@icar.gov.in

V. Kumar
Department of Genetics & Plant Breeding, Rani Lakshmi Bai Central Agricultural University, Jhansi, Uttar Pradesh, India

A. Kumar
Directorate of Education, Rani Lakshmi Bai Central Agricultural University, Jhansi, Uttar Pradesh, India

A. Kumar et al. (eds.), *Millets and Millet Technology*,
https://doi.org/10.1007/978-981-16-0676-2_1

above advantages of millets, the efforts have hastened to collect, conserve, and utilize germplasm of millets in breeding programs. Of late, several private and government agencies have ventured into value addition of millets to manufacture food and non-food products. But, the governments have a key role in formulating policies to promote cultivation and consumption of millets.

Keywords

Millets · C4 plants · Health related disorders · Climate resilient cropping system · Food and non-food products · Government policies · Climate resilience · Chronic disorders

1.1 Introduction

Millets also called small millets are cultivated for their small kernels which are the products of small grassy plants belonging to the Poaceae family. The other name minor millets may indicate them to be minor crops yet are important for their nutritional values, medicinal benefits, feed for animals, and saviors during food crisis (Joshi and Agnihotri 1984; Yenagi et al. 2010). The word "millet" has originated from the French word "Mile" meaning thousand which implies a handful of millets contain thousands of grains. Millets are often grown in semi-arid conditions with very less rainfall and marginal or degraded lands with very low nutrient contents. The crops support the livelihood of people in areas where famine is a regular phenomenon and the millets yield a more dependable harvest compared to other crops in low rainfall areas (Tadele 2016). Millets are C4 plants with very superior photosynthetic efficiency, short duration, higher dry matter production capacity, and a high degree of tolerance to heat and drought. They also easily adapt to degraded saline, acidic and aluminum toxic soils (Yadav and Rai 2013). These extraordinary characters of millets make them suitable crops to tackle the hurdles of climate change and formulate millet-based climate resilient technologies.

The modern sedentary lifestyle associated with several health issues has urged people to seek for healthy and nutritious diets. Small millets satiate these requirements of modern society by being a healthy food choice because millets are a storehouse of nutrient and, in particular, finger millet grains contain remarkably high calcium content (>350 mg/100 g); foxtail millet, barnyard millet, and proso millet are prosperous in protein (>10%); little millet and foxtail millet are well-off in fat (>4.0%); foxtail millet, barnyard millet, and little millet are superior in crude fiber (6.7–13.6%), barnyard millet and little millet contain high amount of iron (9.3–18.6 mg/100 g) in comparison to other major cereals like rice, wheat, barley, maize, and sorghum (Dwivedi et al. 2012; Kam et al. 2016). The increase in demands of millets in both national and international markets have triggered the interests of researchers to collect, conserve, and utilize the germplasm available globally for their important traits in crop improvement, development of genomic resources, and value addition. Recently several private organizations have ventured into value addition and marketing chain of millets that has boosted millet cultivation and consumption. In this process, the government has a role to play by formulating suitable policies in order to motivate cultivation, marketing, and consumption of millets to achieve food and nutritional security.

There are no less than 14 species of millets belonging to 10 genera, that include pearl millet (*Pennisetum glaucum* L.), foxtail millet (*Setaria italica* L. subsp. *italica*), Finger millet (*Eleusine coracana* L.), barnyard millet (*Echinocloa esculenta* A. and *Echinocloa colona* L.), proso millet (*Panicum miliaceum* L. subsp. *miliaceum*), kodo millet (*Paspalum scrobiculatum* L.), and little millet (*Panicum sumatrense* Roth.) that are cultivated widely throughout the world. However, the study of literature on millets is very cumbersome because of different common names and vernacular names given to the same species of millets and they are usually studied as minor cereals. In this section, we will be listing seven important millets cultivated in the world, their centers of origin, common names or vernacular names, along with area and production in India and the world (Table 1.1).

Pearl millet (*Pennisetum glaucum* L.): Bajra or Pearl millet is estimated to be originated as early as 5000 years in Africa (Andrews and Kumar 1992) and was introduced to the Indian subcontinent around 3000 years ago. The crop is well adapted to adverse environmental conditions with rainfall less than 250 mm and temperature of 30 °C and above and mainly grown by subsistence farmers throughout Africa, Asia, and Australia. Recently the crop is gaining importance as a commercial crop in Australia and accounts for almost half of the world's area under millets (National Research Council 1996).

Foxtail millet (*Setaria italica* L. subsp. *italica*): The probable center of origin of Foxtail millet or Italian millet is China; however, the crop is known to be domesticated during Neolithic culture. It is among one of the ancient cereals cultivated in Europe and Asia, with China contributing more than 45% of world production (Jiaju and Yuzhi 1994). The crop is well adapted to cooler climates and matures in less than 70–120 days.

Proso millet (*Panicum miliaceum* L. subsp. *miliaceum*): Broom millet or Proso millet has probably originated in Manchurian region of China (House 1995) and presently cultivated in northwest China, southern and central parts of India, Australia, the USA, and Europe. It is the third most important millet crop cultivated after pearl millet and foxtail millet and it is well adapted to temperate climatic conditions up to altitudes of 3500 m and various soil types (Baltensperger 2002).

Finger millet (*Eleusine coracana* L.): The ear heads of the crop resemble finger of human hand thus giving it the name. The probable origin of Ragi or Finger millet is highlands of Ethiopia and Uganda (National Research Council 1996). Asia and Africa are major centers of production and India is the leading producer in the world. The crop is adapted to tropical climates with an intermediate altitude (500–2400 m) and low to moderate rainfall (500–1000 mm). The crop can thrive under dry and hot conditions up to 35 °C in well-drained soils. The grains of finger millet can be stored for up to 50 years thus serving as a good reserve against famine (National Research Council 1996).

Table 1.1 Important millets cultivated in India and the world, along with their area and production (2016–2017)

Common name	Scientific name	Vernacular names	Area in world (million ha)	Production in world (million tons)	Center of origin	Major countries	Area in India (million ha)	Production in India (million tons)	Major states	Uses
Pearl millet	*Pennisetum glaucum* L.	Bajra, Bulrush millet, cattail millet, Babala,	27.16	23.09	Africa Sahelian zone (western Sudan to Senegal)	India, Western and Central Africa, Eastern and Southern Africa	7.12	10.28	Rajasthan, Uttar Pradesh, Gujarat, and Madhya Pradesh	Grown for food grain and fodder
Finger millet	*Eleusine coracana* L.	Ragi, Wimbi	2.11	3.42	Highlands of Uganda and Ethiopia	India, Nepal, Sri Lanka, Rwanda, Ethiopia, Uganda, Malawi, Burundi	1.14	1.82	Karnataka, Uttarakhand, Tamilnadu, Maharashtra	Grown for food grain and beer making
Barnyard millet	*Echinocloa esculenta* A.	Japanese barnyard millet	0.15	0.15	Japan	India, Japan, China, Malaysia	0.14	0.146	Madhya Pradesh, Chhattisgarh, Uttarakhand, Maharashtra, and Karnataka	Grown for human food and animal fodder
Shama millet	*Echinocloa colona* L.	Sawa millet, Awnless barnyard grass, Jungle rice Corn panic grass, Jungle			India	India				

		ricegrass, Deccan grass								
Kodo millet	*Paspalum scrobiculatum* L.	Creeping paspalum, Indian paspalum, Water couch, Ditch millet	0.20	0.084	India	India	0.19	0.082		Grown for human food and animal fodder
Proso millet	*Panicum miliaceum* L. subsp. *miliaceum*	Broom millet, Panic millet, Common millet, Hog millet	0.94	1.45	Manchuria	The USA, Russia, India, Iran, Ukraine, South Korea, Poland, Belarus, Kazakhstan, France	0.003	0.002		Grown for human food and animal fodder
Foxtail millet	*Setaria italica* L. subsp. *italica*	Italian millet, Foxtail bristle millet	1.06	2.29	China	China, Myanmar, India, Eastern Europe	0.072	0.052		Grown for human food and animal fodder
Little millet	*Panicum sumatrense*	Blue panic, heen	0.25	0.12	Eastern Ghats of India	India	0.25	0.12	Karnataka, Tamilnadu,	Grown for food

(continued)

Table 1.1 (continued)

Common name	Scientific name	Vernacular names	Area in world (million ha)	Production in world (million tons)	Center of origin	Major countries	Area in India (million ha)	Production in India (million tons)	Major states	Uses
		meneri, Sajje							Kerala, Andra Pradesh	grain and fodder

Source: Directorate of Economics & Statistics, DAC&FW *4th Advance Estimates; IIMR estimates based on FAO/DES-GOI data, Taylor and Emmambux 2008

Barnyard Millet or Sawa millet (*Echinocloa esculenta* A. and *Echinocloa colona* L.): The Barnyard millet or Japanese millet has originated in Japan province, whereas sawa millet was domesticated in the Indian subcontinent (House 1995). Both millets belong to the same genus and their morphology is similar. The crops prefer warm climatic conditions but can be cultivated in cold temperatures too. The cultivation of barnyard millet is mainly taken up in Japan, China, Korea, and India and the crop is known for its good storability.

Kodo Millet (*Paspalum scrobiculatum* L.): or Ditch millet is known to be domesticated 3000 years ago and is indigenous to India (House 1995). Kodo millet is majorly produced in India and the production accounts for 90% of total world production (Hedge and Chandra 2005). Although kodo millet is well adapted to tropical and sub-tropical climatic conditions, the crop takes 120–180 days to mature and the grain yields are very low (250–1000 kg/ha).

Little millet (*Panicum sumatrense* Roth.): Eastern Ghats of India are known to be the place of domestication of Little millet as early as 2000 years ago and the crop is majorly cultivated in peninsular Indian states like Andhra Pradesh, Karnataka, Tamil Nadu, and Kerala. The crop is adapted to both dry and humid conditions and can be cultivated in drought-prone areas as well as water-logged conditions, as the crop matures early and withstands adverse conditions. The genetic diversity of this crop is very little because of its restricted cultivation in India, Sri Lanka, Nepal, and Myanmar, with India accounting for more than 98% of the area and production of little millet.

1.2 History of Millets in India and World

Millets are among the oldest crops domesticated and cultivated in the world for human food and animal fodder and their cultivation dates back to 8700–10,300 years ago (Lu et al. 2009a, b). Various millet species were initially domesticated in different parts of the world like South Asia, East Asia, East Africa, and West Africa. However, they spread well beyond their initial area of domestication. Whereas, the earliest record of millets domestication and cultivation is of Foxtail millet and Proso millet from China around 3000–2000 BC and Indian valley of Kashmir is regarded as a place of integrated networks, where major trade of millets took place between Asia, Europe, and Africa (Oelke et al. 1990). The history regarding the origin of millets, their domestication, and cultivation is very vast and difficult to summarize in a small section, so important dates in the history are briefly covered in the current section.

1.2.1 History in India

The cultivation of millets in India is as old as human civilization. The record of cultivation of foxtail millet during Harappan civilization, Pearl millet in the Neolithic period in South India (2000–1200 BC), kodo millet and finger millet during early

Iron age (1200–1000 BC), little millet, native small millet (*Setaria* spp.), and browntop millet during the later Harappan period (2500–2000 BC), Browntop millet and bristly foxtail millet (*Setaria verticillata*) during Neolithic-Chalcolithic period are some of the recorded proofs of Indian history. There are also specific records like

Harappan civilization: the foxtail millet spread from China and its cultivation started during Harappan civilization in India. Around 2500–2200 BC (Harappan levels) the cultivation of foxtail millet started in Shikarpur (Kutch) and around 1900–1400 BC (late Harappan levels) the cultivation began in Punjab.

Yajurveda or Indian Bronze age (1500 BC): The mention of millets foxtail millet (priyangava), proso millet (aanava), and Barnyard millet (shyaamaka) in Indian Sanskrit text Yajurveda's verses, indicated that millet cultivation and consumption was very common in India (Roy 2009).

Ancient Indian texts: There is a mention of millet cultivation in ancient Indian texts like Sushruta Samhita (600–500 BC)—classification of cereals into millets, Charaka Samhita (100–200 AD)—Sorghum, Vishnu Purana (450 AD)—classification of cereals and millets, Abhijnana Shakuntalam (400–500 AD)—foxtail millet and Ramadhanya Charithre (1600 AD)—finger millet.

1.2.2 History of the World

Cultivation of millets in China and Africa cover the major history of world millet cultivation, wherein foxtail millet cultivation started in China during 3000–2000 BC and was later spread to other regions of the world (Lu et al. 2009a, b; Lawler 2009).

The Pearl millet was cultivated as early as 2500 BC in the Sahel region of West Africa and moreover the cultivation of finger millet in highlands of East Africa dates back to 1800 BC (Engels et al. 1991), wherein vast majority of the people were dependent on the cultivation of millets for their livelihood. But, Palaeoethnobotanists could not find any evidence of millets in their surveys in ancient Egyptian tombs (Krishnaswamy 1938).

There are several literatures showing the recent introduction of millets to European countries, i.e., between 1850 and 1950 AD. But, the cultivation of millets in European countries dates back to the Late Bronze age (3000 BC), wherein in the region of Macedonia and northern Greece the millets were growing in wild and their seeds were stored by the farmers and used as food (Nesbitt and Summers 1988). Megasthenes (350–290 BC) the Greek historian revealed about millet cultivation after the summer solstice, along with sorghum and wheat.

1.3 Nutritional Values of Millets

In the last few decades, the incidence of obesity and diabetes has increased exponentially due to excessive consumption of processed junk foods. To combat them there is rise in demand for foods containing high amounts of complex carbohydrates, dietary fibers, and beneficial phytochemicals (Shobana et al. 2007). Although, the research is in progress to biofortify the wholegrain cereals like wheat and rice, with phenolic acids that impart antimutagenic, antiglycemic, and antioxidative properties, the high content of gluten in these cereals is an obstacle to develop healthy foods or nutraceuticals (Friedman 1997; Jones and Engleson 2010; Chandrasekara and Shahidi 2010). Keeping these drawbacks of wholegrain cereals and the growing public awareness of health care and nutrition, there is an immediate need to identify novel sources of nutraceuticals, natural foods, and nutritional materials with desirable functional characters.

Millets are a good substitute for the major cereals and an elixir to overcome health problems like obesity and diabetes. They are distinctive among cereals because of their richness in dietary fibers, antioxidants, phytochemicals, proteins, and polyphenols (Devi et al. 2011; Muthamilarasan et al. 2016). Table 1.2 represents the nutritional value of minor millets and their comparison with major non-millet cereals, along with mineral content, amino acid composition, and vitamin concentration. From Table 1.2, we can conclude that millets are a very rich source of nutrients and are significantly nutritionally richer than widely cultivated and consumed cereals like rice, wheat, maize, and sorghum (Obilana and Manyasa 2002). Millets are a rich source of calcium (10–348 mg/100 g), iron (2.2–17.7 mg/100 g), zinc (32.7–60.6 mg/100 g), and phosphorus (200–339 mg/100 g), vitamins such as thiamine (0.15–0.60 mg/100 g), niacin (0.09–1.11 mg/100 g), and riboflavin (0.28–1.65 mg/100 g), that makes them a perfect energy food (Kumar et al. 2018). The millets are non-glutinous, so easy to digest. The carbohydrate concentration in most millets ranges from 60% to 70% and having a major portion as non-starchy polysaccharides adding to many health benefits of millets (Shivran 2016). The lower carbohydrate content and non-glutinous property make them most suitable diets for diabetic patients. In particular, finger millet is the richest source of calcium (344 mg/100 g), magnesium (137 mg/100 g), potassium (408 mg/100 g), sodium (11 mg/100 g), and phenolic compounds (0.3–3%). The barnyard millet and foxtail millet have the highest mineral nutrient concentration even among millets, while barnyard millet contains six times higher mineral content than rice. The millets are also rich in fats and millet oil is a good source of oleic acid, linoleic acid, palmitic acid, and tocopherols that are known to play a major role in the immune system and defense mechanism in the human body (Liang et al. 2010; Amadou et al. 2011). Moreover, the millets are a rich source of most essential amino acids like methionine, cysteine, and lysine thus providing special dietary benefits to vegetarians dependent on plant foods to complete their dietary requirements (FAO 2012). However, the millets do contain phytates (0.48%), polyphenols, tannins (0.61%), trypsin inhibitory factors, and dietary fiber, which were earlier considered as "anti-nutrient elements" due to their metal chelating and enzyme inhibition activities (Thompson 1993) but the recent studies have termed them as nutraceutical compounds with antioxidant, metal chelating, and metal reducing power (Chandrasekara and Shahidi 2010).

Apart from grains, the seed coat of millets is an edible component that is rich in phytochemicals, such as polyphenols (0.2–3.0%) and dietary fiber (Hadimani and Malleshi 1993; Ramachandra et al. 1977). It is now recognized that polyphenols,

Table 1.2 Nutritional values of minor millets in comparison with major cereals (composition per 100 g edible portion)

Nutrients	Pearl millet	Finger millet	Foxtail millet	Proso millet	Little millet	Barnyard millet	Kodo millet	Rice (white milled, Raw)	Wheat	Sorghum	Maize	Oats
Energy (kcal)	361	336	331	341	329	300	353	345	346	329		
Protein (g)	11.6	7.7	12.3	12.5	7.7	6.2	8.3	6.8	11.8	10.6	12.1	17.1
Fat (g)	5	1.5	4.3	1.1	4.7	2.2	1.4	0.5	1.5	3.5	4.6	6.4
Ash (g)	2.2	2.7	3.3	1.9	1.5	4.4	2.6	0.6	1.5	1.6	1.8	3.2
Crude fiber (g)	2.3	3.6	8.0	7.2	7.6	9.8	9.0	0.2	1.2	2.0	2.3	11.3
Carbohydrate (g)	67.5	72.6	60.9	70.4	67.0	65.5	65.9	78.2	71.2	72.1	62.3	52.8
Phenols (mg/100 g)	51.4	102	106	–	–	–	368	2.51	20.5	43.1	2.91	1.2
Minerals (g)	2.3	2.7	3.3	1.9	1.5	4.3	2.6	0.6	1.5	1.6		
Minerals and trace elements (mg/100 g dry matter)												
Phosphorous (mg)	296	283	290	206	251	340	215	160	298	222	290	380
Magnesium (mg)	137	137	81	153	133	82	166	64	138	165	140	130
Calcium (mg)	42	350	31	14	12	21	31	10	30	13	30	11
Iron (mg)	8	3.9	2.8	0.8	13.9	9.2	3.6	0.7	3.5	3.36	3.1	6.2
Zinc (mg)	3.1	2.3	2.4	1.4	3.5	2.6	1.5	1.3	2.7	1.7	2.0	3.7
Copper (mg)	1.06	0.47	1.60	5.80	1.6	1.30	5.80	–	–	–	–	–
Manganese (mg)	1.15	5.94	0.6	0.6	1.03	1.33	2.9	0.51	2.29	0.78	0.50	0.45
Molybdenum (mg)	0.069	0.102	0.7	–	–	–	–	0.05	0.051	0.039	–	–
Chromium (mg)	0.023	0.028	0.070	0.040	0.240	0.140	0.080	–	–	–	–	–

Sodium (mg)	10.9	11	4.6	8.2	–	–	–	–	17.1	2	3.0	2.0
Potassium (mg)	307	408	250	113	–	–	–	–	284	363	370	470
Amino acid composition (mg per gram protein)												
Isoleucine	256	275	475	405	416	288	188	181	148	173	154	164
Leucine	598	594	1044	762	679	725	419	345	269	523	480	287
Lysine	214	181	138	189	114	106	188	166	130	89	126	166
Methionine	154	194	175	160	142	133	94	103	56	54	85	62
Cysteine	148	163	–	–	–	175	–	101	101	70	97	136
Phenyl alanine	301	325	419	307	297	362	375	222	179	218	199	192
Tyrosine	203	–	–	–	–	150	213	199	125	175	166	160
Threonine	122	191	263	61	49	35	63	144	125	133	148	137
Tryptophan	122	191	61	49	35	63	38	55	43	99	16	39
Valine	345	413	431	407	379	388	238	269	199	221	211	227
Vitamins												
Thiamine (mg)	0.33	0.42	0.59	0.2	0.41	0.30	0.33	0.41	0.45	0.33	0.38	0.77
Riboflavin (mg)	0.25	0.19	0.11	0.18	0.28	0.09	0.10	0.06	0.17	0.096	0.14	0.14
Niacin (mg)	2.3	1.1	3.2	2.3	–	–	0.70	1.9	5.5	3.7	2.80	0.97
Total folic acid (ug)	45.5	18.3	15	–	–	–	–	8	36.6	20	–	–
Vitamin E (mg)	–	22	–	–	–	–	–	–	–	0.5	–	–

Source: Shivran (2016), Kaur et al. (2019), Gopalan et al. (2004), USDA National Nutrient Database for Standard Reference, Release 28 (2016), Muthamilarasan et al. (2016)

phytates, and tannins add to antioxidant activity of millet foods, which is an imperative factor in human health, aging, and metabolic diseases (Bravo 1998). Considering the nutritional richness of millets, they have to be an essential constituent of human diet in order to augment people's diet diversity and maintaining good nutrition.

1.4 Medicinal Values or Health Benefits of Millets

As seen in the previous section, the millets are known for their superior nutritional values along with several health benefits, which are attributed to their richness in polyphenols, dietary fibers, non-glutinous as well as non-starchy carbohydrate content making them superior over major wholegrain cereals (Muthamilarasan et al. 2016). Millets are also documented for their beneficial health effects like antioxidant activity, anti-diabetic, anti-tumorigenic, atherosclerogenic effects, and antimicrobial properties (Yang et al. 2012). Cardiovascular diseases, risk of type II diabetes, gastrointestinal cancers, and a range of other disorders can be kept at bay by regular consumption of whole grain millets and their products (McKeown 2002). Based on several epidemiological studies, consumption of millets improves the digestive system, detoxifies the body, lowers the risk of cancer, increases energy levels, increases immunity in respiratory health, and improves neural and muscular systems. Millet consumption also protects against several degenerative diseases such as Parkinson's disease and metabolic syndrome (Manach et al. 2005; Chandrasekara and Shahidi 2012). Usually, millets are consumed along with seed coat that is rich in minerals, dietary fiber, vitamins and phenolics, which offer better health benefits to humans than other wholegrain cereals (Antony et al. 1996). Some major health benefits of individual millets in our daily diet are furnished in Table 1.3.

1.4.1 Millets Are Cure for Coronary Diseases

The millet grains are the powerhouse of nutrition that help in improving heart health and effectively trim down the coronary blockage. They are enriched with magnesium, potassium, and plant lignins, which effectively reduce blood pressure by acting as a vasodilator and decrease heart attacks and other cardiovascular risks. The high fiber content of millets lowers the cholesterol level thus eliminating LDL (Low Density Lipoprotein) from the system and increases the positive effects of HDL (High Density Lipoprotein) in the body (Kumari and Thayumanavan 1997; Park et al. 2008; Lee et al. 2010).

1.4.2 Millets Can Manage Sugar Level to Tackle Diabetes

Diabetes mellitus is a chronic metabolic disorder characterized by hyperglycemia with alterations in carbohydrate, protein, and lipid metabolism. It is a common

Table 1.3 Important millets and their medicinal values

Millet	Medicinal importance	References
Pearl millet	Turns the gut condition to alkaline and cures stomach ulcers, high amounts of magnesium and potassium control blood pressure and relieve heart diseases, magnesium also reduces respiratory problems and migraine attack, high phosphorous content helps in bone growth and development in kids, high amount of dietary fiber and slow release of glucose maintains blood sugar level and more suitable for diabetic patients, phytic acids reduce the cholesterol levels of body, hypoallergic properties make it a suitable diet for lactating mothers, infants, elderly people, and convalescents.	Shobana et al. (2007), Taylor and Emmambux (2008), Chandrasekara and Shahidi (2010), Lee et al. (2010), Saleh et al. (2013), Muthamilarasan et al. (2016)
Finger millet	High amounts of phenolic acids have anti-ulcerative properties, lower blood sugar level and cholesterol, phenolic compounds are nephron protective and anti-cataractogenic, germinated seeds improved hemoglobin level in infants, protection against epithelialization, mucosal ulceration, increases the synthesis of collagen, activation of fibroblasts and mast cells, tryptophan lowers appetite and keeps weight in control, high amount of calcium strengthens bones, lecithin, and methionine eliminate excess fat from liver and thus reduce cholesterol level in the body, high amount of iron protects from anemia.	Antony et al. (1996), Friedman (1997), Kumari and Thayumanavan (1997), Chandrasekara and Shahidi (2010), Lee et al. (2010), Mohamed et al. (2012), Saleh et al. (2013)
Foxtail millet	Soluble and insoluble bound phenolic extracts present in the seeds show antioxidant, metal chelating, and metal reducing powers, they reduce toxicity caused by xenobiotics and toxins in the body, high amount of proteins and essential amino acids helps in building body tissues and advised for infants and elderly people.	Hedge and Chandra (2005), Devi et al. (2011), Muthamilarasan et al. (2016)
Proso millet	High amount of copper facilitates the body to form red blood cells, helps maintain blood vessels, healthy bones, nerves, and immune function, and contributes to iron assimilation. Sufficient copper in the diet prevents cardiovascular diseases and osteoporosis. Magnesium reduces	Dykes and Rooney (2006), Taylor and Emmambux (2008), Chandrasekara and Shahidi (2012), Saleh et al. (2013), Muthamilarasan et al. (2016)

(continued)

Table 1.3 (continued)

Millet	Medicinal importance	References
	respiratory problems and migraine attack, potassium controls blood pressure and relieves heart diseases.	
Kodo millet	Phenolic compounds have anti-ulcerative properties, lower blood sugar level and cholesterol, Magnesium and potassium control blood pressure and relieve heart diseases, magnesium also reduces respiratory problems and migraine attack.	Shobana et al. (2007), Taylor and Emmambux (2008), Chandrasekara and Shahidi (2012)
Barnyard millet	The richness of phenolic acids, tannins, phytates, and dietary fibers show antimutagenic and anti-carcinogenic properties, high amounts of dietary fiber reduce the risk of colon cancer and oesophageal cancer, phosphorous content helps in bone growth and development in kids.	Bravo (1998), Hedge and Chandra (2005), Devi et al. (2011), Mohamed et al. (2012), Shivran (2016)
Little millet	High amounts of iron help to safeguard many fundamental functions in the body, including general energy and focus, the immune system, gastrointestinal processes, and the regulation of body temperature. Higher amounts of zinc aid in enzymatic reactions, immune function, wound healing, DNA and protein synthesis, and regular growth and development of the body.	Antony et al. (1996), McKeown (2002), Chandrasekara and Shahidi (2010), Lee et al. (2010)

endocrine disorder that adversely affects insulin production; thereby bringing about an imbalance in sugar levels in the body. Millets are a rich source of magnesium that helps in stimulating the level of insulin in the body and thus increase the glucose receptors' efficiency in the body. This in turn helps in a healthy balance of sugar level in the body thus reducing the chance of type I and type II diabetes (Saleh et al. 2013; Kumari and Sumathi 2002; Shahidi and Chandrasekara 2013; Sarita and Singh 2016; Devi et al. 2014).

1.4.3 Millets Promote Digestion and Cure Gastrointestinal Disorders

Regular and excessive consumption of gluten-rich foods may lead to increased nutrient retention and severe gastrointestinal disorders like colon cancer and gastric ulcers. Consumption of fiber and phenolics rich millets along with their seed coat eliminates the disorders like excess gas, bloating, constipation, and cramping in the gastrointestinal tract and shrinks other diseases related to the kidney and liver (Taylor et al. 2006; Taylor and Emmambux 2008; Thompson 2009; Chandrasekara and Shahidi 2010; Catassi and Alarida 2011; Muthamilarasan et al. 2016).

1.4.4 Millets Help in Detoxification (Antioxidant Properties)

Millets are rich in components like curcumin, ellagic acid, quercetin, and various other beneficial catechins that assist the system to clear any toxins and xenobiotics by promoting appropriate excretion and neutralizing enzymatic activity (Tsao 2010). The antioxidant, metal chelating, and metal reducing powers are shown by various soluble and insoluble bound phenolic millet extracts (Chandrasekara and Shahidi 2010). More than 50 phenolic compounds belonging to several classes, namely, phenolic acids and their derivatives, dehydrodiferulates and dehydrotriferulates, flavonols, and flavones, flavan-3-ol monomers and dimmers, etc., are greatly crammed in millets and can be utilized as functional food ingredients and as sources of natural antioxidants (Hedge and Chandra 2005; Dykes and Rooney 2006; Devi et al. 2011; Mohamed et al. 2012).

1.4.5 Millets Protect from Cancer

The richness of millet grains with phenolic acids, tannins, phytates, and dietary fibers show antimutagenic and anti-carcinogenic properties (Watanabe 1999). Regular consumption of millets reduces the risk of colon cancer, breast cancer, and oesophageal cancer (Van Rensburg 1981; Shobana and Malleshi 2007). Few recent researches have revealed that consumption of 30 g dietary fiber every day by women can reduce the chance of breast cancer almost by 50%.

1.5 Important Food and Non-food Products Prepared from Millets

The millets are cultivated in semi-arid conditions on marginal lands by small and marginal farmers and consumed throughout Asia and Africa for more than 3000 years. But, recently millets have gained huge demands in national and international markets because of their superior nutritional values along with several health benefits. There are a large number of traditional millet-based foods and beverages prepared and consumed in different parts of Asia, Africa, the Indian subcontinent, and East Asia. They can be characterized into wholegrain foods, meal or flour-based foods, nonalcoholic and alcoholic beverages (Taylor and Emmambux 2008).

1.5.1 Food Products from Millets

1.5.1.1 Whole Grain Foods

Popping the grains: The popping process involves moistening the grains to 19% moisture content, tempering the grains for several hours, agitating the grains in hot sand bed (240 °C) for a few minutes (Malleshi et al. 1996). The outer pericarp is removed and the popped grains are consumed as a snack or further processed by milling, e.g., pearl millet, finger millet, and foxtail millet.

Germinated seeds: The seeds are soaked overnight and allowed to germinate; the protein, mineral, and vitamin content increases after seed germination. The seeds are then consumed raw or cooked. Germinated seeds are advised for infants and elderly people, e.g., finger millet, little millet, and kodo millet.

Direct cooking: Rice like product called Kichadi is prepared by cooking the whole grains (Subramanian and Jambunathan 1980), e.g., Pearl millet, Foxtail millet, Kodo millet, and Little millet.

1.5.1.2 Foods made from Flour

Flatbreads: These are staple foods in Africa. The millet flours undergo a specialized fermentation treatment with yeast and lactic acid bacteria that yields soft leavened textured bread with acidic flavor (Gashe et al. 1982). The two most famous flatbreads in Africa are kisra and injera that are thin, flexible wafers with a spongy texture and are relished with spicy sauce (Badi et al. 1990; Yetneberk et al. 2004).

Rotis or chapatis: These are the most well-known unfermented flatbreads made from millets and are popular staple foods of India. Rotis or chapatis resemble a soft pancake with a flexible puffed texture. They are usually served with pickles, vegetables, chutney, meat, or sauce (Murty and Kumar 1995).

Dosa and Idli: These products are popular in southern parts of India, the semi-fermented millet flour is used in making dosas and idlis, that are served with sambar or chutney (Murty and Kumar 1995).

Couscous: It is pasta-like culinary prepared from semolina of millets in North Africa. The semolina is steamed and agglomerated stirred with yogurt and consumed. Usually, the couscous products are categorized based on the size of the particle of semolina used.

Dumplings and other dough products: In India and Africa the boiled dough dumplings traditionally prepared of pearl millet flour and finger millet flour are popular (Quin 1959; Malleshi et al. 1996). The dumplings in Africa are called dingwa and in India they are called mudde or ragi balls. In India, ponganumis is a snackfood prepared by frying steamed millet dough (Subramanian and Jambunathan 1980).

Porridges: In Africa porridges are prepared from millet meals and there are an almost countless range of conventional porridges consumed throughout the world. The porridges are usually classified based on their consistency (runny to stiff), solid content of the porridge (10–30%), serving temperature (very hot to cold) and their pH ranges (Alkaline to acidic) (Rooney et al. 1997; Lin et al.

1998), e.g., porridges prepared from flours of foxtail millet (China), pearl millet (South Africa), finger millet (Sahel region of Africa), and little millet (India).

1.5.1.3 Alcoholic Beverages from Millets

Opaque beers: The alcoholic beverages are popular in Africa and India. The traditional African beers are prepared by fermenting pearl millet and finger millet grains. The beers are usually greenish-brown or opaque in color with a milk-like effervescent consistency and a musty, pleasant, bitter-sour taste because of semi-suspended particles of millets, gelatinized starch, and yeast and they are not pasteurized either. They have moderately low alcohol content (up to 3%) and consumed when they are vigorously fermenting (Quin 1959).

Busa or Bouza: In Egypt and Turkey, a thick, pale yellow colored liquid with a characteristic acid-alcoholic aroma is manufactured by fermenting the grains of proso millet and the product is called Busa or Bouza. The alcoholic content ranges from 1% to 7% (Arici and Daglioglu 2002).

Tella and Katikalla: The traditional opaque beers called tella and a spirit called katikalla in Ethiopia are prepared by using finger millet and teff grains as ingredients (Bultosa and Taylor 2004).

Chhang or Jaanr or Jnard: In the Indian Himalayas, finger millet grains are fermented to produce a traditional alcoholic beverage called Chhang or Jaanr or Jnard. The alcohol content ranges from 3% to 8% (Malleshi et al. 1996). Notably, in the preparation of Chhang, the brewing process does not involve malting of grains (Basappa 2002).

Sulim, Burukutu, Dolo, talla, or Pito: In West Africa, the traditional beers are made from pearl millet grains. The fermented product is filtered to form a clear but cloudy liquid called Sulim, Burukutu, Dolo, talla, or Pito. The product is sweetish in taste and fruity in aroma and the alcohol content ranges from 1% to 5% (Demuyakor and Ohta 1993).

1.5.1.4 Nonalcoholic Beverages

There are several nonalcoholic beverages prepared and consumed throughout Africa like Oskikundu which is prepared by fermenting cooked pearl millet and malted sorghum flour popular in Namibia (Belton and Taylor 2004). Togwa is another popular product made from finger millet malt and cornmeal popular in Tanzania and Nigeria (Oi and Kitabatake 2003). Other semi-fermented traditional beverages popular in Zimbabwe are prepared from mile and finger millet flour that gives highly nutritious drinks (Mugocha et al. 2000).

1.5.2 Non-food Products from Millets

Bioethanol: Several millets rich in starch are used for manufacturing bioethanol. Usually, the starch content of major millets ranges from 65% to 73% and are good candidates to be utilized in bio-industries (Shivran 2016). Research on the utilization of sorghum grains, maize grains, and sweet stemmed sorghum for large-scale production of bioethanol has been studied extensively by House et al. (2000) and Schaffert (1995). But, the investigations on the utilization of millet grains and stems for the production of bioethanol are still in infancy.

Starch wet milling: Wet-grinding process can be used for the isolation of Starch from millet grains thus yielding ethanol as a by-product. The isolated starch from millet grains can be used as feed for animals and birds or in other industrial relevances in a similar fashion to corn starch (Munck 1995). But, the starch extracted from millets is of poor quality because of the high content of polyphenolic pigments in the glumes and/or pericarp that stains the starch (Beta et al. 2000).

Biopolymer Films and Bioplastic: Almost all the millet grains contain Kafirin a prolamin storage protein that is highly hydrophobic in nature and a superior choice for the production of bioplastics (Belton et al. 2006; Duodu et al. 2003). The kafirin in combination with glycerol, polyethylene glycol 400 (PEG 400), and lactic acid as a plasticizer yields a stable bioplastic that has lower strain, higher tensile strength, and high water vapor permeability compared to commercial zein plastics from maize (Buffo et al. 1997). These bioplastics prepared from kafirin are commercially used in Africa to coat the pears to reduce stem-end shriveling, increase shelf life, and delay ripening by up to 13 days at 20 °C storage temperature (Taylor et al. 2006).

Bio-coatings: The pericarp of most cereals contains wax that can be used for making coatings and films. Presently the sorghum pericarp wax in combination with carnauba palm wax is used as edible coatings for candies and confections (Weller et al. 1998).

1.6 Millet Genetic Resources

Millets are hardy nature crops that are well adapted for cultivation on marginal and poor soils under semi-arid and arid climatic conditions. These nutri-cereals are cultivated by small and marginal farmers in very small portion of degraded lands and their contribution in global food production is very meager. But, recently these crops are gaining importance because of their nutritionally superior grains that are a healthy dietary choice among present-day modern sedentary lifestyle. Moreover, climate resilient features of millets necessitate enhancing the focus on research and development towards these crops. Recent advances in phenotyping and genome sequencing technologies, together with availability of large germplasms; the projects on improvement of genetic resources of millets have accelerated. Globally large numbers of germplasm accessions are preserved in genebanks for several economically important traits. Moreover, efforts are being made towards collection and conservation of germplasm for important traits, utilization of these germplasms in crop improvement programs, and development of genomic and genetic resources for crop breeding and yield enhancement. The status of cultivated millet germplasms available globally are furnished in Table 1.4 and the target traits identified in different millets that can be exploited in future breeding programs are mentioned

Table 1.4 Status of cultivated millet germplasm available globally

S. No.	Crop	Asia	Africa	USA	Europe	Oceania	Total
1.	Finger millet (*Eleusine coracana*)	28,663	6700	1456	36	18	36,873
2.	Other species of the genus *Eleusine*	256	1628	20	40	22	1966
3.	Foxtail millet (*Setaria italica*)	38,572	166	1145	4548	330	44,761
4.	Other species of the genus *Setaria*	209	976	341	372	9	1907
5.	Barnyard millet (*Echinochloa colona* and *E. crusgalli*)	7444	59	316	53	51	7923
6.	Other species of the genus *Echinochloa*	371	248	71	8	9	707
7.	Proso millet (*Panicum miliaceum*)	12,110	11	1147	15,812	228	29,308
8.	Little millet (*Panicum sumatrense*)	2830	7	226	–	1	3064
9.	Other species of the genus *Panicum*	9599	3853	2161	677	142	16,432
10.	Kodo millet (*Paspalum scrobiculatum*)	4043	356	354	14	13	4780
11.	Other species of the genus *Paspalum*	190	357	2812	41	524	3924
12.	Pearl millet (*Pennisetum glaucum*)	8429	12,131	231	45	8	20,844
13.	Other species of the genus *Pennisetum*	152	586	11	1	–	750

Adopted from http://www.fao.org/wiews-archive/germplasm_query.htm?i_l=EN, accessed in 10 Mar 2018

in Table 1.5. From the above tables, we can conclude that millets have a vast genetic base and genetic diversity in crop cultivars that diminish the risk of vulnerability resulting in crop failure due to infestation by insect pests and diseases or unpredictable climatic effects. An assessment of the entire collection of seven small millets germplasm globally revealed considerable morpho-agronomic variability and considerable variation in cultivated gene pool (Upadhyaya et al. 2014).

1.7 Role of Millets in Climate Smart Agriculture

Climate change is an indubitable fact of present days and its impacts are experienced all over the world by way of changes in precipitation pattern, more rapid melting of the glaciers, and rising air temperature (Knappenberger et al. 2001; IPCC 2007). The present estimates designate an augmentation in global mean annual temperatures of 1 °C by 2025 and 3 °C by the end of the twenty-first century, the frequency and

Table 1.5 Target traits identified in different millets

Crop	Target trait
Pearl millet	Smut, Ergot, Downy mildew, and rust resistance, Drought tolerance, salinity resistance, high seed protein content, yellow endosperm, sweet stalk, and male sterility.
Finger millet	Drought and salinity tolerance, blast resistance, bold grain size, non-lodging, machine harvestable.
Foxtail millet	Bold grain size, blast and sheath blight resistance, non-lodging, strong culm for mechanical harvesting.
Proso millet	Non-shattering type, shoot fly resistance.
Little millet	Bold grain size, shoot fly resistance, non-lodging.
Kodo millet	Head smut and sheath blight resistance, shoot fly, nutrient-response and drought recovery, non-lodging.
Barnyard millet	Bold grain size, sheath blight, shoot fly and grain smut-resistance.

Source: Bennetzen et al. (2012), Zhang et al. (2012), Guo et al. (2017)

amount of rainfall will reduce by more than 30% and the amount of atmospheric CO_2 will double between 2025 and 2070 (UNFCCC 2007). Under this climate change regime, the yield of major staple food crops will drastically shrink and may pose threat to food security among the ballooning global population. For example, an increase in 1 °C temperature could reduce the rice (*Oryza sativa* L.) yield by 10% in the Philippines (Peng et al. 2004), an increase in 6 °C temperature and 300 mm deficit in precipitation in the European union can reduce yields of maize (*Zea mays* L.) up to 36% (Ciais et al. 2005). Although researchers have developed improved varieties to counteract the ill effects of abiotic stress; much of the enhancement in grain production has resulted from better agronomic practices, increase in area under cultivation, and improved pest management practices (Myers 1999). However, the yields of several cereal crops have already become stagnant in developed countries and excessive consumption of rice, wheat, and maize have paved path for new deadly diseases like diabetes, cancer, cardiovascular disorders, obesity, malnutrition as well as over nutrition (Yang et al. 2011; McKeown 2002; Chandrasekara and Shahidi 2012). Under this scenario, the researchers have to formulate strategies to adapt to climate through climate smart and climate resilient agricultural technologies.

Millets are a wonderful alternative to fight this battle against climate change, for their adaptation to a wide range of environmental conditions, better growth, and productivity under low nutrient input conditions, less irrigation requirement, least vulnerable to biotic and abiotic stresses, and less reliant on synthetic fertilizers (Kole et al. 2015). Moreover, millets are nutritionally superior to other cereals as they are rich in vitamins, minerals, essential amino acids, dietary fibers, phenolic compounds, several other bioactive compounds, and gluten-free storage proteins (Amadou et al. 2013). Although millets have a large number of attributes to be chosen for cultivation in semi-arid and arid regions of the world, they are less considered in the fight against climate change when compared to other major cereals that often are considered as model crops to study stress biology and bioenergy characters but rather fail to meet the demands of researchers (Muthamilarasan and Prasad 2015).

1.7.1 Why Millets?

Millets are superior to other cereals in their physiological, morphological, biochemical, and molecular characters that confer them both tolerance and resistance to environmental stress. In particular, the life cycle of millets completes in 84–98 days when compared to rice, wheat, and Maize that require 140–168 days. Several other traits like short plant stature, small leaf area with large leaf area index, erect leaf type with high radiation use efficiency, thickened cell walls, and dense root system helps to circumvent the adverse effects of abiotic stresses (Carberry and Campbell 1985; Ram et al. 1999; Li and Brutnell 2011). Details regarding climate resilient traits of seven millets are furnished in Table 1.6. Above all, the C4 photosynthetic pathway of millets is highly advantageous because the water use efficiency (WUE) and nitrogen use efficiency (NUE) are enhanced as RuBisCO of C4 plants works at elevated CO_2 levels and at warm conditions and confers almost two- to fourfold higher photosynthetic activity than the C3 photosynthetic plants like rice (Aubry et al. 2011; Sage and Zhu 2011). In addition to conferring higher WUE and NUE to plants, the C4 photosynthetic pathway provides secondary benefits like flexible allocation of biomass and reduced hydraulic conductivity per unit leaf area, improved growth, and adaptability to warm temperatures (Sage and Zhu 2011).

Several case studies carried out by researchers show why millets are model plants to study stress biology. For instance, the study conducted by Bidinger et al. (2007) showed that pearl millet adjusts flowering phenology according to rainfall pattern. Ajithkumar and Panneerselvam (2014) reported an increase in root length in little millet during drought conditions. Similarly, the biochemical studies conducted by Lata et al. (2011) in foxtail millets and Ajithkumar and Panneerselvam (2014) in little millets showed enhanced levels of reactive oxygen species, antioxidants and their enzyme scavenging, enzyme activity of catalase and superoxide, and synthesis of osmolytes and other stress-related proteins in response to abiotic stresses. In addition to morphological, physiological, and biochemical traits, there are several genetic and genomic resources like novel genes, alleles, and QTLs (Quantitative trait loci) being identified in millets that confer stress tolerance (Goron and Raizada 2015; Saha et al. 2016). Specific genes, alleles, and QTLs and their functional role in stress tolerance are mentioned in Table 1.7.

The genomic and genetic resources are important tools for improvement of any crop species. However, in case of millets, the information available regarding molecular markers and genetic maps is scarce compared to major cereals but availability of whole-genome sequence of foxtail millet (*Setaria italica*) and its wild progenitor, green foxtail (*S. viridis*) has facilitated the development of several high throughput genome-wide molecular markers (Yadav et al. 2015; Zhang et al. 2014) and integrated marker databases (Muthamilarasan and Prasad 2015). These resources would be of great importance for large-scale genotyping studies including

Table 1.6 Climate resilient traits of important millets

Crop	Duration (days)	Climate resilient traits
Pearl millet	80–95	Highly adaptable wide range of heat and drought can be cultivated in very poor soils and responsive to high input management
Finger millet	90–130	Adapted to wide altitude range, moderately resistant to drought, heat stress, and humidity
Foxtail millet	70–120	Adapted to high altitude and low rainfall conditions
Proso millet	60–90	Short duration crops, adapted to high altitude and low rainfall conditions
Little millet	70–110	Famine food, adapted to poor soils, low rainfall, and can also withstand waterlogging to some extent
Kodo millet	100–140	Very hardy crop with long duration, adapted to low rainfall, poor soils, and shows good response to improved agronomic practices
Barnyard millet	45–60	Short duration crop, well adapted to high altitudes and low rainfall conditions

Source: Stockle et al. (2003), Yadav et al. (2013)

genetic diversity analysis, high density physical and genetic linkage map construction, and mapping of QTL-related nutritional traits (Gupta et al. 2014; Kumar et al. 2015). Considering the above attributes of millets, they hold a great promise in global food security and nutrition amid changing climatic conditions and burgeoning population worldwide. Millets are nutritious, possess additional health benefits, and require significantly low inputs for cultivation and naturally tolerant to biotic and abiotic stresses. These traits of millets accentuate their choice for developing climate smart and climate resilient technologies.

1.7.2 Government Policies to Design Millet-Based Climate Resilient Agricultural Technologies

As seen in the previous sections, millets are packed up with numerous advantages and are superior to other cereals. They are storehouse of nutrition; provide infinite number of health benefits, can be cultivated on the poorest of soils, demand less or no synthetic fertilizers, pest-free, produce multiple securities like food, fodder, fiber, health, and livelihood. The millets can also withstand climate change and are a foremost option in climate resilient agriculture. But, post green revolution the cultivation of millets declined drastically in India, as the main objective of the green revolution was to achieve food security; ignoring nutritional security. The area under millets in the 1960s in India was 36.81 million ha with production of 18.41 million tonnes, while it reduced to 21.31 million ha with production of 17.97 million tonnes in the year 2006 (Balasubramanian et al. 2007). These drawbacks of ignoring cultivation and consumption of millets have lead to an exponential increase in health-related causalities in India and the world. Considering innumerable health benefits and environmental advantages of millets, the need of the hour is to develop

Table 1.7 Differentially expressed abiotic stress-related genes in millets

Gene name	Source of the gene	Test organism (Type)	Reference
SiLEA	Foxtail millet	Increases drought tolerance	Wang et al. (2014)
SiARDP	Foxtail millet	Increased drought tolerance	Li et al. (2016)
Argonaute protein 1 encoding gene	Foxtail millet	Regulate stress response in foxtail millet	Liu et al. (2016)
Abscisic acid stress ripening gene (ASR)	Foxtail millet	Confers tolerance to drought and oxidative stress	Feng et al. (2016)
Autophagy-related gene (ATG)	Foxtail millet	Confers tolerance to nitrogen starvation and drought stress	Li et al. (2015)
Late embryogenesis abundant protein (LEA)	Foxtail millet	Confers tolerance to salt, osmotic, and drought stress	Wang et al. (2014)
ABA-responsive DRE-binding protein (ARDP)	Foxtail millet	Confers tolerance to salt and drought stress	Li et al. (2014)
Dehydration-responsive element-binding protein 2 (DREB2)	Foxtail millet	Role in conferring dehydration tolerance	Lata et al. (2011)
12-Oxophytodienoic acid reductase (OPR1)	Foxtail millet	Confers tolerance to drought	Zhang et al. (2007)
Phospholipid hydroperoxide glutathione peroxidase (PHGPX)	Foxtail millet	Association in conferring salinity tolerance	Sreenivasulu et al. (2004)
Ec-apx1	Finger millet	Expression increased under drought	Bhatt et al. (2013)
Metallothionein	Finger millet	Induced under drought condition	Parvathi et al. (2013)
Farnesylated protein ATFP6	Finger millet	Induced under drought condition	Parvathi et al. (2013)
Farnesyl pyrophosphatesynthase	Finger millet	Induced under drought condition	Parvathi et al. (2013)
Protein phosphatase 2A	Finger millet	Induced under drought condition	Parvathi et al. (2013)
RISBZ4	Finger millet	Induced under drought condition	Parvathi et al. (2013)
NAC transcription factor	Finger millet	Confers tolerance to salinity and drought stress in rice	Rahman et al. (2016)
bHLH transcription factor	Finger millet	Confers tolerance to salinity, oxidative, and drought stress	Babitha et al. (2015)
Dehydrin7	Finger millet	Confers tolerance to drought stress	Singh et al. (2015)
β-Carbonic anhydrase (PgCA)	Pearl millet	Up-regulated when exposed to drought	Achary et al. (2015)

(continued)

Table 1.7 (continued)

Gene name	Source of the gene	Test organism (Type)	Reference
Glutathione reductase	Pearl millet	Putative involvement in stress-responsive pathways	Achary et al. (2015)
Voltage-dependent anion channel (VDAC)	Pearl millet	Tolerance to several abiotic stresses	Desai et al. (2006)
Mt1D	Bacteria	Finger millet expressing mt1D had better osmotic adjustment and chlorophyll retention under drought	Hema et al. (2014)

Source: Tadele (2016), Bandyopadhyay et al. (2017)

millet-based government policies to recognize their contributions in achieving nutritional security and retrieve them back into agricultural production in order to formulate climate resilient cropping systems. In this section, we will be briefly explaining the role of researchers in developing improved varieties, government in promoting cultivation and consumption of millets, stakeholders in increasing area under cultivation, marketing agencies in production and marketing of value-added products (Balasubramanian et al. 2007; Mishra et al. 2014; Ullah et al. 2017).

- To develop a millet-based climate smart cropping system, the role of researchers, farmers, policymakers, and rural agro-service providers are essential to achieve success.
- It is essential to brief farmers and extension agencies regarding basics of climate information including all possible mitigation and adaptation strategies.
- There is a need to focus on socioeconomic situation, resource availability, average production of focus crop, and existing risk management approaches of a given locality, before recommending climate smart cropping system.
- Farmers can be motivated through the supply of high-yielding varieties, range of tools such as hand-held crop sensors, rain gauges, decision support tools, leaf color charts, zero-till machinery, and improved production options in order to study the crop status in field.
- In order to develop a successful climate resilient cropping system, adaptation of weather smart techniques by near-term weather forecasting station is important.
- The farming community must be linked to crop and weather information along with agro-advisory through newspapers, radio talks, television shows, and mobile phone messages.
- There is a need to develop index-based insurance schemes to cover risks associated with changes in rainfall, temperature, and floods to benefit the farming community.
- Resilient water and nutrient management practices to enhance water and nutrient use efficiency and productivity are also important in climate smart interventions.

- In arid and semi-arid areas receiving less rainfall, the use of high-efficiency water management techniques such as rainwater harvesting and water conservation through drip irrigation care imperative under changing climate scenarios.
- Identification and selection of climate resilient millet varieties through participatory variety selections, considering the culinary properties, regional preferences, agronomic attributes, and scientific observations.
- Fixing up minimum support price for millets by the government is necessary to provide assured returns to millet growers.
- The nutritional and health benefits can be communicated efficiently to consumers through print and electronic media in order to increase the market demands for millets and their value-added products.
- Millet cultivation can also be boosted through official procurement of millets by government agencies for food grain distribution under several food security programs.
- There is an urgent need to include millets in Public Distribution System (PDS) to make PDS a food and nutritional security program.
- Space can be created to place millets in the food menus of schools, welfare hostel programs, and public gatherings.

1.8 Conclusion

By now we are familiar with important millets and their superiority over major staple cereals with respect to nutritional values, climate resilient features, germplasm collections, and health promoting properties. Although millets have been referred to as poor man's crop and are a staple food for more than 60% of the world's poor, the major drawback in the cultivation of millets is their productivity, which is to be addressed by increasing efforts to utilize globally available germplasms to develop high yielding varieties with all the favorable morpho-physiological characters and superior agronomic traits. Moreover, there has been considerable innovation by private and government organizations in cultivation, value addition, and marketing of millets in order to achieve both food and nutritional security. Of late, the researchers are also focusing more on exploiting neutraceutical values of millets and bio-fortifying them in order to utilize in the fight against micronutrient deficiency, malnutrition, and other health-related disorders. Even though the demand for millets and millet-based food products is increasing in national and international markets, the area under cultivation is not seeing any rise because of changing climatic conditions, nonavailability of improved high yielding varieties and hybrids that are suitable for mechanized farming, lack of favorable government policies to support cultivation and marketing of millets, noninclusion of millets in public distribution system and not fixing minimum support price for millets. However, various national and international agencies under the Consultative Group on International Agricultural Research (CGIAR) like the International Crop Research Institute for the Semi-Arid Tropics (ICRISAT) are working to tackle these obstacles; but, there is an immediate need for integration among researchers, farmers, policymakers, and rural agro-service providers to formulate climate resilient cropping systems and create more favorable resources and conditions globally that favor cultivation and consumption of millets.

References

Achary VMM, Reddy CS, Pandey P, Islam T, Kaul T, Reddy MK (2015) Glutathione reductase a unique enzyme: molecular cloning, expression and biochemical characterization from the stress adapted C4 plant, *Pennisetum glaucum* (L.) R. Br. Mol Biol Rep 42(5):947–962

Ajithkumar IP, Panneerselvam R (2014) ROS scavenging system, osmotic maintenance, pigment and growth status of *Panicum sumatrense* roth. under drought stress. Cell Biochem Biophys 68 (3):587–595

Amadou I, Amza T, Shi YH, Le GW (2011) Chemical analysis and antioxidant properties of foxtail millet bran extracts. Songklanakarin J Sci Technol 33(5):509–515

Amadou I, Gounga ME, Le GW (2013) Millets: nutritional composition, some health benefits and processing-a review. Emir J Food Agric 25:501–508

Andrews DJ, Kumar KA (1992) Pearl millet for food, feed, and forage. Adv Agron 48:89–139

Antony U, Sripriya G, Chandra TS (1996) Effect of fermentation on the primary nutrients in finger millet (*Eleusine coracana*). J Agric Food Chem 44(9):2616–2618

Arici M, Daglioglu O (2002) Boza: a lactic acid fermented cereal beverage as a traditional Turkish food. Food Rev Int 18(1):39–48

Aubry S, Brown NJ, Hibberd JM (2011) The role of proteins in C3 plants prior to their recruitment into the C4 pathway. J Exp Bot 62:3049–3059

Babitha KC, Vemanna RS, Nataraja KN, Udayakumar M (2015) Overexpression of EcbHLH57 transcription factor from *Eleusine coracana* L. in tobacco confers tolerance to salt, oxidative and drought stress. PLoS One 10(9):e0137098

Badi S, Pedersen B, Monowar L, Eggum BO (1990) The nutritive value of new and traditional sorghum and millet foods from Sudan. Plant Foods Hum Nutr 40(1):5–19

Balasubramanian S, Viswanathan R, Sharma R (2007) Post harvest processing of millets: an appraisal. Agric Eng Today 31(2):18–23

Baltensperger DD (2002) Progress with proso, pearl and other millets. In: Trends in new crops and new uses. ASHS Press, Alexandria, pp 100–103

Bandyopadhyay T, Muthamilarasan M, Prasad M (2017) Millets for next generation climate-smart agriculture. Front Plant Sci 8:1266

Basappa SC (2002) Investigations on Chhang from finger millet (*Eleusine coracana* Gaertn.) and its commercial prospects. Indian Food Ind 21(1):46–51

Belton PS, Taylor JR (2004) Sorghum and millets: protein sources for Africa. Trends Food Sci Technol 15(2):94–98

Belton PS, Delgadillo I, Halford NG, Shewry PR (2006) Kafirin structure and functionality. J Cereal Sci 44(3):272–286

Bennetzen JL, Schmutz J, Wang H, Percifield R, Hawkins J, Pontaroli AC (2012) Reference genome sequence of the model plant *Setaria*. Nat Biotechnol 30:555–561

Beta T, Rooney LW, Marovatsanga LT, Taylor JRN (2000) Effect of chemical treatments on polyphenols and malt quality in sorghum. J Cereal Sci 31:295–302

Bhatt D, Saxena SC, Jain S, Dobriyal AK, Majee M, Arora S (2013) Cloning, expression and functional validation of drought inducible ascorbate peroxidase (Ec-apx1) from *Eleusine coracana*. Mol Biol Rep 40(2):1155–1165

Bidinger FR, Nepolean T, Hash CT, Yadav RS, Howarth CJ (2007) Quantitative trait loci for grain yield in pearl millet under variable post flowering moisture conditions. Crop Sci 47(3):969–980

Bravo L (1998) Polyphenols: chemistry, dietary sources, metabolism, and nutritional significance. Nutr Rev 56:317–333

Buffo RA, Weller CL, Gennadios A (1997) Films from laboratory-extracted sorghum kafirin. Biological systems engineering: Papers and Publications, 105

Bultosa G, Taylor JRN (2004) Functional properties of grain tef (Eragrostistef (Zucc.) Trotter) starch. Starch/Stiirke 56:20–28

Carberry PS, Campbell LC (1985) The growth and development of pearl millet as affected by photoperiod. Field Crops Res 11:207–217

Catassi C, Alarida K (2011) Another brick in the (great) wall: celiac disease in Chinese children. J Pediatr Gastr Nutr 53(4):359–360

Chandrasekara A, Shahidi F (2010) Content of insoluble bound phenolics in millets and their contribution to antioxidant capacity. J Agric Food Chem 58(11):6706–6714

Chandrasekara A, Shahidi F (2012) Bioaccessibility and antioxidant potential of millet grain phenolics as affected by simulated in vitro digestion and microbial fermentation. J Funct Foods 4(1):226–237

Ciais P, Reichstein M, Viovy N, Granier A, Ogée J, Allard V, Chevallier F (2005) Europe-wide reduction in primary productivity caused by the heat and drought in 2003. Nature 437 (7058):529–533

Demuyakor B, Ohta Y (1993) Comparison of the malt characteristics of traditional and improved varieties of Sorghum vulgare from Ghana. J Inst Brew 99(3):227–230

Desai MK, Mishra RN, Verma D, Nair S, Sopory SK, Reddy MK (2006) Structural and functional analysis of a salt stress inducible gene encoding voltage dependent anion channel (VDAC) from pearl millet (*Pennisetum glaucum*). Plant Physiol Biochem 44:483–493. https://doi.org/10.1016/j.plaphy.2006.08.008

Devi PB, Vijayabharathi R, Sathyabama S, Malleshi NG, Priyadarisini VB (2011) Health benefits of finger millet (*Eleusine coracana* L.) polyphenols and dietary fiber: a review. J Food Sci Technol. https://doi.org/10.1007/S13197-011-0584-9

Devi PB, Vijayabharathi R, Sathyabama S, Malleshi NG, Priyadarisini VB (2014) Health benefits of finger millet (*Eleusine coracana* L.) polyphenols and dietary fiber: a review. J Food Sci Technol 51(6):1021–1040

Duodu KG, Taylor JRN, Belton PS, Hamaker BR (2003) Factors affecting sorghum protein digestibility. J Cereal Sci 38(2):117–131

Dwivedi SL, Upadhyaya HD, Senthilvel S, Hash CT, Fukunaga K, Diao X, Prasad M (2012) Millets: genetic and genomic resources. Plant Breed Rev 35:247–375

Dykes L, Rooney LW (2006) Sorghum and millet phenols and antioxidants. J Cereal Sci 44 (3):236–251

Engels JM, Engels JMM, Hawkes JG, Hawkes JG, Worede M (eds) (1991) Plant genetic resources of Ethiopia. Cambridge University Press, Cambridge

Feng ZJ, Xu ZS, Sun J, Li LC, Chen M, Yang GX, Ma YZ (2016) Investigation of the ASR family in foxtail millet and the role of ASR1 in drought/oxidative stress tolerance. Plant Cell Rep 35 (1):115–128

Food and Agricultural Organization [FAO] (2012) Economic and social department: the statistical division. Statistics Division, FAO, Rome

Friedman M (1997) Chemistry, biochemistry and dietary role of potato polyphenols—a review. J Agric Food Chem 45:1523–1540

Gashe BA, Girma M, Bisrat A (1982) Tef fermentation. I. The role of microorganisms in fermentation and their effect on the nitrogen content of tef. SINET 5:69–76

Gopalan C, Ramashastri BV, Balasubramanium SC (2004) Nutritive value of Indian foods. National Institute of Nutrition (NIN), Indian Council of Medical Research, Hyderabad, pp 59–67

Goron TL, Raizada MN (2015) Genetic diversity and genomic resources available for the small millet crops to accelerate a New Green Revolution. Front Plant Sci 6:157

Guo L, Qiu J, Ye C, Jin G, Mao L, Zhang H (2017) *Echinochloa crusgalli* genome analysis provides insight into its adaptation and invasiveness as a weed. Nat Commun 8(1):1031

Gupta S, Kumari K, Muthamilarasan M, Parida SK, Prasad M (2014) Population structure and association mapping of yield contributing agronomic traits in foxtail millet. Plant Cell Rep 33 (6):881–893

Hadimani NA, Malleshi NG (1993) Studies on milling, physico-chemical properties, nutrient composition and dietary fibre content of millets. J Food Sci Technol 30(1):17–20

Hedge PS, Chandra TS (2005) ESR spectroscopic study reveals higher free radical quenching potential in kodo millet (*Paspalum scrobiculatum*) compared to other millets. Food Chem 92:177–182

Hema R, Vemanna RS, Sreeramulu S, Reddy CP, Senthil-Kumar M, Udayakumar M (2014) Stable expression of mtlD gene imparts multiple stress tolerance in finger millet. PLoS One 9(6): e99110

House LR (1995) Sorghum and millets: history, taxonomy, and distribution. American Association of Cereal Chemists, St Paul, MN, pp 1–9

House LR, Gomez M, Murty DS, Sun Y, Verma BN (2000) Development of some agricultural industries in several African and Asian countries. In: Smith CW, Frederiksen RA (eds) Sorghum: origin, history, technology, and production. Wiley, New York, pp 131–190

IPCC CC (2007) The physical science basis. IPCC

Jiaju C, Yuzhi Q (1994) Recent developments in foxtail millet cultivation and research in China. In: Riley KW, Gupta SC, Seetharam A, Mushonga JN (eds) Advances in small millets. International Science Publisher, New York, pp 101–108

Jones JM, Engleson J (2010) Whole grains: benefits and challenges. Annu Rev Food Sci Technol 1:19–40

Joshi PK, Agnihotri AK (1984) Millet production in India: problems and prospects. Agric Situation India 39:329

Kam J, Puranik S, Yadav R, Manwaring HR, Pierre S, Srivastava RK, Yadav RS (2016) Dietary interventions for type 2 diabetes: how millet comes to help. Front Plant Sci 7:1454

Kaur P, Purewal SS, Sandhu KS, Kaur M, Salar RK (2019) Millets: a cereal grain with potent antioxidants and health benefits. J Food Meas Charact 13(1):793–806

Knappenberger PC, Michaels PJ, Davis RE (2001) Nature of observed temperature changes across the United States during the 20th century. Clim Res 17(1):45–53

Kole C, Muthamilarasan M, Henry R, Edwards D, Sharma R, Abberton M, Cai H (2015) Application of genomics-assisted breeding for generation of climate resilient crops: progress and prospects. Front Plant Sci 6:563

Krishnaswamy N (1938) Geography and history of millets. Curr Sci 6(7):355–358

Kumar A, Gaur VS, Goel A, Gupta AK (2015) De novo assembly and characterization of developing spikes transcriptome of finger millet (*Eleusine coracana*): a minor crop having nutraceutical properties. Plant Mol Biol Rep 33(4):905–922

Kumar A, Tomer V, Kaur A, Kumar V, Gupta K (2018) Millets: a solution to agrarian and nutritional challenges. Agric Food Secur 7(1):31

Kumari PL, Sumathi S (2002) Effect of consumption of finger millet on hyperglycemia in non-insulin dependent diabetes mellitus (NIDDM) subjects. Plant Foods Hum Nutr 57 (3–4):205–213

Kumari SK, Thayumanavan B (1997) Comparative study of resistant starch from minor millets on intestinal responses, blood glucose, serum cholesterol and triglycerides in rats. J Sci Food Agric 75(3):296–302

Lata C, Bhutty S, Bahadur RP, Majee M, Prasad M (2011) Association of an SNP in a novel DREB2-like gene SiDREB2 with stress tolerance in foxtail millet [*Setaria italica* (L.)]. J Exp Bot 62(10):3387–3401

Lawler A (2009) Millet on the move. Science 325(5943):942–943

Lee SH, Chung IM, Cha YS, Park Y (2010) Millet consumption decreased serum concentration of triglyceride and C-reactive protein but not oxidative status in hyperlipidemic rats. Nutr Res 30 (4):290–296

Li P, Brutnell TP (2011) *Setaria viridis* and *Setaria italica*, model genetic systems for the Panicoid grasses. J Exp Bot 62(9):3031–3037
Li C, Yue J, Wu X, Xu C, Yu J (2014) An ABA-responsive DRE-binding protein gene from *Setaria italica*, SiARDP, the target gene of SiAREB, plays a critical role under drought stress. J Exp Bot 65(18):5415–5427
Li X, Hao J, Liu X, Liu H, Ning Y, Cheng R, Jia Y (2015) Effect of the treatment by slightly acidic electrolyzed water on the accumulation of γ-aminobutyric acid in germinated brown millet. Food Chem 186:249–255
Li W, Tang S, Zhang S, Shan J, Tang C, Chen Q, Diao X (2016) Gene mapping and functional analysis of the novel leaf color gene SiYGL1 in foxtail millet [*Setaria italica* (L.) P. Beauv]. Physiol Plant 157(1):24–37
Liang S, Yang G, Ma Y (2010) Chemical characteristics and fatty acid profile of foxtail millet bran oil. J Am Oil Chem Soc 87(1):63–67
Lin R, Li W, Corke H (1998) Spotlight on Shanxi province China: its minor crops and specialty foods. Cereal Foods Worm 43:189–192
Liu X, Tang S, Jia G, Schnable JC, Su H, Tang C, Diao X (2016) The C-terminal motif of SiAGO1b is required for the regulation of growth, development and stress responses in foxtail millet (*Setaria italica* (L.) P. Beauv). J Exp Bot 67(11):3237–3249
Lu H, Zhang J, Liu KB, Wu N, Li Y, Zhou K, Ye M, Zhang T (2009a) Earliest domestication of common millet (Panicum miliaceum) in East Asia extended to 10,000 years ago. Proc Natl Acad Sci USA 106(18):7367–7372
Lu H, Zhang J, Wu N, Liu KB, Xu D, Li Q (2009b) Phytoliths analysis for the discrimination of foxtail millet (Setaria italica) and common millet (Panicum miliaceum). PLoS One 4(2):e4448
Malleshi NG, Hadimani NA, Chinnaswamy R, Klopfenstein CF (1996) Physical and nutritional qualities of extruded weaning foods containing sorghum, pearl millet, or finger millet blended with mung beans and nonfat dried milk. Plant Foods Hum Nutr 49(3):181–189
Manach C, Williamson G, Morand C, Scalbert A, Remesy C (2005) Bioavailability and bioefficacy of polyphenols in humans. 1. Review of 97 bioavailability studies. Am J Clin Nutr 81:230S–242S
McKeown NM (2002) Whole grain intake is favorably associated with metabolic risk factors for type 2 diabetes and cardiovascular disease in the Framingham Offspring Study. Am J Clin Nutr 76:390–398
Mishra AK, Muthamilarasan M, Khan Y, Parida SK, Prasad M (2014) Genome-wide investigation and expression analyses of WD40 protein family in the model plant foxtail millet (Setaria italica L.). PLoS One 9(1):e86852
Mohamed TK, Issoufou A, Zhou H (2012) Antioxidant activity of fractionated foxtail millet protein hydrolysate. Int Food Res J 19(1):207
Mugocha PT, Taylor JRN, Bester BH (2000) Fermentation of a composite finger millet-dairy beverage. World J Microbiol Biotechnol 16(4):341–344
Munck L (1995) New milling technologies and products: whole plant utilization by milling and separation of the botanical and chemical components. In: Sorghum and millets: chemistry and technology. American Association of Cereal Chemists, St. Paul, MN, pp 223–281
Murty DS, Kumar KA (1995) Traditional uses of sorghum and millets. In: Dendy DAV (ed) Sorghum and millets: chemistry and technology. American Association of Cereal Chemists, St Paul, MN, p 221
Muthamilarasan M, Prasad M (2015) Advances in Setaria genomics for genetic improvement of cereals and bioenergy grasses. Theor Appl Genet 128(1):1–14
Muthamilarasan M, Dhaka A, Yadav R, Prasad M (2016) Exploration of millet models for developing nutrient rich graminaceous crops. Plant Sci 242:89–97
Myers N (1999) The next green revolution: its environmental underpinnings. CurrSci 76:507–513
National Research Council (1996) Lost crops of Africa, Grains, vol I. National Academies Press, Washington, DC

Nesbitt M, Summers GD (1988) Some recent discoveries of millet (Panicum miliaceum L. and Setaria italica (L.) P. Beauv.) at excavations in Turkey and Iran. Anatol Stud 38:85–97

Obilana AB, Manyasa E (2002) Millets. In: Pseudocereals and less common cereals. Springer, Berlin, pp 177–217

Oelke EA, Oplinger ES, Putnam DH, Durgan BR, Doll JD, Undersander DJ (1990) Millets. Alternative field crops manual. University of Wisconsin-Madison, Madison, WI

Oi Y, Kitabatake N (2003) Chemical composition of an East African traditional beverage, togwa. J Agric Food Chem 51(24):7024–7028

Park KO, Ito Y, Nagasawa T, Choi MR, Nishizawa N (2008) Effects of dietary Korean proso-millet protein on plasma adiponectin, HDL cholesterol, insulin levels, and gene expression in obese type 2 diabetic mice. Biosci Biotechnol Biochem 72(11):2918–2925

Parvathi MS, Nataraja KN, Yashoda BK, Ramegowda HV, Mamrutha HM, Rama N (2013) Expression analysis of stress responsive pathway genes linked to drought hardiness in an adapted crop, finger millet (Eleusine coracana). J Plant Biochem Biotechol 22(2):193–201

Peng S, Huang J, Sheehy JE, Laza RC, Visperas RM, Zhong X, Cassman KG (2004) Rice yields decline with higher night temperature from global warming. Proc Natl Acad Sci U S A 101 (27):9971–9975

Quin PJ (1959) Foods and the feeding habits of the pedi. Witwatersrand University Press, Johannesburg

Rahman H, Ramanathan V, Nallathambi J, Duraialagaraja S, Muthurajan R (2016) Over-expression of a NAC 67 transcription factor from finger millet (Eleusine coracana L.) confers tolerance against salinity and drought stress in rice. BMC Biotechnol 16(1):35

Ram N, Sheoran K, Sastry CVS (1999) Radiation efficiency and its efficiency in dry biomass production of pearl millet cultivars. Ann Agric Sci 20:286–291

Ramachandra G, Virupaksha TK, Shadaksharaswamy M (1977) Relation between tannin levels and in vitro protein digestibility in finger millet (Eleusine coracana Gaertn.). J Agric Food Chem 25 (5):1101–1104

Rooney LW, RD Waniska, and R Subramanian (1997) Overcoming constraints to utilization of sorghum and millet. pp 549–557

Roy M (2009) Agriculture in the Vedic period. Indian J Hist Sci 44(4):497–520

Sage RF, Zhu XG (2011) Exploiting the engine of C4 photosynthesis. J Exp Bot 62(9):2989–3000

Saha D, Gowda MC, Arya L, Verma M, Bansal KC (2016) Genetic and genomic resources of small millets. Crit Rev Plant Sci 35(1):56–79

Saleh ASM, Zhang Q, Chen J, Shen Q (2013) Millet grains: nutritional quality, processing, and potential health benefits. Compr Rev Food Sci Food Saf 12:281–295

Sarita ES, Singh E (2016) Potential of millets: nutrients composition and health benefits. J Sci Innov Res 5(2):46–50

Schaffert RE (1995) Sweet sorghum substrate for industrial alcohol. In: Dendy DAV (ed) Sorghum and millets: chemistry and technology. American Association of Cereal Chemists, St. Paul, MN, pp 365–374

Shahidi F, Chandrasekara A (2013) Millet grain phenolics and their role in disease risk reduction and health promotion: a review. J Funct Foods 5:570–581

Shivran AC (2016) Biofortification for nutrient-rich millets. In: Biofortification of food crops. Springer, New Delhi, pp 409–420

Shobana S, Malleshi NG (2007) Preparation and functional properties of decorticated finger millet (*Eleusine coracana*). J Food Eng 79(2):529–538

Shobana S, Usha Kumari SR, Malleshi NG, Ali SZ (2007) Glycemic response of rice, wheat and finger millet based diabetic food formulations in normoglycemic subjects. Int J Food Sci Nutr 58 (5):363–372

Singh J, Reddy PS, Reddy CS, Reddy MK (2015) Molecular cloning and characterization of salt inducible dehydrin gene from the C4 plant *Pennisetum glaucum*. Plant Gene 4:55–63

Sreenivasulu N, Miranda M, Prakash HS, Wobus U, Weschke W (2004) Transcriptome changes in foxtail millet genotypes at high salinity: identification and characterization of a PHGPX gene specifically up-regulated by NaCl in a salt-tolerant line. J Plant Physiol 161(4):467–477
Stockle CO, Marcello D, Roger N (2003) CropSyst, a cropping system simulation model. Eur J Agron 18:289–307
Subramanian V, Jambunathan R (1980) Traditional methods of processing of sorghum (Sorghum bicolor) and pearl millet (Pennisetum americanum) grains in India. Rep Int Assoc Cereal Chem 10:115–118
Tadele Z (2016) Drought adaptation in millets. In: Shanker AK, Shanker C (eds) Abiotic and biotic stress in plants: recent advances and future perspectives. InTech, Rijeka, pp 639–662
Taylor JR, Emmambux MN (2008) Gluten-free foods and beverages from millets. In: Gluten-free cereal products and beverages. Academic, Cambridge, MA, p 119
Taylor JR, Schober TJ, Bean SR (2006) Novel food and non-food uses for sorghum and millets. J Cereal Sci 44(3):252–271
Thompson LU (1993) Potential health benefits and problems associated with antinutrients in foods. Food Res Int 26:131–149
Thompson T (2009) The nutritional quality of gluten-free foods. In: Gluten-free food science and technology. Blackwell, OXford, pp 42–51
Tsao R (2010) Chemistry and biochemistry of dietary polyphenols. Nutrients 2(12):1231–1246
Ullah A, Ahmad A, Khaliq T, Akhtar J (2017) Recognizing production options for pearl millet in Pakistan under changing climate scenarios. J Integr Agric 16(4):762–773
United Nations Framework Convention on Climate Change (UNFCCC) (2007) Climate change: impacts, vulnerabilities and adaptation in developing Countries. Climate Change Secretariat, INFCCC, Bonn, Germany
Upadhyaya HD, Dwivedi SL, Ramu P, Singh SK, Singh S (2014) Genetic variability and effect of postflowering drought on stalk sugar content in sorghum mini core collection. Crop Sci 54 (5):2120–2130
USDA National Nutrient Database for Standard Reference, Release 28 (2016) US Department of Agriculture, Agricultural Research Service, Nutrient Data Laboratory
Van Rensburg SJ (1981) Epidemiological and dietary evidence for a specific nutritional predisposition to esophageal cancer. J Natl Cancer Inst 67(2):243–251
Wang M, Li P, Li C, Pan Y, Jiang X, Zhu D, Yu J (2014) SiLEA14, a novel atypical LEA protein, confers abiotic stress resistance in foxtail millet. BMC Plant Biol 14(1):290
Watanabe M (1999) Antioxidative phenolic compounds from Japanese barnyard millet (Echinochloa utilis) grains. J Agric Food Chem 47(11):4500–4505
Weller CL, Gennadios A, Saraiva RA, Cuppett SL (1998) Grain sorghum wax as an edible coating for gelatin-based candies. J Food Qual 21(2):117–128
Yadav OP, Rai KN (2013) Genetic improvement of pearl millet in India. Agric Res 2(4):275–292
Yadav SB, Patel HR, Lunagaria MM, Parmar PK, Chaudhari NJ, Karande BI, Pandey V (2013) Impact assessment of projected climate change on pearl millet in Gujrat. In: National Seminar on Climate change impacts on water resources systems. International Crops Research Institute for the Semi-arid Tropics, India, pp 33–38
Yadav CB, Muthamilarasan M, Pandey G, Prasad M (2015) Identification, characterization and expression profiling of Dicer-like, Argonaute and RNA-dependent RNA polymerase gene families in foxtail millet. Plant Mol Biol Rep 33(1):43–55
Yang Q, Li X, Liu W, Zhou X, Zhao K, Sun N (2011) Carbon isotope fractionation during low temperature carbonization of foxtail and common millets. Org Geochem 42(7):713–719
Yang X, Wan Z, Perry L, Lu H, Wang Q, Hao C, Li J, Xie F, Yu J, Cui T, Wang T, Li M, Ge QH (2012) Early millet use in Northern China. Proc Natl Acad Sci U S A 109(10):3726–3730
Yenagi NB, Handigol JA, Ravi SB, Mal B, Padulosi S (2010) Nutritional and technological advancements in the promotion of ethnic and novel foods using the genetic diversity of minor millets in India. Indian J Plant Genet Res 23(1):82–86

Yetneberk S, de Kock HL, Rooney LW, Taylor JR (2004) Effects of sorghum cultivar on injera quality. Cereal Chem 81(3):314–321

Zhang JP, Zhang JP, Liu TS, Zhang JP, Liu TS, Zheng J, Zhang JP (2007) Cloning and characterization of a putative 12-oxophytodienoic acid reductase cDNA induced by osmotic stress in roots of foxtail millet: full length research paper. DNA Seq 18(2):138–144

Zhang G, Liu X, Quan Z, Cheng S, Xu X, Pan S (2012) Genome sequence of foxtail millet (*Setaria italica*) provides insights into grass evolution and biofuel potential. Nat Biotechnol 30:549–554

Zhang S, Tang C, Zhao Q, Li J, Yang L, Qie L, Liu X (2014) Development of highly polymorphic simple sequence repeat markers using genome-wide microsatellite variant analysis in Foxtail millet [Setaria italica (L.) P. Beauv.]. BMC Genomics 15(1):78

Global Scenario of Millets Cultivation

2

Rajendra Prasad Meena, Dinesh Joshi, J. K. Bisht, and Lakshmi Kant

Abstract

Millets are staple food in the developing world, especially in the drylands of Africa and Asia. Most of the millets are indigenous to Africa and later domesticated to other parts of the world. Globally, millets are cultivated in 93 countries and only 7 countries have more than 1 M ha acreage of millets. In general, more than 97% of millets production and consumption is by developing nations. It has been estimated that from 1961 to 2018 around 25.71% area under millets cultivation has been declined across the continents. However, global millet productivity has increased by 36% from 1961 (575 kg/ha) to 2018 (900 kg/ha). The average data of the last 58 years indicated that millet production reduced in most parts of the world, except Africa. The highest increment was recorded in West Africa, almost double than the 1960s. In Asia, although the area under millet cultivation has declined production trend showed a gradual increase, which led to productivity enhancement. In the Indian scenario, millet production was at peak during the 1980s, thereafter decreased gradually due to sharp reduction under cultivated area. India is the largest producer of millets with 37.5% of the total global output followed by Sudan and Nigeria. In terms of trade, the highest global import and export value of millets (155.26 and 127.60 million US$, respectively) were recorded during the year 2011–2017. The continuous downfall in the global area under millets may be attributed due to the area shifting for other crops, changed food habits, assured irrigation facilities, and ensured returns from major commercial crops.

R. P. Meena · D. Joshi (✉) · J. K. Bisht · L. Kant
ICAR-Vivekananda Institute of Hill Agriculture, Almora, Uttarakhand, India
e-mail: Dinesh.Joshi@icar.gov.in

A. Kumar et al. (eds.), *Millets and Millet Technology*,
https://doi.org/10.1007/978-981-16-0676-2_2

Keywords

Millets · Developing nations · Global millet production and productivity · Millets in India · Global millet trade · Food security

2.1 Introduction

The word millet is derived from the French word "mille" which means that a handful of millet contains thousands of seed grains (Taylor and Emmambux 2008). They are broadly categorized into two major groups (1) major millets, viz., sorghum [*Sorghum bicolor* (L.)] and pearl millet [*Pannisetum glaucum* (L.)]; (2) minor or small millets, viz., finger millet [*Eleusine coracana* (L.) Gaertn.], proso millet [*Panicumm iliaceum* (L.)], foxtail millet [*Setaria italica* (L.) Beauv.], kodo millet [*Paspalum scrobiculatum* (L.)], barnyard millet (*Echinochloa* spp.), and little millet [*Panicum sumatrense* Roth ex. Roem. and Schult.] (Table 2.1). Millets are the important staple food of resource for poor farmers in hot and drier regions of the developing world especially in Africa and Asia (McDonough et al. 2000). Globally millets (pearl millet and minor millets) are cultivated in more than 93 countries. Among the millets, sorghum is the most widely grown crop with 42.1 M ha area in 105 countries, whereas production figures are available for pearl millet and other minor millets from 93 countries (Obilana 2003; Fig. 2.1). About 97% of millets are produced and consumed by developing countries especially in Africa and Asia. Globally, India is the largest grower of millets with 26.6% of the world and 83% of Asia's millet cropping area. In India, millets have been an integral part of tribal food in the states of Odisha, Madhya Pradesh, Jharkhand, Rajasthan, Karnataka, and Uttarakhand (Sood et al. 2019). However, in recent years owing to their tremendous nutraceutical potential they are becoming popular in urban areas as well. According to an estimate, there has been a decline of 25.7% in the global area under millets cultivation from 1961 to 2018 (Table 2.2; FAOSTAT 2018). Among the continents, the largest area reduction was observed in Asia (148%), whereas the lowest was observed in Africa (Table 2.2). This decline may be attributed to lack of concentrated crop improvement efforts, shift towards high-value cash crops, lack of government policies, and low farm profitability. The continuous decline in global cultivated area under millets in the last decades have given them the status of minor or underutilized grains.

Millets have agricultural superiority over other commercial crops attributed to their ability to adapt under marginal and less input demanding cultivation. Additionally, the C_4 photosynthetic pathway and ability to withstand environmental stress make them a suitable choice for future agricultural systems. The nutritional superiority over major cereals in terms of balanced micronutrient profile and bioactive flavonoids of diverse pharmaceutical uses makes them highly valuable crops (Sood et al. 2019 missing). By virtue of immense health benefits, there has been an increase in global export and import of millets in the last decade and highest value (155.26 and 127.60 million US$, respectively) was recorded during the year 2011–2017 (Table 2.6; FAOSTAT 2018). Despite their immense agricultural value, global area under millet cultivation and production in the last five decades have declined or

Table 2.1 Millets and their special characteristics

Millet	Common name	Botanical name	Special characteristics	Reference
Sorghum	Great millet, Jowar, Kafir corn, Guinea corn, Kaolin in China, and Milo in Spain	*Sorghum bicolor*	Tolerate moisture stress and high temperature better than any other crops	–
Pearl millet	Bajra, Cattail millet, Black millet, German millet	*Pennisetum glaucum*	Grow in arid and semi-arid region, richest source of folic acid	–
Finger millet	Ragi, Wimbi, Mandua, Nachni, Kapai, Nagli, Marua	*Eleusine coracana*	Wider adaptability, rich source of calcium	Seetharam (1998)
Proso millet	Cheena, Common millet, Broom millet	*Panicumm iliaceum*	Short duration, tolerant to heat and drought	Sahib (1997)
Foxtail millet	Indian paspalum, Kangni, Water couch, Italian millet	*Setaria italica*	Short duration, tolerant to low soil fertility and drought	Jijau (1989)
Kodo millet	Kodo, Ditch millet, Creeping paspalum	*Paspalum scrobiculatum*	Long duration, grown well in shallow and deep soil, rich in folic acid	Hegde and Gowda (1989)
Barnyard millet	Sawan, Jhingora, Kudraivali, Oodalu	*Echinochloa frumentacea*	Fastest growing, voluminous fodder	Gupta et al. (2009)
Little millet	Kutki, Samai, Samalu, Hog millet	*Panicum sumatrense*	Short duration, withstand both drought and waterlogging	Doggett (1989)
Browntop millet	Korale in Kannada	*Brachiaria ramosa*	Rapidly maturing, best suited for catch crop	Sheahan (2014)
Teff	Teff, lovegrass, annual bunch grass, Williams love grass	*Eragrostis tef*	Massive fibrous rooting system, drought tolerant, ephemeral nature	Assefa et al. (2011)
Fonio	Fonio, Acha, Hungry rice	*Digitaria exilis* (White fonio) *Digitaria iburua* (White fonio)	Smallest seeds among millets, fast growing and highly nutritious	NRC (1996)
Job's tears	Adlag, Adlay millet	*Coix laeryma*	Grown in higher areas, used in folk medicine	Duke (1983)
Guinea millet	False signal grass, Babala, Bajra/Bajira	*Urochloa deflexa*	Potential as grain crop	–

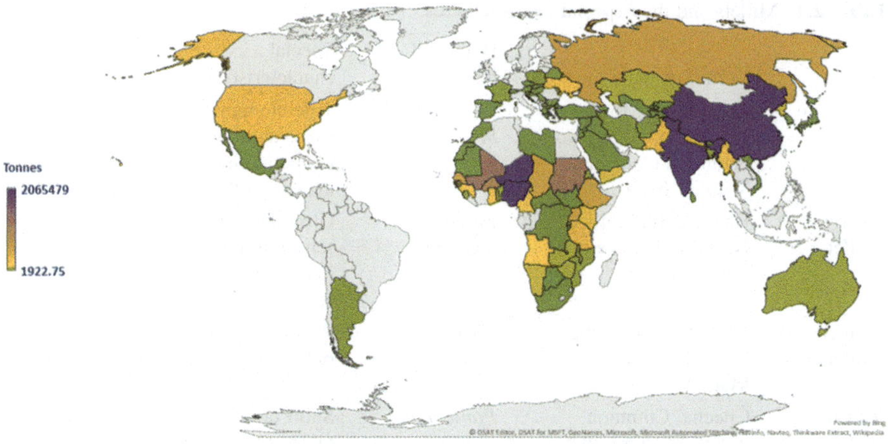

Fig. 2.1 Country wise production of millets (FAOSTAT 2018)

remained stagnated compared to major cereals. This is mainly because appreciable genetic gain through modern plant breeding is yet to be realized in millets.

Furthermore, integration of sustainable and cost-effective crop management practices is the key to project them as golden crops of the future. In the present chapter, we summarize the global scenario of millets and their importance as a healthy alternative food. We also provided an overview of major production constraints associated with millets and suggested future prospects that can accelerate millet production and productivity.

2.2 Cropping Area

At Food and Agriculture Organization (FAO) website (FAOSTAT 2018) the area, production, and productivity data for millets are available in two categories (1) millets (small millets and pearl millet) and (2) sorghum. Therefore, we have described the global scenario of millets (Table 2.2) and sorghum (Table 2.3) separately in this chapter. In the Indian scenario, data is included for sorghum, pearl millet, and finger millet only (INDIASTAT 2020; Table 2.4).

Out of the total 93 millet growing countries of the world (Fig. 2.1), only 7 countries (India, Niger, Sudan, Nigeria, Mali, Burkina Faso, and Chad) have more than 1 M ha harvested area, whereas around 25 countries have more than 0.1 M ha harvested area. Together all contribute around 97% of the total world millet harvested area (34.1 M ha). Among the top seven millet growing countries of the world, India ranked first with 15.29 M ha harvested area followed by Niger (7.03 M ha), Sudan (3.75 M ha), Nigeria (2.7 M ha), Mali (2.15 M ha), Burkina

Table 2.2 Global millets (except sorghum) area and production by region[a] (FAOSTAT 2018)

	Area (lakh ha)							Production (lakh tons)						
	1961–1963	1971–1973	1981–1983	1991–1993	2001–2003	2011–2013	2016–2018	1961–1963	1971–1973	1981–1983	1991–1993	2001–2003	2011–2013	2016–2018
Africa	118.390	133.227	108.751	168.994	197.694	191.280	207.067	69.424	74.512	77.617	109.664	142.483	113.391	140.569
Eastern Africa	16.364	17.758	12.333	13.412	14.382	15.096	15.032	11.566	13.585	12.052	12.065	12.939	16.004	17.570
Middle Africa	8.017	6.804	4.945	8.119	11.430	14.655	15.962	5.147	3.996	2.974	4.021	6.609	7.780	9.356
Northern Africa	4.799	10.489	11.727	12.567	25.402	21.870	30.351	3.077	3.651	3.578	3.342	6.609	7.136	16.698
Southern Africa	0.813	1.600	1.745	1.729	2.836	2.481	2.695	0.275	0.449	0.533	0.453	0.689	0.497	0.602
Western Africa	88.397	96.576	78.002	133.166	143.645	137.178	143.027	49.359	52.831	58.481	89.783	115.637	81.975	96.342
Americas	2.657	2.529	2.457	2.259	2.149	1.670	1.676	3.257	2.983	3.067	3.355	2.885	2.448	3.628
Northern America	0.971	0.977	0.859	1.599	1.996	1.593	1.644	1.157	1.267	1.165	2.314	2.610	2.318	3.578
Central America	0.000	0.000	0.000	0.000	0.005	0.008	0.002	0.000	0.000	0.000	0.000	0.005	0.008	0.003
Caribbean	0.000	0.000	0.000	0.000	0.000	0.000	0.000	0.000	0.000	0.000	0.000	0.000	0.000	0.000
South America	1.686	1.552	1.598	0.661	0.148	0.069	0.030	2.100	1.717	1.902	1.041	0.270	0.122	0.048
Asia	271.746	272.350	229.054	174.644	144.703	121.958	109.255	152.925	181.630	178.191	142.069	137.569	142.501	139.522
Central Asia	0.000	0.000	0.000	6.499	0.725	0.440	0.477	0.000	0.000	0.000	0.680	0.532	0.458	0.530
Eastern Asia	74.310	66.337	40.679	19.771	11.760	8.014	7.490	66.283	79.594	66.976	36.292	20.990	17.772	17.317
Southern Asia	193.676	202.853	184.903	147.468	128.739	109.925	97.864	83.835	99.717	108.229	101.515	113.598	121.086	118.503
South-Eastern Asia	1.452	1.540	1.754	1.955	2.329	2.231	2.415	0.468	0.390	1.512	1.316	1.659	2.198	2.474
Western Asia	2.308	1.621	1.718	1.118	1.151	1.348	1.009	2.339	1.929	1.474	0.680	0.790	0.987	0.698
Europe	40.893	26.865	28.023	22.453	8.182	6.278	4.029	23.776	26.753	21.400	16.277	9.043	8.363	6.237
Eastern Europe	40.631	26.765	27.989	22.440	8.097	6.128	3.881	23.444	26.582	21.319	16.248	8.793	7.870	5.688
Northern Europe	0.000	0.000	0.000	0.000	0.000	0.000	0.000	0.000	0.000	0.000	0.000	0.000	0.000	0.000
Southern Europe	0.226	0.081	0.019	0.013	0.015	0.009	0.014	0.286	0.127	0.034	0.029	0.033	0.016	0.026
Western Europe	0.036	0.018	0.015	0.000	0.071	0.140	0.134	0.047	0.044	0.047	0.000	0.218	0.477	0.524
Oceania	0.294	0.334	0.327	0.295	0.357	0.353	0.351	0.322	0.363	0.317	0.262	0.288	0.358	0.359
Australia and New Zealand	0.294	0.334	0.327	0.295	0.357	0.353	0.351	0.322	0.363	0.317	0.262	0.288	0.358	0.359
World	433.980	435.305	368.613	368.645	353.085	321.539	322.378	249.704	286.242	280.593	271.627	292.268	267.061	290.314

[a]Each figure is an average of 3 years for the respective period, for example, 1961–1963

Table 2.3 Global sorghum harvested area and production by continents (FAOSTAT 2018)

	1961	1971	1981	1991	2001	2011	2018
	Sorghum harvested area (lakh ha)						
Africa	132.14	151.31	138.92	200.57	234.90	265.75	297.11
Americas	58.00	104.95	104.35	72.70	73.28	58.31	53.01
Asia	267.60	237.64	206.21	148.85	118.38	89.08	64.32
Europe	1.32	1.72	2.62	2.49	1.93	2.57	2.36
Oceania	1.03	5.53	6.60	3.78	7.59	6.34	4.64
World	460.09	501.14	458.69	428.39	436.08	422.04	421.43
	Sorghum production (lakh tons)						
Africa	106.92	117.85	135.30	161.99	207.09	239.92	297.82
Americas	143.91	302.21	378.10	237.38	249.18	197.51	192.44
Asia	155.41	180.67	199.60	140.24	114.91	101.94	79.74
Europe	1.45	5.22	7.70	8.34	7.33	9.32	10.79
Oceania	1.63	13.01	12.07	7.52	19.39	19.39	12.62
World	409.32	618.95	732.79	555.47	597.90	568.08	593.42

Faso (1.39 M ha), and Chad (1.22 M ha). India being the largest grower of millets contributes about 26.6% of the global harvested area (FAOSTAT 2018).

Considering the 1961 year as a baseline, the average harvested area of millets from different continents indicated that, Asia was the largest grower of millet (27.1 M ha) followed by Africa (11.8 M ha), Europe (4 M ha), America (2.6 M ha), and Oceania (0.02 M ha) (Table 2.2). However, the recent data of 2016–2018, revealed that Africa recorded the highest area (20.7 M ha) followed by Asia (10.9 M ha), Europe (0.4 M ha), America (0.16 M ha), and Oceania (0.03 M ha). In different regions of the African continent, West Africa recorded the highest millet cropping area with 14.3 M ha followed by North Africa (3 M ha). West Africa alone contributed around 44.3% of the world millet growing area (Table 2.2; FAOSTAT 2018). Interestingly, among the seven countries with the largest harvested area, four countries (Niger, Nigeria, Mali, and Burkina Faso) are from West Africa. In North West Africa, Sudan occupies the largest harvested area under millets whereas Chad is the number one millet growing country in North Central Africa. In the Asian scenario, more than 80% of Asia's millets are grown in India followed by China. In America, United States (0.16 M ha) followed by Argentina are the largest grower of millets. Likewise, Ukraine, Poland, France, and Belarus are the largest millet growing countries of Europe.

In spite of tremendous potential of millets, the cultivated area across the globe has declined at the rate of 22.5 lakh ha area per decade. Worldwide the area has come down by 25% in 2016–2018 compared to 1961–1963 levels (Fig. 2.2). In Asia, around 148% reduction in millet cropping area were reported from 1961–1963 to 2016–2018 (Table 2.2; FAOSTAT 2018). The sharp reduction in millets cropping area in Asia is mainly attributed to shrinking portfolio of food crops, heavy dependence on starchy food such as rice, maize, and potato and lower farm profitability.

Table 2.4 Area, production, and productivity of millets in India (1951–2018)[a] (INDIASTAT 2020)

	Finger millet			Sorghum			Pearl millet			Total millets		
	Area (M ha)	Production (Mt)	Productivity (kg/ha)	Area (M ha)	Production (Mt)	Productivity (kg/ha)	Area (M ha)	Production (Mt)	Productivity (kg/ha)	Area (M ha)	Production (Mt)	Productivity (kg/ha)
1951–1960	2.33	1.70	725.4	17.09	7.65	446	10.66	3.21	300.00	30.08	12.56	300
1961–1970	2.49	1.86	746.8	18.30	9.29	506.9	11.58	4.00	345.00	32.37	15.14	345
1971–1980	2.51	2.41	956.3	16.36	9.75	596.6	11.97	5.35	444.40	30.84	17.51	444.4
1981–1990	2.43	2.57	1059.1	15.83	11.09	701.6	10.94	5.08	460.40	29.20	18.73	460.4
1991–2000	1.85	2.42	1319.5	11.76	9.80	831	10.32	7.33	64.60	23.92	19.55	657.3
2001–2010	1.48	2.07	1395	8.76	7.27	836.9	9.39	7.87	829.50	19.63	17.20	829.5
2011–2020	1.17	1.79	1591.375	6.07	5.07	883.375	8.05	9.02	1130.10	15.29	15.88	1130.1

[a]Each figure is an average of 10 years for the respective period, for example, 1951–1960

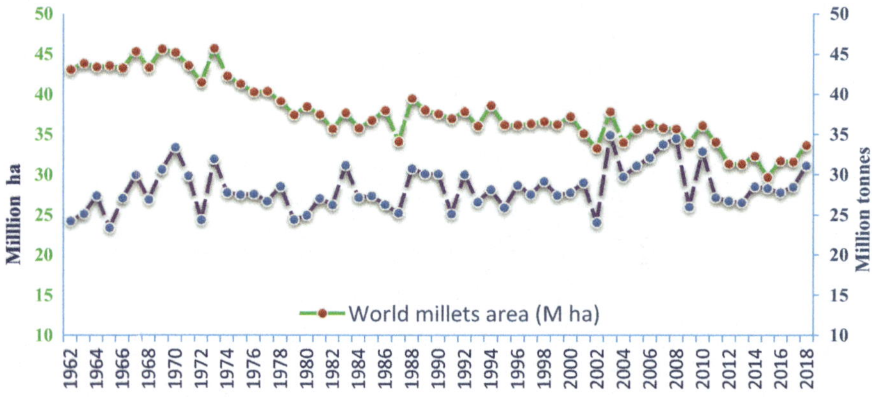

Fig. 2.2 Global trends in cultivation of millets (FAOSTAT 2018). The figure illustrates trends in area under cultivation and production from 1962 to 2018

On the other hand, in Africa an increase in area under millets has been observed by 42% compared to 1961–1963 levels (Fig. 2.4a). Highest increase was observed in Western Africa from 8.8 M ha in 1961–1963 to 14.3 M ha in 2016–2018 (Table 2.2).

Sorghum is cultivated in more than 105 countries with 37 countries having more than 0.1 M ha area and 10 countries (Sudan, Nigeria, India, Niger, America, Burkina Faso, Ethiopia, Mali, Mexico, and Chad) has more than 1 M ha cultivated area. Among the different continents, Africa holds the largest area with 70% of the world sorghum cropping area followed by Asia and America. Sudan is the largest sorghum grower in the world with ~7.1 M ha area followed by Nigeria (6.12 M ha) and India (4.9 M ha). Over the last five decades, sorghum cropped area around the world has reduced at the rate of 15 lakh ha per decade. However, like small millets, the sorghum area has increased significantly (44%) from 1961 to 2018 in Africa (Table 2.3). The steady increase in African countries is attributed to hostile weather patterns in last decades, which has resulted in the replacement of more water demanding crops like maize with small millets and sorghum.

2.3 Production

Developing countries produce and consume around 97% of the world's millets and only a small fraction comes from the rest of the world. The average data of millet production of different continents from 1961 to 1963 indicated that Asia is the largest producer of millets (13.2 Mt) followed by Africa (6.9 Mt), Europe (2.3 Mt), America (0.32 Mt), and Oceania (0.03 Mt) (Table 2.2; Fig. 2.3a). In Asia, millet production concentrates mainly in India, China, and Nepal. With ~37.5% of global output, India is the largest producer of millets followed by Sudan and Nigeria. In India, millet production is concentrated mainly in dry and arid regions where rainfall is low and erratic. The most widely produced millet is pearl millet, which is mainly grown in the states of Rajasthan, Uttar Pradesh,

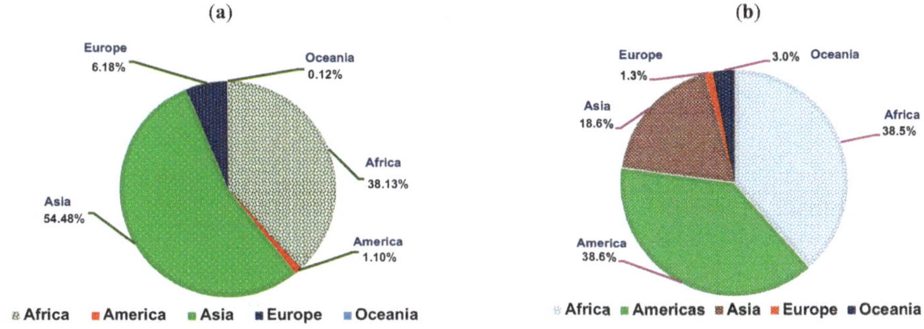

Fig. 2.3 Production share of millets by region (FAOSTAT 2018). (**a**) Production share of Pearl millet and small millets (Average 1961 to 2018). (**b**) Production share of sorghum (Average 1994 to 2018)

Gujarat, Madhya Pradesh, and Haryana which accounts for 56% (9 Mt) of total millets production in India (Table 2.4). Among the minor millets, finger millet is the most widely produced millet in India with a production of 1.79 Mt from the total cropped area of 1.17 M ha (Table 2.4). The main finger millet growing states are Karnataka, Uttarakhand, Maharashtra, Tamil Nadu, Odisha, and Andhra Pradesh contributing to more than 90% of national production. After finger millet, kodo millet is the most widely grown minor millet in India.

Africa contributes to a major share in global millets production because millets are an important staple food in larger part of the continent. The major millet-producing African countries are Nigeria, Sudan, Mali, Guinea, and Ghana. In America, the United States is the major North American millet producing country whereas in South America it is dominated by Argentina. Millets production in Europe is concentrated in the drier regions of Eastern Europe and shares 90% of the regions total. Likewise, in Oceania, Australia and New Zealand are the major producers of millets. During the last 58 years from 1961, millets production enhanced in Africa especially in West African countries but declined in most parts of the world (Fig. 2.2). The most recent production data (2016–2018) revealed that Africa recorded the highest millet production (14 Mt) after overtaking Asia (13.9 Mt) followed by Europe (0.62 Mt), America (0.36 Mt), and Oceania (0.03 Mt). The production increase in Africa was due to area expansion into drier lands due to climatic vagaries which ultimately resulted in area reduction under more water demanding crops like maize. In Asia, millets production peaked during 1980s thereafter decrease gradually due to the reduction of cultivated areas in India and China (Tables 2.4 and 2.5). Interestingly, millet production in Oceania remains almost constant from 1961 to 2018.*Each figure is an average of 3-years for the respective period, for example, 1961-1963

Global production of sorghum during 2018 was 59.3 million tons (Table 2.3). The United States (9.27 Mt) is the world's largest producer of sorghum followed by Nigeria (6.8 Mt), Sudan (4.95 Mt), Ethiopia (4.93 Mt), India (4.8 Mt), Mexico (4.53 Mt), and China (2.19 Mt). These seven countries represent 57% of global sorghum production. In Africa, sorghum production is concentrated mainly in

Table 2.5 Global millets (except sorghum) productivity by region[a] (FAOSTAT 2018)

	Millet grain productivity (kg/ha)						
	1961–1963	1971–1973	1981–1983	1991–1993	2001–2003	2011–2013	2016–2018
Africa	586	559	714	649	720	597	677
Eastern Africa	707	766	987	904	900	1061	1170
Middle Africa	642	586	605	494	580	522	586
Northern Africa	641	360	301	262	260	319	525
Southern Africa	339	281	305	265	249	201	221
Western Africa	558	547	750	674	804	602	673
Americas	1223	1176	1247	1486	1269	1361	2166
Northern America	1192	1297	1356	1448	1214	1326	2177
Central America	0	0	0	1030	909	1000	5595
Caribbean	0	0	0	0	0	0	0
South America	1236	1087	1184	1595	1838	1763	1576
Asia	562	666	777	811	938	1171	1276
Central Asia	0	0	0	523	760	1025	1104
Eastern Asia	889	1195	1644	1846	1786	2219	2308
Southern Asia	433	489	584	686	864	1104	1210
South-eastern Asia	322	254	801	673	712	985	1025
Western Asia	1013	1180	858	601	682	732	689
Europe	583	982	761	765	1079	1301	1517
Eastern Europe	578	979	759	764	1057	1249	1424
Northern Europe	0	0	0	0	0	0	0
Southern Europe	1257	1551	1792	2252	2212	1727	1869
Western Europe	1293	2372	3129	0	3050	3400	3909
Oceania	1087	1067	975	855	814	1015	1022
Australia and New Zealand	1087	1067	975	855	814	1015	1022
World	575	656	761	736	823	832	900

[a]Each figure is an average of 3-years for the respective period, for example, 1961–1963

Nigeria, Sudan, Ethiopia, Burkina Faso, Chad, Mali, and Niger. With 23% of total output, Nigeria is the largest producer of sorghum in Africa. In America, United States, Mexico, Brazil, and Argentina are the major sorghum-producing countries. In Asia, India and China are the major producers with 88.5% of the region's total (FAOSTAT 2018). Likewise, France, Italy, Russian Federation, Albania, Italy, Romania, Spain, and Ukraine are the major sorghum-producing countries of Europe.

World sorghum grain production achieved its peak during 1981 with ~73.2 Mt of grains, nearly 44.4% higher than the production levels recorded in 1960s (Table 2.3; Fig. 2.3b). During the 1980s–2018 global sorghum production was declined by 19% and the cropping area also declined during the same period by 8.1%. From the year 1961, production increased mainly in the African continent but decreased in other parts of the globe. Highest negative growth was recorded in Asia, which is mainly attributed to a sharp decline in the cultivated area in India and China. The sharp decline is evident by the fact that in 1950s, India harvested 7.65 Mt of sorghum from the total cropped area of 17 M ha. Currently, the area and production are reduced to 6 M ha and 6 Mt respectively (Table 2.4).

In general, production figures of all the millets from 1961 to 2018 revealed that global production slightly increased from 249 lakh tons to 290 lakh tons despite a sharp reduction in total cropping area. Loss of area might have been compensated by increasing production per unit area through adopting high-yielding varieties and good management practices. The major increment was recorded in Africa due to expansion in the cropping area (Fig. 2.4b).

2.4 Productivity

Millet productivity has increased by 36% in the last seven decades (Fig. 2.4c). Comparative account of productivity across the continents from 1961–1963 to 2016–2018 revealed that highest increment was observed in Europe (583–1517 kg/ha; 62%) followed by Asia (562–1276 kg/ha; 56%), America (1223–2166 kg/ha; 44%), and Africa (586–677 kg/ha; 13%).

However, Oceania has recorded a negative growth rate in millet productivity (1087–1022 kg/ha; minus 6%) (Fig. 2.4a; Table 2.5). In Africa, after the 1960s, yield level increased gradually and reached a peak in the 1980s, before declining in the 1990s. Thereafter, productivity (720 kg/ha) in 2001–2003 reached the highest level which was 18.6% higher compared to the 1960s (Table 2.5). After that again negative growth rate in productivity was observed. However, in Eastern Africa yield level increased continuously from 1961–1963 to 2016–2018 by 40% which is almost comparable to America. Among different regions of Asia, South-Eastern Asia recorded the highest millet productivity by 69% from 1961–1963 (322 kg/ha) to 2016–2018 (1025 kg/ha). However, in 2016–2018, the highest productivity (2308 kg/ha) was recorded in Eastern Asia followed by Southern Asia. Among different countries, Mexico has recorded the highest millets productivity (15.7 t/ha) followed by Uzbekistan (6.76 t/ha), Austria (4.4 t/ha), and Switzerland (4 t/ha) (FAOSTAT 2018). In the Indian scenario, millets productivity has increased by more than 73% in the last seven decades. In the last decade, average millet productivity of 1130 kg/ha was recorded in the country (Table 2.4). Among different millets finger millet has the highest productivity (1591 kg/ha) followed by pearl millet (1130 kg/ha) and sorghum (883 kg/ha) (Table 2.4). Improvement in productivity may be attributed to the development of input responsive millet cultivars, higher input use efficiency, and adoption of better crop and resource management practices.

Over the years, yield enhancement in sorghum has been observed in different sorghum-growing countries. This is attributed to the development of input responsive varieties and hybrids and better agronomic management practices. In India, its productivity varies among regions due to variability in soil type, rainfall, and seasons. In India, sorghum is grown twice a year, during the rainy season and post rainy season. Rainy season crop gives higher grain yield than post rainy season crop (Rakshit and Wang 2016). Globally, Oman has reported highest sorghum grain yield

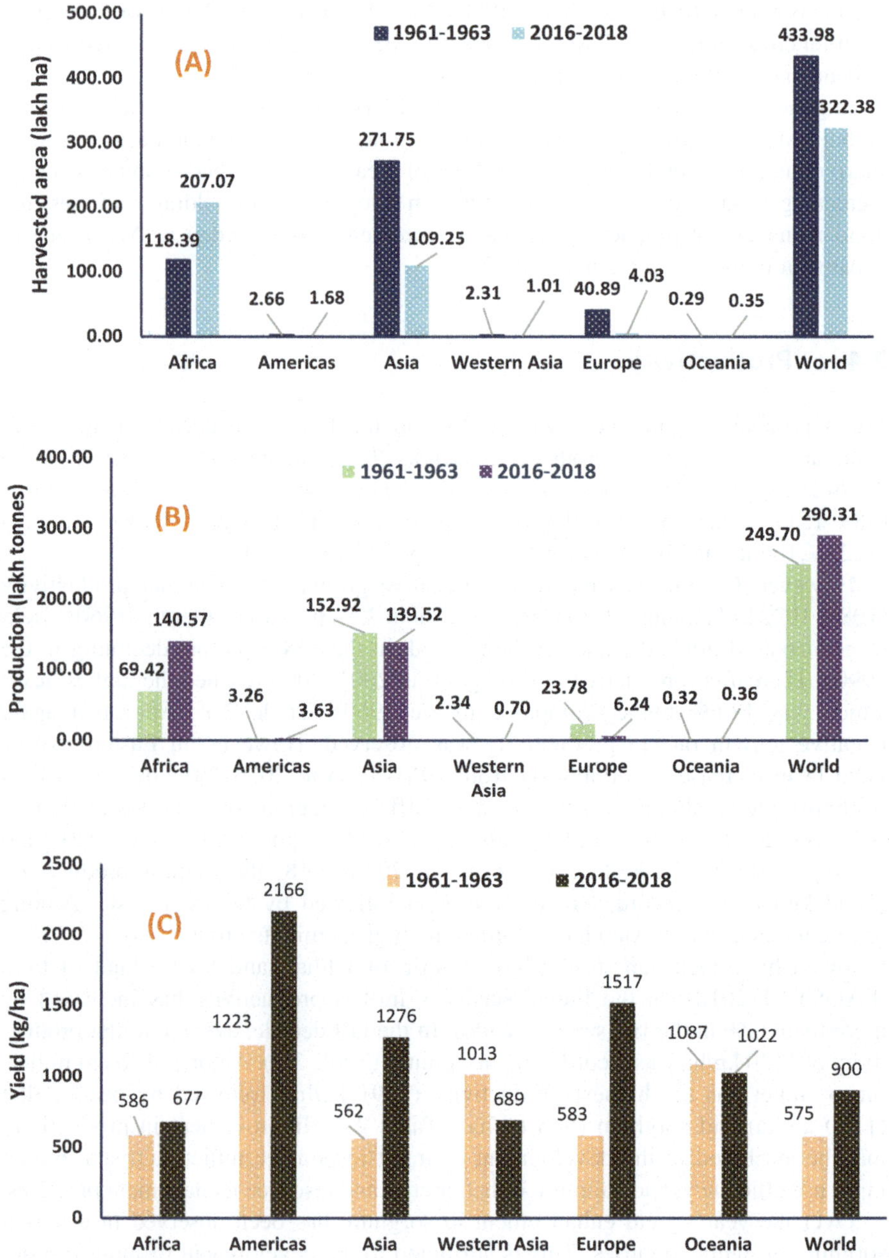

Fig. 2.4 Changes in harvested area (**a**), production (**b**), and productivity (**c**) of millets in different continents over the last 58 years (FAOSTAT 2018)

(28.1 t/ha) followed by Jordan (22.1 t/ha) and Algeria (13.3 t/ha) (FAOSTAT 2018). In Asia, China has recorded highest grain yield of sorghum (4.5 t/ha).

2.5 International Trade

Most of the millets grains are consumed, where they are produced as 97% of millets are cultivated by developing nations especially by resource poor and marginal farmers. However, its import and export increased gradually from 1960s to 2017 by 25.4% and 25.9%, respectively. In the last decade, the import and export of millets grain were 374.5 and 376.4 thousand tons, respectively (Table 2.6). The sharp rise in import was observed during the 1970s; it was mainly attributed to high import of millets in Europe. Global import and export value of millets also increased and highest value (155.26 and 127.60 million US$, respectively) were recorded during 2011–2017 (Table 2.6). Average data of 2010–2017 revealed that Asia is the largest importer of millet grains; it alone shares more than 65% of global import. Similarly, America is the largest exporter of millet; it alone shares more than 83% of global millet export. Among different countries, India, the United States, Argentina, and China together contribute more than 33% of millets export (FAOSTAT 2018). India contributes a major role in pearl millet export and the United States in proso millet. Similarly, China is the largest exporter of foxtail millet.

Globally, sorghum trade is mainly dependent on the demand for animal feed and price differences between sorghum and maize (Hariprasanna and Rakshit 2016). Highest expansion was observed between 1960s and 1980s especially after the mid-1960s (Table 2.6). Import and export declined from the 1990s onward, remained at 7.7 and 7.6 Mt, respectively, and then started to increase further. The highest global value of sorghum import (2290.29 million US$) and export (1926.80 million US$) were recorded during 2010–2017 (Table 2.6). In global sorghum export, around 90% share comes from Argentina, Australia, China, and the United States (Hariprasanna and Rakshit 2016). China has set a record of largest exporter and importer of sorghum during the mid-1980s (FAOSTAT 2018). India started exporting large amount grains mostly after 2002 and before that export was inconsistent.

2.6 Production Constraints

Millet farming is mainly concentrated in developing nations where average productivity is still below the world average (Sood et al. 2020). In most of these countries, markets for millet grains are not well established resulting in poor economic returns to the farmers. Furthermore, millets seed supply in most of the developing countries is dependent on informal seed chain. This results in nonavailability of improved seeds and large-scale cultivation of less productive and heterogeneous landraces or local cultivars (Rakshit and Wang 2016). However, in developed nations and some developing nations such as India and China, farmers have better socioeconomic

Table 2.6 Importer and exporters of millets and sorghum by continents[a] (FAOSTAT 2018)

	Major millet importer continents ('000 tons)						World import value (million US$)
	Africa	Americas	Asia	Europe	Oceania	World	
1961–1970	105.30	1.20	37.53	135.24	0.00	279.28	20.58
1971–1980	162.81	4.01	68.10	172.57	0.18	407.67	67.84
1981–1990	38.43	10.65	46.53	136.40	1.00	233.01	56.61
1991–2000	15.25	18.66	106.23	146.61	0.96	287.71	69.45
2001–2010	60.88	20.54	122.13	124.37	1.31	329.23	99.49
2011–2017	46.73	27.50	177.23	121.07	2.03	374.56	155.26
	Millet export quantity ('000 tons)						World export value (million US$)
	Africa	Americas	Asia	Europe	Oceania	World	
1961–1970	99.52	143.71	21.69	2.61	11.06	278.58	15.72
1971–1980	46.75	129.27	34.59	14.35	23.49	248.46	32.15
1981–1990	18.75	125.87	10.85	27.07	20.23	202.76	39.06
1991–2000	54.52	86.55	50.23	62.80	16.09	270.19	55.72
2001–2010	24.35	45.60	129.46	103.87	7.22	310.49	84.73
2011–2017	21.35	67.41	112.67	172.70	2.28	376.42	127.60
	Major sorghum importer continents ('000 tons)						World import value (million US$)
	Africa	Americas	Asia	Europe	Oceania	World	
1961–1970	75.63	72.56	2496.11	1949.32	0.00	4593.62	272.00
1971–1980	129.42	1207.76	5565.52	2354.67	1.49	9258.85	1104.97
1981–1990	318.73	3024.48	5325.60	2240.61	13.23	10,922.65	1424.16
1991–2000	373.43	3607.20	3129.20	607.32	52.43	7769.57	1002.23
2001–2010	744.47	3326.81	1686.60	1132.47	68.26	6958.61	1270.37
2011–2017	846.32	1587.97	5473.41	392.69	53.45	8353.83	2290.29

(continued)

Table 2.6 (continued)

	Sorghum export quantity ('000 tons)						World export value (million US$)
	Africa	Americas	Asia	Europe	Oceania	World	
1961–1970	175	4681	48	109	27	5040	237.80
1971–1980	291	8329	169	409	697	9896	1008.33
1981–1990	252	9372	510	205	763	11,102	1293.72
1991–2000	202	6714	210	256	307	7690	877.56
2001–2010	61	5738	126	255	425	6604	987.81
2011–2017	140	6848	133	316	822	8260	1926.80

[a]Each figure is a 10 years average for the respective period, for example, 1961–1970, except 2011–2017

conditions, well-developed marketing system, better accessibility to inputs including improved varieties. Together these factors have a positive impact on the millet production scenario in these countries resulting in higher productivity compared to Africa. Many minor millets are not adapted to modern agroecosystems and mechanization. This is mainly because of some inherent problems like high seed shattering and unsynchronized maturity. Besides these basic traits, grain size is also an important yield component as the very small seeds of small millets are causing difficulties for mechanical planting and harvest and ultimately for their commercialization. Seeds of minor millets are subjected to dehulling before human consumption. The traditional methods of dehulling followed in developing countries are labor intensive and time-consuming (Sood et al. 2015). The drudgery involved in manual processing is an important factor in reduced consumption and commercialization of small millets at a large scale.

Climatic factors such as rainfall pattern and distribution, edaphic factors such as soil type, soil fertility, agronomic management, and moreover socioeconomic status of farming communities are equally important for better performance of millet production system (Sood et al. 2019). Incidence of diseases, insect-pests, parasitic nematodes, birds, parasitic plants, and weeds are the most important biotic constraints associated with millets. The important diseases of millets are downy mildew (sorghum and pearl millet), blast (finger millet), grain mold (sorghum), smut (foxtail millet, barnyard millet, teff, and sorghum), rust (sorghum, teff, and foxtail millet), ergot (pearl millet and sorghum) and charcoal rot (sorghum) (Strange and Scott 2005; Das 2013). Weed infestation is also considered as a major constraint in the global millet production as more than 29% reduction in millet grain yield is associated with weed infestation only (Burkill 1985). The poor initial vigor of small millets promotes excessive growth of weeds resulting in more competition for sunlight, nutrient, space, and water in early growth stage, which ultimately reduce crop productivity (Lall and Yadav 1982). Striga, a semi-root parasitic weed is one of the major constraints for millet production in Africa causing huge yield losses in millets. Yield reduction due to striga is higher in sorghum and pearl millet than other crops (Ejeta 2007). Bird damage is also considered as a major biotic threat for millet growers, yield reduction may reach 100% in isolated crop fields (Sood et al. 2015). The manual weed management in the absence of robust preemergence weedicide coupled with manual bird scaring increases cost of quality seed production in minor millets.

Abiotic constraints of millet production are mainly associated with environmental and soil factors such as moisture stress, nutrient stress, salinity, alkalinity, acidity, and heat stress. Among all, moisture stress is considered the most important constraint for millet production, as millets are mostly grown by resource poor farmers in drylands. Drought may occur at any physiological growth stage of millets. In African countries, drought is considered as one of the most important stress for millet production (Matanyaire 1996; Gebretsadik et al. 2014). In India, millets (sorghum and finger millet) are cultivated in two seasons, in the rainy and post rainy season. The low productivity of post rainy season sorghum is mainly associated with terminal drought stress (Patil 2007). As millets are grown on marginal land having low soil fertility and low organic carbon, salinity, and alkalinity that leads to low productivity of millets. Soil salinity and poor drainage severely affect the crop during the seedling emergence stage (Macharia et al. 1994). Changing food habits and consumer preferences have led to the shifting of land for the cultivation of other high-value cereal grains thereby lowering the production of millets. For instance, in India, the millet cropping area is reduced to 2.3 M ha during 2011–2012 compared to 8 M ha during the late 1940s, this reduction was mainly associated with shifting of millet cultivated area to other cereals grains (Seetharam 2015).

2.7 Conclusion

By virtue of their unique nutritional profile, tremendous health benefits and C_4 photosynthetic pathway, millets are well-suited crops to diversify cropping systems for climate resilient agriculture. Since ages, millets are being grown by resource-poor farmers of drylands and tribal communities inhabiting less productive and fragile ecosystems. However, the growing awareness about their potential health benefits and industrial uses resulted in the renaissance of millets. Shrinking of global millet cropping area is the main concern associated with millet production. Lack of improved cultivars, agricultural inputs, and policy support are major limiting factors associated with lower productivity of millets and shrinking area. Well-planned and long-term public sector investment for multidisciplinary research activities, jointly by major growing countries are required for projecting millets as golden crops of the future. For instance, in India, the government is setting in place an Initiative for Nutritional Security through Intensive Millet Promotion (INSIMP). Recognizing their immense nutraceutical potential and climate resilient nature, the government of India has launched a national nutraceutical mission. The national nutraceutical mission is an arching national strategy, which has prioritized eight millets (sorghum, pearl millet, finger millet, barnyard millet, foxtail millet, proso millet, kodo millet, and little millet) and two pseudocereals (amaranth and buckwheat) and termed them as nutri-cereals. The government of India declared the year 2018 as the national year of millets to boost up the indigenous production of millets. The UN Food and Agriculture Organization (FAO), Rome has declared the year 2023 as the international year of millets, upon the request from the Indian government. Similar national and international multidisciplinary public sector initiatives are required by other major millet growing countries for their promotion and enhanced consumption. Furthermore, linking small millets to the industry through value addition will fetch higher returns to marginal farmers of Asia and Africa. On the whole, policy support in conjunction with concentrated crop improvement efforts and public awareness on nutritive values will help in regaining the lost cultivated area under millets.

References

Assefa K, Yu JK, Zeid M, Belay G, Tefera H, Sorrells ME (2011) Breeding tef [*Eragrostis tef* (Zucc.) trotter]: conventional and molecular approaches. Plant Breed 130:1–9

Burkill HM (1985) The useful plants of west tropical Africa, vol 2. Royal Botanic Gardens, Kew, UK, p 648

Das IK (2013) Disease management in grain, forage and sweet sorghum. In: Chapke RR, Bhagwat VR, Patil JV (eds) Sorghum cultivation for value-added diversified products and sweet sorghum perspectives. Directorate of Sorghum Research, Hyderabad, pp 99–104

Doggett H (1989) Small millets – a selective overview. In: Seetharam A et al (eds) Small millets in global agriculture. Oxford & IBH Publishing Co., Delhi, India, pp 59–70

Duke JA (1983) Handbook of energy crops. http://www.hort.purdue.edu/newcrop/dukeenergy/arundodonax

Ejeta G (2007) The Striga scourge in Africa–a growing pandemic. In: Ejeta G, Gressel J (eds) Integrating new technologies for striga control towards ending the witch-hunt. World Scientific Publishing Co, Singapore, pp 3–16

FAOSTAT (2018) Production-yield quantities of millets in world + (total) 1962–2018. http://www.fao.org/faostat/en/#data/QC/visualize. Accessed 25 May 2020

Gebretsadik R, Shimelis H, Laing MD, Tongoona P, Mandefro N (2014) A diagnostic appraisal of the sorghum farming system and breeding priorities in striga infested agro-ecologies of Ethiopia. Agric Syst 123:54–61

Gupta A, Mahajan V, Kumar M, Gupta HS (2009) Biodiversity in the barnyard millet (*Echinochloa frumentacea* Link: Poaceae) germplasm in India. Genet Resour Crop Evol 56:883–889

Hariprasanna K, Rakshit S (2016) Economic importance of sorghum. In: Rakshit S, Wang Y-H (eds) The sorghum genome. Springer International Publishing, Singapore, pp 1–25

Hegde BR, Gowda BKL (1989) Cropping systems and production technology for small millets in India. In: Seetharam A et al (eds) Small millets in global agriculture. Oxford & IBH Publishing Co, Delhi, India, pp 209–235

INDIASTAT (2020) Statistical information about India. https://www.indiastat.com/agriculture data/2/stats.aspx. Accessed 6 Apr 2020

Jijau C (1989) Importance and genetic resources of small millets with emphasis on foxtail millet (*Setaria italica*) in China. In: Seetharam A et al (eds) Small millets in global agriculture. Oxford & IBH Publishing Co, Delhi, India, pp 93–100

Lall M, Yadav LNS (1982) Critical time of weed removal in finger millet. Indian J Weed Sci 14:85–88

Macharia JM, Kamau J, Gituanja JN, Matu EW (1994) Effects of sodium salinity on seed germination and seedling root and shoot extension of four sorghum [*Sorghum bicolor* (L.) Moench] cultivars. Int Sorghum Millet Newsl 35:124–125

Matanyaire CM (1996) Pearl millet production system(s) in the communal areas of northern Namibia: priority research foci arising from a diagnostic study. In: Leuschner K, Manthe CS (eds) Drought-tolerant crops for southern Africa. In: Proceedings of the SADC/ICRISAT regional sorghum and pearl millet workshop, 25–29 July 1994

McDonough CM, Rooney LW, Serna-Saldivar SO (2000) The millets, food science and technology: handbook of cereal science and technology, 2nd edn. CRC Press, Boca Raton, FL, pp 177–210

National Research Council (NRC) (1996) Lost crops of Africa, Grains, vol I. The National Academies Press, Washington, DC, p 408. https://doi.org/10.17226/2305

Obilana AB (2003) Overview: importance of millets in Africa. In: Belton PS, Taylor JRN (eds) Proceeding of the workshop on the proteins of sorghum and millets: enhancing nutritional and functional properties for Africa, 2–4 April 2003, Pretoria, South Africa. http://www.afripro.org.uk/papers/Paper02Obilana.pdf. Accessed 25 May 2020

Patil SL (2007) Performance of sorghum varieties and hybrids during post-rainy season under drought situations in Vertisols in Bellary, India. J SAT Agric Res 5:1–3

Rakshit S, Wang Y-H (2016) The Sorghum genome. Springer International Publishing, Singapore, p 284

Sahib KH (1997) Importance of Proso millet in Indian agriculture. National Seminar on Small Millets, 23–24 April 1997, Coimbatore, India. pp 11–12

Seetharam A (1998) Small millets research: achievement during 1947–97. Indian J Agric Sci 68:431–438

Seetharam A (2015) Genetic improvement in small millets. In: Tonapi VA, Patil JV (eds) Millets: ensuring climate resilience and nutritional security. Daya Publishing House, New Delhi, p 649

Sheahan CM (2014) Plant guide for browntop millet (Urochloaramosa). USDA-Natural Resources Conservation Service, Cape May Plant Materials Center, Cape May

Sood S, Khulbe RK, Gupta AK, Agrawal PK, Upadhyaya HD, Bhatt JC (2015) Barnyard millet – a potential food and feed crop of future. Plant Breed 134:135–147

Sood S, Joshi DC, Chandra AK, Kumar A (2019) Phenomics and genomics of finger millet: current status and future prospects. Planta 250:731–751

Sood S, Joshi DC, Pattanayak A (2020) Breeding advancements in barnyard millet. In: Gosal SS, Wani HS (eds) Accelerated plant breeding, Cereal crops, vol 1. Springer Nature, Switzerland, pp 391–410

Strange RN, Scott PR (2005) Plant disease: a threat to global food security. Annu Rev Phytopathol 43:83–116

Taylor JRN, Emmambux MN (2008) Gluten-free cereal products and beverages. In: Arendt EK, Bello FD (eds) Gluten-free foods and beverages from millets. Elsevier, Amsterdam, p 464

Minor Millets: Profile and Ethnobotanical Scenario

3

Laxmi Rawat, A. K. Karnatak, T. S. Bisht, and Akshit Kukreti

Abstract

Minor millets are an agronomic community of genetically diverse species of cereal grasses, well adapted to a range of marginal growing conditions where major cereals are relatively ineffective, such as wheat, rice, and maize. Minor millets are grown in various soils in India, in varying rainfall regimes, and in areas where thermal and photographic cycles vary widely. Seven cultivated species, viz., finger millet, barnyard millet, foxtail millet, proso millet, little millet, kodo millet, and browntop millet represent minor millets. These millets provide millions of households with highly nutritious food and livelihood security, especially small and marginal farmers and residents of rainfed areas, particularly in remote tribal areas. They are now no longer referred to as coarse cereals but as nutricereals or nutraceutical crops, and are considered as a plausible answer to combat malnutrition and secret hunger worldwide. Indian tribal groups have a special link to minor millets as these crops have been an integral component of their agricultural systems and operations. Minor millets are being used by them from time immemorial not only to fight hunger but also for ethnomedical uses. This chapter emphasizes on the potential of minor millets for combating hunger, malnutrition and for ensuring food and nutritional security for tribal communities and discusses the initiatives being taken by the government and civil societies to

L. Rawat (✉) · A. Kukreti
Plant Pathology Division, College of Forestry, Rancihauri, V.C.S.G. Uttarakhand University of Horticulture and Forestry, Bharsar, Uttarakhand, India

A. K. Karnatak
V.C.S.G. Uttarakhand University of Horticulture and Forestry, Bharsar, Uttarakhand, India

T. S. Bisht
Department of Horticulture, School of Agriculture and Allied Sciences, HNB Garhwal University, Srinagar Garhwal (A Central University), Srinagar Garhwal, Uttarakhand, India

A. Kumar et al. (eds.), *Millets and Millet Technology*,
https://doi.org/10.1007/978-981-16-0676-2_3

promote millets-based farming system in India. Since these small millets have the capacity to cope up with the situation of current climate aberrations, it is imperative to increase the quality and productivity of these crops not only for tribal people living in harsh and difficult terrains but also for other masses living in suitable areas.

Keywords

Minor millets · Cereal grasses · Nutricereals · Malnutrition · Tribal people

3.1 Introduction

Advancement in agricultural research, innovation, and policy continue to put a firm focus on increasing production of the three principal staple crops, rice, wheat, and maize, while other crops remain low in public and private investment (Grovermann et al. 2018; Verma et al. 2018). Worldwide food and nutrients transition make a speciality of precise solution to feed the expected population of nine billion by 2050, equitably, healthily, and sustainably (Pradhan et al. 2019). Each wheat and rice are fed on as delicate flour and polished rice; and due to their smooth appearance and flavor fullness, our traditional grains like bajra (pearl millet), jowar (sorghum), ragi (finger millet), and rajgira (amaranth) have taken a back seat. In fact, the millets were called as "orphan crops" by the Father of Green Revolution, Dr. M. S. Swaminathanas. Millets used to represent 40% of the cultivated grains before the inexperienced green revolution, which after 1965 dropped to about half (Batra 2019).

Even though all grains, cereals in addition to millets, broadly speaking include starch (60–65%), but the concentrations of protein, dietary fiber, and minerals substantially vary among them. Wheat grains comprise 70%, while rice grains are composed of 90% starch. Protein content material tiers from 13% to 15% in wheat but is basically lowest in rice among all major cereals. The fiber content, as well as the minerals composition of each of the privileged grains, is also poor. Thinking about the present modern sedentary lifestyle, the quantity of refined carbohydrates we consume without an awful lot physical hobby is taking a toll on our health, as is clear by using growing malnutrition numbers (Batra 2019). The World Summit on Food Security, 2009, anticipated that as a minimum 70% more food production is needed by 2050 to feed the ever-growing population. It would require an annual boom of approximately 44 million tons that is 38% above present-day annual growth in food production (Tester and Langridge 2010). Similarly, a big populace in developing countries is laid low with dietary imbalance and prevalence of a couple of micronutrient deficiencies (Singh et al. 2015). Crops such as wheat, maize, and rice provide only food security, but crops such as minor millets account for a variety of securities like grain, fodder, fiber, nutrition, health, environment, and livelihood at minimal cost, offering great food and nutrition security opportunities (Gupta et al. 2017).

One feasible solution is identifying and enhancing the yield of conventional or native vegetation which are particularly adaptive to neighborhood climate, have excessive nutrient cost, and may efficiently resist biotic and/or abiotic stresses. Those plants in large part fed on with the aid of indigenous communities across the globe are frequently referred to as coarse cereals or minor millets (Gull et al. 2014; Pradhan et al. 2019). The unexploited millets known as "nutri cereal" crops are slowly being rediscovered and researched through the agricultural research and development community (Kumar 2016; Grovermann et al. 2018).

Millet grains are having better nutritional fiber and resistant starch. These slowly digestible nutritional elements provide satiety feeling for a longer period of time, and help to prevent constipation by accelerating meal movement through the gastrointestinal tract. They also bind to toxins and eliminate them from the gut thereby protecting the colon mucosa from cancers. The recorded contention is that millets have been effectively developed for centuries, which could demonstrate both their flexibility to an assortment of conditions, yet additionally some natural characteristics that merited the valuation for endless ages (Rawat et al. 2020). Millets is an aggregate term used to portray various little grained cereal grasses and oldest among the developed addition crop (Kumar 2016; Rawat et al. 2019). The large number of individuals living in the dry and semi-dry tropical regions relies upon millets as a wellspring of food in Asia and Africa landmass (Doggett 1989; Maloles et al. 2011).

Millets are commonly subdivided into two most important groups of species, major millets and minor millets. The crops underneath the umbrella of minor millets are finger millet (*Eleusine coracana*), barnyard millet (*Echinochloa frumentacea*), proso millet (*Panicumm iliaceum*), foxtail millet (*Setaria italica*), kodo millet (*Paspalum scrobiculatum*), little millet (*Panicum sumatrense*), and browntop millet (*Brachiaria ramose*) which can be cultivated as food and fodder crops (Fig. 3.1),

Fig. 3.1 Seven cultivated species of minor millets in India, viz., 1. finger millet, 2. barnyard millet, 3. foxtail millet, 4. proso millet, 5. little millet, 6. kodo millet, and 7. browntop millet

more often grown on poor and marginal lands in dry areas of temperate, subtropical, and tropical areas throughout the globe (Dwivedi et al. 2012; Gupta et al. 2017; Kukreti et al. 2017). Millets mature rapidly, a significant trait for rainfed farming, and need relatively little input compared to major cereals (Verma et al. 2018). Minor millets are also the staple food of the people in millet-producing areas in many Asian and African countries and are used to prepare various traditional foods and beverages such as idli, dosa, papad, chakli, porridges, breads, baby, and snack foods (Chandrasekara and Shahidi 2012). Minor millets are an important source of carbohydrates and proteins for people living in semi-arid zones.

They also contribute to national food security and potential health benefits, and hence millet grain is now receiving rising interest from scientists, technologists, and nutritionists. They are doing groundbreaking research on processing and adding value to minor millets to provide ready-to-eat and ready-to-cook products so big masses can take advantage of it. Value addition to minor millet grain also offers rural and tribal farmers a good opportunity for income generation, and supporting development and marketing may contribute to food and nutritional protection, job creation, and revenue generation (Verma et al. 2018). These healthy "super-foods," as the new health industry christened, have become part of modern daily diet of health-conscious people (Behera 2017; Rao et al. 2011).

In India, tribal population have largely lived on different types of protein-rich minor millets that have helped them sustain their health (Behera 2017). The resilience exhibited by using these crops facilitates them to regulate to distinctive forms of ecological niches and features made them pretty indispensable to rainfed, tribal, and hill agriculture in which crop substitution is tough. The protection of these important traditional minor millets in their natural environment will be an effective approach for ensuring future generations' food security.

3.2 Finger Millet (*Eleusine coracana*)

3.2.1 Origin and Distribution

The generic name *Eleusine* derives from the Greek cereal goddess, "*Eleusine*" while the common name finger millet suggests "finger-like" panicle branching. Finger millet is widely referred to as "nutritious millet" since grains are nutritionally better than many cereals, that provide people with equal quantities of proteins, minerals, calcium, and vitamins (Devi et al. 2014; Kumar 2016).

The finger millet belongs to the Poaceae family and was domesticated in East Africa (possibly Ethiopia) about 5000 years ago and introduced to India 3000 years ago. *Eleusine coracana* subsp. *africana* is the nearest wild relative to finger millet which was originated in Africa (Hilu and Johnson 1992). Scientific research has focussed the crop on its extraordinary ability to grow under high temperatures, low humidity, and weak soils (Shukla et al. 2015; Gull et al. 2014; Lata 2015).

3.2.2 Botanical Description

Finger millet is an annual tufted crop, which grows to a height of 30–150 cm and matures in 75–160 days. Leaves are small, grassy, and can produce several tillers and nodal branches. The panicle consists of a group of digitally arranged spikes which are also called fingers. The spikelets are composed of 4–10 florets serially arranged on the finger. The grain is oblong to circular and oval, in color it appears as reddish brown with finely corrugated surface (Duke 1979).

3.2.3 Edible Uses

The finger millet is milled with testa, which is typically rich in dietary fiber and micronutrients for the preparation of flour, and the whole meal is used in the preparation of traditional foods such as roti (unleavened bread), ambali (thin porridge), and mudde (dumplings) (Devi et al. 2014). Finger millet is eaten in Uttarakhand as roti, and a savory porridge cooked in buttermilk. Zan is the most popular porridge recipe made from finger millet and vegetables by the Monpa tribes of Arunachal Pradesh (Bhat et al. 2019). For preparations like cakes, pancakes, porridges, and even local Indian cuisines like bhakri, dhebra, papad, etc., the whole grain or refined finger millet flour is used. In addition to having a decent shelf life, the flour is free from gluten and fat, easy to digest, and needs little cooking (Saturni et al. 2010) even the finger millet straw is excellent as an animal feed with up to 60% digestible nutrients in total (Gupta et al. 2017). Hence, finger millet has retained high socioeconomic significance in the Indian and African semi-arid tropical regions subsistence farmers background (Gull et al. 2014).

3.2.4 Medicinal Uses

The products derived from finger millet are useful for the growth of bone mass in growing children, as well as for the prevention of osteoporosis and other bone disorders in adults and aging populations (Kumar et al. 2016). The seed is astringent and used to treat fever, biliousness, and hepatitis (Duke and Ayensu 1985). The seed coat is rich in phytochemicals like polyphenols and nutritional fiber and is also very excessive in minerals specially calcium (Devi et al. 2014). Chethan and Malleshi (2007) confirmed upto 2.3 ± 0.3 gallic acid equivalents in wholemeal and upto 6.4 ± 1.5 inside the seed coat of finger millet grains. The abundance of these minerals in finger millet may be in a roundabout way linked with its capability to alleviate type-2 diabetes hazard. Due to excessive polyphenolic content inside the seed coat of finger millet reduces the threat of most cancers and diabetes and excessive fiber that promotes sluggish digestion and blood sugar balance (Devi et al. 2014).

3.3 Barnyard Millet

3.3.1 Origin and Distribution

Barnyard millet (*Echinochloa frumentacea*) is an ancient millet crop grown in warmer and temperate parts of the world, growing since 5000 B.C. in India and then 3000 B.C. in China. *Echinochloa* species have very few cultivable forms and are consequently cultivated by marginal farmers as minor millet. *Echinochloa frumentacea* (Roxb.) Link; syn. *E. colona* var. *frumentacea* (allohexaploid, $2n = 6x = 54$), commonly known as Indian barnyard millet, originated from wild *E. colona* (L.) (Jungle rice), and shows an evolutionary parallel in India and Africa. *E. frumentacea* has four races that are widely cultivated in Central Africa, India, Malawi, Nepal, and Tanzania, namely *stolonifera*, *intermedia*, *robusta*, and *laxa* (Doggett 1989; Upadhyaya et al. 2014).

3.3.2 Botanical Description

Echinochloa frumentacea is an annual grass with rather steep and erect stems, up to 242 cm tall, leaf length and arc width 15–40 cm long and 1–2.5 cm wide, mostly green plants, but violet tinges are also found in vegetative and reproductive parts, leaf blades are smooth and glabrous, culms slender to strong. The inflorescence is a contract green to violet, typically erect and compact, 1–28 cm long, with numerous 20–70 and 1–3 cm long racemes, seldom drooping, and awnless. The spikelet present on the panicle is small, unbranched, tightly clustered, 2–4 mm long, acute, and awnless on the rachis (Renganathan et al. 2020). The spikelet is filled with stiff bristles and crowded (You et al. 2011; Bicha and Noguchia 2012). But this plant is easily distinguished by its lack of hairiness, even when very young stage (Chauhan 2013).

3.3.3 Edible Uses

Barnyard millet dehusked grain is cooked like rice, and eaten or turned into porridge. In Uttarakhand, barnyard millet is eaten as paleu or chencha, a savory cooked porridge in buttermilk. The grain is processed very close to the rice parboiling in some parts of South India (Bhat et al. 2019).

3.3.4 Medicinal Uses

Barnyard millet sprouted seed is astringent, acidic, emollient, and stomachic. It is used for the treatment of abdominal dyspepsia, impaired digestion, and nutritional stagnation. White seeds are refrigerant and are used in cholera and fever care. Green seeds are diuretic and make virility stronger (Chauhan and Jhonson 2011).

3.4 Proso Millet (*Panicumm iliaceum*)

3.4.1 Origin and Distribution

Archeological researchers suggest that proso millet domestication took place at the beginning of the Holocene when global temperatures were warmer and new plants and habitats became introduced to hunter-gatherers (Habiyaremye et al. 2017). The wild ancestor of proso millet has yet to be identified (Miller et al. 2016) though weedy types of millet are found throughout Eurasia, which may have a wild offspring (Zohary et al. 2012). The allotetraploid form of proso millet, with or similar to *Panicum capillare*, and *Panicum repens* as ancestors, is revealed by chromosomal in situ hybridization with genomic DNA and phylogenetic evidence (Hunt et al. 2014). Evidence suggests important changes in the cultivation of proso millet on the Tibetan plateau before it was abandoned in eastern Tibet (Habiyaremye et al. 2017). Proso millet was later mostly substituted for wheat and barley on the Tibetan plateau (Boivin et al. 2012).

3.4.2 Botanical Description

Proso millet or common millet (*Panicumm iliaceum*), with low moisture requirements, is a relatively short-term emergency or fast-season irrigated crop. It is well adapted to many types of soil and climate conditions. Proso millet is a short seasonal crop compared to all millets and can complete its life cycle within 60–100 days of planting. It is an annual summer herb, most widely cultivated as a late-seeded summer crop (Rao 1989; Williams et al. 2007; Baltensperger 2002). Grains are round, approximately 3 mm long and 2 mm wide, and contained in a smooth shell, usually white or creamy-white, yellow or red in color, but may be gray, brown or black (Hardman 1990; Changmei and Dorothy 2014). Proso millet, with few tillers and an adventive root system, ranges from 30 to 100 cm tall (Baltensperger 2002). Historically, this millet was cultivated in Russia, China, the Balkan countries, and Northern India, later was replaced by rice and other cereals in most areas (Bhat et al. 2019).

3.4.3 Edible Uses

The protein content of proso millet, compared to wheat, is approximately 11% (dry base) and richer in essential amino acids (leucine, isoleucine, methionine) (Kalinova and Moudry 2006).

It can be cooked in various ways and prepared. Similar to rice, the grains can be boiled, steamed to make salad, or cooked completely. Proso millet can be a substratum in distilled liquors and beers, and is used in Africa and Asia to make fermented beverages (Habiyaremye et al. 2017).

3.4.4 Medicinal Uses

The intake of proso millet and other millets is associated with a decreased risk of type 2 diabetes mellitus because whole grains are a rich source of magnesium. Since the incidence of migraine headaches and heart attacks can also be decreased by magnesium, people with atherosclerosis and heart disease benefit from it (Shobana and Malleshi 2007; Gelinas et al. 2008).

3.5 Foxtail Millet (*Setaria italica*)

3.5.1 Origin and Distribution

One of the oldest cultivated millet crops is foxtail millet (*Setaria italica*), which needs warm weather and limited water for rapid ripening during the hot and dry months of the year. In global millet production, it comes second after pearl millet (Food and Agricultural Organization 2005; http://faostat.fao/org/). *Setaria italim* (L.) was the Roman "panicum," and *Setaria italim* (L.) is currently cultivated worldwide. The cultivated form of *Setaria italica* is its wild ancestor *S. viridis* (Wang et al. 1995; Le Thierry d'Ennequin et al. 2000). The most primitive archeological remnants of foxtail millet have been discovered in the ruins of Cishan and Peiligang in Northern China's Yellow River Valley, almost 7400 and 7935 years ago, respectively (Li and Wu 1996; Doust et al. 2009). Its domestication may have taken place anywhere in the area that stretches from Europe to Japan.

3.5.2 Botanical Description

Foxtail millet is a C4 monocot annual with slender, erect, leafy stems that can grow to heights of 90–180 cm. It has a thick root system, with adventitious roots usually small and lanky. The leaves are arc-broad and lack hairiness while the culms with hollow internodes are vertical and slender (Lata et al. 2012). It has the structure of a distinctive domesticated plant consisting of a single stalk or a limited number of tillers, and large inflorescences that mature more or less uniformly (Doust et al. 2009). The inflorescence is a constricted panicle due to its small branches, which sometimes nodes at the top and looks like a needle. The spikelets (one yellow pistilled flower per spikelet) are packed and combined with firm bristles (normally lengthy sterile branches and regularly reddish or purplish in shade) giving the panicle the advent of a foxtail (http://www.plantguide.org/foxtail-millet.html). Bristles are normally lengthy and regularly reddish or purplish. They deliver the panicle a look of a foxtail, which is the not unusual call for cultivated millets belonging to the genus *Setaria*. Each spikelet consists of most effective one flower with a yellow pistil (Lata et al. 2012).

3.5.3 Edible Uses

A significant amount of nutritional components are contained in foxtail millet, mainly starch, protein, vitamins, minerals, and crude fiber. All these nutritional components have made foxtail millet an essential ingredient in Chinese noodle preparation, nourishing gruel or soup, brewing alcoholic drinks, cereal porridges, and pancakes (Sharma and Niranjan 2018). Typically, foxtail millet grain is cooked whole like rice (millet rice) or made into meal. It is also eaten in India as stiff porridge called sargati or as leavened bread known as roti after the dehulled grain was milled into flour (Bhat et al. 2019). It is also possible to sprout the seed until it is used as it becomes sweeter (Duke and Ayensu 1985).

3.5.4 Medicinal Uses

Foxtail millet is also a good source of crude fiber, assists in the digestive process, and helps to stimulate bowel movement, producing a laxative effect that supports a balanced digestive system, like most millets. Foxtail millet also shows many health benefits such as cancer prevention, hypoglycemic effects, and hypolipidemic effects (Sharma and Niranjan 2018).

3.6 Kodo Millet (*Paspalum scrobiculatum*)

3.6.1 Origin and Distribution

Kodo millet (*Paspalum scrobiculatum*) is grown extensively in poor soil and it is considered to be highly resilient, resistant to drought, and grows on stony or gravelly soils that would not sustain other crops. The length is comparatively long, that takes 5 or 6 months to mature compared with 2–4 months for the other millets (Bhat et al. 2019). Kodo millet, a tropical African native, was thought to have been domesticated in India around 3000 years ago (de Wet et al. 1983). Cow grass, rice grass, ditch millet, Native *Paspalum*, or Indian crown grass are also known as kodo millet (*Paspalum scrobiculatum*). It is cultivated in India, Philippines, Indonesia, Vietnam, Thailand, and West Africa. In India, kodo millet grown in the Deccan region (Gujarat, Karnataka, and elements of Tamil Nadu), some regions of Maharashtra, Odisha, West Bengal, Rajasthan, and Uttar Pradesh and is historically consumed as healthy and vitality meals particularly in rural India (Hegde and Chandra 2005; Deshpande et al. 2015).

3.6.2 Botanical Description

There are more than 400 species in the *Paspalum* genus, usually an annual crop, but some cultivars emerge at the nodes and culminate after mature plant flowers and mature inflorescence. Probably, some of the species are perineal (de Wet et al. 1983). Kodo millet is resistant to drought and typically cultivated without intercultural operations in semi-arid areas. Kodo is a monocot and has very small seeds and ellipsoidal, about 1.5 mm in width and 2 mm in length; ranging in color from light brown to dark gray. Kodo millet has a superficial root system that may be suitable for intercropping. The grain is enclosed in permanent, rough, and corneous husk (Deshpande et al. 2015).

3.6.3 Edible Uses

Using traditional or industrial methods, unprocessed or processed kodo millet grains can be cooked whole or decorated, and where necessary ground to flour. In India, kodo millet grains are used by tribal people as rice and ground seeds used for making chapattis as flour. Traditionally Kodo flour is used to make novel food products such as idli, dosa, chapatti, pongal, puttu, idiyappam, kozhukattai, boli, cutlet, biscuits, bread, cookies, laddoo, etc. (Kalpana and Koushikha 2013; Deshpande et al. 2015).

3.6.4 Medicinal Uses

Kodo has water-soluble fiber and this property can be used to maintain or minimize blood glucose response in patients with diabetes and cardiovascular disease, glycemic load (GL) reflecting both the quality and quantity of carbohydrate in food and allows comparison of the possible glycemic impact of practical food portions and low glycemic index foods such as kodo glucose tolerance has been shown to increase in both healthy and diabetic subjects (Riccardi et al. 2008; Thakur et al. 2018).

3.7 Little Millet (*Paspalum sumatrense*)

3.7.1 Origin and Distribution

Paspalum sumatrense, a small millet, is native to India so it is also called Indian millet. The name of the species is based on a collected specimen from Sumatra (Indonesia) (de Wet et al. 1983). It is grown mainly in India, China, eastern Asia, and Malaysia. Little millet is suited to both the tropical and temperate climates. Currently, the crop is almost limited to some hilly areas in India. It is an important catch crop and cultivated by Indian tribals (Kalaisekar et al. 2017).

3.7.2 Botanical Distribution

Little millet grown mostly within the drought susceptible place, semi-arid and in slight rainfall location. It is grown in the most important parts of Madhya Pradesh, Chhattisgarh, and Tamil Nadu. Little millet is also grown in many different states; however, on a confined scale. Little millet is regarded as a cash crop in many tribal regions as it earns much higher fees than rice. Little millet possesses husk and bran much like rice. So, they require dehusking and debraning previous to utilization.

3.7.3 Edible Uses

In India, little millet finds place in Madhya Pradesh and Chhattisgarh where it is consumed as a bhagaar food during fasting. Many other typical small millet foods are made from popped flour that is mixed with sugar/jaggery/ghee/milk/butter milk and salt. A vast array of traditional snacks, bhat, kheer, dosa, upma, paddu, masala idli, and halwa are made from small millet in many rural households (Yenagi 2004). For various food uses, milled millet can be further processed, such as flakes, fast food cereals, ready-to-eat snacks, supplementary foods, extrusion cooking, malt-based items, weaning foods, and, above all, health foods.

3.7.4 Medicinal Uses

Little millet is used for patients with diabetic and cardiovascular disorders to control or reduce the blood glucose response (Riccardi et al. 2008). Little millet is also a good source of nutraceuticals such as phenolics, butyric gama-amino acid (GABA), lignans, starch resistant, sterols, and phytats. In millet-based foods, the additive and synergistic effects of these bioactive nutraceuticals can give many health benefits (Prathapan et al. 2011).

3.8 Browntop Millet (*Brachiaria ramosa*)

3.8.1 Origin and Distribution

Browntop millet (*Brachiaria ramosa* (L.) Stapf; *Panicum ramosum* L.) is an introduced annual grass of South-East Asian origin (Clayton et al. 2006). Browntop millet is drought-resistant and heat tolerant but it can also be planted in low flooding areas. It was launched in the United States from India in 1915 (Oelke et al. 1990). In India, browntop millet is grown particularly in rainfed tracts of Karnataka state districts of Tumakuru, Chitradurga, and Chikkaballapura. The crop is common in this area and is mainly used by people from economically weaker sectors for food purposes. This millet seed is cultivated in a number of soils and climates. It is a hardy crop like other millets, and well suited for dry land.

3.8.2 Botanical Distribution

Browntop millet is an annual species growing 1–3 ft high in the warm season. There are pubescent nodes in the smooth stems and can stand up from a decumbent base or ascend. It grows with either a compact or open panicle and can have spikelets that are either shattering or indehiscent. The leaves are 2.2–18 cm long and 6–18 mm wide, and smooth on both surfaces. In browntop millet, the inflorescence is indeterminate, open, with simple axis and stalked flowers spreading out. It has inflorescences of 3–15 and has white flowers. Seed is ellipsoid and tan in color; it matures in about 60 days (Sheahan 2014).

3.8.3 Edible Uses

Browntop millet grains are used to make unleavened bread or porridge (Nesbitt 2005). Not only is it nutritious but it is also very delicious, rich in essential nutrients, and gluten-free. It is a rich source of natural fiber, producing quality fodder compared to other grains and straw, by which cattle can enjoy it. Browntop millet tends to be ground into flour and used for making flat breads (roti, dosa) or to make gruel (anna, kheer) polished and boiled. In religious practices, some of these foods are used that may partially account for their longevity in cultivation.

3.8.4 Medicinal Uses

Browntop millet contains about 12.5% of fiber, which is why it is used as a medicine for the treatment of lifestyle diseases. Among those who regularly consume millets, lower incidence of cardiovascular diseases, duodenal ulcer, and hyperglycemia (diabetes) are recorded. It is seen as a solution to health issues such as diabetes, arthritis, and heart disease that are prevalent. However, it is yet to gain popularity due to a lack of consumer awareness, and disinterest among food processing companies. Browntop millet has a great scope for consumption growth. The farmer can benefit from its low input and maintenance requirements, and hence low risk (Ashoka and Sunitha 2020).

3.9 Ethnobotany: What and Why?

Ethnobotany is coined with two words, i.e., "ethno"-human study and "botany" means plant study; it is the study of the relationship between plants and humans. It is the branch of ethnobiology (study of plants as treated or used by different traditional communities) and is a multidisciplinary science described as plant-to-people interaction. The relationship between plants and human cultures is not restricted to using vegetation for food, apparel, and refuge however also includes their use for religious ceremonies, ornamentation, asthenic and fitness care, and a part of belief and taboo of tribal human beings (Schultes 1992). People commenced their life inside the forest as a vital part of the forest atmosphere. Dwelling near nature he has acquired particular expertise about the ambient biodiversity by means of intuition, trial or blunders and experimentation, and used a ramification of flora in nature to meet his crucial necessities like food, remedy, gasoline, fiber, and so on (Pandey and Tripathi 2017).

Millets were eaten as staple cereals in Asia, Africa, and Europe, and were fermented since prehistoric times. In the "Hoe Era" they may have been among the first crops to be cultivated prior to the "Plow Era." In India and the rest of the globe, millets are among the oldest cultivated crops. In rainfed conditions, millets are traditionally grown particularly by the marginal farmers and tribals (Phondani et al. 2010; Sahu and Sharma 2013; Dileep and Nair 2015). Because of its drought-resistant growth adaptations, the millet has proved to be a crucial staple food in African and Asian culture. Proso millet is the first domesticated cereal grain and one of the oldest human foods that was commonly accepted, domesticated, and cultivated simultaneously in Asia and Africa during the Neolithic Period over 7000 years ago, and then spread all over the world as a staple food. The earliest archeological plant, found in regions of India, Mexico, China, and Africa, has documented millets in many parts of the world. In Indian, Chinese Neolithic, and Korean Mumun societies, millets also formed essential parts of the prehistoric diet. While millet cultivation has followed the introduction of domesticates in some regions, in other regions it appears to be an independent mechanism preceding the introduction of crops from other regions, such as South India and West Africa (Anonymous 2018).

In India, millets are the staple food among the tribal communities and they are cultivating since a long period of time. The millets are utilized as dehusked grains which are cooked and consumed like rice, popped grains are used for making kheer and laddoos (Punia et al. 2003; Jaybhaye et al. 2014), the grain as well as the plant parts are used for curative of many diseases also; however, the plants are used as fodder for the domestic animals and seed for the feed for animals including ruminants, pigs, poultry, and pet birds (Dileep and Nair 2015).

In twenty-first century, the importance of these millets has increased due to good remediation properties in millets for many human diseases as well as used as nutraceuticals. The ethnobotanical importance of these millets for the human community is not much revealed rather than the tribal of the Indian Subcontinent (Phondani et al. 2010; Sahu and Sharma 2013; Dileep and Nair 2015). The Indian Anthropological Survey under the "People of India Project" identified 461 ethnic/indigenous (Tribal) groups in India (Xaxa 1999). These cultures, here called indigenous societies, follow many diverse lifestyles and practices. The information on ethnobotanical, utilized by the tribal peoples, is congregated for better utilization of minor millets for the human community.

3.9.1 Millets and Tribal Agriculture: Understanding the Connections

India's tribal groups have a special relationship with millets. It is eaten by the local tribal groups as the staple food and fermented drink. Millets were an important part of their agricultural systems and agricultural operations. For instance, Soligas, a Biligiriranga Hills tribe in Karnataka district of Chamarajanagar celebrates RagiHabba (Festival) at the time of harvesting millets. Similarly, for most Madhya Pradesh tribals, minor millets of "Kodo" and "Kutaki" have been the largest single crop grown and constituted the staple food (Behera 2017).

Patil et al. (2015) also set out the role of tribals in millet conservation. Millets contribute 50% of the total germplasms retained by Maharashtra district Nandurbar tribes according to a study undertaken by them.

Tribal agriculture in India has been mostly affected by the Green revolution. The traditional approach of tribals to farming and land use has changed a lot mostly because of the influence of state policies, which undoubtedly encouraged selective crop cultivation such as paddy and wheat (Behera 2017).

3.9.2 Ethno-Medical Approach of Minor Millets

Millets play an important role in Indian subcontinents' tribal life. Not only are they used as food but they are also used by tribal communities for certain other purposes (Table 3.1). The following are the tribals' documented medicinal practices and other uses of small millets, living in Bastar Plateau Zone of Chhatisgarh, India (Sahu and Sharma 2013).

1. The wall of the houses are made by mixing the straw of the kodo millet with mud, the resulting mud wall becomes very strong and moreover resistant to termites. At the same time they reported that when the straw of the kodo millet is burnt and the ash is spread in the field of onion, it results in higher yield of the crop.
2. The tribal people usually mix and store the husk of the finger millet with the green gram and pigeon pea seeds and this helps in protecting against pests during storage. The tradition of using kodo millet straw for baking earthen pots tribal people is also very common as they believe the pots become stronger and bake better by burning the kodo millet straw.
3. Kodo millet straw is used by Bastar tribals to provide immediate relief in cattle for tympany disorder, which is a fatal disease in which the animal's stomach swells enormously and the cattle may die if adequate care is not given in time, they also use 3–4-year old kodo millet grains as an instant-relief medicine for poultry ranikhet disease, which is again a very contagious and dangerous foul disease, leading to heavy toll if not effectively and timely managed.
4. Kodo straw is also used in paddy fields by tribal people to control leaf folders and blast diseases while mixing millet grains in cattle feed helps to increase milk production and reduce feed costs.

Table 3.1 Ethnobotanical utilization of minor millets by different tribes of Indian Subcontinent

Use	Plant used	Plant part or method of preparation	Tribe	Region	References
For the cure of measles	Proso millet (*Panicum miliaceum*)	Make a chaval (bhat) for consumption	Bhotiya	Niti Valley, Uttarakhand	Phondani et al. (2010)
For the cure of jaundice	Barnyard millet (*Echinochloa frumentacea*)	Make a chaval (bhat) and consume it	Bhotiya	Niti Valley, Uttarakhand	Phondani et al. (2010)
For cuts and wounds	*Finger* millet (*Eleusine coracana*)	Flour mixed with ghee	Tolcha, Marcha sub-communities of Bhotiya	Dhauli Ganga, Uttarakhand	Kandari et al. (2012)
For tympani disorder in cattle	Kodo millet (*Paspalum scrobiculatum*)	Straw	Tribal farmers	Bastar, Chattisgarh	Sahu and Sharma (2013)
For ranikhet disease of poultry	Kodo millet (*Paspalum scrobiculatum*)	4 years old grains	Tribal farmers	Bastar, Chattisgarh	Sahu and Sharma (2013)
Leaf folder and blast diseases of rice	Kodo millet (*Paspalum scrobiculatum*)	Straw	Tribal farmers	Bastar, Chattisgarh	Sahu and Sharma (2013)
For high yield in onion	Kodo millet (*Paspalum scrobiculatum*)	Straw ash	Tribal farmers	Bastar, Chattisgarh	Sahu and Sharma (2013)
For minimizing the burning sensation and quick healing of wound	Finger millet (*Eleusine coracana*)	Flour	Tribal farmers	Bastar, Chattisgarh	Sahu and Sharma (2013)

(continued)

Table 3.1 (continued)

Use	Plant used	Plant part or method of preparation	Tribe	Region	References
For storage pests of pigeon pea and green gram	Finger millet (*Eleusine coracana*)	Husk	Tribal farmers	Bastar, Chattisgarh	Sahu and Sharma (2013)
For baking of earthen pots	Kodo millet (*Paspalum scrobiculatum*)	Straw	Tribal farmers	Bastar, Chattisgarh	Sahu and Sharma (2013)
Shoots are eaten as vegetable	*Echinochloa colona*	Whole plant	Paniya tribe	Wayanad, Kerala	Dileep and Nair (2015)
For the cure of dysentery; for wound healing; measles; pleurisy and smallpox fevers; ulcer; diabetes	Finger millet (*Eleusine coracana*)	Fruit, leaves	Adiyan, Irular, Kuruman, and Vitolia	Andhra Pradesh, Karnataka, Kerala, and Tamil Nadu	Latheef et al. (2014), Sreeramulu et al. (2013)
Used as antivenom (snakebite); fever; for reducing blood sugar level; dysentery; prevent soil erosion; antiabortion	*Eleusine indica*	Whole plant	Adiyan, Irular, Kuruman and Vitolia	Andhra Pradesh, Karnataka, Kerala, and Tamil Nadu	Dey and De (2012), Srithi et al. (2009)
For the cure of diuretic; diabetes; wound healing; bowel cleanser; reduces inflammation; vitamin supplement	Kodo millet (*Paspalum scrobiculatum*)	Stem, rhizome, root, seed	Adiyan, Irular, Kuruman, and Vitolia	Andhra Pradesh, Karnataka, Kerala, and Tamil Nadu	Dileep and Nair (2015), Maloles et al. (2011)
Whole plant used against shivering; diuretic; dyspepsia; poor digestion; bone fracture	Foxtail millet (*Setaria italica*)	Whole plant	Adiyan, Irular, Kuruman, and Vitolia	Andhra Pradesh, Karnataka, Kerala, and Tamil Nadu	Maloles et al. (2011)
For the cure of jaundice	Barnyard millet (*Echinochloa frumentacea*)	Seeds, stem	Kharka and Maira/Chani	Jakholi Block, Uttarkhand	Singh et al. (2017)

Used in body ache due to exhaustion and gastric problem	Finger millet (*Eleusine coracana*)	Fermented seeds	Lepcha	Dzongu valley, North Sikkim	Pradhan and Badola (2008)
For the cure of stomach disorder, cold	Finger millet (*Eluesine coracana*)	Grains	Apatani	Ziro valley, Arunachal Pradesh	Kala (2005)
Used as poultry feed	*Panicum repens*	Seeds	Kurumas Tribe	Wayanad, Kerala	Dileep and Nair (2015)
To cure stomach ache and gastric disorders	*Panicum repens*	Rhizome	Kurichya Tribe	Wayanad, Kerala	Dileep and Nair (2015)
To cure diabetes and wound healing	Kodo millet (*Paspalum scorbiculatum*)	Rhizome and root	Kurichya Tribe	Wayanad, Kerala	Dileep and Nair (2015)
Used for making mudde and porridge	Kodo millet (*Paspalum scorbiculatum*)	Grains	Kattuanaicka, Kurichya, Kuruma, Paniya Tribes	Wayanad, Kerala	Dileep and Nair (2015)
Used as feed for birds and fodder for livestock	Foxtail millet (*Setaria itallica*)	Entire panicle	Kattuanaicka, Kurichya, Kuruma, Paniya Tribes	Wayanad, Kerala	Dileep and Nair (2015)
Used for the treatment of dyspepsia, poor digestion, and bone fracture	Foxtail millet (*Setaria itallica*)	Plant	Kattuanaicka, Kurichya, Kuruma, Paniya Tribes	Wayanad, Kerala	Dileep and Nair (2015)

5. In the case of burns, the use of thick finger millet flour paste serves as an effective medication to reduce the burning pain and is also very effective in rapid wound healing.
6. Tribal people cover finger millet grains in a piece of cloth with ash and salt to tie up in the neck as Taabiz, which they believe serves as an effective measure to check the bad omen.
7. Minor millets are also associated with lots of faiths among tribal people. Gaadi is a major Bastar tribe festival, where the Prasaad given is finger millet grains and if those finger millet grains received as the Prasaad of Gaadi festival are sprinkled on the nonbearing mango and/or tamarind trees, the very next season they begin to bear fruit.

Pradhan et al. (2008) conducted a major study in Bastar, Chhattisgarh, surveyed and selected 18 people with diabetes problems and giving them prepared ragi and wheat multigrain flours with ragi between 30% and 70%. They suggested replacing their roti (chapatti) with chapatti with multigrain rice. Observations of capillary blood glucose were reported every third day. There were no changes to their normal food feeds or regulated diets. Blood glucose decreased as humans continued to eat multigrain flour. It has been found in all situations, whether the diabetic belonged to rural or urban areas blood glucose was dramatically reduced, whereas for some time people who did not take ragi due to some inevitable circumstances were found high levels of glucose in their blood during the period.

There are more numbers of tribes in the mid and hilly areas of Uttarakhand and the minor millets are the only crops these tribes take on very bad soil where other crops cannot or are very difficult to cultivate. In India, there are 645 distinct tribes, whereas in Uttarakhand there are six tribes in existence, viz., Bhotias, Buksa, Jaunsari, Khas, Raji, and Tharu (Fig. 3.2). The Uttarakhand tribes have maintained traditional ways of living through their generation and they reflect the characteristic cultures and qualities of primitive life (Farswan 2017). On very bad soil, the tribes take the crops of small millets and consume those crops as they are late in digestion. According to these tribes, they need to take the food twice or thrice if they are taking rice, but if they are taking kodo, one time food is enough.

Foxtail millet (local "Shol" or "Kangni") and proso millet (local "Ping" or "China") were cultivated throughout Kashmir till several decades ago and are presently grown in negligible areas. Husked grains of these crops are hard to cook and usually eaten as porridge. In some places, both these millets are grown in mixed cropping for fodder. The Doda and Kishtwar districts have a wide range of local high-value crop cultivars among which prominent are drought-tolerant millets like shol and ping which are grown in low rainfall areas, in the month of May (Bhat et al. 2019).

Fig. 3.2 (**a**) Bhotia tribe of Uttarakhand. (**b**) Tribal woman preparing porridge from millet. (**c**) Tribal/Gujjars woman of Kashmir preparing food. (**d**) Dhar tribe of Kashmir. (Source: Photographs by https://uttarakhand.pscnotes.com, https://www.thehansindia.com and Lakhera 2017)

3.10 Ethnobotanical Significance of Minor Millets with Respect to Tribal/Remote Farmers

Mishra and Chaudhury (2012) studied the tribal communities of Koraput district, Odisha. These tribal communities have Traditional Healthcare Practitioners (THP) of their own that came from heredity, taking care of their well-being from herbal medications. Every village/community has its own THP. THPs usually have typical palm leaf books, known locally as Pothis, where plant species names, plant description, medicine preparation methods, and doses are written down. Koraput is renowned for its small millets. The paroja, gadaba, and bhumia communities cultivate a range of millet species with varying lengths and grains with different sizes, forms, and colors in lower slopes of hills and unbundled rainfed uplands largely beneath the multiple method for cropping. Short, medium, or long duration ragi (*Eleusine coracana* (L.) Gaertn.), suan black and white (*Panicum sumatrense* Roth. ex Roem. et Schult.), kangu black, white, and red (*Setaria italica* (L.) P. Beauv.), still cultivated by them. Small and marginal farmers depend more on short duration ragi harvested in September/October for a short period that is considered as a lean time. Ragi (*Eleusine coracana* (L.) Gaertn.) gruel with rice is an important part of tribal cuisine as well as being used for special occasions to make delicious foods and country liquors. Manithottam and Francis (2007) studied the ethnobotanical

Fig. 3.3 The various activities (**a**) Finger millet seed stored for next season (**b**) Winnoing of seed of finger millet (**c**) Traditional method for dehusking of finger millets seed (**d**) Powdering of finger millet seed in millstone of tribal people for finger millet utilization

significance of finger millet in Muthuvan community of Idukki District of Kerala. Muthuvans adopt the splash and burn method for the cultivation of finger millet. They called kepa for finger millets and grow three varieties which are having 3, 4, and 6 months maturity. The grains are separated from straw, sun-dried for 3–4 days, and are kept in gunny bags. Various activities like storage, winnowing, dehusking and powdering of finger millet seeds is done manually by the tribal people (Fig. 3.3). For the next season, seeds are selected from the fully matured and healthy plants. The panicles along with straw are stalked and hanged vertically over the fireplaces above the cheru (a platform with perforations weaved with bamboo splits). The smoke and heat from fireplaces protect, preserve, and extend the viability of seed that will remain for the 5 years. The once cultivated area is abandoned by them for 7 years to restore the fertility of that land. They also prepare a unique pudding by powder of finger millet which is known as Katty (1 kg powder + 4 L water).

Drinking locally fermented drinks and distilled alcoholic beverages is a common practice for many Himalayan ethnic people. In Sikkim, the northern part of Arunachal Pradesh and Ladakh, alcoholic beverages locally called kodo ko jaanr or alterations made from fermented finger millets or barley are popular. Finger millet is eaten in Uttarakhand as roti, barnyard millet as paleu or chencha, a savory, buttermilk-cooked porridge. Zan is the most common porridge recipe made from finger millet and vegetables by the Monpa tribes of Arunachal Pradesh (Bhat et al. 2019).

Nayogi (2018) visited the villages in Dindori district of Madhya Pradesh for collecting information on the utilization of Sikiya millet by Baiga tribe (Fig. 3.4). Sikiya is a crabgrass finger millet. It has been part of tribal culture all along. Baigas have nursed sikiya as food for decades, which they use to prepare kheer (porridge). Other groups and communities do not know these millets as well as these millets have yet not included in the central government's nutri cereals scheme.

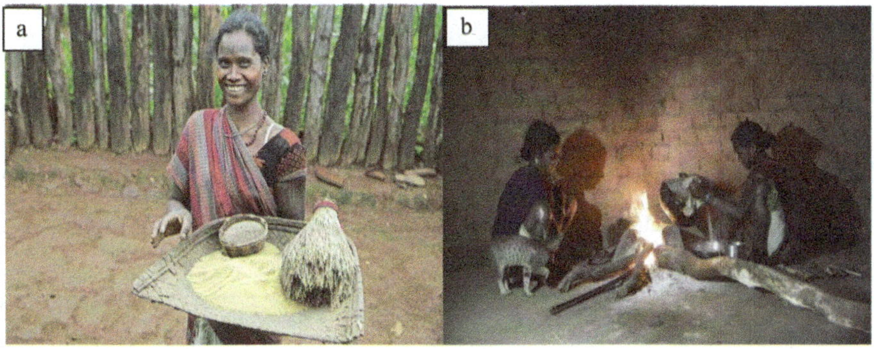

Fig. 3.4 (**a**) Woman of Baiga tribe in Madhya Pradesh showing sikiya as a plant, and its whole and dehusked grains. (**b**) Baiga tribe women preparing rice from sikiya millet in Dindori district, Madya Pradesh. (Source: Reproduced from Nayogi 2018, Photographs by Vikas Choudhary)

3.11 Few Evidences of Minor Millets Coping with Changing Climatic Conditions Around that in India

There are plenty of evidences indicating the climate resilience of these minor millets and their potential to cope up with the changing situations and a few of them are mentioned below as per the study made by Behera (2017).

1. A particular millet-growing belt survived the famine in 2016 in Medak district, Telangana, which records the highest number of farm suicides in the state, when most of the farmers who stuck to input intensive crops such as sugarcane, cotton, and paddy were in deep distress.
2. After witnessing a steady decline in groundnut yields for more than four decades, a farmer from Kogira village in Andhra Pradesh's south-western district of Ananthapuramu chose to grow foxtail millet for the first time in 2012 and achieved a gross profit of 81%.
3. Many farmers in Odisha's tribal districts such as Kandhamal and Kalandi have benefited immensely from the cultivation of millet, especially during the food crisis due to enormous paddy losses as a result of the scarce rainfall of 2015–2016. For the whole year, even after the entire paddy was lost, the paddy growers with very small patches left for millet crops were able to fulfill their food requirements through millets.

There are many traditional recipes which are made of minor millets. Roy (2020) collected the information related to the traditional recipes out of finger millet, collected from 100 sample households, from tribals of Dasamantpur Block of Koraput, Odisha. It was seen that most of them consume one recipe most that is Jau (finger millet mixed with warm water). About 35% household sample found to consume Jau, Mudde, and Tampa (1 Rice:2 Ragi flour) whereas few consumed pitha

Fig. 3.5 The various cuisine (**a–f**) made from finger millet by the tribals of Dasamantpur Block of Koraput, Odisha: (**a**) Mandia Anda (ragi flour with broken rice), (**b**) Mandia Tampa (ragi flour with broken rice and warm water), (**c**) Mandia Kandul Raav (ragi flour with whole arhar dal with some masala), (**d**) Mandia Roti (ragi flour roti), (**e**) Mandia Pitha (ragi flour in between banana leaves), (**f**) Mandia Sukua Raav (ragi with dried fish and masala). (Source: Roy 2020 and www.awesomecuisine.com/recipes/356/ragi-roti.html)

occasionally (Fig. 3.5). It is interesting to note that they also consume ragi in between Jackfruit leaf (Water + Ragi Flour + Salt + Jaggery + Chili). Cakes are made by wrapping ragi flour in maize husks or banana leaves and then roasted.

3.12 Initiatives of Government of India and Civil Societies to Promote Millets Framing in India

A number of programs to encourage millet cultivation have been initiated by the Government of India, particularly in the last decade (Behera 2017).

1. The most important of the current schemes are Initiative for Nutritional Security through Intensive Millets Promotion (INSIMP) and Rainfed Area Development Programme (RADP) as part of Rashtriya Krishi Vikas Yojana (RKVY) and Integrated Cereals Development Programmes in Course Cereals Based Cropping Systems Areas (ICDP-CC) under Macro Management of Agriculture (MMA).
2. M. S. Swaminathan Research Foundation (MSSRF) has been working through the collectives of women in Kolli Hills for quite a long time to encourage traditional millet farming. It has sponsored more than 35 Self Help Groups (SHGs) in the Kolli Hills, with more than 386 members (of whom 214 are women), in institutionalizing the various millet operations including the value

chain for millets. The SHGs have formed the Kolli Hills Agrobiodiversity Conservers' Federation (KHABCoFED), a network of SHGs, trained in operating mills, processing, packing, and labeling.

3. Millet Network of India (MINI) is a pan India alliance of over 65 institutions supporting different millet varieties. More than 50,000 farmers are part of the alliance that has spread activities to Uttarakhand, Nagaland, and Odisha.
4. In Odisha, a volunteer organization named "Nirman" restored 12 varieties of traditional millets with the aid of the Kutia Kandha community, which were almost extinct and preserved in seven villages in Kandhamal district involving tribal members. One of those 12 millets is Guruji, a tiny-sized millet.

3.13 Minor Millets: Present Scenario in Mid and High Hills of Uttarakhand, India

Uttarakhand has long been a stronghold of millets. In the state of Uttarakhand, these crops are being grown by poor, marginal and tribal farmers since time immemorial (Fig. 3.6). Uttarakhand has a distinct range of landforms from hills to plains. Hilly

Fig. 3.6 Minor millets: a staple diet of farmers in hilly/tribal areas of Uttarakhand

area is characterized by gravel and light-textured soils which do not retain water for a long time and suits small millets. The resilience exhibited by these crops to different kinds of ecological niches has been the reason that these crops are being grown since time immemorial in hilly/tribal areas of Uttarakhand.

A number of initiates are currently also being conducted to raise awareness among farmers that it is good to eat, which in fact sometimes has more to do with taste than nutrition. A number of NGOs and Government schemes are prevailing in Uttarakhand to uplift the area, production, and productivity under millets cultivation. In a "maiden initiative" by the Uttarakhand Government aimed to enhance millets farmers' income, the State-owned MandiParishad (wholesale market) will directly buy farm millet from them. This is the first of its kind initiate, and the move would directly benefit the farmers as they are expected to get adequate price for their farm produce millets. The scheme was initiated with two districts Almora and Chamloi and will be extended to other districts in the next phase. One of these millet's food grains has been proposed to be a part of the midday meal scheme meant for schools.

These schemes inspire farmers to produce more and more millets, as it is not a simple task to expect farmers in Uttarakhand—resource-poor, marginal, or otherwise—to grow millets without strong policy support. Millets are often the last standing crop in times of climate change and are therefore a successful risk management strategy for resource-poor marginal farmers.

Despite all the benefits, these millets are encountered with several production constraints. As policymakers, there is a need to reflect and we really have to drive the progressive farmers to go for these crops. The area under minor millets has declined terribly including the production. Climate change concerns surrounding these crops need to be discussed. It is time to understand the value of millets, and how to popularize them. The role of government is very important and should be rewarded to the farmers who are involved in producing these precious seed materials. The millet is not a mere seed crop, but it is a sign of prosperity and hope.

It is time to think seriously about the initiatives whereby we can improve small millet productivity. As to be honest, there is no chance for increasing area under small millets cultivation but new technologies are needed to develop to enhance the productivity levels. Out of total developed technologies, only 30–38% of technologies are being taken to the farmers and the rest are lying as such. On one side, the area is going down and proportionally the land conditions are totally deteriorated. If we can aware the farmers and take all the technologies in the right order, it is possible to increase the productivity levels in millets systems.

The state contribution of agriculture to the state's domestic product is about 23.4% and the population dependent on agriculture for their livelihood is about more than 70% (Sakamma et al. 2018; Anbukkani et al. 2017). However, underweight and child malnutrition is a major problem in state Uttarakhand. According to the National Family Health Survey (2018), one-third (34%) of children under 5 years age is stunted, or too short for their age which indicated that they have been undernourished from some time. Twenty percent are undernourished, or too thin, less height which may have resulted from inadequate food intake. These two problems can be solved by developing nutria-cereal model villages in tribal/hilly areas. There is a need to popularize small millets among the people for their nutritive and medicinal benefits and the development of a remunerative market. To make hilly/tribal villages into self-sufficient units in terms of nutrition and income through the utilization of available resources and awareness regarding health education, there is a need to establish connection between agriculture, nutrition, and farmers. Institute–farmer partnership (Participatory breeding programme) through the development of Nutria-cereal model villages in hilly tribal areas is need of the hour for in situ conservation of germplasm and for the development of acceptable crop varieties, improvement of economic as well as the health status of poor tribal/hilly farmers. At the same time, small millet cultivation can be made sustainable for farming communities through the novel approach, i.e., "Development of nutria-cereal seed villages."

3.14 Future Strategies

The increasing demand for food that is nutritious, the effectiveness and efficiency of the consumed product and the increased knowledge of the consumer and healthcare industries are the major factors that have led to the creation of a nutraceutical market that is expected to expand several folds in the years to come. Therefore, the Food and Drug Administration has also issued legislation promoting this new industry, encouraging scientific research. Therefore, an urgent aim should be to recognize health-benefiting factors that increase the essential nutrient levels in staple crops to have a major effect on human nutrition around the world. Through emerging biotechnology tools and techniques, the targeting of nutritionally essential genes and proteins will lead to the development of "smart" biofortified crops. Products from these value-added crops can help with many issues, such as reducing protein-energy malnutrition. Research can evaluate the effect of these products on the body's absorption, protection, homeostasis, and nervous system control, and then dive into hypoallergenic foods and modern approaches to nutraceutical development. Initial research has shown that in the nutraceutical industry, finger millet has a promising future and provides a theoretical foundation for its use as an economically viable store of nutrients for chronic disease depreciation. But the properties of millets and their agricultural and nutritional potential are yet largely uninvestigated.

3.15 Conclusion

Setting up minor millet as a nutraceutical, from the point of view of the customer, transcend the usual wait, commitment, and cost inputs to bring traditional healthcare to the market and provide "self-care" for their satisfaction. We need to take initiatives to encourage small-scale millet farming and consumption, beginning at the very basic village level.

More intakes of these grains will result in a decrease in health problems leading to lower spending on medical bills and greater availability of funds for other purposes.

Additionally, the utilization of its rich nutritional value on a global scale takes on significance in providing developing countries with food security, agricultural production, self-dependence, and economic enhancement. With the growing awareness about minor millet's nutraceutical properties, the day is not far away when this crop and its various products will find their place in every individual's daily menu.

In India, the government is setting up an Initiative for Nutritional Protection through Intensive Millet Promotion (INSIMP), while those ancient grains have appeared on specialized shop shelves in the developed world. To meet the food and nutrition demands of the nation while ensuring that natural resources are used sustainably, the farming system we need today is more robust and diversified. The GoI and state governments have taken several progressive steps over the past decade to promote mission-based millet farming and raise awareness among the population, particularly urban Indians, of increased millet consumption. As a result, millets are becoming more attentive and have been given greater importance.

References

Anbukkani P, Balaji SJ, Nithyashre ML (2017) Production and consumption of minor millets in India - a structural break analysis. Ann Agric Res 38(4):1–8

Anonymous (2018) The story of millets. Karnataka State Department of Agriculture, Bengaluru, India with ICAR-Indian Institute of Millets Research, Hyderabad, India, pp 26–35

Ashoka P, Sunitha NH (2020) Review on browntop millet-a forgotten crop. J Exp Agric Int 42 (7):54–60

Baltensperger DD (2002) Progress with proso, pearl and other millets. In: Janick J, Whipkey A (eds) Trends in new crops and new uses. ASHS Press, Alexandria, pp 100–103

Batra GA (2019) Can millets be the answer to India's nutritional problems? Inno Health Magazine

Behera MK (2017) Assessment of the state of millets farming in India. MOJ Eco Environ Sci 2 (1):16–20

Bhat BV, Arunachalam A, Kumar D, Tonapi VA, Mohapatra T (2019) Millets in the Indian Himalaya. Indian Council of Agricultural Research, New Delhi. 84p

Bicha T, Noguchia HK (2012) Allelopathic potential of two aquatic plants, duckweed (*Lemna minor* L.) and water lettuce (*Pistia stratiotes* L.), on terrestrial plant species. Aquatic Bot 103:30–36

Boivin N, Fuller DQ, Crowther A (2012) Old world globalization and the Columbian exchange: comparison and contrast. World Archaeol 44:452–469

Chandrasekara A, Shahidi F (2012) Antioxidant phenolics of millet control lipid peroxidation in human LDL cholesterol and food systems. J Am Oil Chem Soc 89(2):275–285

Changmei S, Dorothy J (2014) Millet-the frugal grain. Int J Sci Res Rev 3:75–90

Chauhan BS (2013) Shade reduces growth and seed production of *Echinochloa colona*, *Echinochloa crus-galli*, and *Echinochloa glabrescens*. Crop Protect 43:241–245

Chauhan B, Johnson DE (2011) Ecological studies on *Echinochloa crusgalli* and the implications for weed management in direct-seeded rice. Crop Prot 30:1385–1391

Chethan S, Malleshi NG (2007) Finger millet polyphenols: optimization of extraction and the effect of pH on their stability. Food Chem 105:862–870

Clayton WD, Vorontsova MS, Harman KT, Williamson H (2006) Grass base–the online world grass flora. Royal Botanic Gardens, Kew, UK

de Wet JM, Brink DE, Rao KP, Mengesha MH (1983) Diversity in kodo millet, *Paspalum scrobiculatum*. Econ Bot 37(2):159–163

Deshpande S, Mohapatra SD, Tripathi MK, Sadvatha RH (2015) Kodo millet-nutritional value and utilization in Indian foods. J Grain Process Storage 2(2):16–23

Devi PB, Vijayabharathi R, Sathyabama S, Malleshi NG, Priyadarisini VB (2014) Health benefits of finger millet (*Eleusine coracana* L.) polyphenols and dietary fiber: a review. J Food Sci Technol 51:1021–1040

Dey A, De JN (2012) Traditional use of plants against snakebite in Indian subcontinent: a review of the recent literature. African J Tradit Complement Altern Med 9(1):153–174

Dileep P, Nair GG (2015) Taxonomic and ethnobotanical studies of grasses used by tribals of Wayanad District, Kerala, South Western Ghats of India. J Global Biosci 4(5):2212–2235

Doggett H (1989) Small millets-a selective overview. In: Seetharam A, Riley KW, Harinarayana G (eds) Small millets in global agriculture. Oxford and IBH Publ. Co. Pvt. Ltd, Janpath, New Delhi, pp 3–18

Doust AN, Kellogg EA, Devos KM, Bennetzen JL (2009) Foxtail millet: a sequence-driven grass model system. Plant Physiol 149:137–141

Duke JA (1979) Ecosystematic data on economic plants. Quart J Crude Drug Res 17:91–110

Duke JA, Ayensu ES (1985) Medicinal plants of China. Reference Publications Inc, Algonac, MI. ISBN-13:978-0917256202

Dwivedi SL, Upadhyaya HD, Senthilvel S, Hash CT, Fukunaga K, Diao X, Prasad M (2012) Millets: genetic and genomic resources. In: Janick J (ed) Plant breeding reviews. Wiley, Hoboken, NJ, pp 247–374

Farswan DS (2017) Tribes in Uttarakhand: status and diversity. Int J Multi Discip Res Dev 4:89–93

Food and Agricultural Organization (2005) Economic and social department: the statistical division. https://archive.gramene.org/species/setaria/foxtailmillet_maps_and_stats.html

Gelinas P, McKinnon CM, Mena MC, Mendez E (2008) Gluten contamination of cereal foods in Canada. Int J Food Sci Technol 43:1245–1252

Grovermann C, Umesh KB, Quiédeville S, Kumar BG, Moakes S (2018) The economic reality of underutilised crops for climate resilience, food security and nutrition: assessing finger millet productivity in India. Agriculture 8:131. https://doi.org/10.3390/agriculture8090131

Gull A, Jan R, Nayik GA, Prasad K, Kumar P (2014) Significance of finger millet in nutrition, health and value added products: a review. J Environ Sci Comput Sci Eng Technol 3:1601–1608

Gupta SM, Arora S, Mirza N, Pande A, Lata C, Puranik S, Kumar J, Kumar A (2017) Finger millet: a "certain" crop for an "uncertain" future and a solution to food insecurity and hidden hunger under stressful environments. Front Plant Sci 8:643. https://doi.org/10.3389/fpls.2017.00643

Habiyaremye C, Matanguihan JB, Dalpoim GJ, Ganjyal GM, Whiteman MR, Kidwell KK, Murphy KM (2017) Proso millet (*Panicum miliaceum* L.) and its potential for cultivation in the Pacific Northwest, US: a review. Front Plant Sci 7:1961

Hardman LH (1990) Varietal trials of farm crops. Minnesota Report 24. University of Minnesota, Minneapolis, MN

Hegde PS, Chandra TS (2005) ESR spectroscopic study reveals higher free radical quenching potential in kodo millet (*Paspalum scrobiculatum*) compared to other millets. Food Chem 92 (1):177–182

Hilu KW, Johnson JL (1992) Ribosomal DNA variation in finger millet and wild species of *Eleusine* (Poaceae). Theor Appl Genet 83:895–902

Hunt HV, Badakshi F, Romanova O, Howe CJ, Jones MK, Heslop-Harrison JSP (2014) Reticulate evolution in *Panicum* (Poaceae): the origin of tetraploid broom corn millet, *P. miliaceum*. J Exp Bot 65:3165–3175

Jaybhaye RV, Pardeshi IL, Vengaiah PC, Srivastav PP (2014) Processing and technology for millet based food products: a review. J Ready To Eat Food 1(2):32–48

Kala CP (2005) Ethnomedicinal botany of the Apatani in the eastern Himalayan region of India. J Ethnobiol Ethnomed 1:11. https://doi.org/10.1186/1746-4269-1-11

Kalaisekar A, Padmaja PG, Bhagwat VR, Patil JV (2017) Introduction. In: Kalaisekar A, Padmaja PG, Bhagwat VR, Patil JV (eds) Insect pests of millets. Academic, London, pp 1–25. https://doi.org/10.1016/B978-0-12-804243-4.00001-X

Kalinova J, Moudry J (2006) Content and quality of protein in proso millet (*Panicum miliaceum* L.) varieties. Plant Foods Hum Nutr 61:45–49

Kalpana CA, Koushikha NM (2013) Development and evaluation of varagu incorporated recipes. Paper presented National Seminar on Recent advances in processing, utilization and nutritional impact of small millets. Madurai symposium, Thamukkam Grounds, Madurai, 13 September 2013

Kandari LS, Phondani PC, Payal KC, Rao KS, Maikhuri RK (2012) Ethnobotanical study towards conservation of medicinal and aromatic plants in upper catchments of Dhauli Ganga in the Central Himalaya. J Mt Sci 9:286–296

Kukreti A, Kurmanchali N, Rawat L, Bisht TS (2017) Evaluation of fluorescent *Pseudomonas* spp. against *Pyricularia grisea*, *Rhizoctonia solani* and *Sclerotium rolfsii* causing blast, sheath blight and foot rot diseases of finger millet (*Eleusine coracana* L.) crop in mid hills of Uttarakhand: *in vitro* study. Bull Environ Pharmacol Life Sci 6:29–35

Kumar B (2016) Status of small millets diseases in Uttarakhand. Int J Plant Protect 9:256–263

Kumar A, Metwal M, Kaur S, Gupta AK, Puranik S, Singh S, Singh M, Gupta S, Babu BK, Sood S, Yadav R (2016) Nutraceutical value of finger millet [*Eleusine coracana* (L.) Gaertn.], and their improvement using omics approaches. Front Plant Sci 7:934

Lakhera S (2017) Food and spices of the Rung community in Uttarakhand. https://www.sahapedia.org/food-and-spices-of-the-bhotia-community-uttarakhand

Lata C (2015) Advances in genomics for enhancing abiotic stress tolerance in millets. Proc Indian Natl Sci Acad 81:397–417

Lata C, Gupta S, Prasad M (2012) Foxtail millet: a model crop for genetic and genomic studies in bioenergy grasses. Crit Rev Biotechnol:1–16

Latheef A, Kumar SP, Remashree AB (2014) Ethnomedicine used for treating cuts and wounds by the tribes of Attappady, Kerala. Int J Herbal Med 2(2):1–8

Le Thierry d'Ennequin M, Pan Maud MO, Toupance B, Sarr A (2000) Assessment of genetic relationships between Setaria italica and its wild relatives S. viridis using AFLP marker. Theor Appl Genet 100:1061–1066

Li Y, Wu SZ (1996) Traditional maintenance and multiplication of foxtail millet (*Setaria italica* (L) P. Beauv.) landraces in China. Euphytica 87:33–38

Maloles JR, Berg K, Ragupathy S, Nirmala BC, Althaf KA, Palanisamy VC, Newmaster SG (2011) The fine scale ethnotaxa classification of millets in Southern India. J Ethnobiol 31(2):262–287

Manithottam J, Francis MS (2007) Ethnobotany of finger millet among Muthuvan tribes of Idukki district, Kerala. Indian J Tradit Knowl 6:160–162

Miller NM, Spengler RN, Frachetti M (2016) Millet cultivation across Eurasia: origins, spread, and the influence of seasonal climate. Holocene 26:1566–1575

Mishra S, Chaudhury SS (2012) Ethnobotanical flora used by four major tribes of Koraput, Odisha, India. Genet Resour Crop Evol 59:793–804

Nayogi DG (2018) Madhya Pradesh's century-old millet Sikiya finds few revivalists. Down to Earth. DIALOG. https://www.downtoearth.org.in/news/food/madhya-pradesh-s-century-old-millet-sikiya-finds-few-revivalists-61394. Accessed 16 Aug 2020

Nesbitt M (2005) Grains. In: Prance G, Nesbitt M (eds) The cultural history of plants. Routledge Press, New York

Oelke EA, Oplinger ES, Putnam DH, Durgan BR, Doll JD (1990) Millets. In: Alternative field crops manual. Univ of Wisc-Ext Serv, Univ of Minn ExtServ and Univ of Minn. CAPAP, Madison

Pandey AK, Tripathi YC (2017) Ethnobotany and its relevance in contemporary research. J Med Plants Stud 5(3):123–129

Patil S, Patil KS, Sawarkar P, Kulkarni DK (2015) Germplasm conservation of maize, sorghum, millets and vegetables from Dhadgaon and Akkalkuwa tribal block of Nandurbar district, Maharashtra State. Sci Res Rep 5(2):137–146

Phondani PC, Maikhuri RK, Rawat LS, Farooquee NA, Kala CP, Vishvakarma SCR, Rao KS, Saxena KG (2010) Ethnobotanical uses of plants among the Bhotiya tribal communities of Niti Valley in Central Himalaya, India. Ethnobot Res Appl 8:233–244

Pradhan BK, Badola HK (2008) Ethnomedicinal plant use by Lepcha tribe of Dzongu valley, bordering Khangchendzonga Biosphere Reserve, in North Sikkim, India. J Ethnobiol Ethnomed 4:22. https://doi.org/10.1186/1746-4269-4-22

Pradhan A, Nag SK, Tomar NS, Sharma RL (2008) Ragi controls diabetes. Kurukshetra 56(9):47–48

Pradhan A, Panda AK, Bhavani RV (2019) Finger millet in tribal farming systems contributes to increased availability of nutritious food at household level: insights from India. Agric Res 8 (4):540–547

Prathapan A, Singh MK, Anusree SS, Soban Kumar DR, Sundaresan A, Raghu KG (2011) Antiperoxidative, free radical scavenging and metal chelating activities of *Boerhaavia diffusa* l. J Food Bio-Chem 35:1548–1554

Punia D, Dalal A, Sindhu S (2003) Nutritional evaluation of kangini (*Setaria italica*): an under utilised millet and sensory evaluation of value added products from kangini. In: Recent trends in millet processing and utilization. CCS Hisar Agril. Univ, Hisar, India, pp 32–37

Rao MV (1989) Small millets in global agriculture. The small millets: their importance, present status and outlook. Delhi, Oxford and IBH Publishing Co Pvt Ltd, pp 9–12

Rao BR, Nagasampige MH, Ravikiran M (2011) Evaluation of nutraceutical properties of selected small millets. J Pharm Bioallied Sci 3:277–279

Rawat L, Prasad S, Bisht TS, Naithani DC, Tiwari A (2019) An impact assessment of front line demonstrations on yield and economics of finger millet and barnyard millet under rainfed conditions of Uttarakhand. Int J Pure Appl Biosci 7:408–414

Rawat L, Karnatak AK, Nautiyal BP, Bisht TS, Nautiyal A (2020) Management of shoot fly damage in barnyard millet by seed treatment for higher monetary return in hills of Uttarakhand. J Entomol Zool Stud 8(3):1762–1767

Renganathan VG, Vanniarajan C, Karthikeyan A, Ramalingam J (2020) Barnyard millet for food and nutritional security: current status and future research direction. Front Genet 11:500

Riccardi G, Rivelluse AA, Giacco R (2008) Role of glycemic index and glycemic load in the healthy state, in prediabetics and in diabetes. Am J Clin Nutr 87:269S–274S

Roy TC (2020) Traditional millet recipes of the tribals of Koraput and their perception. TCR connecting agriculture. Dialog. https://www.tcrconnectingagriculture.com/2020/03/traditional-millet-recipes-of-tribals.html. Accessed 16 Aug 2020

Sahu RK, Sharma ML (2013) Medicinal and other uses of small millets by the tribal farmers of the Bastar Plateau Zone of Chhattisgarh. Agric Update 8(4):596–599

Sakamma S, Umesh KB, Girish MR, Ravi SC, Satishkumar M, Bellundagi V (2018) Finger millet (*Eleusine coracana* L. Gaertn.) production system: farmer's welfare. J Agric Sci 10:162–179

Saturni L, Ferretti G, Bacchetti T (2010) The gluten-free diet: safety and nutritional quality. Nutrients 2:16–34

Schultes RE (1992) Ethnobotany and technology in the Northwest Amazon: a partnership. In: Plotkin M, Famolare L (eds) Sustainable harvest and marketing of rain forest products. Island Press, Covelo, CA, pp 45–76

Sharma N, Niranjan K (2018) Foxtail millet: properties, processing, health benefits, and uses. Food Rev Int 34(4):329–363

Sheahan CM (2014) Plant guide for browntop millet (*Urochloa ramosa*). USDA-Natural Resources Conservation Service, Cape May Plant Materials Center, Cape May

Shobana S, Malleshi NG (2007) Preparation and functional properties of decorticated finger millet (*Eleusine coracana*). J Food Eng 79:529–538

Shukla A, Lalit A, Sharma V, Vats S, Alam A (2015) Pearl and finger millets: the hope of food security. Appl Res J 1:59–66

Singh B, Bahuguna A, Bhatt A (2015) Small millets of Uttarakhand for sustainable nutritional security and biodiversity conservation. Int J Manag Soc Sci Res 4(8):26–30

Singh A, Nautiyal MC, Kunwar RM, Bussmann RW (2017) Ethnomedicinal plants used by local inhabitants of Jakholi block, Rudraprayag district, western Himalaya, India. J Ethnobiol Ethnomed 13:49. https://doi.org/10.1186/s13002-017-0178-3

Sreeramulu N, Suthari S, Ragan A, Raju VS (2013) Ethno-botanico-medicine for common human ailments in Nalgonda and Warangal districts of Telangana, Andhra Pradesh, India. Ann Plant Sci 2(7):220–229

Srithi K, Balslev H, Wangpakapattanawong P, Srisanga P, Trisonthi C (2009) Medicinal plant knowledge and its erosion among the Mien (Yao) in northern Thailand. J Ethnopharmacol 123 (2):335–342

Tester M, Langridge P (2010) Breeding technologies to increase crop production in a changing world. Science 327:818–822

Thakur J, Kanwar RR, Kumar P, Salam JL, Kar S (2018) Studies of genetic parameters for yield and yield attributing traits of kodo millet (*Paspalum scrobiculatum* L). Int J Curr Microbiol Appl Sci 7(09):278–287

Upadhyaya H, Dwivedi SL, Singh SK, Sube S, Vetriventhan M, Sharma S (2014) Forming core collections in barnyard, kodo, and little millets using morpho agronomic descriptors. Crop Sci 54:2673–2682

Verma VC, Verma VC, Singh A, Agrawal S (2018) Ethnobotanical study of small millets from India: prodigious grain for nutritional and industrial aspects. Int J Chem Stud 6(4):2155–2162

Wang RL, Wendel JF, Dekker JH (1995) Weedy adaptation in *Setaria* spp. I. Isozyme analysis of genetic diversity and population genetic structure in *Setaria viridis*. Am J Bot 82:308–317

Williams MM, Boydston RA, Davis AS (2007) Wild proso millet (*Panicum miliaceum*) suppressive ability among three sweet corn hybrids. Weed Sci 55:245–251

Xaxa V (1999) Tribes as indigenous people of India. Econ Polit Wkly:3589–3595

Niramala B. Yenagi (2004) Value adding strategies for conservation and sustainable use of indigenous minor millets. Presented at First National Convention on "Science and tradition of food - India's heritage of 5000 years" during 25–27 July 2004 organised by Academy of Sanskrit Research Melkote, India, and CFTRI, Mysore. p 80

You L, Wang P, Kong C (2011) The levels of jasmonic acid and salicylic acid in a rice-barnyard grass coexistence system and their relation to rice allelochemicals. Biochem Systemat Ecol 39:491–497

Zohary D, Hopf M, Weiss E (2012) Domestication of plants in the old world: the origin and spread of domesticated plants in South West Asia, Europe and the Mediterranean basin. Oxford University Press, New York

Millets: Malnutrition and Nutrition Security 4

Sarita Srivastava and Chhavi Arya

Abstract

Globally malnutrition is an emerging challenge in the context of rising uncertainty of food supplies. The latest estimates of the Food and Agriculture Organization revealed that up to the year 2017 the number of undernourished people in the world has increased to 821 million. Stunting, wasting, and anemia are the most prevalent form of malnutrition in low-income and food-deficit countries. Hidden hunger or micronutrient deficiency is yet another malnutrition problem faced globally. Together all these factors put children and women at a greater risk of dying as they become susceptible to common infections and life-threatening diseases. With changing lifestyle and food habits noncommunicable diseases have emerged as major health problems worldwide more so in developing countries. It is predicted that noncommunicable diseases will increase substantially in low- and middle-income countries because of lifestyle, transition associated with increasing urbanization, and globalization. Millets are the ancient heritage grains grown on poor soils without the use of chemicals therefore in a way these are organic grains. Owing to their exceptional nutritional profile in terms of micronutrients concentration, fiber content, gluten free nature, resistant starch, and various phytochemicals of therapeutic uses they are known as miracle grains. In the present chapter, a brief account of malnutrition statistics and the role of millets in achieving nutritional security to combat hidden hunger has been furnished.

S. Srivastava (✉)
Department of Food and Nutrition, College of Home Science, G.B. Pant University of Agriculture and Technology, Pantnagar, Uttarakhand, India

C. Arya
Department of Home Science, DSB Campus, Kumaun University, Nainital, Uttarakhand, USA

A. Kumar et al. (eds.), *Millets and Millet Technology*,
https://doi.org/10.1007/978-981-16-0676-2_4

Keywords

Millets · Malnutrition · Micronutrients · Lifestyle · Nutrition security

4.1 Malnutrition: The Global Scenario

Malnutrition is a serious problem in today's world of changing climate and rising uncertainty of food supplies for both the developing as well as developed countries (UNICEF 2018). It refers to deficiency, excess or imbalance in the intake of energy and nutrient by a person (WHO 2020). According to FAO (2018) a total of 821 million people are malnourished (those facing chronic food deprivations) across the world. As both deficiency and excess coexist in the world, it is seen that 109 billion adults are overweight and another 462 million are underweight (WHO 2020). One-third of the reproductive age women are anemic and 20 million babies are born underweight (UNICEF 2018).

The impact of malnutrition is severe in developing countries especially in economically deprived masses. For instance, out of all deaths occurring in children less than 5 years of age, nearly half are attributed to malnutrition (UNICEF 2020). Malnutrition leads to high rate of mortality, impaired body development and physical growth of a person which ultimately results in reduced economic productivity of the country (Masoud et al. 2018). Hidden hunger, the most prevalent form of malnutrition is reported to be very high in sub-Saharan Africa and South Asian countries. Hidden hunger might not be felt in the belly but it strikes at the core of health and vitality (Gautam 2014). The global nutrition report revealed that hidden hunger and other forms of malnutrition alone lead to a reduction of 11% in GDP every year in Africa and Asia (IFPRI 2016). Globalization and industrialization have severely impacted the lifestyle including food habits of people resulting in week immunity against noncommunicable diseases (Anand et al. 2016; Singh et al. 2018). Economically deprived people may face serious challenges related to their food and nutrition security due to the lack of funds for preventing malnutrition due to global economic slowdown (Braun Von 2020).

4.2 Malnutrition: The Indian Scenario

In the Indian scenario, the proportion of malnourished people is quite high among children (<5 years), adolescent girls, pregnant women, and lactating mothers in both rural and urban masses (Narayan et al. 2019). Children are hit the hardest by malnutrition in India. It has been estimated that India has the maximum number of malnourished children across the globe and one in every three children is malnourished in the country (CRY 2020). The severe condition is evident by the fact that in the year 2017 more than 65% of children (<5 years) died because of malnutrition (CRY 2020). Likewise, in the same year, the percent of children with low birth weight, wasting, anemia, and anemic adolescent girls in the country were 21.4%,

39.3%, 15.7%, 59.7%, and 54.4%, respectively. All forms of malnutrition varied between the states in India (Swaminathan et al. 2019). India ranked 102nd out of 117 countries on the Global Hunger Index (2019).

In general, the percentage of people lacking essential micronutrients and vitamins in their diet is quite high in South Asian countries (Harding et al. 2018). Of the two billion people suffering from hidden hunger globally, nearly half live in India. There are a widespread (more than 80% of the total Indian population) risk of deficiencies of calcium, vitamin A, vitamin B12, and folate. Availability of the amino acids, viz., lysine and leucine is limited in the average Indian diet. Besides these, other localized deficiencies of iron, zinc, and vitamin B6 are also prevalent. A major reason for these implicated deficiencies is lack of dietary diversity and dominated monotonous cereal diet and therefore insufficient food intake (Ritchie et al. 2018; Kumar and Kumar 2020). The acute deficiency of essential micronutrients and vitamins may lead to severe health complications including poor mental and physical growth and even death (Biro and Menon 2014). These deficiencies ultimately lead to poor working efficiency of vulnerable section of the society which ultimately hampers the economic growth of developing countries (Bhandari and Banjara 2015). For instance, the short-term cost of micronutrient malnutrition in India amounts to 0.8–2.5% of the GDP (Qaim 2007).

Epidemiological studies conducted across the country revealed an overall zinc deficiency of 43.8% with a high deficiency rate in the states of Orissa, Uttar Pradesh, Madhya Pradesh, and Karnataka (Kapil and Jain 2011). From 2006 to 2016 anemia declined in India but it remains highly prevalent in children, pregnant, and nonpregnant women (Nguyen et al. 2018a, b). A study conducted by Didzun et al. (2019) in India reported the prevalence of anemia in men to be 28.2%.

Yet another form of malnutrition is obesity. Overweight and obesity are emerging as serious challenges in developing countries including India. It has been observed that obesity among women and children are on the rise and with the proportion of problem becoming serious it needs to be addressed soon. According to an estimate, nearly 135 million people are suffering from obesity and associated problems in India (Ahiwar and Mondal 2019). It is known that a sedentary lifestyle and unhealthy eating habits with too much reliance on processed and fast food are the root causes of rising obesity in India. Luhar et al. (2020) estimated that the prevalence of overweight and obesity will reach 30.5% among men and women by 2040 especially in the older population and rural residents. Obesity is linked with several life-threatening diseases including diabetes, cardiovascular diseases, cancer, osteoarthritis, liver, and kidney diseases). Furthermore, it is known to negatively impact reproductive health. Obesity affects the quality of life and is a persisting risk factor leading to morbidity and mortality. Among all, diabetes and cardiovascular diseases are the most common and serious consequences of obesity.

India is second after China in the prevalence of diabetes mellitus with a total number of 69.1 million diabetic people (Tripathi et al. 2017). According to the World Health Organization diabetes was the major cause of death in the year 2016. Similarly, cardiovascular disease which is a noncommunicable disease is becoming a major cause of mortality in India because a quarter of all mortality is attributed to it

(Prabhakaran et al. 2016). There are large gaps in knowledge about the association of macronutrients to cardiovascular disease in lower and middle-income countries (Anand et al. 2016). Indian cities and metros have already witnessed a substantial change in the dietary pattern with more dependence on processed food. The frequency of consuming junk food and the tendency of skipping home-cooked meal is picking up.

Furthermore, the current COVID 19 pandemic in India and all over the world is devastating and has aggravated the nutrition security situation. In India, survival of a large number of factory workers, daily wage earners, and laborers as well as food security has been threatened (FAO 2020). Coronavirus infection has changed the scenario of communicable versus noncommunicable diseases globally. Nevertheless, the current situation demands more attention on immunity and health of people as a matter of their survival. In these situation, millets seem to be the best alternative staple food for sustaining food security and building immunity.

4.3 Millets

Millets are small-seeded grains that have been domesticated as food grains since ages (Weber and Fuller 2008). These have been part of the diets of the people belonging to ancient civilizations (Pokharia et al. 2014) and have also well stood the test of time (Green and Hemming 2014) as they continue to be used even today. Thus, they have a long cultivation and usage history. Diets with millets as staples ensured survival and nourishment of ancient people. Millets are the native of Africa and then introduced to other parts of the world (Bhat et al. 2018). They are divided into two broad categories (1) major millets which include sorghum and pearl millet and (2) minor millets which include finger millet, kodo millet, barnyard millet, proso millet, little millet, and foxtail millet (Behera 2017). They are widely utilized for food, feed, fodder, and fuel. Just five to six decades ago different types of millets were being cultivated and an integral part of people's diet in various parts of India. These were used as flour, partially broken, or as a whole grain in various local recipes.

In an attempt to increase food grain production, the food policies of the government relied heavily on wheat, rice, and maize while neglecting other small grains which had roots embedded in our tradition. The green revolution favored the production of major cereals but at the same time production of other minor crops and millets declined (Nelson et al. 2019). The green revolution though a boon for the country gave the much-needed agricultural, financial, and research attention to wheat and rice. As a result, these crops became popular while other coarse cereals and millets lost out to them (Dhan Foundation 2012).

4.3.1 The Sustainability and Agricultural Advantages of Millets

Millets are the hardiest and climate-resilient crops. Millets adapt well to a wide range of agro-climatic conditions and even on poor soils. Millets are extremely drought-tolerant compared to other crops such as wheat, rice, and maize. Millets have a short growing season and are pest-resistant in field and store (Passi and Jain 2014). It is notable that very few food crops possess these qualities which make them golden crops of the future.

In India, millets are traditionally grown on poor soils without the use of chemical fertilizers and pesticides therefore in a way these are organic grains. Marginal farmers generally grow them with the use of farmyard manure. Millets can be stored safely well for years and during this duration they remain practically pest-free. Pearl, foxtail, proso, and barnyard millets are some of the fastest maturing millets. Compared to major cereals, sorghum, and millets have low carbon footprints. By virtue of their unique C4 photosynthetic pathway millets utilize water in a highly efficient manner and require low inputs (Kumar et al. 2018). It has been predicted that due to accelerated climate change production of major cereals will decline in the future therefore drought tolerant and low input demanding crops like millets should be promoted in an extensive way to mitigate the effects of climate change and food insecurity (Saxena et al. 2018).

Thus, millets offer several agricultural advantages which make them crops which are less demanding on agricultural inputs yet are organic and environment friendly. Under conditions of environmental stress, millets can thrive and yet provide nutritious food grains. Therefore, millets are the key alternative to the expected food insecurity problem.

4.3.2 Role of Millets in Nutritional Security and Combating Malnutrition

"Food and nutrition security exists when all people at all times have physical, social and economic access to food, which is consumed in sufficient quantity and quality to meet their dietary needs and food preferences, and is supported by an environment of adequate sanitation, health service and care, allowing healthy and active life" (UNSCN 2013). Nutrition security is an essential element of food security, as sound nutrition requires more than just enough energy for every man, woman, and child (Hwalla et al. 2016). According to the UNICEF, three determinants of nutritional security are (1) easy access to sufficient food, (2) care and feeding practices, and (3) hygiene and good health to all the sectors of the society (UNSCN 2013). In short, nutrition security requires simultaneously "food," "health," and "care." So there is no way to achieve nutrition security without food security at household level (FAO 2009).

Nutritional insecurity is a major concern to that section of the society which is highly dependent on carbohydrate-rich diet and deficient in essential vitamins and micronutrients (Vinoth and Ravindhran 2017). Therefore, supplementation of

Table 4.1 Proximate composition of millets and other cereals (per 100 g)

Millet/cereal	Calorific value (kcal)	Carbohydrate (g)	Protein (g)	Fat (g)	Crude fiber (g)	Minerals (g)
Sorghum	349	72.6	10.4	1.9	1.6	1.6
Pearl millet	361	67.5	11.6	5.0	1.2	2.3
Finger millet	328	72.0	7.3	1.3	3.6	2.7
Foxtail millet	331	60.9	12.3	4.3	8.0	3.3
Proso millet	341	70.4	12.5	1.1	2.2	1.9
Barnyard millet	307	65.5	6.2	2.2	9.8	4.4
Kodo millet	309	65.9	8.3	1.4	9.0	2.6
Little millet	341	67.0	7.7	4.7	7.6	1.5
Rice (raw milled)	345	78.2	6.8	0.5	0.2	0.6
Wheat flour (whole)	341	69.4	12.1	1.7	1.9	2.7
Maize (dry)	342	66.2	11.1	3.6	2.7	1.5

Source: Gopalan et al. 2016

carbohydrate-rich diet with micronutrients and vitamins-rich crops would be a potential strategy to improve dietary diversification to combat malnutrition. Being one of the rich reservoirs of micronutrients, dietary fiber, vitamins, and phytochemicals of diverse therapeutic uses millets has the huge potential to work as alternative grains for ensuring food and nutritional security in most parts of the world (Konapur et al. 2014; Kumar et al. 2018). By virtue of their excellent nutritional composition, there has been an increased interest in the consumption of millets in the form of various functional food products across the world (Among the minor millets finger millet alone was found to have higher nutritional value than rice and contains more than ten times the average amount of calcium than major cereals and is high in iron, magnesium, phosphorus, and potassium (Kumar et al. 2016)). Further absence of gluten made protein quality of small millets very high compared to conventional food sources like wheat, barley, and maize (Kumar et al. 2016). Thus, nutraceutical characteristics of small millets must be harnessed (Tables 4.1 and 4.2) to develop finger millet as novel and functional food for nutritional security.

4.3.3 Macronutrients Present in Millets

The macronutrient profile of millet grains is comparable to major cereals (Table 4.1). A perusal of Table 4.1 revealed that compared to cereals millets are the rich and cheap source of energy especially for tribal and resource-deprived poor communities (Devi et al. 2011; Nambiar et al. 2011; Verma and Patel 2012). The energy content of millet varies from 307 to 361 kcal being lowest for barnyard millet and highest in pearl millet (Table 4.1).

Table 4.2 Mineral composition of millets and other cereals (mg/100 g)

Millet/cereals	Iron	Calcium	Zinc	Phosphorus	Potassium	Magnesium
Sorghum	4.1	25	1.6	222	131	171
Pearl millet	8.0	42	3.1	296	307	137
Finger millet	3.9	344	2.3	283	408	137
Foxtail millet	2.8	31	2.4	290	250	81
Proso millet	0.8	14	1.4	206	113	153
Barnyard millet	5.0	20	3.0	280	–	82
Kodo millet	0.5	27	0.7	188	144	147
Little millet	9.3	17	3.7	220	129	133
Rice (milled)	0.7	10	1.4	160	–	90
Wheat flour (whole)	4.9	48	2.2	355	315	132
Maize (dry)	2.3	10	2.8	348	286	139

Source: Gopalan et al. 2016

Millets are a major source of carbohydrates and proteins for people living in semi-arid tropics of Africa and Asia (Saleh et al. 2013). In the Indian diet, the major carbohydrate is starch mainly derived from cereals such as rice, wheat, finger millet, and sorghum (Urooj et al. 2006). The carbohydrate content of millets is comparable with major cereals and it ranges from 60.9 to 72.6 g/100 g. Finger millet contains 72% carbohydrates, a high proportion of which is in the form of slow digestive starch and resistant starch which helps in the prevention of constipation and lowering blood glucose (Vidyavati et al. 2004). Starch content among pearl millet genotypes ranges from 62.8 to 70.5 g/100 g. By virtue of its slow digestive nature and low cost, it can be used in the development of functional food products suitable for infants and people suffering from obesity and diabetes (Suma and Urooj 2015). Further, higher swelling power and solubility above 65 °C indicates its usefulness as a thickener. Likewise, the rich carbohydrate profile of foxtail millet (60.9 g/100 g) is an ideal source of energy for sick people, young children, and pregnant and lactating women (Hariprasanna 2016). Similarly, the slowly digestible carbohydrate profile of barnyard millet is ideal for people engaged in sedentary activities (Verma et al. 2015).

In sub-Saharan Africa millets especially sorghum accounts for nearly half of the cereal production and act as a staple food and also the primary source of protein (Belton and Taylor 2004). The protein content of sorghum and most of the millet compares well with that of major cereals. However, the major advantage over cereal protein especially wheat is that it is gluten-free so ideal for the consumption of celiac patients (Santra and Rose 2013). Foxtail millet, barnyard millet, pearl millet, and sorghum contain 12.3%, 12.0%, 11.6%, 10.4% protein in their grains (Surekha et al. 2013; Gopalan et al. 2016), respectively, which is more than the protein present in milled rice (6.8%). However, the protein quality of millets is comparatively better than the cereal protein. For instance, the amount of essential amino acid lysine in sorghum, pearl millet, and finger millet proteins is reported to be 1.77%, 2.56%, and 3.38%, respectively (Malleshi et al. 2004). Similarly, the amount of essential amino acids in protein composition of pearl millet grains is 27–32% higher than major cereals (Davis et al. 2003). Likewise, the seed storage protein of finger millet and foxtail millet is enriched with essential amino acids including sulfur-containing amino acids methionine and cystine (Vidyavati et al. 2004; Mbithi-Mwikya et al. 2000; Mohamed et al. 2009).

Millets are low in fat and contain mono and polyunsaturated fatty acids. Pearl millet has high oil content (up to 4.2%) which consists of 50% unsaturated fatty acids (Davis et al. 2003). Among the major cereals and millets, pearl millet contains highest amount of fat. Predominant fatty acids present in finger millet are oleic (49%), linoleic (25%), and palmitic acids. Characterization of fatty acid profile of finger millet revealed that it is mainly comprised of triglycerides, which reduces the risk of duodenal ulcer (Gull et al. 2016). Likewise, linoleic (66.5%) and oleic (13.0%) are the major fatty acids of foxtail millet bran oil (Liang et al. 2010). Millets are non-glutinous and easily digestible grains (FSSAI 2019). Several millet-based food preparations such as *ambali* (fermented semi-liquid product) and *hurihittu* (flour of popped finger millet) were traditionally used in India to meet the dietary needs of pregnant women (Lakshmi 2018). Finger millet malt is a popular and traditional energy-rich and low viscosity weaning food utilized for feeding the infants (Chandrashekhar 2010; Verma and Patel 2012). The weaning food prepared from 30% each of foxtail millet and barnyard millet flour is reported to contain 18.37% protein and 398 kcal energy per 100 g (Thathola and Srivastava 2002).

Millets can act as an ideal supplement of cereal-based diet to reduce the protein energy malnutrition in vulnerable masses of society (Saleh et al. 2013; Kumar et al. 2018). In a study conducted on adolescent children revealed significant improvement in their body mass index and stunting after continuously feeding them millet-based lunch for an interval (Anitha et al. 2019).

4.3.4 Micronutrients of Millets

Micronutrients and trace elements are nutrients that are required in very small quantities but perform essential metabolic functions in human body and provide resistance against various diseases. These are required for normal growth and good health. Each millet individually is a rich source of one or the other mineral (Table 4.2). Finger millet is an exceptionally rich source of calcium (344 mg/100 g) and it also contains a good amount of iron. In comparison to milled rice same quantity of finger millet supplies approximately 34 times more calcium. Finger millet is the richest source of calcium among cereals with up to tenfold higher calcium content than brown rice, wheat, or maize and three times than that of milk (Kumar et al. 2016). Pearl millet is an excellent source of iron (8.0 mg/100 g), phosphorous (296 mg/100 g), zinc (3.1 mg/100 g), and magnesium (137 mg/100 g). Foxtail millet is a good source of phosphorous and zinc. Little millet is a rich source of iron (9.3 mg/100 g) and zinc (3.7 mg/100 g). Barnyard millet too is a rich source of iron (5.0 mg/100 g) and zinc (3.0 mg/100 g) and phosphorus (280 mg/100 g) (Gopalan et al. 2016). However, phytic acid present in cereals and millets reduces the bioavailability of minerals. Various processing techniques are known to reduce the concentration of anti-nutritional factors such as phytic acid content in millets (Elhag et al. 2002; Makokha et al. 2002).

Calcium is an essential element which plays a major role in bone mineralization, cell signaling, and homeostasis (Puranik et al. 2017). Calcium deficiency coupled with inadequate intake of vitamin D with associated genetic and environmental factors is the major reason for rickets (Thacher et al. 2006). Rickets is a condition that results in weak or soft bones in children is highly prevalent in Africa and tropical Asia (Jideani 2012; Chappalwar et al. 2013). In India, the dietary intake of calcium in both rural and urban masses is reported to be very low than the recommended daily intake of calcium for a normal human (Harinarayan et al. 2007). Supplementation of carbohydrate-rich cereal-based diet with calcium-rich millet-based food is one sustainable approach to reduce calcium deficiency in developing countries (Ekbote et al. 2017). Among all the millets, finger millet is the richest reservoir of calcium (Table 4.2). Its grains have nearly 30-fold higher concentration of calcium compared to the rice grains and can be used as a model plant system for deciphering the molecular genetics underlying calcium biosynthesis and accumulation (Sharma et al. 2016). Finger millet grains can act as an ideal dietary supplement of calcium especially in case when families are unable to afford sufficient milk and milk products or if people suffer from lactose intolerance. Apart from finger millet, pearl millet, foxtail millet, kodo millet, and barnyard millet are also a good reservoir of calcium.

Millet grains are rich in phosphorus, which is required for energy production in the form of ATP molecules during various metabolic pathways (Himanshu et al. 2018). Among the millets pearl millet, finger millet, and foxtail millet are the rich sources of phosphorus (Table 4.2).

Anemia caused by iron deficiency is the major malnourishment concern, which has affected nearly 2.36 billion people globally (GBD 2015). Anemia alone leads to 3.7% and 12.8% of maternal deaths during pregnancy in Africa and Asia, respectively (Khan et al. 2006). Reports have indicated that though the overall rate of anemia has reduced in the last decade in India but still prevalent in children and pregnant women). In India about 89 million children are anemic and children less than 5 years are particularly vulnerable to anemia (Singh and Patra 2014). Pearl, proso, barnyard, and finger millet are good sources of iron. The inclusion of these millets in the diet of pregnant women and children can help reduce iron deficiency anemia. Millet-based weaning food can increase the supply of iron and calcium in the diet of infants and toddlers. A low-cost pearl millet-based weaning food fortified with iron and vitamin A has been developed by Sihag et al. (2016). This well-developed product could serve as a potential delivery system of iron and vitamin A and could be an effective tool to deal with the micronutrient deficiencies in infants. Studies on finger millet cookies showed that they retained a good amount of iron. The standardized recipes contained more amounts of calcium and iron than those available in the market (Kazi and Auti 2017). Biofortification of staple food items with iron-rich millets is one sustainable and effective approach to reduce the risk of micronutrient-related disorders, especially anemia. For instance, among the millets pearl millet is the richest source of iron and can act as an ideal dietary supplement to tackle iron deficiency. In this connection, hybrids of pearl millet possessing high micronutrient density in their grains have been developed and released in India (Vinoth and Ravindhran 2017). Dietary diversification with iron biofortified pearl millet grains is reported to enhance the iron status of secondary school children (Finkelstein et al. 2015). In another study, Kodkany et al. (2013) observed higher absorption of iron and zinc from biofortified pearl millet grains than the control grains by young children of 22–35 months.

Zinc is another vital micronutrient required for maintaining sound health. It plays an essential role in biosynthesis and metabolism of important proteins, lipids, and nucleic acids, which are ultimately require for maintaining reproductive functions, immunity boost up, wound repairs, and other complementary activities in human body. It has been reported that approximately 17% of the global population and 30% of people living in south Asian countries are suffering from zinc deficiency (Akhtar 2013). Among all the age groups, dietary deficiency of zinc severely affects the children resulting in stunted growth in their first 5 years of life (Maxfield and Crane 2019). Adolescent girls and pregnant women are also severely hampered by dietary deficiency of zinc in developing countries.

Millet grains are the rich source of all the trace elements including zinc required for maintaining normal physiological functions of human body. Almost all millets comprise excellent zinc concentration in their grains with maximum in pearl millet (3.29 ± 0.47 mg/100 g) followed by finger millet (3.13 ± 0.67 mg/100 g), foxtail millet (2.29 ± 0.13 mg/100 g), and proso millet (2.22 ± 0.21 mg/100 g) (Shankaramurthy and Somannavar 2019). Many studies have confirmed that processing treatments like malting enhance the bioavailability of micronutrients profile (including zinc) of millets. For instance, upon malting zinc concentration of pearl millet grains increased from 2.04 mg/100 g to 5.25 mg/100 g. Likewise, concentration of zinc in malt was reported to be 3.14 mg/100 g compared to the raw finger millet grains (2.15 mg/100 g) (Rao and Deosthale 1983).

Magnesium is another vital micronutrient for maintaining physiological processes and metabolic activities. Its critical role in maintaining various molecular processes in human body is evident by the fact that it is a cofactor of more than 300 enzymes regulating various biosynthetic pathways (Himanshu et al. 2018). Adequate dietary intake of magnesium is known to reduce the risk of heart stroke, atherosclerosis, respiratory problems, and migraine (Rao et al. 2017; Ambati and Sucharitha 2019). Among the millets, kodo millet grains (147 mg/100 g) are the richest reservoir of magnesium followed by finger millet (137 mg/100 g) and little millet (133 mg/100 g) (FAO 1995).

Millets are also a good source of B vitamins—thiamine, riboflavin, niacin, and folic acid. These B vitamins are essential for energy metabolism in the body and they function as coenzymes in a vast array of catabolic and anabolic enzymatic reactions (Sarita and Singh 2016). They are important for proper brain functioning, energy production, nucleic acid synthesis, and synthesis of various neuro-chemicals (Kennedy 2016). Compared to major cereals like rice (0.04 mg/100 g) and wheat (0.1 mg/100 g), concentration of essential vitamins in little millet (0.28 mg/100 g), pearl millet (0.21 mg/100 g), and foxtail millet (0.19 mg/100 g) is very high (FAO 1995).

Folic acid is another essential trace element required for proper physiological functions in the body of adolescent girls and pregnant women. It plays an important role in the biosynthesis of nucleic acids and red blood cells. Proper dietary intake of folic acid can reduce the risk of anemia, neural disorders in infants, and cardiovascular diseases in women particularly in developing countries (Krishnaswamy and Nair 2001). Pearl millet, sorghum, and kodo millet are good sources of folic acid and can easily be incorporated into the diets, especially that of women and children. The folate content of sorghum, pearl millet, finger millet, little millet, and kodo millet is 39.42, 36.11, 34.66, 36.20, and 39.49 μg/100 g, respectively, while that of milled rice is 9.32 μg/100 g (NIN 2007). In general, it is seen that millets contain a good amount of important vitamins and minerals which can help to reduce deficiency diseases and thus contribute to mitigate hidden hunger.

4.3.5 Fiber and Phytochemicals of Millets and Their Role in Disease Prevention

There is much more use of refined wheat flour, polished rice, high sugar, and saturated fats in the typical Indian urban diet than it used to be five to six decades ago. Fiber in the form of whole grains is decreasing in terms of quantity from the urban Indian diet. The incidence of diabetes and obesity are increasing globally and to combat them there is a raised demand for food containing complex carbohydrates with higher levels of dietary fiber and phytochemicals (Shobana et al. 2007). On the other hand, degenerative or noncommunicable diseases are on an increase worldwide (Mohan et al. 2019).

Millets are a rich source of dietary fiber which plays a role in the prevention of many diseases. Crude fiber content of small millets ranges from 2.2 to 9.8 g/100 g (Gopalan et al. 2016). These millets contain much more fiber than milled rice and wheat. Dietary fiber has health benefits like good bowel movement and reduction in blood cholesterol and sugar. High fiber diet is important in prevention and management of obesity, type 2 diabetes, heart disease, and cancer (Kendell et al. 2010). Dietary fiber is important for its hypoglycaemic, hypolipidemic, and hypocholesterolemic effects. Hence, it helps in the prevention of atherosclerosis. It also shows antitoxic and anti-cancerous effects (Bisoi et al. 2012). With too much inclusion of refined cereals in the diet, constipation, obesity, and other noncommunicable disease are on the increase.

Fiber is very important for proper functioning of the gastrointestinal tract and providing relief in constipation. Fibers decrease transit time, increase fecal bulk, have good water holding capacity, and result in the formation of softer feces and relieves constipation. Owing to its rich fiber content, pearl millet can be recommended in the treatment of celiac disease, constipation, and several noncommunicable diseases (Nambiar et al. 2011). The high fiber content in millets prevents gallstone formation (Deshpande et al. 2015).

The incidence of obesity is on the rise globally and the demand for food, containing complex carbohydrates with a higher amount of dietary fiber and health beneficial phytochemicals has been in demand (Shobana et al. 2007). Animal model studies are also important and provide in vivo experimental evidence. Murtaza et al. (2014) studied the supplementation of finger millet whole grain and bran to LACA mice for 12 weeks and reported that inclusion of finger millet bran at 10% (w/w) had beneficial effects and it prevented body weight gain, improved lipid profile and anti-inflammatory status, alleviated oxidative stress and regulated the expression level of several obesity-related genes. Likewise, Mice fed normal foxtail millet protein diet for 3 weeks showed an increase in HDL cholesterol (Choi et al. 2005).

Millets have a beneficial role in the prevention and nutritional management of diabetes mellitus. Millet fiber and phytochemicals are thought to play a beneficial role, especially in type 2 diabetes mellitus. Whole-grain consumption is associated with a reduced risk of type 2 diabetes (Montonen et al. 2003). Post et al. (2012) concluded that an intervention involving fiber supplementation for type 2 diabetes mellitus can reduce fasting blood glucose and HbA1c and also suggested that increasing dietary fiber in the diet of patients with type 2 diabetes is beneficial.

Kumari and sumathi (2002) in their study evaluated the effect of consumption of finger millet-based diet on hyperglycemia on six NIDDM subjects showed that consumption of finger millet-based diets resulted in significantly lower blood glucose response curve which might have been due to the higher fiber content of finger millet. Foxtail millet has relatively low starch digestibility and moderate glycaemic index. Intake of 50 g foxtail millet/day for 12 weeks brought about a decrease in mean fasting blood glucose of the subjects. The study concluded that foxtail millet could improve the glycaemic control in free-living subjects with impaired glucose tolerance. The increase in the consumption of foxtail millet might be beneficial for type 2 diabetic individuals (Ren et al. 2018). A study by Jin et al. (2019) concluded that the interaction between endogenous proteins and lipids with starch plays a significant role in the hypoglycaemic properties of foxtail millet.

An animal model study showed that finger millet feeding to diabetic animals for 4 weeks controlled the blood glucose and improved the antioxidant status which hastened the dermal wound healing process (Rajasekharan et al. 2004). Sluijs et al. (2010) in their study showed a positive correlation between higher GI food and diabetes. The fiber intake inversely correlated with diabetes. A study by Jali et al. (2012) on type 2 diabetic patients fed with foxtail millet for 90 days showed improved glycemic control. Thathola et al. (2011) concluded that foxtail millet as a low glycemic index food product leads to modest improvement in long-term glycemic control in type 2 diabetics. Another study by Chhavi and Sarita (2012) concluded that millet flour incorporated bread showed low glycemic indices in normal subjects. Similar results have also been reported by Shukla and Srivastava (2014) with finger millet incorporated noodles.

Another noncommunicable disease with high morbidity and mortality rate is cardiovascular disease. Zhu et al. (2018) in their study highlighted that foxtail millet bran is a good source of dietary fiber and foxtail millet bran dietary fiber (FMBDF) showed good functional properties. In the study, it was found that FMBDF exhibited good absorption capabilities to lipophilic substances such as lard (3.34 g/g), peanut oil (2.32 g/g), and cholesterol (5.19 mg/g). Its bile salt absorption capacity (143.03 μmol/g) for sodium cholate and (76.65 μmol/g) for sodium taurocholate indirectly reflected its cholesterol lowering effect. Soluble fiber part of the dietary fiber of millets is also known for its cholesterol lowering effect.

A study undertaken by Ugare et al. (2014) to access health benefits of barnyard millet in type 2 diabetic patients reported that feeding intervention of 28 days revealed a significant reduction in glucose, LDL cholesterol, and VLDL cholesterol. Marginal decrease of triglycerides and increase of HDL were registered in diabetic groups. In one of the studies by Lee et al. (2010) on hyperlipidemic rats, it was concluded that foxtail millet and proso millet may prevent cardiovascular disease by reducing plasma triglycerides in hyperlipidemic rats. This study was undertaken to investigate the hypothesis that whole grain consumption would have beneficial effects on lipid profiles of hyperlipidemic rats. However, in the same study, it was found that sorghum increases total cholesterol, HDL cholesterol, and LDL cholesterol concentrations.

After cardiac disease, cancer has emerged as an important cause of morbidity and mortality in India with an estimate of 1.45 million cases of cancer in 2016 (Singh et al. 2018). A diet which provides sufficient dietary fiber and the vital nonnutritive substances such as polyphenols and antioxidants has a preventive and protective effect against various types of cancers. Millet grains are a rich source of bioactive phytochemicals, particularly phenolic acids and flavonoids. Phenolic compounds of millet grains are bioaccessible and possess bioactivities against several pathophysiological conditions. The use of millets as a source of nutraceuticals and specialty food in disease reduction and overall health and wellness is important (Shahidi and Chandrasekara 2013). Sripriiya et al. (1996) showed high (94%) DPPH radical quenching of brown finger millet extract and it also had high phenolic content. Finger millet polyphenols could be used as a natural source of antioxidants for minimizing the risk of diseases arising from oxidative deterioration (Banerjee et al. 2012).

Finger millet grains contain high phenolic content and antioxidant activities compared to proso and foxtail millets (Kumari et al. 2017). The bran fraction of millets is particularly rich in antioxidants. For instance, the whole flour and bran-rich fractions of foxtail millet are reported to have high antioxidant potency (Suma and Urooj 2012). Likewise, the phenolic extract of kodo millet hull exhibit three times higher inhibition activities against oxidation of LDL cholesterol and liposome compared to pearl millet (Chandrasekhara and Shahidi 2011).

Phytochemicals and antioxidants present in millets could play a protective role in the proliferation of mutagenic cells. Zhang et al. (2014) studied the antioxidant activity and anti-proliferative properties of three varieties of proso millet. Results exhibited a differential and possible selective anti-proliferative property of proso millet. Similarly, a novel protein with anti-cancerous activities was extracted and purified from foxtail millet bran (Shan et al. 2014). It inhibited cell proliferation and may serve as a therapeutic agent against colon cancer.

4.4 Conclusion

Malnutrition is a serious global and national problem. Both under-nutrition and over-nutrition are present at the same time in many countries. The burden of malnutrition is particularly high in India. Of the two billion people suffering from "hidden hunger" globally nearly half live in India. To further aggravate the situation noncommunicable diseases are on a rise. For instance, the current COVID 19 pandemic in India threatened the food and nutrition security of many poor people of the country. Under the circumstances of widely prevalent malnutrition and climatic vagaries on one hand and an urgent need to increase food grain production on the other, C4 millet crops are one of the most feasible options. Compared to polished rice and refined wheat nutritionally millets are the storehouse of many vital nutrients. Studies have suggested that the presence of a good amount of dietary fiber and phytochemicals in millets has a preventive role in many degenerative diseases such as obesity, type 2 diabetes, cardiovascular disease, and cancer.

The nutritional potential of millets can be utilized by the commercialization of millet-based foods which would cater to the need of many who are suffering from under-nutrition and degenerative diseases. Dietary intervention should be brought about by the inclusion of locally available millets in the diets of people especially women and children to help mitigate the widespread problem of under-nutrition and hidden hunger. Review of literature reveals that value addition of small millets has been done for diversified uses. Millets incorporated convenience mixes, viz., ready to cook as well as ready to use have been developed. Research has also been done to incorporate millets in Indian traditional food products. Refined wheat flour replacement by millets in various baked products and junk foods has resulted in nutritious food products with functional properties. In India, some low glycemic index functional foods are available in the market but their availability is limited to some parts of urban areas only with almost negligible variety. There is a need to popularize millet food products by commercializing them and by increasing awareness among the public through nutrition education for incorporation of millets in the diet. Thus, use of millets in the diet in a traditional way and also through use of unconventional food products can provide multiple benefits for ensuring health and nutrition security.

References

Ahiwar R, Mondal PR (2019) Prevalence of obesity in India: a systematic review. Diabet Metab Syndr Clin Res Rev 13:318

Akhtar S (2013) Zinc status in South Asian population - an update. J Health Population Nutr 31:139–149

Ambati K, Sucharitha KV (2019) Millet review on nutritional profiles and health benefits. Int J Recent Sci Res 10I:33943–33948

Anand SS, Hawker C, de Souza RJ, Meile A, Dehgoin M et al (2016) Food consumption and its impact on cardiovascular disease: importance of solutions focused on the globalized food system. J Am Coll Cardiol 66:1590–1614

Anitha S, Potaka JK, Tsusaka TW, Tripathi D, Upadhayay S et al (2019) Acceptance and impact of millet based mid day meal on the nutritional status of adolescent school going children in a peri urban region of Karnataka state in India. Nutrients 11(9):2077

Banerjee S, Sanjay KR, Chethan S, Malleshi NG (2012) Finger millet (*Eleusine coracana*) polyphenols: investigation of their antioxidant capacity and antimicrobial activity. Afr J Food Sci 6:362–374

Behera MK (2017) Assessment of the state of millets farming in India. MOJ Ecol Environ Sci 2:00013. https://doi.org/10.15406/mojes.2017.02.00013

Belton PS, Taylor RN (2004) Sorghum and millets: protein source for Africa. Trends Food Sci Technol 15(2):94–98. https://doi.org/10.1016/j.tifs.2003.09.002

Bhandari S, Banjara MR (2015) Micronutrient deficiency, a hidden hunger in Nepal: prevalence, causes, consequences and solution. Int Sch Res Notices 2015:276469

Bhat BV, Rao DB, Tonapi VA (2018) The story of millets. Karnataka State Department of Agriculture in association with ICAR, Indian Institute of Millet Research, Hyderabad, India

Biro E, P Menon (2014) Global hunger index, 2014, addressing the challenge of hidden hunger. www.globalhungerindex.org

Bisoi PC, Sahoo G, Mishra SK, Das KL (2012) Hypoglycemic effect of insoluble fiber rich fraction of different cereals and millets. J Food Process Technol 3:1–11

Braun Von J (2020) Forecast 2020, Financial meltdown and malnutrition. UN chronicle. www.un.org

Chandrasekhara A, Shahidi F (2011) Bioactivities and antiradical properties of millet grains and hulls. J Agric Food Chem 59:9563–9571

Chandrashekhar A (2010) Finger millet: *Elusine coracana*. Adv Food Nutr Res 59:25–62

Chappalwar VM, Peter D, Bobde H, John SM (2013) Quality characteristics of cookies prepared from oats and finger millet based composite flour. Eng Sci Technol 3:667–683

Chhavi A, Sarita S (2012) Evaluation of composite millet breads for sensory and nutritional quality and glycemic response. Malaysian J Nutr 18:89–101

Choi YY, Osada K, Ito Y, Nagasawa T, Choi RM (2005) Effects of dietary protein of Korean foxtail millet on plasma adiponectin, HDL-cholesterol, and insulin levels in genetically type 2 diabetic mice. Biosci Biotechnol Biochem 69:31–37

CRY (2020) Child rights and you, malnutrition in India. www.cry.org

Davis AJ, Dale NM, Ferreira FJ (2003) Pearl millet as an alternative feed ingredient in broiler diets. J Appl Poultry Res 12:137–144

Deshpande SS, Mohapatra D, Tripathi MK, Sadvatha RH (2015) Kodo millet nutritional value and utilization in Indian foods. J Grain Process Storage 2:16–23

Devi BP, Vijayabharathi R, Sathyabama S, Malleshi NG, Priyadarshini VB (2011) Health benefits of finger millet (*Eleusine coracana* L.) polyphenols and dietary fiber: a review. J Food Sci Technol. https://doi.org/10.1007/s13197-011-0584-9

DHAN Foundation 2012 WASSAN, supporting millets in India, policy review and suggestions for action, revalorising small millets in rainfed regions of South Asia. Supported by IDRC, CRDI, Canada International Development Agency

Didzun O, De Neve JW, Awasthi A, Dubey M, Theilmann M, Barnighausen T et al (2019) Anaemia among men in India: a nationally representative cross-sectional study. Lancet Glob Health 7:1685–1694

Ekbote VH, Khadilkar VA, Khadilkar VV, Chiplonkar SA, Mughal Z (2017) Dietary pattern with special reference to calcium intake in 2–16-year-old urban western Indian children. Indian J Public Health 16:188–193

ElHag ME, ElTinay AH, Yousif NE (2002) Effect of fermentation and dehulling on starch, total polyphenols, phytic acid content and *in vitro* protein digestibility of pearl millet. Food Chem 77:193–196

FAO (1995) Sorghum and millets in human nutrition. Rome, Italy. http://www.fao.org/docrep/T0818e/T0818E00

FAO (2009) Food Security and Nutrition Security-What is the problem and what is the difference. Summary of the FSN Forum Discussion No 34, Proceedings. http://kmfao.org/fileadmin/user
FAO (2018) The state of food security and nutrition in the world, building climate resilience for food security and nutrition. FAO, Rome. Licence-CC BY-NC-SA3.0IGO
FAO (2020) Local food systems and COVID-19. A glimpse on India's responses city region food system, programme by Pothan PE. www.fao.org
Finkelstein JL, Mehta S, Udipi SA, Ghugre PS, Luna SV et al (2015) A randomized trial of iron biofortified pearl millet in school children in India. J Nutr 145:1576–1581
FSSAI (2019) FSSAI publishes guidance note on Millets – Nutri Cereals. https://foodsafetyhelpline.com
Gautam CK (2014) Addressing the challenge of hidden hunger. IFPRI. http://www.ifpri.org
GBD (2015) Disease and injury incidence and prevalence collaborators, global, regional, and national incidence, prevalence and years lived with disability for 310 diseases and injuries, 1990–2015: a systematic analysis for the global burden of disease study 2015. Lancet 388:1545–1602
Global Hunger Index (2019) Global Hunger Index by severity. http://www.globalhungerindex.org
Gopalan C, Ramashastri BV, Balasubramanium SC (2016) Nutritive value of Indian foods. National Institute of Nutrition, ICMR, Hyderabad
Green P, Hemming C (2014) Grain power: over 100 delicious gluten-free ancient grains and super blend recipes. Penguin Canada. https://books.google.co.in
Gull A, Ahmad GN, Prasad K, Kumar P (2016) Technological processing and nutritional approach of finger millet (*Eleusine coracana*) – a mini review. J Food Process Technol 7:593
Harding LK, Aguayo VM, Webb P (2018) Hidden hunger in South Asia: a review of recent trends and persistent challenges. Public Health Nutr 21:785–795
Harinarayan VC, Ramalakshmi T, Prasad VU, Sudhakar D, Sriniwasrao PVLN, Sharma VS, Kumar TGE (2007) High prevalence of low dietary calcium, high phytate consumption and vitamin D deficiency in healthy south Indians. Am J Clin Nutr 85:1062–1067
Hariprasanna K (2016) Foxtail millet –nutritional importance and cultivation aspects. Indian Farm 65:25–29
Himanshu CM, Sonawane SK, Arya SS (2018) Nutritional and nutraceutical properties of millets: a review. Clin J Nutr Diet 1:1–10
Hwalla N, Labban SE, Bahn RA (2016) Nutrition security is an integral component of food security. Front Life Sci 9:167–172
IFPRI (2016) Global Nutrition Report: malnutrition becoming the "new normal" across the globe press release. www.ifpri.org. Sited 5 May 2020
Jali MV, Kamatar MY, Jal SM, Hiremath MB, Naik RK (2012) Efficacy of value added foxtail millet therapeutic food in the management of diabetes and dyslipidamea in type 2 diabetic patients. Recent Res Sci Technol 4:3–4
Jideani IA (2012) Non conventional cereal grains with potentials. Sci Res Essay 7:3834–3843
Jin Z, Bai F, Chen Y, Bai B (2019) Interactions between protein, lipid and starch in foxtail millet flour affect the in vitro digestion of starch. CyTA J Food 17:640–647
Kapil U, Jain K (2011) Magnitude of zinc deficiency amongst under five children in India. Indian J Pediatr 78:1069–1072
Kazi T, Auti S (2017) Calcium and iron recipes of finger millet. J Biotechnol Biochem 3:64–68
Kendell WC, Esfahani A, Jenkins DJA (2010) The link between dietary fiber and human health. Food Hydrocoll 24:42–48
Kennedy OD (2016) B vitamins and the brain: mechanism, dose and efficacy - a review. Nutrients 27:68. https://doi.org/10.3390/nu8020068
Khan KS, Wojdyla D, Say L, Gulmezoglu AM, Look FPV (2006) WHO analysis of causes of maternal death: a systematic review. Lancet 367:1066–1074
Kodkany SB, Bellad RM, Mahantshetti NS, Westeott JE, Krebs NF, Kemp JF, Hambridge KM (2013) Biofortification of pearl millet with iron and zinc in a randomized controlled trial

increases absorption of these minerals above physiological requirements in young children. J Nutr 143:1489–1943
Konapur A, Gavaravarapu SM, Gupta S, Nair KM (2014) Millets in meeting nutrition security: issues and way forward for India. Ind J Nutr Dietetics 51:306–321
Krishnaswamy K, Nair M (2001) Importance of folate in human nutrition. Br J Nutr 85:124
Kumar N, Kumar A(2020) Biofortification: a plausible antidote to India's hidden hunger problem. www.ifpri.org. IFPRI Blog: Issue post
Kumar A, Metwal M, Kaur S, Gupta AK, Puranik S, Singh S, Singh M, Gupta S, Babu K, Sood S, Yadav R (2016) Nutraceutical value of finger millet (*Eleusine corcana* L Gaertn) and their improvement using omics approaches. Front Plant Sci. https://doi.org/10.3389/fpls.2016.00934
Kumar A, Tomer V, Kaur A, Kumar V, Gupta K (2018) Millets: a solution to agrarian and nutritional challenges. Agric Food Secur 13. https://doi.org/10.1186/s40066-018-0183-3
Kumari LP, Sumathi S (2002) Effects of consumption of finger millet on hyperglycemia in non-insulin dependent diabetes mellitus (NIDDM) subjects. Plant Foods Hum Nutr 57:205–213
Kumari D, Madhujith T, Chandrasekara A (2017) Comparison of phenolic content and antioxidant activities of millet varieties grown in different locations in Srilanka. Food Sci Nutr 5:474–485. https://doi.org/10.1002/fsn3.415
Lakshmi (2018) Traditional foods of Indian origin in pregnancy. MOJ Food Proces Technol 6:280–281
Lee HS, Chung IM, Cha SY, Park Y (2010) Millet consumption decreased serum concentration of triglyceride and C – reactive protein but not oxidative status in hyperlipidemic rats. Nutr Res 3:290–296
Liang S, Yang G, Ma Y (2010) Chemical characters and fatty acid profile of foxtail millet bran oil. J Am Oil Chemists Soc 87(1):63–67
Luhar S, Timaeus IM, Jones R, Curningham S, Patel SA, Kinra S et al (2020) Forecasting the prevalence of overweight and obesity in India to 2040. PLoS One 15(2):e022438. https://doi.org/10.1371/journal.pone.0229438
Makokha AO, Oniang'o RK, Simon MN, Kamar OK (2002) Effect of traditional fermentation and malting on phytic acid and mineral availability of sorghum (*Sorghum bicolor*) and finger millet (*Eleusine coracana*) grain varieties grown in Kenya. Food Nutr Bull 23(3_Suppl 1):241–245
Malleshi NG, Reddy PV, Klopenstein CF (2004) Milling trials of sorghum, pearl millet and finger millet in quadrumat junior mill and experimental roll stands and the nutrient composition of milling fractions. J Food Sci Technol 41:618–622
Masoud S, Menon P, Bhutta ZA (2018) Handbook of famine, starvation and nutrient deprivation. Addressing Child Malnutrition in India, Springer Reference Live
Maxfield L, Crane JS (2019) Zinc deficiency. In StatPearls [Internet]. StatPearls Publishing
Mbithi-Mwikya S, Ooghe W, Van Camp J, Nagundi D, Huyghebaert A (2000) Amino acid profile after sprouting. Autoclaving and lactic acid fermentation of finger millet (*Elusine coracona*) and kidney beans (*Phaseolus vulgaris* L.). J Agric Food Chem 48:3081–3085
Mohamed TK, Zhu K, Issoufou A, Fatmata T, Zhou H (2009) Functionality, *in vitro* digestibility and physicochemical properties of two varieties of defatted foxtail millet protein concentrates. Int Mol Sci 10:522–438
Mohan P, Brahmawar M, Dutta M (2019) Communicable or noncommunicable diseases? Building strong primary health care systems to address double burden of disease in India. J Fam Med Prim Care 8:326–329
Montonen J, Kenkt P, Jarvinen R, Aromaa A, Reunanen A (2003) Whole-grain and fiber intake and the incidence of type 2 diabetes. Am J Clin Nutr 77:622–629
Murtaza N, Baboota RK, Jagtap S, Singh DP (2014) Finger millet bran supplementation alleviates obesity-induced oxidative stress, inflammation and gut microbial derangements in high-fat diet-fed mice. Br J Nutr 112:1447–1458
Nambiar VS, Dhaduk JJ, Sareen N, Shahu T, Desai R (2011) Potential functional implications of pearl millet (*Pennisetum glaucum*) in health and disease. J Appl Pharmaceut Sci 01:62–67

Narayan J, John D, Ramdas N (2019) Malnutrition in India: status and government initiatives. J Public Health Policy 40:126–141

Nelson ELRA, Ravichandran K, Antony U (2019) Impact of green revolution on indigenous crops of India. J Ethnic Foods 6:1–10. https://doi.org/10.1186/s42779-019-0011-9

Nguyen PH, Scott S, Avula R, Tran ML, Menon P (2018a) Trends and drivers of change in the prevalence of anaemia among million women and children in India 2006 to 2016. BMJ Global Health 3:e001010

Nguyen PH, Scott S, Avula R, Tran ML, Menon P (2018b) Trends and drivers of change in the prevalence of anaemia among 1 million women and children in India, 2006 to 2016. BMJ Glob Health 3:e001010

NIN (2007) Indian food composition tables, NIN and Nutritive value of Indian Foods. NIN

Passi SJ, Jain A (2014) Millets: the nutrient rich counterpart of wheat and rice. Press Information Bureau, Government of India, 17 Jul 2014

Pokharia AK, Kharkwal JS, Srivastava A (2014) Archaeobotanical evidence of millets in the Indian subcontinent with some observation son their role in the Indus civilization. J Archaeol Sci 42:442–445

Post ER, Mainous AG, King ED, Simpson NK (2012) Dietary fiber for the treatment of type 2 diabetes mellitus: a meta-analysis. J Am Board Fam Med 25:16–23

Prabhakaran D, Jeemon P, Roy A (2016) Cardiovascular disease in India. Current epidemiology and future directions. Circulation 133:1605–1620

Puranik S, Kam J, Sahu PP, Yadav R, Srivastava RK, Ojulong H, Yadav R (2017) Harnessing millet to combat deficiency in humans: challenges and prospects. Front Plant Sci 26. https://doi.org/10.3389/fpls.2017.013389

Rajasekharan ND, Nithya M, Rose C, Chandra TS (2004) The effect of finger millet feeding on the early response during the process of wound healing in diabetic rats. Biochim Biophys Acta Mol basis Dis 1689:190–201

Rao SSD, Deosthale YG (1983) Mineral composition, ionisable iron and soluble zinc in malted grains of pearl millet and ragi. Food Chem 11:217–223

Rao DB, Bhaskarachary K, Christina ADG, Devi SG, Tonapi AV (2017) Nutritional and health benefits of millets. ICAR_Indian Institute of Millet Research (IIMR), Rajendranagar, Hyderabad, p 122

Ren X, Yin R, Hou D, Xue Y, Zhang M, Diao X et al (2018) The Glucose-lowering effect of foxtail millet in subjects with impaired glucose tolerance: a self-controlled clinical trial. Nutrients 10:1509

Ritchie H, Reay DS, Higgins P (2018) Quantifying, projection and addressing India's hidden hunger. Front Sustain Food Syst 2:11

Saleh SMA, Zhang Q, Chen J, Shen Q (2013) Millet grains: nutrition quality processing and potential health benefits. Compr Rev Food Sci Food Saf 12:281–295

Santra KD, Rose JD (2013) Alternative uses of proso millet, NebGuide, University of Nebraska – Lincoln Extension, Institute of Agriculture and Natural Resources – Lincoln and the United States Department of Agriculture issued October

Sarita S, Singh E (2016) Potential of millets nutrient composition and health benefits. J Sci Innov Res 5:46–50

Saxena R, Vanga KS, Wang J, Orsat V (2018) Millet for food security in the context of climate change: a review. Sustainability 10:2228. https://doi.org/10.3390/su10072228

Shahidi F, Chandrasekara A (2013) Millet grain phenolics and their role in disease risk reduction and health promotion: a review. J Funct Food 5:570–581

Shan S, Li Z, Newton IP, Zhoo C, Li Z, Guo M (2014) A novel protein extracted from foxtail millet bran display anti-carcinogenic effects in human colon cancer cells. Toxicol Lett 227:129–138

Shankaramurthy KN, Somannavar MS (2019) Moisture, carbohydrate, protein, fat, calcium and zinc content in finger, foxtail, pearl and proso millets. Indian J Health Sci Biomedical Res 12:228–232

Sharma D, Gautam J, Singh UM, Sood S (2016) Calcium biofortification: three pronged molecular approaches for dissecting complex trait of calcium nutrition in finger millet (*Eleusine corcana*) for devising strategies of enrichment of food crops. Front Plant Sci 17(2028). https://doi.org/10.3389/fpls.2016.02028

Shobana S, Usha Kumari SR, Malleshi NG, Ali SZ (2007) Glycemic response of rice, wheat and finger millet based diabetic food formulations in normoglycemic subjects. Int J Food Sci Nutr 58:363–372

Shukla K, Srivastava S (2014) Evaluation of finger millet incorporated noodles for nutritive value and glycemic index. J Food Sci Technol 51:527–534

Sihag KM, Sharma V, Goyal A, Arora S (2016) Development of an alternative low cost cereal-based weaning food fortified with iron and vitamin A. Indian J Anim Sci 86:478–484

Singh RK, Patra S (2014) Extent of anaemia among preschool children in EAG states, India: a challenge to policy makers. Anaemia 9:868752

Singh M, Prasad CP, Singh DT, Kumar L (2018) Cancer research in India: challenges and opportunities. Indian J Med Res 148:362–365

Sluijs I, Van der Schouw YT, Van den Daphne D, Spijkesman AM, Hu FB, Grobbee DE et al (2010) Carbohydrate quantity and quality and sick of type 2 diabetes in the European perspective investigation into cancer and nutrition Netherland (EPIC-NL) study. Am J Clin Nutr 92:905–911

Sripriiya G, Chandrashekharan K, Murty VS, Chandra TS (1996) ESR spectroscopic studies on free radical quenching action of finger millet (*Eleusine coracana*). Food Chem 57:537–540

Stein AJ, Qaim M (2007) The human and economic cost of hidden hunger. Food Nutr Bull 28:125–134. https://doi.org/10.1177/156482650702800201

Suma FP, Urooj A (2012) Antioxidant activity of extracts from foxtail millet (*Setaria italica*). J Food Sci Technol 49:500–504

Suma PF, Urooj A (2015) Isolation and characterization of starch from pearl millet (*Pennisetum typhodium*) flours. Int J Food Prop 18:2675–2687

Surekha N, Naik RS, Mythri S, Devi R (2013) Barnyard millet (*Echinochloa Frumentacea*), cookies development, value addition, consumer acceptability, nutritional and shelf life evaluation. J Environ Sci Toxicol Food Technol 7:01–10

Swaminathan S, Hemlatha R, Pandey A, Kassebaum NJ, Laxmaiah A et al (2019) The burden of child and maternal malnutrition and trends in its indicators in the states of India: the global burden of disease study 1990–2017. Lancet Child Adolesc Health 3:855–870

Thacher TD, Fischer PR, Strand MA, Pettifor JM (2006) Nutritional rickets around the world causes and future direction. Ann Trop Paediatr 26:1–16

Thathola A, Srivastava S (2002) Physicochemical properties and nutritional traits of millet-based weaning food suitable for infants of the Kumaun hills. Asia Pacific J Clin Nutr 11:28–32

Thathola A, Srivastava S, Singh G (2011) Effect of foxtail millet (*Setaria italica*) supplementation on serum glucose, serum lipid and glycosylated haemoglobin in type 2 diabetics. Diabetol Croatica 40:23–28

Tripathi PJ, Thakur JS, Saran R (2017) Prevalence and risk factors of diabetes in a large community-based study in North India: results from a STEPS survey in Punjab, India. Diabetol Metab Syndr 9:8

Ugare R, Chimmad B, Naik R, Bharati P, Itagi S (2014) Glycemic index and significance of barnyard millet (*Echinochloa frumentacaea*) in type 2 diabetics. J Food Sci Technol 51:392–395

UNICEF (2018) For every child. Press release 28 November 2018. unicef.org

UNICEF (2020) Malnutrition, in children UNICEF data. Unicef.org

UNSCN (2013) Food and Nutrition Security, UNSCN Meeting of the minds, Nutrition impact of food systems, United Nations Systems, Standing Committee on Nutrition, 25–28 March 2013. Presented by Dr Wustefeld M. https://www.unscn.org

Urooj A, Rupashrik K, Puttaraj S (2006) Glycaemic responses to finger millet based Indian preparations in non-insulin dependent diabetic and healthy subjects. J Food Sci Technol 43:620

Verma V, Patel S (2012) Nutritional security and value added products from finger millet. J Applicable Chem 1:485–489

Verma S, Srivastava S, Tiwari N (2015) Comparative study on nutritional and sensory quality of barnyard and foxtail millet food products with traditional rice products. J Food Sci Technol 52:5147–5155

Vidyavati HG, Begum MJ, Vijayakumari J, Gokavi SS, Begum S (2004) Utilization of finger millet in the preparation of papad. J Food Sci Technol 41:379–382

Vinoth A, Ravindhran R (2017) Biofortification in millets: a sustainable approach for nutritional security. Front Plant Sci. https://doi.org/10.3389/fpls.2017.00029

Weber SA, Fuller DQ (2008) Millets and their role in early agriculture. Paper presented in the International seminar on the 'First Farmers in Global Perspective', Lucknow, India, 18–20 January

WHO (2020) Factsheets- malnutrition. WHO. 1 Apr 2020

Zhang L, Liv R, Niv W (2014) Phytochemical and antiproliferative activity of proso millet. PLoS One 9:e104058

Zhu Y, Chu J, Lu Z, Fengxia LV, Zhang B, Zhao H (2018) Physiochemical and functional properties of dietary fiber from foxtail millet (*Setaria itatica*) bran. J Cereal Sci 79:456–461

Nutritional Composition of Millets

5

Manoj Kumar Tripathi, Debabandya Mohapatra, Rajpal S. Jadam, Sharad Pandey, Vaishali Singh, Vishnu Kumar, and Anil Kumar

Abstract

Millets are drought-tolerant crops, extensively grown in Asia and semi-arid tropics of Africa. Millets are important nutritious minor cereal food crops and can ensure nutritional security. However, a decline is seen in the consumption of minor millets over the last few years. Millets are rich sources of minerals, vitamins, proteins, fatty acids, fiber, and other phytonutrients. Millet proteins have a balanced amount of essential amino acids, especially sulfur (S) containing amino acids. Millets are also enriched with several positive health attributing phytochemicals, including lignans, phytosterols, polyphenols, phyto-oestrogens, and phytocyanins. Processing and milling of millets remove the germ and bran and layers that are rich sources of fiber and phytochemicals. The millets are rich sources of antioxidants like glycated flavonoids and phenolic acids. Millets also serve as prebiotics and can enhance the effectiveness of probiotics. The nutritional significance of different millet cultivars are very useful in developing

M. K. Tripathi (✉) · D. Mohapatra · R. S. Jadam
ICAR-Central Institute of Agricultural Engineering, Bhopal, Madhya Pradesh, India

S. Pandey
Himgiri Zee University, Dehradun, Uttarakhand, India
e-mail: sharad.pandey@hzu.edu.in

V. Singh
SHUATS, Prayagraj, Uttar Pradesh, India

V. Kumar
Department of Genetics & Plant Breeding, Rani Lakshmi Bai Central Agricultural University, Jhansi, Uttar Pradesh, India

A. Kumar
Directorate of Education, Rani Lakshmi Bai Central Agricultural University, Jhansi, Uttar Pradesh, India

A. Kumar et al. (eds.), *Millets and Millet Technology*,
https://doi.org/10.1007/978-981-16-0676-2_5

value-added products. Presences of highly valuable nutraceuticals components in millets are supportive in prevention of various lifestyle illnesses such as cancer, cardiovascular diseases, low and high blood pressure, and diabetes. Because of their significant involvement in nutritional security and possible increasing health effects, millet is now addressing an important area of research for food scientists. The nutritional quality of millets can be further improved by adopting suitable and effective processing methods. Value addition and product development from millets might be helpful in avoidance and management of nutrition-related disorders. Advancement in these high-value and nutritious products will increase the immunity, health, and socioeconomic status of the consumers.

Keywords

Millets · Phytochemical · Antinutrients · Lifestyle diseases · Value addition

5.1 Introduction

Small seed millets are grown across the world for grains and human food due to their nutritional advantages and ability to withstand adversities. Millets are preferred over other grains due to their short life cycle, low input requirement, and ensured production with in-built genetic tolerance for abiotic stresses. The term "*Millet*" has evolved from the word "mil or thousand," indicating that large number of seeds can be generated from one or single seed. Pearl millet is one of the widely grown millet, which is an important crop in the dry parts of India and Africa.

Millets differ from each other for agronomic traits, morphology, nutritional composition, genetic constitution, etc. The minor millets include Finger millet (*Eleusine coracana*), Foxtail millet (*Setaria italica*), Barnyard millet (*Echinochloa* spp.), Proso millet or white millet (*Panicum miliaceum*), Kodo millet (*Paspalum scrobiculatum*), Little millet (*Panicum miliare*), Browntop millet (*Urochloa ramosa*), Guinea millet (*Brachiaria deflexa*), Job's tears (*Coix lacrima-jobi*), and Fonio (*Digitaria exilis*) (Desai et al. 2010).

Ecological crop alternates are desired to fulfill the world hunger and also to support the farmers' economy. Millets crop has the potential to achieve food and nutritional security (Fig. 5.1). Millets possess vigorous nutrients and protein content. Millets' role is prominent in the formulation of nutri-rich foods like multigrain and gluten-free foods. High proportion of polyphenols and other bioactive compounds in millets enhance the rate of absorption of fats and also increase the time of the release of sugars thereby reducing the glycemic index. This property proves a boon for diabetic patients and patients with cardiovascular diseases. Millets are categorized as photosynthetic C_4 cereals, which utilize a higher amount of CO_2 from the environment and transform more oxygen and also have better water uptake efficiency, and considered more eco-friendly. As C_4 plants have a low CO_2 compensation point, they are more efficient in utilizing CO_2 present in the atmosphere and thereby increase Carbon fixation in plants.

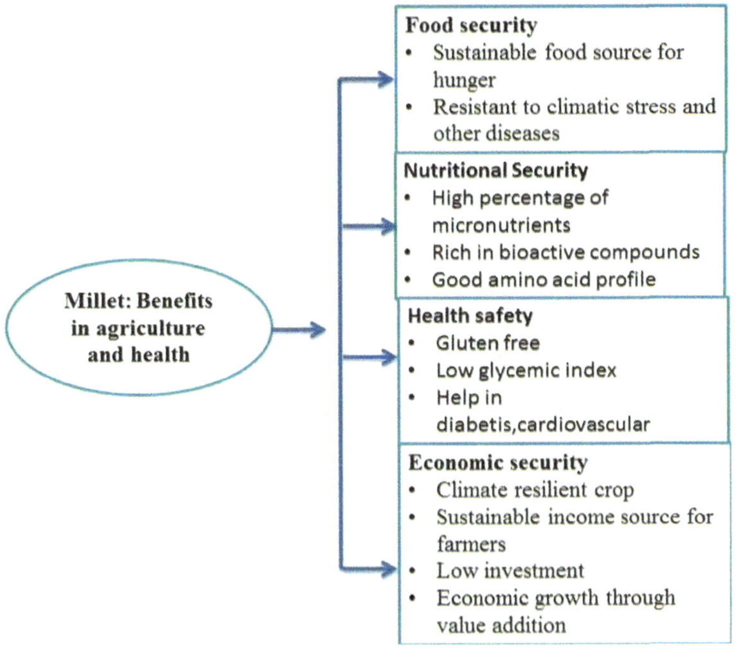

Fig. 5.1 Millet: Benefits in agriculture and health

Coarse cereals are grown in different parts of the country under various ecologies and different agro-climatic conditions, mostly as rainfed crops. The minor millets like finger millet (ragi), little millet, kodo millet, proso millet, foxtail millet, and barnyard millet altogether with pearl millet, sorghum, and maize are considered as nutri-cereals because of their excellent nutritional potential.

Millets were hardy, nutritious, and low input requiring crops since ancient times but discarded in favor of staple food crops like wheat and rice with the development of irrigation facilities, open markets, and with urbanization. Millets are now being relooked as a better option for healthy foods due to their anti-diabetic and anti-hypertension properties as these ailments are spreading faster with more and more urbanization, low physical activity, and modern lifestyle. The regular intake of millets keeps our colon hydrated and resultantly avoids the problem of constipation. The higher level of serotonin is produced by tryptophan in millets, which has soothing effects on our minds. Magnesium (Mg) in millets can be helpful in reducing the effects of cardiac attack and migraine, whereas, vitamin B3 (niacin) is helpful in lowering blood cholesterol levels. Regular millet intake is helpful in decreasing C-reactive proteins and triglycerides and therefore prevents coronary diseases. All millets are gluten-free and have a good amount of antioxidant activities coupled with nonallergenic effects.

Millets are rich in fatty acids, proteins, vitamins, minerals, polyphenols, and dietary fibers. The protein present in millets is mainly rich in essential amino acids, specifically sulfur containing amino acids such as cysteine and methionine (Amadou et al. 2011). Sorghum, pearl millet, finger millet, barnyard, and kodo millet are the major commercially available millets in the market.

5.2 Nomenclature of Millets

The botanical names and local names of millets are presented in Table 5.1.

5.2.1 Distribution of Millets

India is the largest producer of millets. These crops are mostly cultivated under rainfed conditions and hence require less water for cultivation. These crops are grown either as a sole crop or as an intercrop with pulses, oilseeds, spices, and condiments.

Sorghum (*Sorghum bicolor* (L.) Moench) is classified into four different groups, i.e., grain sorghum, forage sorghum, grass, and Sudan sorghum. It is generally a warm weather crop with fair resistance to serious pests and diseases. The fruit is a caryopsis made up of pericarp, endosperm, and germ (Macrae et al. 1993).

Ragi or Mandua (Finger millet) is a cereal grass and serves as a staple food in many South Asian and African countries (FAO 2012; Quattrocchi 2006). The pericarp is loosely bound and can be easily separated from the seed coat. De Wet (2006) reported that the grain of finger millet is nutritious, easily digestible, and versatile. Sprouted grains of finger millet are recommended for elderly people and infants. Pearl millet (*Pennisetum glaucum*) is a hardy crop and grown worldwide in

Table 5.1 The millets, common, scientific, and local names

S. No.	Common name	Scientific name	Local name (Hindi)
(A) Millets			
1	Sorghum	*Sorghum bicolor* (L.)	Jowar
2	Pearl millet	*Pennisetum glaucum* (L.)	Bajra
3	Finger millet	*Eleusine coracana* (L.)	Ragi/Mandu
4	Barnyard millet	*Echinochloa species*	Sanwa/Jhangora
5	Proso millet	*Panicum miliaceum* (L.)	Cheena
6	Foxtail millet	*Setaria italic*	Kakun/Kangni
7	Kodo millet	*Paspalum scrobiculatum* (L.)	Kodo
8	Little millet	*Panicum sumatrense*	Kutki
(B) Lesser known millets			
1	Browntop millet (signal grass)	*Brachiaria ramose*	–
2	Crap grass (finger grass)	*Digitaria sanguinalis*	–
(C) Pseudocereals			
1	Purple amaranthus	*Amaranthus cruentus*	Chaulai
2	Buckwheat	*Fagopyrum esculentum*	Kuttu ka aata

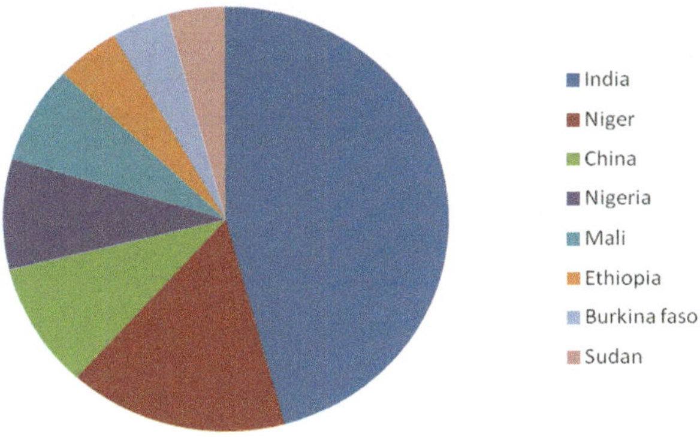

Fig. 5.2 Contribution of major millet growing countries in the world

dry regions. The cultivation of Pearl Millet is not new as it was domesticated as a food crop several 1000 years ago. Pearl millet is well adapted for adverse climatic conditions like low fertility, drought, and high-temperature regimes. Some millets like Proso millet (*Panicum miliaceum* L.) is used as a forage crop also.

Kodo millet (*Paspalum scrobiculatum* L.) is used widely across the world. The seeds of kodo millet are rich in protein and fiber content 11% and 14%, respectively with low fat 4.2% content. It is easy to digest and contains good amount of lecithin, which depicts positive effects on the human nervous system. Vitamins B3, B6, folic Acid, and minerals such as Ca, Fe, K, Mg, and Zn are available in good amounts in this millet. Kodo millet is also good food for people having gluten intolerance.

One of the world's primitive grown crops is Foxtail millet (*Setaria italica* (L.) P. Beauvois) and has been reviewed as a native of China. It is an early maturing crop and thereby has drought escape mechanism. The grains of foxtail millets are also used in poultry feed and for other avians. Barnyard millet (*Echinochloa crusgalli* L.) is a very good food and animal feed. It is also known as sawan, sanwa, ooda, oodalu, and sanwank. Barnyard millet has low carbohydrate content which is slowly digestible and makes it highly suitable for the person involved in sedentary activities. Linoleic acid, palmitic, and oleic acids which are regarded as good fatty acids are major fatty acids in barnyard millet. Therefore, it is good for patients with diabetes and cardiovascular ailments as it lowers high lipid and glucose levels. One millet known as Little millet (*Panicum sumatrense*) has smaller seeds than the other millets. The contribution of major millet growing countries is given in Fig. 5.2.

5.3 Nutrient Composition of Millets

5.3.1 Macronutrients

India shares the world's largest undernourished population accounting for 15.2% of the total population. Among 119 countries, India ranked 100th in Global Hunger Index, 2017 (Von Grebmer et al. 2017). Protein energy malnutrition (PEM) along with iodine, iron, and vitamin A deficiency has been a major cause of illness and

Table 5.2 Proximate composition of different millets (per 100 g) (Mohapatra et al. 2019)

Crop	Protein (g)	Carbohydrates (g)	Fat (g)	Crude fiber (g)	Mineral matter (g)
Millets					
Pearl	11.6	67.5	5.0	1.2	2.3
Finger	8.0	72.0	1.3	3.62	2.1
Barnyard	11.6	74.3	5.8	14.7	4.7
Foxtail	12.3	60.9	4.3	8.0	3.3
Proso	11.5	75.2	2.3	2.2	1.9
Little	8.7	75.7	5.3	8.6	1.7
Kodo	8.3	65.9	1.4	9.0	2.6
Sorghum	10.4	72.6	1.9	1.6	1.6

deaths (Lozano et al. 2012). Obesity is one of the major health problems in India (11% and 15% in men and women, respectively). Millets are the source of primary food in many parts of the world with sixth place in terms of world agricultural production. These are a very good source of phytonutrients, vitamins, and minerals for combating nutrient deficiencies.

Millets are nutritionally comparable or better than many major cereal grains. Millets have high fiber content, antioxidants, gluten-free proteins, low glycemic index, and fulfilled with biologically active compounds (Kannan et al. 2017). The nutritional profile of these small seeded crops depicts that these are an excellent source of energy, protein, and minerals including trace elements. Millet grains are a good source of vitamins namely, niacin (B3) thiamine, riboflavin, and folic acid. The grains constitute nearly 65% carbohydrates, majorly in the form of soluble carbohydrates and dietary fiber. Amylose and amylopectin (72–84%) is the major fraction of polysaccharides in millets. They are rich in phytochemicals such as polyphenols (0.2–0.3%) and dietary fiber (Hadimani and Malleshi 1993). Antioxidant activity of millets is mainly due to polyphenols, phytates, and tannins and these compounds have a role in controlling aging (Bravo 1998). Calcium content (344 mg/ 100 g) is found maximum in Finger millet among all cereals. It also contains high amount of phytates, polyphenols, and tannins (Thompson 1993).

Jowar (Sorghum) has a good amount of moisture content (11.9%) with nearly 10.4% and 1.9% of protein and fat content, respectively. The fiber and mineral content of grain sorghum amounts to 1.6%. The grain of Sorghum is a good source of energy (349 kcal) with 72.6% of carbohydrates (Gopalan et al. 1996). Cellulose and hemicellulose are other carbohydrates present in sorghum. A good amount of dietary fiber (14.3%) and elements such as Ca (25 mg), P (222 mg), and Fe (4.1 mg) is present per 100 g of edible portion, respectively in Sorghum (Hosmani and Chittapur 1997).

Little millet and kodo millet have 37–38% of dietary fiber and regarded as nutraceutical crops (Hegde and Chandra 2005). Millets are regarded as a complete food because of their nutritional composition and are popularly used as snacks, baby food, processed foods, etc. Sorghum and millets play a major role in ensuring nutritional security in developing countries. The millets are also known as miracle grains and *nutri*-cereals. The elemental composition of millets is given in Table 5.2.

5.3.2 Micronutrients

Minerals play a vital role in bone development, blood clotting, cell signaling, heartbeat control, cell energy metabolism, oxygen transport, metabolism of proteins and fats, and functions as coenzymes in body immunity and nerve conduction system (Soetan et al. 2010). The minerals (1.7–4.3 g/100 g) present in millets is several folds higher than other cereals such as rice (0.6%) and wheat (1.5%). Iron is found in ample amounts in Pearl Millet and Barnyard millet which is useful to meet out the Fe requirement in anemia. Foxtail millet contains the maximum content of Zn (4.1 mg/100 g) and also serves as an enriched source of Fe (Chandel et al. 2014). Both the minerals play a significant role in immunity boosting. Millets are a good source of water-soluble vitamins like niacin (B3), riboflavin (B2), and folic acid which are not stored in the body and have to be supplied from outside. Adding millets to the diet can be a real delight as it improves nutritional balance and security. Platel (2013) proposed for iron and zinc fortification by adding millet flour in India. The detail of micronutrient content of millets has been presented in Tables 5.3 and 5.4.

One of the key components of micronutrients in millets is carotenoids that are found in a large quantity in millets. The details of carotenoids profile of millets are given in Table 5.5.

5.4 Availability of Nutrients

There are several reports which indicate the role of industrial and home processing techniques as milling, roasting, soaking, popping, puffing, malting, fermentation, and germination enhances the bioavailability of nutrients by reducing the antinutrient factors. The nutrient content of millets decreases after milling but the bioavailability of some nutrients in case of millet improves considerably. Bioavailability of iron increases if weaning foods are prepared by grain roasting. Popping in case of finger millet has also been found to enhance protein and carbohydrate digestibility. In majority of cases soaking millet flour prior to heating enhances the zinc availability (Kamaraddi and Shanthakumar 2003). Protein digestibility, Protein efficiency ratio (PER), and Net Protein Utilization were also increased after malting in case of finger millet. It has also been reported that protein content and level of antinutrients were reduced after malting in pearl millet and finger millet; however, PER and bioavailability were increased.

Fermentation process also improves the Net Protein Utilization (NPU), Biological Value (BV), and contents of thiamine (B1), riboflavin (B2), and niacin (B3) in case of fermented finger millet. Sprouting process in millet has been found to increase methionine and cysteine. Fermentation of finger millet flour also found a

Table 5.3 Micronutrients content of millets (mg/100 g) (Li et al. 2008; Maha Ali et al. 2003)

Crop	Calcium content	Iron content	Phosphorus content	Zinc content	Thiamine content	Niacin content	Riboflavin content
Millets							
Pearl	35	10.3	339	–	0.30	1.11	1.4
Finger	348	4.2	250	36.6	0.40	0.80	0.6
Foxtail	31	3.5	300	60.6	0.60	0.55	1.6
Kodo	32.3	3.1	300	32.7	0.15	0.09	2.0
Barnyard	18.3	17.4	–	57.4	0.33	0.10	4.2
Proso	10	2.2	200	–	0.41	4.54	0.2
Sorghum	35.2	5.2	266	3.01	0.28	5.19	–

Table 5.4 Some essential amino acid composition in different millet (mg/g) (Gopalan et al. 2009; Geervani and Eggum 1989)

	Millets					
Parameters	Finger	Proso	Foxtail	Little	Pearl	Kodo
Lysine	220	190	140	110	190	150
Arginine	300	290	220	250	300	270
Tryptophan	100	050	060	060	110	050
Histidine	130	110	130	120	140	120
Phenylalanine	310	310	330	330	290	430
Tyrosine	220	–	–	–	200	–
Cysteine	140	–	100	090	110	110
Methionine	210	160	180	180	150	180
Isoleucine	400	410	480	370	260	360
Valine	480	410	430	350	330	410
Leucine	696	754	1032	763	748	648
Threonine	243	149	193	191	239	197

Table 5.5 Carotenoids and biological profile of sorghum and millets (in μg) (Dayakar et al. 2017; Geervani and Eggum 1989)

Sorghum and millets	β Carotene	Lutein	Zeaxanthin	Net protein utilization (%)
Sorghum	08.2	09.0	7.4	–
Pearl millet	28.2	29.6	9.3	84.5
Little millet	01.9	07.8	5.2	96.3
Finger millet	01.5	25.5	1.4	–
Kodo millet	01.4	59.4	3.9	95.6

Geervani P, Eggum BO (1989) Nutrient composition and protein quality of minor millets. Plant Foods Hum Nutr 39:201–208

significant reduction of phytate, tannins, trypsin inhibitor activity increase in mineral availability of Ca, P, Fe, and Zn. In case of longer duration of fermentation the level of phytic acid reduces. Different processing methods also affect the nutraceutical properties (Saleh et al. 2013). Sorghum contains about 3.5% lipids, 10% protein while finger millet contains 2–5% lipids and 12–16% protein and is rich source of minerals, phytonutrients, and vitamins. Prolamin is the major portion of sorghum protein and has a better amino acid profile.

5.5 Phytochemicals and Antinutrients in Millets

Millets contain many anti-nutritional compounds. These are mainly classified as tannins, phytates, polyphenols, trypsin inhibitors, and dietary fiber (Mugocha et al. 2000). Most of the phytochemicals including polyphenols and dietary fiber are mainly present in the seed coat of millets (Kaur et al. 2012). These polyphenols, which are complex mixture of cinnamic acid derivatives and benzoic acid, are considered as having major health benefits. Polyphenols such as phenolic acid and tannins are found in more quantity in cereals as compared to flavonoids. These compounds do not have any established role in nutrition (Malleshi and Hadimani 1993) but have other important advantages like anti-mutagenic, anti-carcinogenic, anti-oestrogenic, antiviral effects, and anti-inflammatory activities.

There is an abundance of tannin reported in the testa (seed coat) of millets. The deeper layers of the seed including Bran contain high levels of phenolic compounds and their level decreases during milling of the seed. The levels of phenolic compounds are present in higher amount in finger millet (Liu et al. 2012). Insoluble bound phenolic fraction forms are formed of hydroxycinnamic acids and their derivatives (HCAS), except in finger millets. The phenolics present in millets are pH sensitive and heat stable and these are highly unstable under basic conditions (Chandrasekara and Shahidi 2012). The total phenolic compounds of seed coat of millet showed the presence of kaempferol, glycoside, catechin/epicatechin, apigenin, phloroglucinol, trans-feruloyl-malic acid, catechin gallates, and daidzein (Liu et al. 2012). Some important flavonoids found in leaves of millets are isoorientin, orientin, isovitexin, vitexin, violanthin, saponarin, lucenin-1, and tricin and some major flavonoids reported in grains are glucosylorientin, goitrogenic, glucosyl vitexin, and vitexin isolated from pearl millet grains. Finger millet contains a good amount of proanthocyanidins, which are also known as condensed tannins. It is reported that high fiber foods delay nutrient absorption and finally increases the fecal bulk, increase the fecal transit time, lower the blood lipids, and inhibit colon cancer. The resistant starches (RS) are also involved in some health benefits and also contribute towards dietary fiber. Ragi (Finger Millet) also contains a good amount of residual starch (RS).

Millets and sorghum contain different phytochemicals such as phytosterols, phenolic acids, pinacosanols, tannins, and anthocyanins which are basically classified as secondary metabolites which are more beneficial for humans than the plant itself. Sorghum and other millet grain have been reported to possess high antioxidant activity than other cereals and fruits (Thompson 1993). Millets possess a good amount of phenolic compounds including phenolic acids, flavonoids, and condensed tannins. It was also reported that condensed tannins are absent in sorghum; however, almost all millets contain phenolic acids. Pigmented sorghum contains exceptional anthocyanins that could be possibly used as natural food colorants.

5.5.1 Condensed Tannins

Tannins constitute important secondary metabolites in plants. The tannins present in different varieties of Sorghum are mostly phenolic compounds (Hahn et al. 1984). This compound confers some resistance towards molds (fungal infection) and avoids grain damage and maintains the quality of the grain (Waniska et al. 1989). Condensed tannins are proanthocyanidins, also classified as polyphenols having high molecular weight. Processing of sorghums into food products greatly affects the Phenol levels. It has been reported that tannin content decreased by 52% and

72% when sorghum was processed to produce cookies and breads, respectively (Awika et al. 2003).

5.5.2 Phenolic Acids

Phenolic compounds present in millets are majorly phenolic acids, tannins, and flavonoids. These compounds are chiefly concentrated in the pericarp, aleurone layer, testa, and endosperm of the seed (Hahn et al. 1984; McDonough et al. 1986). Phenolic acids reported in sorghum and millet grains are presented in Table 5.6.

Anthocyanins, also called accessory pigments, are 3-deoxyanthocyanins and are devoid of hydroxyl group at the 3-position of the C-ring which accounts for increased stability at high pH compared to other anthocyanins (Awika et al. 2004). This property of millets is used as potential natural food colorants in industry. It was also reported that large quantities of tannins are present in outer layers of grain in some cultivars (Serna-Saldivar and Rooney 1995). Very few reports are available for levels of anthocyanins in millets and other cereals. The most prevalent anthocyanin in sorghum grains is 3-deoxyanthocyanidins (Gous 1989). Therefore, sorghum can be used as a commercial source of anthocyanins. Various health supplements such as tannins, phenolic acids, plant sterols (PS), anthocyanins, and policosanols (PC) are reported to be present in Grain sorghum and DDGS Sorghum (distillers dried grain with soluble (DDGS)) (Awika and Rooney 2004; Hwang et al. 2002). Nutraceutical and Functional foods with high antioxidant activities have increased over the years to eradicate free radicals by lowering oxidative stress associated with the development of ailments like neuron degeneration, cardiovascular, cancer, diabetes, aging, and hypo-cholesterolemia (Grundy et al. 2004; Wu et al. 2004).

Table 5.6 Some common phenolic acid compounds in millet grains

Compounds (phenolic acid)	Report
Hydroxybenzoic acids: Gallic	Subba Rao and Muralikrishna (2002)
Protocatechuic	Subba Rao and Muralikrishna (2002)
p-Hydroxybenzoic	McDonough et al. (1986)
Gentisic	Waniska et al. (1989)
Syringic	Waniska et al. (1989)
Hydroxycinnamic acids	Subba Rao and Muralikrishna (2002)
Caffeic	Subba Rao and Muralikrishna (2002)
p-Coumaric	Subba Rao and Muralikrishna (2002)
Cinnamic	McDonough et al. (1986)
Sinapic	Waniska et al. (1989)
Salicylic acids	Waniska et al. (1989)
Vanillic acids	Hahn et al. (1983)

5.5.3 Flavonoids

Flavonoids are natural substances found in a wide variety of plants and their parts. These compounds have immense health benefits. Sorghum grains are the best source of flavonoids and they are mainly isolated and identified from sorghum grains (Gous 1989; Wu and Prior 2005; Dykes and Rooney 2006; Seitz 2004; Gujer et al. 1986). Plant sterols have the capacity to lower cholesterol levels (Ostlund 2002). In spite of high levels of phytochemicals and diversity in sorghum, still a lot of research is in waiting in this crop. There is limited incorporation of sorghum fractions in foods for greater nutrition and health. Different reports indicate reduced weight gain of animals like rats, pigs, rabbits, and poultry when fed on sorghum having high tannin content.

Tannins present in sorghum reduce nutritive value mainly by binding several proteins and some important carbohydrates into insoluble complexes, which are difficult to be broken down by digestive enzymes. Most of the lipids are present in the pericarp, germ, and aleurone layers of the millet grains (Rooney 1978). The constitution of free lipids (5.6–7.1%) is more than bound lipids (0.57–0.90%) in case of pearl millet (Lai and Varriano Marston 1980). Hydrocarbons, triglycerides, monoglycerides, diglycerides, and free fatty acids are major components of free lipids and the bound lipid fraction is mainly made up of lecithin (Rooney 1978). Oleic acid and Linoleic acid which are regarded as good fatty acids are the major unsaturated fatty acids present in pearl millet in addition to palmitic acid which is a saturated fatty acid (Lai and Varriano Marston 1980; Rooney 1978). The fatty acid profile of pearl millet indicates that oleic, linoleic, stearic, and palmitic acid are the major fatty acids. High levels (70%) of unsaturated fatty acids are present in the free lipid portion as compared to the bound portion (52%) (Lai and Varriano Marston 1980). High lipid components usually interfere with the storage quality of pearl millet (Carnovale and Quaglia 1973).

Lecithin is another important phospholipid which is present in large amount in the brain cells. This is present in large amounts in Kodo millet and hence strengthens the nervous system. Kodo Millet is also rich in fiber and low in fat content (Itagi 2003). Finger millet is a valuable source of essential amino acids which have diverse health-promoting functions like

1. Improves nitric oxide synthesis in the body known for its role in dilating blood vessels.
2. Amino Acids like arginine and lysine are actively involved in the release of growth hormone (GH).
3. Improves Immunity and reproductive ability.
4. Healing time of injuries is reduced.
5. Lowers Blood Pressure and reduces cardiac ailments.
6. Increases the flow of blood.
7. Improves the action of insulin.
8. Improves sperm production and alleviates male infertility and motility.

Sodium and Potassium are important macronutrients involved in maintaining electrical balance. Potassium is available in good amounts in Sorghum while whole grains are good sources of magnesium (Mg), iron (Fe), zinc (Zn), and copper (Cu). Finger Millet is a very good source of dietary calcium (Ca) which is helpful in bone development.

5.5.4 Phytic Acid

Millets contain important amount of inositol hexaphosphates (IP6) that are mainly known as phytic acid or phytates. Phytate is recognized as a major anti-nutritional factor that influences the bioavailability of trace elements (Mbithi-Mwikya et al. 2000). Several processes such as malting, decoration, and germination reduce the level of antinutrients. Several studies show that millets when consumed raw contain a higher amount of antinutrients as compared to that of processed millets.

Soaking of millets for 12 h and further germination lowers the phytic acid levels upto 451.2 mg/100 g. It is also observed that autoclaving of the sample reduced the phytic acid to half the amount of control.

5.5.5 Amylase Inhibitor

Millets possess Amylase inhibitor activity which was reduced by soaking and coarse grinding. Germination also reduces amylase inhibitor activity. Autoclaving of grains also reduces the amylase inhibitor activity as compared to non-autoclaved samples (Sharma and Kapoor 1996). The level of antinutrients can be reduced by several processes such as malting, decoration, and germination. It is also reported in several studies that the raw millet contains a higher amount of antinutrients as compared to that of processed millets.

5.6 Effect of Processing on Antinutrients

Reports indicate that several processing methods in millets help in reducing the anti-nutritional factors which are utilized in the food industry. The bioavailability of nutrients is also favored by processing methods. Many processing methods that have been applicable in millet processing are roasting/popping, soaking, germination, and fermentation (Jaybhaye and Srivastav 2015). These methods improve the nutritional value of the grain and make it more consumer-friendly. Malting increases the bio-accessibility of iron by 300% and of manganese by 17% (Platel et al. 2010).

The phytic acid content in millets was found to decrease during germination as there is hydrolysis of phytate phosphorus into inositol monophosphate. The amount of tannins also reduces during soaking and germination (Handa et al. 2017). Boiling and pressure cooking of grains is also found to reduce the tannins content. Antinutritional factors are also reported to reduce during fermentation of millets and also increases protein digestibility. Irradiation enhances protein digestibility and has shown an inhibitory effect against antinutrients (Pushparaj and Urooj 2011). Reports show that extrusion cooking or high-temperature short time (HTST) processing has been found to decrease the antinutrients like tannins, phytates, and increase the bioavailability of minerals (Nirmala et al. 2000).

5.7 Nutritional and Nutraceutical Potential of Millets

Millets are reported to have great nutritive value as they are rich in micronutrients and phytochemicals. Millet proteins have a good composition of essential amino acids lacking threonine and lysine but are relatively high in methionine (Singh and Raghuvanshi 2012). Antioxidant properties of some major millet have been examined and it was found that they are a good source of antioxidants. Phenolic compounds present in several varieties of whole grain millets (pearl millet, kodo, foxtail, finger, proso, and little millets) are also involved in metal chelating, high antioxidant activity, and reducing powers (Chandrasekara and Shahidi 2010). Finger millet has a strong free radical-scavenging activity as compared to other common cereals including wheat and rice (Dykes and Rooney 2006). Defatted foxtail millet bran is also reported to contain phenolic compounds with significant antioxidant activity (Amadou et al. 2011). Resistant starch (RS), lipids, oligosaccharides, antioxidants such as avenanthramides, phenolic acids, and flavonoids; hormonally active compounds including phytosterols, lignans, and antinutrients such as tannins and phytic acid are examples of few such compounds (Miller 2001; Edge et al. 2005). It is well documented that millets are rich sources of nutraceutical components such as polyphenols, antioxidants, and crude fibers and can be used for nutritional quality improvement of processed products with several increasing health benefits like preventing cardiac diseases, reducing tumor incidence and cancer, lowering blood pressure, affect absorption of fat, postponing gastric issues, and providing gastrointestinal substance (Truswell 2002; Gupta et al. 2012) as given in Table 5.7.

Changing lifestyle and sedentary habits have given rise to a number of health problems worldwide. Several studies show that diets supplemented with plant phytonutrients not only make our life easy but also protect against several diseases such as cardiovascular ailments, diabetes, cancer, metabolic syndrome, and Parkinson's disease (Chandrasekara and Shahidi 2012). Several Food science experts believe that whole grains complement fruits and vegetables when taken together as they contain several important phytonutrients (Liu 2007). Polyphenols are the major group of phytochemicals in plants having numerous health benefits. Therefore, they have gained much attention among food scientists, nutritionists, and consumers owing to their roles in human health (Tsao 2010).

Table 5.7 Nutritional and Nutraceutical potential of millets (Malik 2015)

S. No.	Disease	Benefits	Positive factors
1	Diabetes	Treating diabetes	With low glycemic index
2	Anemia	Help in increase in hemoglobin	High iron and zinc concentration
3	Constipation	Help in reducing constipation	High fiber
4	Cancer	Anti-cancer property	Inhibit the tumor formation
5	Bone growth, development, and repair	High phosphorus is essential for bone growth Production of ATP, that is, energy currency in body	Due to the presence of phosphorus
6	Stomach ulcer	For the treatment of stomach ulcers. Convert the stomach alkaline and prevent the formation of ulcers	Prevent formation of excess acidity
7	Heart health	• Phytonutrients and lignin act as a strong antioxidant thus prevent heart disease • High content of magnesium useful in controling blood pressure and heart stress	As strong antioxidant
8	Respiratory problems for asthma patients	• High amount of magnesium reduces respiratory problems (asthma patients) • Reduce migraine attacks	High amount of magnesium
9	Weight loss (obesity)	• Useful in weight loss contains high fiber content • Reduces the overall consumption of food	High fiber content
10	Prevention of gall stones	• Reduces the risk in the gall bladder stone • Reduces the production of excessive bile juice	High fiber content
11	Anti-allergic properties	Highly digestible Low allergic response	Due to hypo allergic property

5.8 Conclusion

Owing to the fast lifestyle and sedentary habits, the importance of healthy nutrition is lacking especially among children leading to malnutrition and many health ailments like diabetes, cardiovascular problems, obesity, celiac disease, cancer, etc. Sorghum is widely cultivated for food and feed grain, while millets are grown entirely for food. Millet farming has many advantages as they are not only remunerative crops for farmers but also improves the health of the community as a whole. Community farming and the availability of food processing industries open new avenues and opportunities for farmers as they get sure and competitive prices for their produce. Sorghum and millets altogether constitute a major source of protein and calories for millions of people in Asia and Africa. Millets and sorghum have gluten-free protein, low glycaemic index, and are rich in B-vitamins, minerals, and antioxidants. The incorporation of millet-based food products in international, national, and state-level community feeding programs will certainly help to overcome the existing nutrient deficiencies of protein, calcium (Ca), iron (Fe), and zinc (Zn) in developing countries. Sorghum and Millets comprise of several major and minor nutrients, antioxidants compounds, and phytochemicals. Several anti-nutritional compounds (e.g., protease inhibitors, phytates, lectins, galacto-oligosaccharides, tannins, ureases, phenolics, and saponins) present in sorghum and other millets also play a significant role in the biological functions of plants. The effect of these anti-nutritional compounds on human and animal health is partly negative because they can lower down the digestibility of nutrients and the absorption of minerals. These anti-nutritional factors can be inactivated by processing methods like steeping, cooking, roasting, malting, germination, and fermentation. Integration of millets with other food sources of protein would make up for essential amino acids like lysine which are deficient in other cereals. Bioavailability of the micronutrients can be improved by the novel processing and preparation methods for millet-based diets.

References

Amadou I, Gbadamosi OS, Guo-Wei L (2011) Millet-based traditional processed foods and beverages - a review. Cereal Food World 56(3):115–121

Awika JM, Rooney LW (2004) Sorghum phytochemicals and their potential impact on human health. Phytochemistry 65:1199–1221

Awika JM, Rooney LW, Wu X, Prior RL, Cisneros-Zevallos L (2003) Screening methods to measure antioxidant activity of sorghum (Sorghum bicolor) and sorghum products. J Agric Food Chem 46:5083–5088

Awika JM, Rooney LW, Waniska RD (2004) Anthocyanins from black sorghum and their antioxidant properties. Food Chem 90:293–301

Bravo L (1998) Polyphenols: chemistry, dietary sources, metabolism and nutritional significance. Nutr Rev 56:317–333

Carnovale E, Quaglia GB (1973) Influence of temperature and humidity controlled preservation on chemical composition of milling products from millet. Ann Technol Agric 22:371–380

Chandel G, Kumar M, Dubey M, Kumar M (2014) Nutritional properties of minor millets: neglected cereals with potentials to combat malnutrition. Curr Sci 107(7):1109–1111

Chandrasekara A, Shahidi F (2010) Content of insoluble bound phenolics in millets and their contribution to antioxidant capacity. J Agric Food Chem 58:6706–6714

Chandrasekara A, Shahidi F (2012) Bioaccessibility and antioxidant potential of millet grain phenolics as affected by simulated in vitro digestion and microbial fermentation. J Funct Foods 4:226–237

Dayakar RB, Bhaskarachary K, Arlene Christina GD, Sudha Devi G, Tonapi A (2017) Nutritional and health benefits of millets. ICAR, Indian Institute of Millets Research (IIMR), Rajendranagar, Hyderabad, p 112

De Wet JMJ (2006) Eleusine coracana (L.) Gaertn. record from protabase. In: Brink M, Belay G (eds) PROTA (Plant Resources of Tropical Africa/Ressources végétales de l'Afriquetropicale). Backhuys Publishers, Wageningen, Netherlands

Desai AD, Kulkarni SS, Sahoo AK, Ranveer RC, Dandge PB (2010) Effect of supplementation of malted ragi flour on the nutritional and sensorial quality characteristics of cake. Adv J Food Sci Technol 2(1):67–71

Dykes L, Rooney LW (2006) Sorghum and millet phenols and antioxidants. J Cereal Sci 44:236–251

Edge MS, Jones JM, Marquart L (2005) A new life for whole grains. J Am Diet Assoc 105 (12):1856–1860

FAO (2012) Grassland index. A searchable catalogue of grass and forage legumes. FAO, Rome, Italy

Geervani P, Eggum O (1989) Nutrient composition and protein composition of minor millets. Plant Foods Hum Nutr 39:201–208

Gopalan C, Ramasastri BV, Balasubramanian SC (1996) Nutritive value of Indian foods. National Institute of Nutrition, Indian Council of Medical Research, Hyderabad, India

Gopalan C, Rama Sastri BV, Balasubramanian SC (2009) Nutritive value of Indian foods. National Institute of Nutrition, Indian Council of Medical Research, Hyderabad, India

Gous F (1989) Tannins and phenols in black sorghum. PhD Dissertation, Texas A & M University, College station, TX

Grundy SM, Cleeman JI, Merz CN, Brewer HB, Clark LT, Hunninghake DB, Pasternak RC, Smith SC, Stone NJ (2004) Implications of recent clinical trials for the National Cholesterol Education Program Adult Treatment Panel III guidelines. Circulation 110:227–239

Gujer R, Magnolato D, Self R (1986) Glucosylated flavonoids and other phenolic compounds from sorghum. Phytochemistry 25(6):1431–1436

Gupta N, Srivastava AK, Pandey VN (2012) Biodiversity and nutraceutical quality of some Indian millets. Proc Natl Acad Sci India B Biol Sci. https://doi.org/10.1007/s40011-012-0035-z. http://www.springerlink.com. Accessed 30 May 2012

Hadimani NA, Malleshi NG (1993) Studies on milling, physicochemical properties, nutrient composition and dietary fiber content of millets. J Food Sci Technol 30:17–20

Hahn DH, Faubion JM, Rooney LW (1983) Sorghum phenolic acids, their high performance liquid chromatography separation and their relation to fungal resistance. Cereal Chem 60:255–259

Hahn DH, Rooney LW, Earp CF (1984) Tannins and phenols of sorghum. Cereal Foods World 29:776–779

Handa V, Kumar V, Panghal A, Suri S, Kaur J (2017) Effect of soaking and germination on physicochemical and functional attributes of horsegram flour. J Food Sci Technol 54 (13):4229–4239

Hegde PS, Chandra TS (2005) ESR spectroscopic study reveals higher free radical quenching potential in kodo millet (Paspalum scrobiculatum) compared to other millets. Food Chem 92:177–182

Hosmani MM, Chittapur BM (1997) Sorghum production technology. Sarasijakshi MH, Dharwad, India

Hwang KT, Cuppett SL, Weller CL, Hanna MA (2002) Properties, composition and analysis of grain sorghum wax. J Am Oil Chem Soc 79:521–527

Itagi S (2003) Development and evaluation of millet based composite food for diabetes. Master Thesis, University Agricultural Science, Dharwad

Jaybhaye RV, Srivastav PP (2015) Development of barnyard millet ready-to-eat snack food: Part II. Food Sci 6(2):285–291

Kamaraddi V, Shanthakumar G (2003) Effect of incorporation of small millet flour to wheat flour on chemical, rheological and bread characteristics. In: Recent trends in millet processing and utilization. CCS Hisar Agril. Univ, Hisar, India, pp 74–81

Kannan SM, Thooyavathy RA, Kariyapa RT, Subramanian K, Vijayalakshmi K (2017) Seed production techniques for cereals and millets. In: Vijayalakshmi K (ed) Seed node of the revitalizing rainfed agriculture network Centre for Indian knowledge systems (CIICS), pp 1–39. http://www.ciks.org/downloads/seeds/5.%20Seed%20Production%20Techniques%20for%20Cereals%20and%20Millets.pdf. Accessed 29 Dec 2017

Kaur KD, Jha A, Sabikhi L, Singh AK (2012) Significance of coarse cereals in health and nutrition: a review. J Food Sci Technol. https://doi.org/10.1007/s13197-011-0612-9. http://wwwspringerlinkcom. Posted 25 Jan 2012

Lai CC, Varriano Marston E (1980) Lipid content and fatty acid composition of free and bound lipids in pearl millets. Cereal Chem 57:271–274
Li J, Chen Z, Guan X, Liu J, Zhang M, Xu B (2008) Optimization of germination conditions to enhance hydroxyl radical inhibition by water soluble protein from stress millet. J Cereal Sci 48:619–624
Liu RH (2007) Whole grain phytochemicals and health. J Cereal Sci 46(3):207–219
Liu J, Tang X, Zhang Y, Zhao W (2012) Determination of the volatile composition in brown millet, milled millet and millet bran by gas chromatography/mass spectrometry. Molecules 17:2271–2282
Lozano R, Naghavi M, Foreman K (2012) Global and regional mortality from 235 causes of death for 20 age groups in 1990 and 2010: a systematic analysis for the Global Burden of Disease Study 2010. Lancet 380:2095–2128
Macrae R, Robinson RK, Sadler MJ (1993) Encyclopedia of food science, food technology and nutrition. Academic, London
Maha Ali MAM, Tinay AH, Abdalla AH (2003) Effect of fermentation on the in vitro protein digestibility of pearl millet. Food Chem 80:51–54
Malik S (2015) Pearl millet-nutritional value and medicinal uses. IJARIIE 1(3):414–418
Malleshi NG, Hadimani NA (1993) Nutritional and technological characteristics of small millets and preparation of value-added products from them. In: Riley KW, Gupta SC, Seetharam A, Mushonga JN (eds) Advances in small millets. Oxford and IBH Publishing Co Pvt. Ltd, New Delhi, pp 271–287
Mbithi-Mwikya S, Van Camp J, Yiru Y, Huyghebaert A (2000) Nutrients and antinutrients changes in finger millet (*Eleusine coracana*) during sprouting. LWT–Food Sci Technol 33:9–14
McDonough CM, Rooney LW, Earp CF (1986) Structural characteristics of Eleusine coracana (finger millet) using scanning electron and fluorescence microscopy. Food Microstruct 5:247–256
Miller G (2001) Whole grain, fiber and antioxidants. In: Spiller GA (ed) Handbook of dietary fiber in human nutrition. CRC Press, Boca Raton, FL, pp 453–460
Mohapatra D, Patel AS, Kar A, Deshpande SS, Tripathi MK (2019) Effect of different processing conditions on proximate composition, anti-oxidants, anti-nutrients and amino acid profile of grain sorghum. Food Chem 271:129–135
Mugocha PT, Taylor JRN, Bester BH (2000) Fermentation of a composite finger millet-dairy beverage. World J Microbiol Biotechnol 16:341–344
Nirmala M, Rao MS, Muralikrishna G (2000) Carbohydrates and their degrading enzymes from native and malted finger millet (Ragi, Eleusine corcana, Indaf-15). Food Chem 69:175–180
Ostlund RE (2002) Phytosterols in human nutrition. Ann Rev Nutr 22:533–549
Platel K (2013) Millet flours as a vehicle for fortification with iron and zinc. In: Preedy VR, Srirajaskanthan R, Patel VB (eds) Handbook of food fortification and health. Springer, New York, pp 115–123
Platel K, Eipeson SW, Srinivasan K (2010) Bioaccessible mineral content of malted finger millet (*Eleusine coracana*), wheat (*Triticum aestivum*), and barley (*Hordeum vulgare*). J Agric Food Chem 58:8100–8103
Pushparaj FS, Urooj A (2011) Influence of processing on dietary fiber, tannin and in vitro protein digestibility of pearl millet. Food Nutr Sci 2:895–900
Quattrocchi U (2006) CRC world dictionary of grasses: common names, scientific names, eponyms, synonyms, and etymology. CRC Press, Taylor and Francis Group, Boca Raton, FL
Rooney LW (1978) Sorghum and pearl millet lipids. Cereal Chem 55(5):584–590
Saleh ASM, Zhang Q, Chen J, Shen Q (2013) Millet grains: nutritional quality, processing, and potential health benefits. Compr Rev Food Sci Food Saf 12:281–295
Serna-Saldivar S, Rooney LW (1995) Structure and chemistry of sorghum and millets. In: Dendy DAV (ed) Structure and chemistry of sorghum and millets. American Association of Cereal Chem, St Paul, MN, pp 69–124

Sharma A, Kapoor AC (1996) Levels of anti nutritional factors in pearl millet as affected by processing treatments and various types of fermentation. Plant Foods Hum Nutr 49:241–252
Singh P, Raghuvanshi RS (2012) Finger millet for food and nutritional security. Afr J Food Sci 6:77–84
Soetan KO, Olaiya CO, Oyewole OE (2010) The importance of mineral elements for humans, domestic animals and plants—a review. Afr J Food Sci 4(5):200–222
Subba Rao MVSST, Muralikrishna G (2002) Evaluation of the antioxidant properties of free and bound phenolic acids from native and malted finger millet (Ragi, Eleusine coracana Indaf-15). J Agric Food Chem 50(4):889–892
Thompson LU (1993) Potential health benefits and problems associated with anti nutrients in foods. Food Res Int 26:131–149
Truswell AS (2002) Cereal grain and coronary heart disease. Eur J Clin Nutr 56(1):1–4
Tsao R (2010) Chemistry and biochemistry of dietary polyphenols. Nutrients 2:123–146
Von Grebmer K, Bernstein J, Hossain N, Brown T, Prasai N, Yohannes Y (2017) Global hunger index: the inequalities of hunger. International Food Policy Research Institute. http://www.globalhungerindex.org/pdf/en/2017.pdf. Accessed 21 Jan 2018
Waniska RD, Poe JH, Bandyopadhyay R (1989) Effects of growth conditions on grain molding and phenols in sorghum caryopsis. J Cereal Sci 10:217–225
Wu X, Prior RL (2005) Identification and characterization of anthocyanins by high-performance liquid chromatography-electrospray ionization-tandem mass spectrometry in common foods in the United States: vegetables, nuts, and grains. J Agric Food Chem 53(8):3101–3113
Wu X, Beecher GR, Holden JM, Haytowitz DB, Gebhardt SE, Prior RL (2004) Lipophilic and hydrophilic antioxidant capacities of common foods in the United States. J Agric Food Chem 52 (12):4026–4037

Millet Starch: Current Knowledge and Emerging Insights of Structure, Physiology, Glycaemic Attributes and Uses

6

Vivek Chandra Verma, Sanjeev Agrawal, Manoj Kumar Tripathi, and Anil Kumar

Abstract

The risk of chronic diseases like diabetes, myocardial infarction, and obesity can be reduced by dietary changes and advance research in nutritional biology. Starch composition of millet grains is very high which almost accounts for 60–70% of millet grains and it tremendously affects the quality of millet-based products. Compared to the major cereals, the hypoglycaemic properties and other superior attributes of millet starch could make it a promising ingredient for the functional food industry. However, scanty knowledge on structural and functional attributes of millet starch seriously hinders further improvement of millets as functional and nutraceutical food. Here, we discussed the current knowledge and emerging insights of the isolation, structure, chemical composition, physicochemical properties, interaction with other constituents, and uses of millet starch. The aim of this chapter is to enlighten the benefits of millet starch and its uses in relation to reduce the risk of chronic disorders and enhance nutritional security.

Keywords

Functional food · Hypoglycaemic · Millets · Nutrition starch

V. C. Verma (✉)
DIHAR-DRDO, Chandigarh, India

S. Agrawal
Department of Biochemistry, G.B. Pant University of Agriculture and Technology, Pantnagar, Uttarakhand, India

M. K. Tripathi
ICAR-Central Institute of Agricultural Engineering, Bhopal, Madhya Pradesh, India

A. Kumar
Directorate of Education, Rani Lakshmi Bai Central Agricultural University, Jhansi, Uttar Pradesh, India

A. Kumar et al. (eds.), *Millets and Millet Technology*,
https://doi.org/10.1007/978-981-16-0676-2_6

6.1 Introduction

Globally, the chronic diseases like diabetes and obesity are increasing in an exponential manner. This has demanded advanced and rapid research in nutritional biology to develop food-based nutraceuticals over the last decade. Dietary quality of food is an important criterion for improving health as well as in the prevention of many chronic diseases (Dev et al. 2011). In the context of nutrition, millets are prodigious grain with their richness in starch (60–70%), protein (7–11%), crude fiber (2–7%), fat (1.5–5%), and flavonoids but vastly ignored due to lack of awareness. Besides, millets are gluten-free, rich in minerals (Fe, Ca, K, Zn, and Mg), vitamins, essential fatty acid, and amino acid (Majid and Poornima Priyadarshini 2019). Millets are small-seeded cereal crops which play a major role in the food security and economy of many Asian and African countries (Fig. 6.1). With advancement in agronomic practices millets can be considered as a sustainable food source because of their ability to flourish in harsh climatic conditions including alkaline soil with pH as high as 11, at an altitude of 2500 m from sea level with average rainfall ranging from 800 to 1200 mm, adaptation to a wide range of climatic and ecological conditions with resistance to pests and diseases (Gull et al. 2015; Bora et al. 2019). The importance of millets as an ingredient in multigrain and gluten-free cereal products is gradually increasing its demand as a form of highly exteriorize nutraceuticals or therapeutic agents. The major component of millet flour is nutritionally rich starch and food prepared from it exerts a positive effect on human health. Among the important health are decrease in glycaemic and insulin indexes coupled with the decrease in serum concentration of cholesterol (Lee et al. 2010). Therefore, the focus of this chapter is to provide a detailed overview of millet starch structure, physiochemical properties, starch interaction with other bioactive compounds in relation to glycaemic attributes and its wide-scale utilization.

6.2 Starch Isolation and Yield

The wet milling-based methods have been widely adopted for isolation of starch from millets. In this approach, millet flour is steeped in aqueous solution which is neutral/near neutral, alkaline, or acidic for several hours to facilitate the separation of starch from other components. Then the slurry is repeatedly washed with water and allowed to remove protein and fibrous components. The separated starch is recovered by centrifugation before drying. The addition of small amount of sodium azide (0.01%)/mercury (II) chloride (0.01 M) to clog bacteria growth/amylase activity has been used during starch extraction from pearl, proso, foxtail, barnyard, kodo, and little millets (Fujita and Fujiyama 1993). The isolation solution can greatly influence the chemical composition and properties of isolated starch (Kumari and Thayumanavan 1998). It should be noted that neutral/near neutral, alkaline, or acidic solution may either cause starch degradation or give low recovery of total starch from the flour. The yield percentage of millet starch ranges from 52% to 68.2% and amylose content 6% to 38.6% (Annor et al. 2014). Extraction techniques based on

Fig. 6.1 Commonly used millets in India

dimethyl sulfoxide (DMSO) dissolution (Carpita and Kanabus 1987) and ethanol/methanol precipitation (Klucinec and Thompson 1998) for laboratory analysis may be developed to overcome these issues. DMSO method is a rapid and sensitive method for extraction of plant tissues such as leaves of soybean, corn starch, potato tuber, and cell suspension culture of proso millet. Plant tissue solubilized in 20% DMSO allows complete hydrolysis of starch in 40 min followed by sonication or stirring for 24 h. As number of extraction and sonication time increases, the yield of starch increases upto 30–50%. Recovery of starch fraction by differential alcohol precipitation in common starch and high-amylose starch ranges 80–84% (w/w).

6.3 Morphology and Crystallinity of Starch Granules

The seed coat is firmly bound to the aleurone layer (a layer between the seed coat and endosperm) and the starchy endosperm which is further divided into corneous and floury regions. The corneous region has highly organized starch granules within the cell walls, while the floury region has loosely packed starch granules (McDonough et al. 1986). The starch granules in the floury endosperm of millets are generally bigger compared to the ones present in the corneous endosperm and hence more susceptible to enzymatic digestion (FAO 1995). Starch granules have a complex and highly ordered semi-crystalline structure. Starch granules size ranges between 0.1 and 200 μm and has different shapes like smooth, angular, ellipsoidal, oval, and spherical on the basis of botanical source (Table 6.1) (Buleon et al. 1998; Hoover 2001; Singh et al. 2003). Scanning Electron Microscope studies have revealed the shape and size of starch granules (Fig. 6.2). Proso millet and foxtail millet starch

Table 6.1 Morphology of millet starch in comparison with major cereals

Name of cereal	Shape of starch granule	Size (μm)	References
Wheat	Small spherical; polygonal	5.9–14.6	Karwasra et al. (2017)
Rice	Polygonal and irregular	10–150	Reddy and Bhotmange (2013)
Maize	Round and irregular	20–40	Agama-Acevedo et al. (2008)
Oat	Polygonal, oval, semi-spherical and semi-polygonal	<10	Binqiang et al. (2016)
Barley	Roughly discoid and spherical	15–25	Gubatz et al. (2010)
Proso Millet	Small spherical; large spherical (rare) large polygonal	1.3–8.0	Kumari and Thayumanavan (1998)
Foxtail Millet	Small spherical; small polygonal (especially pentagonal) large pentagonal	0.8–9.6	Kumari and Thayumanavan (1998)
Barnyard Millet	Small spherical; large spherical large polygonal	1.2–10.0	Kumari and Thayumanavan (1998)
Kodo Millet	Large polygonal; rarely small spherical and polygonal	1.2–9.5	Kumari and Thayumanavan (1998)
Little Millet	Small spherical; large spherical	1.0–9.0	Kumari and Thayumanavan (1998)

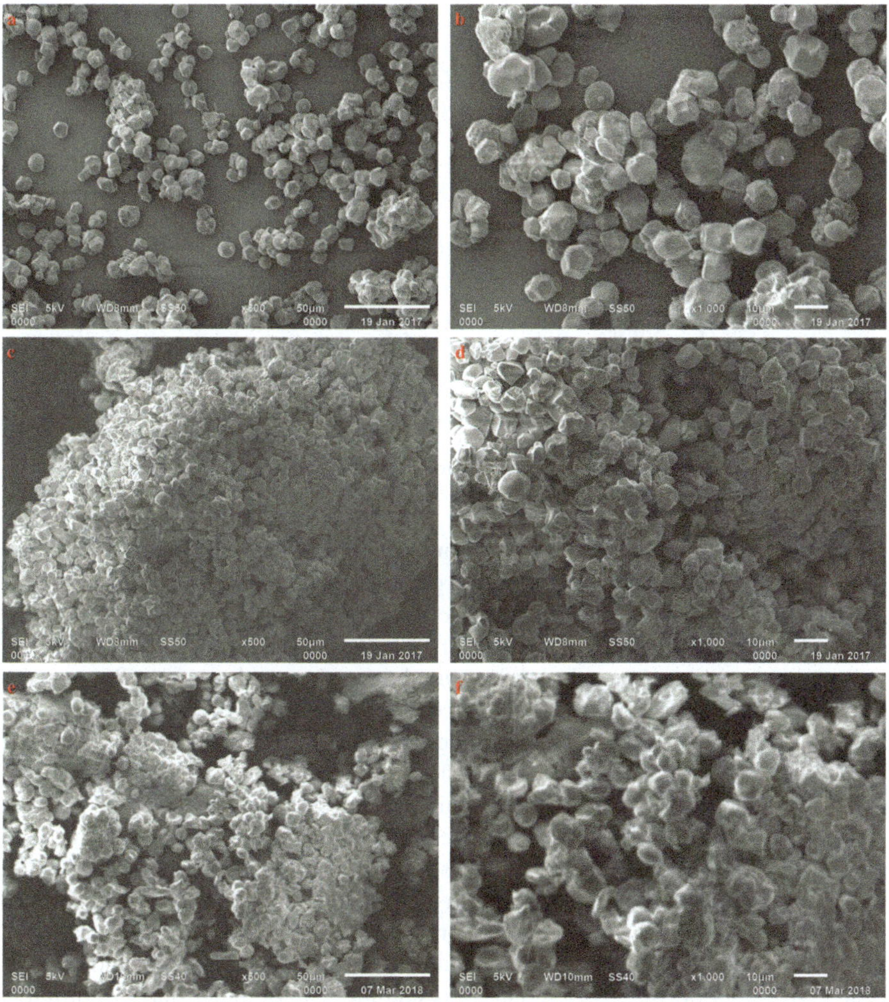

Fig. 6.2 Scanning electron microscope images of barnyard (**a**, **b**), finger (**c**, **d**), and foxtail millet (**e**, **f**) at 500×, 1000×, respectively. (Source: Verma et al. 2018)

granules have small spherical and large polygonal shapes. The size and shape of starch granules vary between different millet species. Starch granules of proso millet starch range from 1.3 to 8 µm, while in foxtail millet starch granules size ranged from 0.8 to 9.6 µm (Kumari and Thayumanavan 1998). Foxtail millet starch contains more small spherical granules than proso millet. Barnyard millet starch is characterized by the presence of large polygonal granules; large and small spherical granules. The number of large spherical granules was quite high in barnyard millet compared to other minor millet. The size ranged from 1.2 to 10 µm which is slightly bigger than

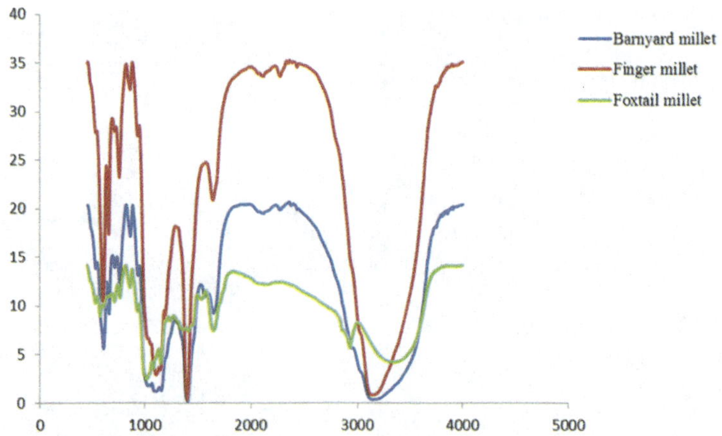

Fig. 6.3 FTIR spectra of barnyard, finger, and foxtail millet. (Source: Verma et al. 2018)

other millet starch (Kumari and Thayumanavan 1998). Kodo millet starch granules were mostly polygonal and the size of the granules ranged from 1.2 to 9.5 μm. Large polygonal, large spherical, and small spherical granules were present in little millet starch. The size of the granules varied from 1 to 9 μm. The starch granules of finger millet are spherical, polygonal, and rhombic shapes whereas the pearl millet granules were exclusively spherical in shape and contained a relatively higher proportion of bigger granules. In large polygonal granules, dense packing of the protein bodies and endosperm results in indentations (Malleshi et al. 1986; Kumari and Thayumanavan 1998; Annor et al. 2014). X-ray crystallography study better understood the granular crystalline and amorphous regions of starch. The crystalline and amorphous parts of starch showed sharp and dispersive peaks respectively (Karwasra et al. 2017). The amylopectin unit chains are crystallized in the form of a double helix arranged in two ways to give A-type and B-type polymorphs (Pérez and Bertoft 2010). Like most other cereal starches, millet starch granules showed the A-type polymorph (Choi et al. 2004; Annor et al. 2014). Relative crystallinity (%) of millet starch of different species ranged from 27% to 30% (Annor et al. 2014). Kumar and Khatkar (2017) studied native starch and A-type and B-type granules of starch by Fourier transform infrared spectroscopy (FTIR). The FTIR spectrum of starch granules showed ordered and amorphous structure at 1047 and 1022 cm^{-1}. The absorbance ratio of FTIR spectra peaks at 1047/1022 and 1022/995 are important in relation to gives the indexes of short range order of double helices of amylose chain present in starch. The absorbance ratio of FTIR spectra peaks of 1047/1022 and 1022/995 are important in relation to gives the indexes of short-range order of double helices of amylose chain present in starch (Sevenou et al. 2002; Shingel 2002). The FTIR spectra of finger millet, foxtail millet, and barnyard millet are shown in Fig. 6.3.

Starch contains two major molecular components, i.e., amylose and amylopectin, which are linked to the quality of starch. Both of these glucan polymers are composed of linear chains of D-glucose linked by α-(1,4) linkage. These linear chains are interconnected by α-(1,6)-glucosidic linkage due to which branching occurs in polymer. Both amylose and amylopectin can form helical complex with iodine molecules in solution giving rise to distinctive blue and red colors, respectively (Zhu 2014). Mostly, amylopectin is the major component of starch granules, while amylose is about 15–30% of total starch. The linear amylose chains form a compact structure that limits enzyme accessibility and rate of amylolysis. On the other hand, amylopectin with its branched structure and several terminal endings is less ordered than amylose and therefore more easily digested (Fardet et al. 2006). Thus, high-amylose/amylopectin ratio influences starch digestion attributing towards resistance starch and affect the glycaemic response (Hallstrom et al. 2011). Along with amylose and amylopectin, starch contains intermediate component in various forms which are available in some mutant plants (Jiang et al. 2010a). These intermediate are poly-glucans whose structures lie between amylopectin and amylose. In recent periods, intermediate components are well studied because these are synthesized in mutant plants during altered biosynthesis (Han et al. 2017).

6.4 Physiochemical Properties of Millet Starch

6.4.1 Swelling Power and Solubility

Swelling power and solubility of starch indicates non-covalent bonding and assess the extent of interaction between amorphous and crystalline domains of the starch granules (Mepba et al. 2009). During heating, the swelling of granules in water is characterized by swelling power (SP), whereas the granular solubilization is quantified by solubility (%) (S). When starch is heated with excessive amount of water the granules absorb water and swell while starch components such as amylose leach out and solubilize. The increase in temperature leads to the granules swelling and break down eventually. This process is influenced by various factors such as molecular structure of amylose and amylopectin, physical associations of chemical components in the granules, size distribution of granules, presence of lipid-amylose complex, and phosphorous groups (Hoover 2001). The swelling power and solubility of millet starches such as pearl, foxtail, proso, and finger millets are mostly reported in the temperature range of 50–90 °C (Lorenz and Hinze 1976; Malleshi et al. 1986; Hoover et al. 1996). In general, millet starch had lower swelling and solubilization values than potato at higher temperature range of ~90 °C (Hoover 2001). Low amylose is associated with higher starch granule which disintegrates easily at 65 °C and attributes higher glycaemic response (Suma and Urooj 2015).

6.4.2 Rheological Property

In relation to significant use of starch in food industries as thickener, characterization of rheological property of starch is remarkably important. Starch granules are insoluble in water but after heating at temperature above the gelatinization, structural integrity of the granules weakens resulting in loss of crystalline structure, water absorption, swelling, and development of significant viscosity. During swelling, the amylose exudes out of the granule and becomes highly hydrated along with the amylopectin (Shinoj et al. 2006). Rheological characteristics of starch depend on the chemical structure of starch, starch concentration, pasting conditions, amylose exudation rate, swollen starch granules, granular fragmentation, colloidally, and molecularly dispersed starch molecules (Sharma and Niranjan 2018). Starch pasting properties can be measured by using Brabender visco amylograph (Kumari and Thayumanavan 1998), Bohlin rheometer (Shinoj et al. 2006), and rapid visco analyzer (Liu et al. 2011). The pasting property of starch granules are remarkably influenced by its minor components such as lipids and phosphate monoester derivatives (Ai and Jane 2015). Kumari and Thayumanavan (1998) reported that barnyard millet showed the highest cold paste viscosity (1165 BU) which differed significantly from other small millets. Viscosity was increased during cooling which can be related to the retrogradation tendency of starch. High paste viscosity indicates a high degree of retrogradation of amylose and formation of resistant starch leading to low glycaemic index. Shinoj et al. (2006) observed viscosity values for little millet (0.058–0.288 Pa s), barnyard millet (0.249–0.419 Pa s), finger (0.333–0.705 Pa s), kodo millet (0.209–0.953 Pa s), proso millet (0.001–0.493 Pa s), and foxtail millets (0.055–0.220 Pa s).

6.4.3 Gelatinization and Retrogradation

Heating or cooling and processing of starch with adequate amount of water at wide range of temperature causes order-disorder phase transition of starch granules termed as gelatinization. The amorphous region of starch uptakes water leading to the swelling of granules, disruption of crystalline region with the breakdown of double helices are the characteristics of gelatinization (Hoover 2001). It occurs in two ways, short-term development of gel structure by amylose crystallization and long-term reordering of amylopectin involves recrystallization of the outer branches (Ring et al. 1987). The gelatinization of millet starch can be assessed by DSC method which involves onset (To), peak (Tp), and conclusion (Tc) temperatures and enthalpy change (ΔH). The finger millet starch granules showed a remarkable change in the gelatinization parameters for starch to water ratio at 1:2 and 1:3 (Premavalli et al. 2005). Starches from various genotypes of foxtail millet and proso millet had ranged from 62.4 to 75 °C and 65.8 to 80.2 °C while ΔH was 8.2–13.5 J/g and ΔH 6.4–11.4 J/g, respectively. Hence, gelatinization properties reveal tremendous genetic diversity between diverse species and among genotypes of the same species (Fujita and Fujiyama 1993). The unlikeness in gelatinization

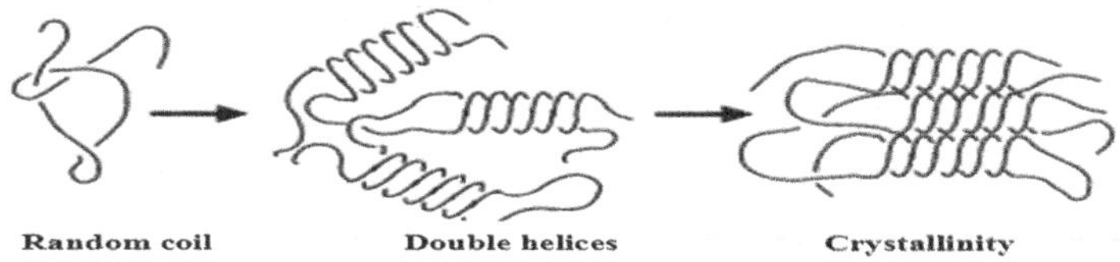

Fig. 6.4 Proposed mechanism of retrogradation (Haralampu 2000)

may be associated to the molecular structure, amylose-amylopectin content, presence of amylose-lipid complex, physical association, and orientation of chemical constituent in the granules (Srichuwong and Jane 2007). The changes in the gelatinization and enthalpy of amylose-lipid complex of starches from two pearl millet genotypes, quantified by DSC may be linked to the amount and complexity of lipid and amylase (Gaffa et al. 2004).

The retrogradation process involves heating of starch in water followed by cooling at 4 °C. The amylose and amylopectin chains are disintegrated on heating while on cooling they reassociate slowly into a distinct ordered structure (Fig. 6.4) (Hoover et al. 2010). The tapioca starch diminished retrogradation of finger millet starch during cooling, whereas xanthan gum may increase retrogradation (Ojijo and Shimoni 2007). Therefore, the blending of tapioca starch to finger millet starch attributing high glycaemic index while intermixing of xanthan gum is associated with low glycaemic index.

6.5 Factors Attributing Glycaemic Attributes of Millet Starch

The characterization of millet starch composition has been proven as one of the inferences for their low blood glucose level. Millet starch granules have polygonal and spherical shapes which are very well discussed in the previous section. The surface of millet starch granules has pores or pinholes which access starch hydrolyzing enzymes into the starch granules. However, these pinholes are not present on starch granules of finer millet and hence starch hydrolyzing enzymes are unable to enter into the starch granules leading to slow digestion of millet starch (Kaur et al. 2007). The rate of starch hydrolysis of different millets follows the sequence: finger millet < pearl millet < proso < foxtail. The in vitro starch hydrolysis by α amylase *wrt* decreasing resistance follows the order: finger millet > potato > chickpea > rice > sorghum > green gram > wheat > tapioca > waxy rice > maize (Singh et al. 2006). The starch hydrolysis was greatly influenced by the interaction of millet starch with other constituents.

6.5.1 Effects of Lipids on Millet Starch Hydrolysis

In millets, palmitic, oleic, and linoleic acids are mainly present fatty acids. The susceptibility of starch to starch degrading enzymes affected by starch-lipid complexes, resulting in slower digestion (Annor et al. 2015), usually involves slower degradation of amylose-lipid complex and rapid degradation of areas which are not associated with amylose–lipid interaction (Jane et al. 1994). Starch-complex with oleic (18:1 Δ^9) and lauric acid (12:0) has very impressive in decreasing starch hydrolysis rates whereas complex formation with linoleic acid (18:2 $\Delta^{9,12}$) has very less effect on hydrolysis rates due to instability of complex (Kawai et al. 2012). From this, it was very clear that the degree of unsaturation affects starch hydrolysis rates which influenced the glycaemic response.

6.5.2 Effects of Proteins on Millet Starch Hydrolysis

Starch–protein interaction forms a physical barrier between the starches and their degrading enzymes affects the starch hydrolysis rates. The starch granules are entrapped by soluble protein fractions, (Albumin, globulin, glutenins) act as a barrier for starch hydrolyzing enzymes such as amylases (Hamaker and Bugusu 2003). The starch digestibility of wheat is greatly influenced by interaction with seed protein resulting in the decrease of glycaemic response (Jenkins et al. 1987). However, it was also reported that the removal of proteins from kodo millet starch increases glycaemic response (Annor et al. 2013).

6.5.3 Effects of Fiber on Millet Starch Hydrolysis

Millet grains are rich sources of fiber (7–21%), have special nutritional significance due to their hypoglycaemic and hypocholesterolemic characteristics. A harmony between phenolic and dietary fiber may play a vital role in amylase inhibition and probably manage type 2 diabetes (Devi et al. 2014).

6.5.4 Effects of Polyphenols on Millet Starch Hydrolysis

Polyphenols are well-known phytoconstituent which have tremendous health-boosting properties and combating oxidative stress generated by free radicals, prevent cancer, anti-diabetic, anti-inflammatory, cardiovascular disease, and control hypertension (Taylor and Duodu 2015). Millet polyphenols such as ferulic acid, vanillic acid, caffeic acid, gentisic acid, and protocatechuic acid may be exploited in the prevention and control of diabetes (type 2) due to their inhibitory effect on starch digestive enzymes such as α glycosidase and pancreatic α amylase (Shobana et al. 2009).

6.6 Health Benefits, Nutraceutical Properties, and Uses of Millet Starch

6.6.1 Health Benefits and Nutraceutical Importance

Millet flour is found protective against several degenerative diseases such as diabetes, cardiovascular diseases, colon cancers, metabolic syndromes, and Parkinson's disease (Fardet et al. 2008; Devi et al. 2014). Polyphenols from millets are acknowledged for the prevention of nonenzymatic glycosylation of collagen and cross linking that is induced during diabetes and aging (Fu et al. 1992). The diabetic conditions showed a deleterious effect on the wound healing process because free oxygen radicals damage the cells. Antioxidants from millet flour have important nutraceutical properties that prevent tissue damage and stimulate the wound healing process (King 2001). Millet starch consumption reduces various diseases and its relation to boost human health condition. Millet flour and its derived product are consumed by a small population of the world but wide interest explores the health benefits as well as nutraceutical properties of millets as they are rich in health-promoting nutrients and phytochemicals. The nutrients present in millets are believed to be responsible for many health benefits which include resistant starch, oligosaccharides, lipids, antioxidants such as phenolic acids, flavonoids, lignans, and phytosterols that are discussed in detail in the following sections (Saleh et al. 2013).

Millet for diabetics: Hyperglycaemia is caused due to variations in carbohydrate, lipid, and protein metabolism which results in a chronic metabolic disorder termed as diabetes mellitus. In present days, diabetes mellitus has evolved as the most common endocrine disorder which is caused due to defective insulin production (type 1) or due to resistance against insulin action and insulin-secretary response (type 2) (Shobana et al. 2009). The enzyme alpha-glycosidase and pancreatic amylase increase glucose level leading to postprandial hyperglycemia which is controlled by chemical synthetic inhibitors of that enzymes. Natural inhibitors such as functional food in diet are potentially safer than chemical inhibitors. Therefore, incorporation of whole food grains in diet is beneficial for the prevention of diabetes mellitus. It has been observed population consuming millet has lower incidences of diabetes (Kim et al. 2011). The anti-nutritional factors present in millet starch are known to decrease starch digestibility and absorption resulting in lower glycaemic response (Kumari and Sumathi 2002). Processing such as dehulling and thermal treatment lowers glycemic index of barnyard millet which is beneficial for type 2 diabetic persons (Ugare et al. 2014). The consumption of multigrain flour-based chapatti results in lowering of blood glucose level (Pradhan et al. 2010). Thus, millet cereals have the potentials to be useful in controlling diabetes.

Prevention of cardiovascular diseases: Physical inactivity, obesity, and grabbing of junk food lead to the risk of cardiovascular diseases. The native and modified starch of barnyard millet showed lowest glycaemic index, lower serum cholesterol and triglycerides (storage lipids) compared to rice and other cereals when given to rats under controlled feeding experiment (Kumari and Thayumanavan 1997). Therefore, millet starch-based diet may prevent cardiovascular disease and provide protection against stokes or heart attack by reducing plasma triglycerides and lipid peroxidation (Lee et al. 2010). The fermentation product of finger millet is also an important source to provide important metabolites like statin and sterol which are commonly used in hypercholesteraemic and obesity therapy (Venkateswaran and Vijayalakshmi 2010). The metabolites statin serves as enzyme inhibitor of the cholesterol biosynthetic pathway while sterol helps in decreasing serum low density lipoprotein (bad cholesterol) without disturbing high density lipoprotein (good cholesterol) (Ostlund Jr 2002).

Anti-celiac diseases: People suffering from celiac disease and overall growing demand of cereal products made from gluten-free grains and other intolerances to rye, barley, and wheat has developed a new commercial market having cereal products made from millet starch (Angioloni and Collar 2013). They have tremendous value in foods and fermented products such as beverage that can be suitable for persons suffering from celiac disease (Chandrasekara and Shahidi 2011).

Cataractogenesis inhibition and prevention of bone ailments: The risk factor of retinopathy and cataract are induced by postprandial hyperglycemia or diabetes. Diabetes-induced cataract leads to accumulation of sugar alcohol sorbitol formed from the reduction of glucose by the key enzyme aldose reductase (AR) (Chauhan et al. 2018). Aldose reductase-mediated sugar-induced cataract involves binding of glucose to protein molecule, nonenzymatic glycation induced during diabetes (Chethan et al. 2008). Apart from that millet starch or flour-based products or derived products can be useful in the development of bone mass in growing children as well as for preventing osteoporosis and other bone ailments in adults and aging community. Thus, all the nutritional benefits of finger millet or other millet starch must be transformed to nutraceutical development and applied to other cereal crops for their optimum enrichment (Kumar et al. 2016).

6.6.2 Glycaemic Response

The millet diet is acknowledged for health potential and is generally recommended for diabetics. Millet starch is slowly assimilated and digested than other cereals starch. Generally, starch is classified into rapidly digestible starches, slowly digestible starches, and resistant starches based on the results of in vitro digestion (Englyst et al. 1992). According to Englyst et al. (1982) a fraction of starch that is slowly digested and escapes digestion in small intestine is fermented by the gut microflora after reaching the large intestine and produces small chain of fatty acids and termed as resistant starch. There are currently five types of resistant starches (Table 6.2). Finger millet reduces the risk of diabetes mellitus and gastrointestinal tract disorders on regular consumption due to the presence of polyphenols and dietary fiber in high concentration. Phenolic extracts from finger millet seed coats partially inhibit amylase and α-glycosidase activity during enzymatic hydrolysis of starch and delays the

Table 6.2 Types of resistant starch

Types	Description	Example	References
Resistant starch I	Starch granules are encapsulated by protein	Coarsely ground or whole-kernel grains	Englyst et al. (1992)
Resistant starch II	Granular starch with the B- or C-polymorph	High-amylose maize starch, raw potato, raw banana starch	Englyst et al. (1992), Jiang et al. (2010b)
Resistant starch III	Retrograded starch	Cooked and cooled starchy foods	Ma et al. (2020)
Resistant starch IV	Chemically modified starches	Succinylation, acetylation, esterification of starch with organic acid	Ačkar et al. (2015)
Resistant starch V	Amylose-lipid complex	Monoacylglycerol complexed high-amylose starch	Panyoo and Emmambux (2017).

absorption of glucose (Devi et al. 2014). Formulations and preparations of finger millet-based food showed lower glycaemic index (Shukla and Srivastava 2014). The risk of diabetes (type 2) reduces by dietary intake of calcium and magnesium (van Dam et al. 2006). Thus, the abundance of these minerals in finger millet could be indirectly connected with its ability to mitigate type 2 diabetes risk. The plasma glucose level is lowered by consumption of multigrain flour containing 30–35% proportion of finger millet (Pradhan et al. 2010).

6.6.3 Value-Added Product of Millet Flour

A schematic diagram for the preparation of various value-added products from millet flour is furnished in Fig. 6.5. Novel approaches and models are required to make value-added products for enhanced millet consumption (Fig. 6.6). For instance, probiotics- and prebiotics-based millet food products can help to repopulate the colon bacteria diminished by antibiotics, chemotherapy, or diseases. *Koko, Mangisi/Masvusvu, Kunu-zaki, Uji, Jandh, Fura, and Dambu* are potential probiotic products of millets (Amadou et al. 2011) represented in Table 6.3. Prebiotics are considered as resistant starch that beneficially affects the host and stimulate the growth of bacteria in the colon (Laminu et al. 2011; Salem et al. 2012). Malting is known to induce important beneficial biochemical changes in the millet grain. According to Lei et al. (2006), *Koko* sour wort could help in reducing diarrhea of young children. The production of millet and millet-based products by households, entrepreneurs, self-help groups, small-scale industries, and large-scale industries can raise their income level and health status of masses in developing countries.

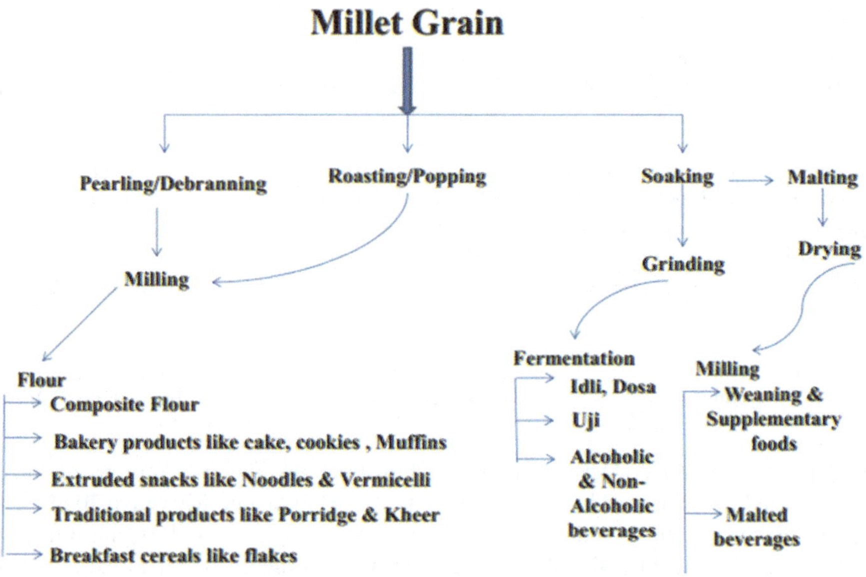

Fig. 6.5 Schematic diagram for developing millet-based value-added foods. (Source: Kumar et al. 2018)

6.7 Decline and Fall of Millets

Urbanization and change in dietary habits are the main factors for the limited production of millets. Even migration of rural population to urban areas also gave millets a status of underutilized crops. After the green revolution, a systematic decline in the production of millets occurred due to the enhancement in area and production of major cereals like rice and wheat under intensive farming and irrigated conditions. Millets possess typical grain texture and hard seed coat that makes them difficult to process as well as cook in convenient form. Due to the lack of awareness as well as appropriate processing technologies, ready-to-eat or value-added products are the major limiting factors for their diversified food uses and better economic status with respect to other cereals. The major constraints for widespread utilization of millet are its coarse fibrous seed coat, colored pigments, astringent flavor, and poor keeping quality of the processed products (Kulkarni et al. 2018). Apart from this, anti-nutritional factors present in millets are a major obstacle in its utilization. For instance, phytic acid has strong chelating ability and readily forms complexes with monovalent cation of potassium and multivalent cations of calcium, iron, zinc, and magnesium, reducing their bioavailability and creating a deficit in their absorption (Simwemba et al. 1984; Raboy 2009). Besides phytic acid, presence of goitrogenic polyphenols might be responsible for some health problems. Epidemiologic evidences indicated that a diet based on millet as a staple food, such as that occurring in rural villages of Africa and Asia, plays a role in the genesis of endemic goiter in

Fig. 6.6 Value-added product from millets (**a**) Laddu, (**b**) Vermicelli Kheer, (**c**) Millet Rawa, (**d**) Ragi Halwa, (**e**) Millet Roti, (**f**) Ragi Papad, (**g**) Biscuits, (**h**) Bread, (**i**) Cake, (**j**) Pizza Base, (**k**) Pasta, and (**l**) Noodles. (Source: Dayakar Rao et al. 2016)

Fig. 6.6 (continued)

these areas (Boncompagni et al. 2018). Advancement in processing technologies (decortication, soaking, germination, fermentation, puffing, and cooking) reduces the levels of anti-nutritional factors such as tannins and phenols thereby increase the bioavailability of starch and minerals. Therefore, robust and cost-effective processing technologies, overcoming the barriers for utilization and industrialization of millets flour for the production of value-added products are desperately needed.

Table 6.3 Millet-based fermented products

Derived product	Form	Country	References
Koko (Pearl millet)	Millet porridge	West Africa, Ghana	Lei and Jakobsen (2004)
Kunu-Zaki (Millet or Millet and other cereals)	Nonalcoholic fermented beverage	Northern part of Nigeria	Agarry et al. (2010)
Mangisi and Masvusvu (Finger millet)	Sweet beer	Africa, Zimbabwe, Uganda	Zvauya et al. (1997)
Uji (Finger millet in combination with Sorghum and Maize)	Porridge	East Africa (Kenya, Uganda, Tanzania)	Onyango et al. (2005)
Jandh (Finger millet)	Slightly acidic and sweet beverage, beer	Nepal	Dahal et al. (2005)
Fura (Pearl millet)	Porridge	Nigeria, Abuja	Jideani et al. (2010)
Dambu (Millet flour)	Dumplings	Africa	Jideani et al. (2010)

6.8 Conclusion

The growing demand for food and nutritional security increases public awareness that challenges the commercial enterprises to develop novel food products with extraordinary health-enhancing characteristics that emphasize the opening of new avenue for research in nutrition biology. In the industry, millet starch or flour is subjected to a combination of different cereal starch or flour to achieve desirable functional products. For processing of millet starch, it is better to understand their natural variation like chemical composition, morphology, structures, and physico-chemical properties from diverse millet species. The study of the shape and surface details of various cereal starches is performed by scanning electron microscopy, while X-ray crystallography provides deep insights into the granular crystalline and amorphous region of starch. Starch physiochemical properties like gelatinization, retrogradation, and rheological properties impart its industrial application. The dietary fiber, minerals, proteins, lipids, and secondary metabolites such as polyphenols and flavonoids in millets are known to offer several health benefits alone and by interacting with its starch or flour by diminishing starch hydrolysis rate resulting in hypoglycemic effect. Several processing technologies were found to improve nutritional characteristics of millet starch and flour. Value-added products from millet starch can improve the socioeconomic status of economically deprived masses as well as combat hidden hunger. This chapter provides a scientific rationale in various aspects of millet starch including its use as a therapeutic and health-promoting food.

References

Ačkar Đ, Babić J, Jozinović A, Miličević B, Jokić S, Miličević R et al (2015) Starch modification by organic acids and their derivatives: a review. Molecules 20(10):19554–19570

Agama-Acevedo E, De La Rosa APB, Méndez-Montealvo G, Bello-Pérez LA (2008) Physicochemical and biochemical characterization of starch granules isolated of pigmented maize hybrids. Starch-Stärke 60(8):433–441

Agarry OO, Nkama I, Akoma O (2010) Production of Kunun-zaki (a Nigerian fermented cereal beverage) using starter culture. Int Res J Microbiol 1(2):18–25

Ai Y, Jane JL (2015) Gelatinization and rheological properties of starch. Starch-Stärke 67 (3–4):213–224

Amadou I, Gbadamosi OS, Le GW (2011) Millet-based traditional processed foods and beverages—a review. Cereal Foods World 56(3):115

Angioloni A, Collar C (2013) Suitability of oat, millet and sorghum in breadmaking. Food Bioprocess Technol 6(6):1486–1493

Annor GA, Marcone M, Bertoft E, Seetharaman K (2013) In vitro starch digestibility and expected glycemic index of kodo millet (Paspalum scrobiculatum) as affected by starch–protein–lipid interactions. Cereal Chem 90(3):211–217

Annor GA, Marcone M, Bertoft E, Seetharaman K (2014) Physical and molecular characterization of millet starches. Cereal Chem 91(3):286–292

Annor GA, Marcone M, Corredig M, Bertoft E, Seetharaman K (2015) Effects of the amount and type of fatty acids present in millets on their in vitro starch digestibility and expected glycemic index (eGI). J Cereal Sci 64:76–81

Binqiang T, Chao W, Lan W, Bijun X (2016) Granule size and distribution of raw and germinated oat starch in solid state and ethanol solution. Int J Food Prop 19(3):709–719

Boncompagni E, Orozco-Arroyo G, Cominelli E, Gangashetty PI, Grando S, Zu TTK et al (2018) Antinutritional factors in pearl millet grains: phytateand goitrogens content variability and molecular characterization of genes involved in their pathways. PLoS One 13(6):e0198394

Bora P, Ragaee S, Marcone M (2019) Characterisation of several types of millets as functional food ingredients. Int J Food Sci Nutr 70(6):714–724

Buleon A, Colonna P, Planchot P, Ball S (1998) Starch granules: structure and biosynthesis. Int J Biol Macromol 23:85–112

Carpita NC, Kanabus J (1987) Extraction of starch by dimethyl sulfoxide and quantitation by enzymatic assay. Anal Biochem 161(1):132–139

Chandrasekara A, Shahidi F (2011) Determination of antioxidant activity in free and hydrolyzed fractions of millet grains and characterization of their phenolic profiles by HPLC-DAD-ESI-MSn. J Funct Foods 3(3):144–158

Chauhan M, Sonawane SK, Arya SS (2018) Nutritional and nutraceutical properties of millets: a review. Clin J Nutr Dietet 1(1):1–10

Chethan S, Dharmesh SM, Malleshi NG (2008) Inhibition of aldose reductase from cataracted eye lenses by finger millet (Eleusine coracana) polyphenols. Bioorg Med Chem 16 (23):10085–10090

Choi H, Kim W, Shin M (2004) Properties of Korean amaranth starch compared to waxy millet and waxy sorghum starches. Starch 56(10):469–477

Dahal NR, Karki TB, Swamylingappa B, Li Q, Gu G (2005) Traditional foods and beverages of Nepal—a review. Food Rev Intl 21(1):1–25

Dayakar Rao B, Vishala AD, Arlene Christina GD, Tonapi VA (2016) Technologies of millet value added products. Centre of Excellence on Sorghum, ICAR-Indian Institute of Millets Research, Rajendranagar, Hyderabad

Dev R, Kumar S, Singh J, Chauhan B (2011) Potential role of nutraceuticals in present scenario: a review. J Appl Pharm Sci 1:26–28

Devi PB, Vijayabharathi R, Sathyabama S, Malleshi NG, Priyadarisini VB (2014) Health benefits of finger millet (*Eleusine coracana* L.) polyphenols and dietary fiber: a review. J Food Sci Technol 51(6):1021–1040

Englyst H, Wiggins HS, Cummings JH (1982) Determination of the non-starch polysaccharides in plant foods by gas-liquid chromatography of constituent sugars as alditol acetates. Analyst 107 (1272):307–318

Englyst HN, Kingman SM, Cummings JH (1992) Classification and measurement of nutritionally important starch fractions. Eur J Clin Nutr 46:S33–S50

Fardet A, Leenhardt F, Lioger D, Scalbert A, Rémésy C (2006) Parameters controlling the glycaemic response to breads. Nutr Res Rev 19(1):18–25

Fardet A, Rock E, Rémésy C (2008) Is the in vitro antioxidant potential of whole-grain cereals and cereal products well reflected in vivo? J Cereal Sci 48(2):258–276

Food and Agricultural Organisation (FAO) of the United Nations. (1995) Sorghum and millets in human nutrition. FAO Food and Nutrition Series, No. 27. FAO, Rome. ISBN 92-5-103381-1

Fu MX, Knecht KJ, Thorpe SR, Baynes JW (1992) Role of oxygen in cross-linking and chemical modification of collagen by glucose. Diabetes 41(Suppl 2):42–48

Fujita S, Fujiyama G (1993) The study of melting temperature and enthalpy of starch from rice, barley, wheat, foxtail- and proso-millets. Starch-Stärke 45(12):436–441

Gaffa T, Yoshimoto Y, Hanashiro I, Honda O, Kawasaki S, Takeda Y (2004) Physicochemical properties and molecular structures of starches from millet (Pennisetum typhoides) and sorghum (Sorghum bicolor L. Moench) cultivars in Nigeria. Cereal Chem 81(2):255–260

Gubatz S, Shewry PR, Ullrich S (2010) The development, structure, and composition of the barley grain. In: Barley: production, improvement, and uses, vol 11. Wiley, Ames, IA, p 391

Gull A, Nayik GA, Prasad K, Kumar P (2015) RETRACTED ARTICLE: Nutritional, technological, and medical approach of finger millet (Eleusine coracana). Cogent Food Agric 1 (1):1090897

Hallstrom E, Sestili F, Lafiandra D, Bjorck I, Ostman E (2011) A novel wheat variety with elevated content of amylose increases resistant starch formation and may beneficially influence glycaemia in healthy subjects. Food Nutr Res. https://doi.org/10.3402/fnr.v55i0.7074

Hamaker BR, Bugusu BA (2003) Overview: sorghum proteins and food quality. In: Workshop on the proteins of sorghum and millets: enhancing nutritional and functional properties for Africa [CD], Pretoria: South Africa

Han W, Zhang B, Li J, Zhao S, Niu M, Jia C, Xiong S (2017) Understanding the fine structure of intermediate materials of maize starches. Food Chem 233:450–456

Haralampu SG (2000) Resistant starch—a review of the physical properties and biological impact of RS3. Carbohydr Polym 41(3):285–292

Hoover R (2001) Composition, molecular structure, and physicochemical properties of tuber and root starches: a review. Carbohydr Polym 45(3):253–267

Hoover R, Swamidas G, Kok LS, Vasanthan T (1996) Composition and physicochemical properties of starch from pearl millet grains. Food Chem 56(4):355–367

Hoover R, Hughes T, Chung HJ, Liu Q (2010) Composition, molecular structure, properties, and modification of pulse starches: a review. Food Res Int 43(2):399–413

Jane JL, Kasemsuwan T, Leas S, Zobel H, Robyt JF (1994) Anthology of starch granule morphology by scanning electron microscopy. Starch-Stärke 46(4):121–129

Jenkins DJ, Thorne MJ, Wolever TM, Jenkins AL, Rao AV, Thompson LU (1987) The effect of starch-protein interaction in wheat on the glycemic response and rate of in vitro digestion. Am J Clin Nutr 45(5):946–951

Jiang H, Campbell M, Blanco M, Jane JL (2010a) Characterization of maize amylose-extender (ae) mutant starches: Part II. Structures and properties of starch residues remaining after enzymatic hydrolysis at boiling-water temperature. Carbohydr Polym 80(1):1–12

Jiang H, Lio J, Blanco M, Campbell M, Jane JL (2010b) Resistant-starch formation in high-amylose maize starch during kernel development. J Agric Food Chem 58(13):8043–8047

Jideani VA, Oloruntoba RH, Jideani IA (2010) Optimization of fura production using response surface methodology. Int J Food Prop 13(2):272–281

Karwasra BL, Gill BS, Kaur M (2017) Rheological and structural properties of starches from different Indian wheat cultivars and their relationships. Int J Food Prop 20(Suppl 1):S1093–S1106

Kaur L, Singh J, McCarthy OJ, Singh H (2007) Physico-chemical, rheological and structural properties of fractionated potato starches. J Food Eng 82(3):383–394

Kawai K, Takato S, Sasaki T, Kajiwara K (2012) Complex formation, thermal properties, and in-vitro digestibility of gelatinized potato starch–fatty acid mixtures. Food Hydrocoll 27 (1):228–234

Kim JS, Hyun TK, Kim MJ (2011) The inhibitory effects of ethanol extracts from sorghum, foxtail millet and proso millet on α-glucosidase and α-amylase activities. Food Chem 124 (4):1647–1651

King L (2001) Impaired wound healing in patients with diabetes. Nursng Stand 15(38):39

Klucinec JD, Thompson DB (1998) Fractionation of high-amylose maize starches by differential alcohol precipitation and chromatography of the fractions. Cereal Chem 75(6):887–896

Kulkarni DB, Sakhale BK, Giri NA (2018) A potential review on millet grain processing. Int J Nutr Sci 3:1–8

Kumar R, Khatkar BS (2017) Thermal, pasting and morphological properties of starch granules of wheat (Triticum aestivum L.) varieties. J Food Sci Technol 54(8):2403–2410

Kumar A, Metwal M, Kaur S, Gupta AK, Puranik S, Singh S et al (2016) Nutraceutical value of finger millet [*Eleusine coracana* (L.) Gaertn.], and their improvement using omics approaches. Front Plant Sci 7:934

Kumar A, Tomer V, Kaur A, Kumar V, Gupta K (2018) Millets: a solution to agrarian and nutritional challenges. Agric Food Secur 7(1):31

Kumari PL, Sumathi S (2002) Effect of consumption of finger millet on hyperglycemia in non-insulin dependent diabetes mellitus (NIDDM) subjects. Plant Foods Hum Nutr 57 (3–4):205–213

Kumari SK, Thayumanavan B (1997) Comparative study of resistant starch from minor millets on intestinal responses, blood glucose, serum cholesterol and triglycerides in rats. J Sci Food Agric 75(3):296–302

Kumari SK, Thayumanavan B (1998) Characterization of starches of proso, foxtail, barnyard, kodo, and little millets. Plant Foods Hum Nutr 53(1):47–56

Laminu HH, Modu S, Numan AI (2011) Production, in vitro protein digestibility, phytate content and acceptability of weaning foods prepared from pearl millet (*Pennisetum typhoideum*) and cowpea (*Vigna unguiculata*). Int J Nutr Metab 3(9):109–113

Lee SH, Chung IM, Cha YS, Park Y (2010) Millet consumption decreased serum concentration of triglyceride and C-reactive protein but not oxidative status in hyperlipidemic rats. Nutr Res 30 (4):290–296

Lei V, Jakobsen M (2004) Microbiological characterization and probiotic potential of koko and koko sour water, African spontaneously fermented millet porridge and drink. J Appl Microbiol 96(2):384–397

Lei V, Friis H, Michaelsen KF (2006) Spontaneously fermented millet product as a natural probiotic treatment for diarrhoea in young children: an intervention study in Northern Ghana. Int J Food Microbiol 110(3):246–253

Liu C, Liu P, Yan S, Qing Z, Shen Q (2011) Relationship of physicochemical, pasting properties of millet starches and the texture properties of cooked millet. J Texture Stud 42(4):247–253

Lorenz K, Hinze G (1976) Functional characteristics of starches from proso and foxtail millets. J Agric Food Chem 24(5):911–914

Ma Z, Hu X, Boye JI (2020) Research advances on the formation mechanism of resistant starch type III: a review. Crit Rev Food Sci Nutr 60(2):276–297

Majid A, Poornima Priyadarshini CG (2019) Millet derived bioactive peptides: a review on their functional properties and health benefits. Crit Rev Food Sci Nutr 60(19):3342–3351

Malleshi NG, Desikachar HSR, Tharanathan RN (1986) Physico-chemical properties of native and malted finger millet, pearl millet and foxtail millet starches. Starch-Stärke 38(6):202–205

McDonough CM, Rooney LW, Earp CF (1986) Structural characteristics of Eleusine corocana (finger millet) using scanning electron and fluorescence microscopy. Food Struct 5(2):9

Mepba HD, Eboh L, Eko CB, Ukpabi UJ (2009) Composition and pasting properties of starch from two cocoyam cultivars. J Food Qual 32(4):522–537

Ojijo NK, Shimoni E (2007) Influence of xanthan gum and tapioca starch on the retrogradation and gelation of finger millet (Eleusine coracana L. Gaertner) starch pastes. J Texture Stud 38 (1):100–115

Onyango C, Noetzold H, Ziems A, Hofmann T, Bley T, Henle T (2005) Digestibility and antinutrient properties of acidified and extruded maize–finger millet blend in the production of uji. LWT Food Sci Technol 38(7):697–707

Ostlund RE Jr (2002) Phytosterols in human nutrition. Annu Rev Nutr 22(1):533–549

Panyoo AE, Emmambux MN (2017) Amylose–lipid complex production and potential health benefits: a mini-review. Starch-Stärke 69(7–8):1600203

Pérez S, Bertoft E (2010) The molecular structures of starch components and their contribution to the architecture of starch granules: a comprehensive review. Starch-Stärke 62(8):389–420

Pradhan A, Nag SK, Patil SK (2010) Dietary management of finger millet (Eleusine coracana L. Gaerth) controls diabetes. Curr Sci 98(6):763–765

Premavalli KS, Jagannath JH, Majumdar TK, Bawa AS (2005) Studies on phase transition in finger millet starch in relation to gelatinisation. J Food Sci Technol Mysore 42(4):336–340

Raboy V (2009) Approaches and challenges to engineering seed phytate and total phosphorus. Plant Sci 177(4):281–296

Reddy DK, Bhotmange MG (2013) Isolation of starch from rice (Oryza sativa L.) and its morphological study using scanning electron microscopy. Int J Agric Food Sci Technol 4(9):859–866

Ring SG, Colonna P, I'Anson KJ, Kalichevsky MT, Miles MJ, Morris VJ, Orford PD (1987) The gelation and crystallisation of amylopectin. Carbohydr Res 162(2):277–293

Saleh AS, Zhang Q, Chen J, Shen Q (2013) Millet grains: nutritional quality, processing, and potential health benefits. Compr Rev Food Sci Food Saf 12(3):281–295

Salem MH, Hippen AR, Salem MM, Assem FM, El-Aassar M (2012) Survival of probiotic Lactobacillus casei and Enterococcus fecium in domiati cheese of high conjugated linoleic acid content. Emirates J Food Agric 24:98–104

Sevenou O, Hill SE, Farhat IA, Mitchell JR (2002) Organisation of the external region of the starch granule as determined by infrared spectroscopy. Int J Biol Macromol 31(1–3):79–85

Sharma N, Niranjan K (2018) Foxtail millet: properties, processing, health benefits, and uses. Food Rev Intl 34(4):329–363

Shingel KI (2002) Determination of structural peculiarities of dexran, pullulan and γ-irradiated pullulan by Fourier-transform IR spectroscopy. Carbohydr Res 337(16):1445–1451

Shinoj S, Viswanathan R, Sajeev MS, Moorthy SN (2006) Gelatinisation and rheological characteristics of minor millet flours. Biosyst Eng 95(1):51–59

Shobana S, Sreerama YN, Malleshi NG (2009) Composition and enzyme inhibitory properties of finger millet (Eleusine coracana L.) seed coat phenolics: mode of inhibition of α-glucosidase and pancreatic amylase. Food Chem 115(4):1268–1273

Shukla K, Srivastava S (2014) Evaluation of finger millet incorporated noodles for nutritive value and glycemic index. J Food Sci Technol 51(3):527–534

Simwemba CG, Hoseney RC, Varriano-Marston E, Zeleznak K (1984) Certain B vitamin and phytic acid contents of pearl millet [Pennisetum americanum (L.) Leeke]. J Agric Food Chem 32 (1):31–34

Singh N, Singh J, Kaur L, Sodhi NS, Gill BS (2003) Morphological, thermal and rheological properties of starches from different botanical sources. Food Chem 81:219–231

Singh V, Ali SZ, Somashekar R, Mukherjee PS (2006) Nature of crystallinity in native and acid modified starches. Int J Food Prop 9(4):845–854

Srichuwong S, Jane JL (2007) Physicochemical properties of starch affected by molecular composition and structures: a review. Food Sci Biotechnol 16(5):663
Suma PF, Urooj A (2015) Isolation and characterization of starch from pearl millet (Pennisetum typhoidium) flours. Int J Food Prop 18(12):2675–2687
Taylor JR, Duodu KG (2015) Effects of processing sorghum and millets on their phenolic phytochemicals and the implications of this to the health-enhancing properties of sorghum and millet food and beverage products. J Sci Food Agric 95(2):225–237
Ugare R, Chimmad B, Naik R, Bharati P, Itagi S (2014) Glycemic index and significance of barnyard millet (Echinochloa frumentacae) in type II diabetics. J Food Sci Technol 51 (2):392–395
Van Dam RM, Hu FB, Rosenberg L, Krishnan S, Palmer JR (2006) Dietary calcium and magnesium, major food sources, and risk of type 2 diabetes in US black women. Diabetes Care 29 (10):2238–2243
Venkateswaran V, Vijayalakshmi G (2010) Finger millet (Eleusine coracana)—an economically viable source for antihypercholesterolemic metabolites production by Monascus purpureus. J Food Sci Technol 47(4):426–431
Verma VC, Kumar A, Zaidi MGH, Verma AK, Jaiswal JP, Singh DK, Sing A, Agrawal S (2018) Starch isolation formed different cereals with variable amylose/amylopectin ratio and its morphological study using SEM and FT-IR. Int J Curr Microbiol Appl Sci 7(10):211–228
Zhu F (2014) Structure, physicochemical properties, and uses of millet starch. Food Res Int 64:200–211
Zvauya R, Mygochi T, Parawira W (1997) Microbial and biochemical changes occurring during production of masvusvu and mangisi, traditional Zimbabwean beverages. Plant Foods Hum Nutr 51(1):43–51

Product Development from Millets

7

Sumedha Deshpande, Manoj Kumar Tripathi,
Debabandya Mohapatra, and Rajpal S. Jadam

Abstract

Whole grains are used widely from ancient times in human diet. They are universally recommended as an important source of energy, carbohydrate, protein, and fiber in human diet. In India millets and sorghum are preferred as a major income-generating crops for the majority of poor farmers exclusively in some farming regions. Millets are the main crop in some regions of the world and specifically in both seasons of Kharif and Rabi. Sorghum plays a major role in nutrient uptake which is about 35% of the total intake of calories, protein, iron, and zinc in the intake zones. Nutritional change and health scenarios indicate a high rate of lifestyle diseases both in developed and developing countries. Due to high nutritional composition millets are categorized as coarse cereals and used for the production of nutrient-rich food development. All millets are very rich in minerals and also have high antioxidant activity. Hence, millets are being consumed as a source of nutraceutical components for nutritional improvement of processed food products to augment their proposed health benefits. Incorporation of millet in various low-cost food formulations intended for adults and children could be used to alleviate malnutrition and other deficiency disorders and can serve as nutritionally dense value-added products.

Keywords

Millet · Malnutrition · Complemenary food · Health promotion

S. Deshpande · M. K. Tripathi (✉) · D. Mohapatra · R. S. Jadam
ICAR-Central Institute of Agricultural Engineering, Bhopal, Madhya Pradesh, India
e-mail: Sumedha.Despande@icar.gov.in

A. Kumar et al. (eds.), *Millets and Millet Technology*,
https://doi.org/10.1007/978-981-16-0676-2_7

7.1 Introduction

Making food for everyone is a major task of government and scientific communities. Sorghums and millets are considered as historic crops which were cultivated many times ago from civilization. Some evidence suggests that farming of sorghum and millets was initiated 4000 years ago (Shahidi and Chandrasekara 2013). These crops are categorized as drought-resistant plants mainly grown in hot climates. In some African and Asian countries, the major sources of daily nutrition of persons living in rural areas are millets. The biochemical composition of millet indicates a high amount of nutraceuticals in comparison to other groups of cereals. Presence of high amounts of nutraceuticals and health-promoting compounds are very helpful in diseases such as blood pressure, diabetic conditions, and cardiovascular problems. Millets are also rich in fiber content which reduces several health problems. The consumption of complete millet grains is very effective for controlling several-health related problems and liable for avoidance of cardiac problems, diabetic problems, obesity problems, breast cancer problems, and some cases of premature death (Balasubramanian 2013; Saleh et al. 2013).

Millets grains are generally of little size and can be easily ready for food use in native form. Millets are very primitive crops and are reported to be cultivated in very dry conditions in China and some African countries. Some evidence also suggests that in the middle age millet was widely consumed in comparison to wheat. The detailed description was also found in the Bible which also indicates their origin. The role of sorghum and millet crops is established in many developing countries which is very useful in their malnutrition and economy. Sorghum and millets show a very important part in the form of food and nutrition security and both types of crops include up to 10% of coarse grain in Asian countries. In case of India, the production of sorghum and millet comprises more than 80% of total Asian country's production but the crop yield in India is comparatively low.

The form of grain utilization in food purpose differs in many countries for both sorghum and millets. In the case of African countries, sorghum and millets are largely used as food grain and directly consumed as a major nutrition source by poor publics. In current times the utilization of sorghum and millet for beer manufacturing is increasing and supports the economy of African countries.

7.2 Features of the Millet Plant

Millet grain size is very small, grown annually, and belongs to the grass family. Millets are very tolerable in hot and extreme conditions. The appearance of millet grains depends upon crop variety. They habitually have rough stems, opaque bunches, leaves similar to grass, and mainly grow over 6 ft long.

The crop seeds are enclosed with colorful hulls. The grain hulls are very flexible in color depending on the crop variety (white, red, yellow, brown, or striped). In case of millet seed which is covered with hulls and hulls have not good digestive quality, it is necessary to remove the hull portion before utilizing it for food purposes.

Dehulling of millet seed is a very difficult process and after dehulling the color and appearance of the seed are changed and grains look like yellow balls.

7.2.1 Millets and Practices

Sorghum crops are broadly cultivated in different regions of India mainly Madhya Pradesh, Uttar Pradesh, Maharashtra, Andhra Pradesh, and Karnataka. Sorghums and millet are observed as significant coarse cereals for food and nutrition. Sorghum contains a very good amount of minerals such as potassium, phosphorus, calcium, and small amounts of iron and sodium. Presence of some micronutrients (iron and zinc) in grain makes it more valuable for reducing the problems of micronutrient deficiency and micronutrient malnutrition worldwide. Regular and proper use of sorghum in different food forms in the meals of pregnant ladies fulfills the need for nutrients and micronutrients.

Pearl millet is an important millet which is widely grown by farmers. The history of farming of pearl millet in Africa and India is not new but is reported from ancient times.

Pearl millet is tall, grows up to 8–16 ft in height, and can also grow in nutrient deficient, sandy soil conditions in low rainfall zones. It contains a high amount of phytochemicals and micronutrients as folic acid, vitamins E, B complex, magnesium, copper, and zinc which are suitable for heart and cholesterol-related issues. It has also a very high energy value in comparison to any other flours. Presence of high calcium and unsaturated fats in grains are responsible for good health.

Proso Millet is also known as common millet and treated as the most healthy and wonderful millet. It is a short season plant and mature within 60–70 days after planting and fit for various soil types and weather circumstances. Its seeds are round with a smooth and silky hull. The grain of Proso Millet is rich in indigestible fiber in comparison to other millet. The seeds are surrounded in the hulls and due to this, it is very difficult to remove by any conventional milling procedures.

Finger millet is a tillering plant which bears finger-like terminal and minor seeds. Maturity period varies from 3 to 6 months depending on millet varieties, climate, and other growing environments. It is mainly grown in dry regions, particularly the southern part of India. It is rich in essential amino acids, vitamin A, vitamin B, and phosphorus and also comprises a great quantity of calcium. In some regions of India, Finger millet flour is used in the preparation of several nutritious foodstuffs as ragi balls. The presence of a high amount of fiber is useful in digestive problems, blood cholesterol, and cancer.

Foxtail millet grain is one of the oldest cultivated millet and is known for gluten-free grain. It can grow in low watery conditions with a narrow root. The maturity period of this crop is short and it matures in 65–75 days. Seeds of this crop are spike-like structures with flattened panicles and show resemblance with rice. Seeds also contain an outer thick husk layer. Little millet can grow in very adverse conditions of drought and waterlogging. It is commonly grown as a mix cropped with pulses, oilseeds, and millets. The seeds of this millet are not large and small as compared to other common millets.

Kodo Millet (*Paspalum scrobiculatum*) is a native of tropical Africa and is known as one of the ancient millet, and reported to origin in India before 3000 years back (De Wet et al. 1983). Kodo Millet is a key food in some parts of India and is used usually as nutritious and healthy foods in rural zones of India (Hegde and Chandra 2005).

7.2.2 Nutritive Summary

Millets are rich in nutritional compounds and have a higher value than wheat and rice specifically in terms of minerals and nutraceutical contents. In case of all the millets, the total fiber contents are very high than wheat and rice. In case of foxtail and little millet, the iron content of grain are also higher than rice. Most of the micronutrients which cause deficiency problems and malnutrition are rich in millets. In case of β-carotene mostly obtained from pharmaceutical sources is abundantly found in millets. In the perspective of micronutrients, each millet is very important and every millet is reported to be higher level to wheat and rice, therefore millet may be a good option for mineral malnutrition for majority of people of India.

7.3 Role of Processing for Food Uses

The seed coat of millets is very tough and mainly linked with flavor (Malleshi et al. 1986) and non-convenience of processing methods. The availability of processed millet is limited which is a major reason for little recognition of millets (Table 7.1). There are many types of machinery available for the processing of many cereals but poor availability of proper machines for millets processing.

The nutritional and functional properties of millet are required urgently for value addition and development of processed millet for nutritional and healthy food development and to satisfy the needs of consumers. In the current scenario, technology and skill for developing varieties of suitable food products specifically healthy food are changing. Proper processing methods with minimal nutritional loss are

Table 7.1 Parameters and impacts for food development

Parameters	Impacts
Enhancement of nutritional availability	• Make more digestible nutrient form • Nutritional fortification, e.g., thiamin
Ready to Eat (RTE) and Convenience	• Nutrient-rich easy and ready to eat options
Digestibility of millets	• Formulate food grade with good digestibility products
Safe food	• Without any harmful compounds • Acceptable microbiological level

Table 7.2 Some important biochemical value for dehulled millets (%)

Millet	True digestibility (%)	Biological value (%)	Net protein utilization (%)	References
Pearl millet	94.6	58.8	55.7	Singh et al. (1987)
Foxtail millet	95.0	48.4	46.3	Geervani and Eggum (1989)
Little millet	97.7	53.0	51.8	Geervani and Eggum (1989)
Kodo millet	96.6	56.5	54.5	Geervani and Eggum (1989)
Proso millet	99.3	52.4	52.0	Geervani and Eggum (1989)
Barnyard millet	95.3	54.8	52.2	Geervani and Eggum (1989)

required by the manufacturer. In case of millet during processing partial modification of some ingredients is possible which affects the edible portion and cost of products.

Currently, many suitable old methods are familiar in some regions of semi-arid tropics where millets are primarily utilized as a food source. In this process some are manual and some are labor-intensive. These methods are not suitable for complete decortication of millet grains and for food consumption after processing. Some manufacturers are not able to process the grain and they directly dry-mill the whole grains and utilize it for the development of several food products (Table 7.2).

7.3.1 Millet-Based Complementary Food

The availability of locally formulated complementary food is limited. Staple foods and components are required all-time for preparation of all types of food products (Samson 1993). The available complementary food with cereal-base are poor in protein quantity and limited in essential amino acids essentially lysine and tryptophan. Use of available legumes in cereal mixture increases the protein content and protein quality specifically essential amino acids. The food formulated at local level or home level must follow the following conditions: Great nutritional value, suitability, low cost, and use of local food items (FAO/WHO/UNICEF 1971; Dewey and Brown 2003).

Several studies found that in case of micronutrient malnutrition the case of iron and zinc deficiency is predominant in children. Due to being rich in micronutrients, utilization of millets will be effective in development of complementary foods (Pelto et al. 2003; Getahun et al. 2001). Some important processing methods have been found to be effective in reduction of anti-nutritional compounds in millet.

Use of the germination process particularly has been found to increase the nutritive value in pearl millet seeds. Germination process increases the accessibility of crucial nutrients and reduces the quantities of anti-nutritional compounds. Millet flour is the best and suitable material for different types of food preparations (Hassan et al. 2006). Therefore, use of millet grains for the development of value-added and health food is a suitable and best approach which nowadays results in high demand for millet-based food in urban areas and also for nontraditional millet users (Obilana 2013). Lack of suitable complementary food and feeding process are major reasons for increasing the problem of under and malnutrition (Villapando 2000; Daelmans and Saadeh 2003). Millet-based complementary food formulation can play a very useful alternative for infant-based food for the low-income people. The formulated food product can be beneficial in avoidance of protein and energy malnutrition due to use of nutrient-rich components. Accordingly continued practice of millet grains in place of other cereal flour, cereal-based mixture, formulated complementary formula, and other blends will be used for development of high quality, safe and longer duration of food products at local level and also helpful in marketing of millet and millet-based formula.

7.4 Utilization of Sorghum and Millets in Food Application

Different regions of the country have their own food habits. The important approach for fighting malnutrition problems and hunger is to design and modify the local recipes as per the populations. Value addition in traditional types of existing recipes using sorghum and millet will be the best strategy for nutrition security. Some important and popular recipes are selected for value addition through millets, fruits, vegetables for enrichment of quality protein, micronutrients, and vitamins.

7.4.1 Sorghum and Millet

Several researches concluded about sorghum biochemical composition and indicated that sorghum has a good amount of quality protein, vitamin B1, B2, niacin, and micronutrients as iron and zinc. In some findings, it is found that sorghum is inexpensive source of iron and zinc after pearl millet (Parthasarathy Rao et al. 2006). The uses of sorghum in different regions of the world vary as per the choice of populations. In Asian countries, it is mainly utilized in food and industries and in African countries it is mainly utilized for food purposes. In European countries, North America, and Australia the sorghum grain is mainly utilized for the purpose of feed. In some developing countries, sorghum is used as livestock feed in some months when the weather is dry.

In the majority of developing countries, millets are preferably used for food purposes. Millet grains are very rich in energy, quality nutrients, and nutraceutical compounds. Currently, millet plays a major role in formulation and development of high market demand for healthy and baby foods. Millet grains are also used as feed in some of the regions of some countries. In the current scenario, the utilization of millet is changing and some countries also shifted for some more uses as alcohol production, feed for poultry purposes and livestock, etc. Millet utilization pattern is altering in different countries and they are shifted to some other applications as alcohol manufacture, livestock, and poultry feed. In some regions of Africa and Asia millet is significant source of feed in post-monsoon seasons.

7.4.2 Uses in Food

Sorghum crop is one of the staple crops in African countries and also in some important regions of India. In these regions, it is consumed as traditional food. The uses of sorghum for food purposes are highest in case of African countries. In Asian countries, the trends of consumption of sorghum as food are decreasing due to increase in income, economic development, and consumer liking for food. In some urban regions of India although it is a staple crop its use and acceptance are decreasing for some years. The major reasons for this decline are substitutions of sorghum grain with some fine cereal grains, increasing pattern of income, and some government policies which favor the use of other grains.

The utilization of millet grains for food purposes in African and Asian countries is not favorable; therefore, use in food is limited. In the current scenario, the demand and use for food purposes increases and favors and increases in demand from African countries in the past two decades. The case of utilization of millet for food uses and demand in Asian countries has been decreasing for several years. In some countries, these trends affect the economy and production.

7.5 Value-Added Products of Sorghum and Millets

7.5.1 Conventional Food Products

Sorghum and millets are grains for very high energy which are exclusively suggested to all age groups. Small millet flour is very similar to rice in many cooking properties and can be used as a good alternative for the preparation of several types of food. In India, several types of food products are prepared at home in rural and urban regions.

7.5.1.1 Roti

Some important food products prepared using millets are roti, mudde, and porridge (Devi et al. 2014). Millet grains are gluten-free protein; therefore, it is not suitable for complete food product formulations and can be used only as a component for preparation of bakery products. In case of preparation of some important products as roti, the hot water is used with millet flour which favors the partial gelatinization of starch. This process is useful in necessary binding and involves rolling the thin leaves. In case of millet flour put off in cold water containing a little buttermilk and gone overnight which supports mild fermentation. After overnight we can use the slurry for preparation of porridge.

7.5.1.2 Multigrain Flour

Flours developed using several grains as multigrain are also known as composite flours. Composite flour is prepared by blending processed and unprocessed grains and pulses and this fulfills the major need of nutrition. The utilization of processed sorghum in sorghum enriched multigrain flour formulation is a very good option for enhancing taste, nutritional, and nutraceuticals value of roti (Rao et al. 2014) (Fig. 7.1). It is found that preparation of flour mixture using finger millet and wheat in the ratio of 3:7 develops favorable semi-finished products which are suitable for making chapatti. In finger millet fortified chapattis, it is found to increase taste as well as antidiabetic properties and found effective in diabetic patients (Ravinder et al. 2008). Composite flour with high fiber content is efficiently useful in constipation problems (Cade et al. 2007).

7.5.1.3 Fermented Foods

Fermentation is an important process which plays several major roles in food processing sectors. In this process, several anti-nutritional compounds which are responsible for nonacceptance of the grains are affected and their level reduces which improve the nutrition value, taste, and other food properties (Varma and Patel 2012). Several fermented foods that are very familiar in most parts of India specifically in South India are very useful for breakfast purposes some of them are dosa and idli. These fermented foods are good examples of fermented food where we can replace rice during preparation. During preparation, millet and black gram are mixed and ground and the resulting mixture is kept for fermentation overnight. The overnight fermented mixtures are steamed and prepared as dosa. Some other sorghum and millet fermented ready-to-cook food products are developed which are given in Figs. 7.2, 7.3 and 7.4.

7.5.1.4 Parboiled Millets Products

Parboiling is a familiar process mainly used in the processing of rice grain which improves the biochemical, milling properties, and yield. In some research, it was also

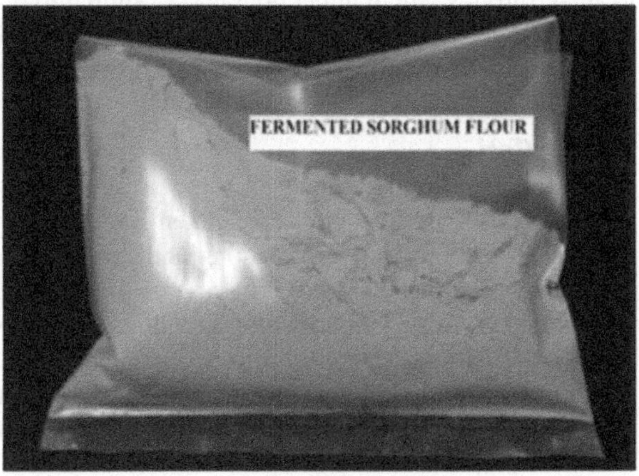

Fig. 7.1 Sorghum flour

Fig. 7.2 Sorghum masala

Fig. 7.3 Upma Mix

Fig. 7.4 Kodo Halwa mix

found that parboiling also affects the phenolic and antioxidant properties of the grain. This process is also beneficial to conserve nutritional compounds of grain from outer coating and bran of the grain. In some research, it was also stated steam treatment process of millet specifically finger millet supports the endosperm part and helps in grits production. Shreshta (1972) found that parboiling of kodo millet upgraded its milling quality. Similar to parboiled rice millets can also be used for development of ready-to-cook products.

7.5.1.5 Papad

In the current scenario, development of several processed food products for food uses are very fast growing some of them which are very popular such as chakli, papad, and idli. Several varieties of sorghum and millets are developed in India and have high yield and biochemical ingredients. Sorghum and millet might be very useful with higher demand in the upcoming time for development of the above types of food products for local and industrial uses.

Papad is one of the very popular food products in India used traditionally from a long time. Begum (2007) reported that adding finger millet flour for preparation of papad maximum up to 55–60% gives proper texture. Papad preparation methods include cooking millet flour in water to gelatinization, preparing thin pieces, rolling the dough, and drying of pieces to achieve moisture maximum up to 6%. Papad is a ready-to-cook product whose dark color changes to lighter color during frying (Varma and Patel 2012).

7.5.2 Nonconventional Food Products

There are several reports available which describe utilization of sorghum and millets in several traditional and healthy foods. Sorghum grains are used successfully in development of several snack foods as cookies, pasta, cake-like food products. Developments of such types of food products are very challenging without wheat and rice. Some additives are required during preparation of such food materials as starches, fat, and hydrocolloid for improvement of quality in case of some products. Sorghum grain might be significant for production of bioethanol and other industrial goods. Many research workers are trying to develop various demanding food products as flaked products, extruded products, popped products, fermented, malted, and composite flours; weaning foods, etc.

7.5.2.1 Flakes Food Products

The demand for sorghum and millet-based flakes products are increasing due to nutritional and nutraceutical behavior of such types of products. Sorghum and millets are exploring for development of ready-to-cook (RTC) flakes products to fulfill the desires of present users. Sorghum and millet flakes are prepared by using the following steps; debranne the millet, add in hot water, and cook for 10 min. Proper cooking procedures are adopted for all types of millet flakes (Rao et al. 2016).

The comparatively lesser mass and rapid hydration process of millets are responsible for preparation of suitable flake products.

7.5.2.2 Popped Products

Popping process is very common in the majority of cereals industries to prepare ready-to-eat (RTE) types of food products. After popping, the grain becomes crunchy and porous and also with improved taste and flavor. Popped products developed using finger millet have very acceptable flavor. In some of the regions of India popped products are available for marketing at small level food industries. In some study, it was found that a grain with moisture of around 18% and popping temperature up to 250 °C will be suitable for the preparation of complete expanded millet products (Malleshi and Desikachar 1981). The important factors responsible for optimum flattening of the grains are their shape and moisture. In case of decorticated finger millet, high temperature and quick time are desirable for expanded products and finally ready-to-eat products (Ushakumari et al. 2007).

7.5.2.3 Weaning Food

Most developed weaning foods are used as complementary food. The major aim of development weaning foods is to provide proper nutrition, taste, and safety aspects of baby food. The important cereals used in India are mainly staple food that provide energy and nutrients in human diets and comprise major components in daily diet intake. For baby complementary food it is essential to formulate energy-rich, nutritionally balanced, highly digestible, and with functional components. Millet, specifically Pearl Millet, is a staple food and favors the majority of the population in Asia and Africa. They are not only rich in protein and energy but also provide enough essential micronutrients.

Use of malting process is very common in some regions of India (Chandrasekhara and Swaminathan 1953). It is also established that the amylase activity in finger millet is higher than sorghum and other millets (Senappa 1988). Malleshi and Desikachar (1986) also found finger millet is suitable for development of weaning food because malt of finger millet is highly acceptable with high value of starch digesting enzymes. The higher level of enzyme activity is achieved at a very short time of 3–5 days of germination. Due to high levels of micronutrients specifically calcium and essential amino acids it can be used as a suitable ingredient for development of weaning food (Malleshi and Desikachar 1986). Malting processes not only increase the digestibility and nutritional properties but also reduces the level of anti-nutritional compounds (Desai et al. 2010).

7.5.2.4 Noodles-Vermicelli

Development of value-added products is one of the major challenges in food industries. Some novel process as extrusion technology is developed for restoring the ingredients in value-added products. Kurkure is one of the important and common products widespread in children who are developing through this extrusion process. Children and teenaged populations are very sensitive and choosy group who like such type of products and therefore demand of millet and millet-based noodles due to nutritional awareness in worldwide is very high.

Millet-based noodles are prepared by using a mixture of legumes and millet flours and can be introduced in weaning foods programs for energy and nutrition balance. Development of such types of food products is very economical at cottage level industry because all the ingredients and machines are very simple with very low amount of investment (Kumate 1983).

7.5.2.5 Bakery Products

Several bakery products available in the market such as biscuit, muffins, and bread can be prepared by using millet flour. Although millet flours are gluten-free which is not favorable for use of entirely pure millet ingradients for preparation of bakery and noodle products. Use of millet flour in preparation of bakery products enriches them in fiber and micronutrients which enhance the overall quality and value of products. In some studies, it was found that the addition of malted finger millet flour improves the nutritional and function properties of cake-like products (Desai et al. 2010). Many studies were performed for development of convenient food using millets and sorghum variety (Singh and Raghuvanshi 2012). Several studies used for formulation of food products using millets found that sorghum, oat, and millet can be used as suitable alternatives (Angioloni and Collar 2013). Eneche (1999) formulated biscuits using legume and millet flour and found higher acceptability of the product.

7.6 Process of Preparation of Some Millet Products

In the current scenario, it is found in several studies that small millets are superior to any conventional food because of the presence of several value-added compounds. It contains a higher level of some complex carbohydrates and starch (resistant) which make it favorable for diabetic patients. Millets are also rich in β-glucans which help in glucose metabolic pathways. The important factor which limits their utilization is flavors and shelf life of flour-based products. Therefore, use of suitable processing conditions, use of pulses, vegetables, fruits, and spices during preparation may be useful to overcome the problems of flavor and improvement in nutritional and functional properties of the products.

7.6.1 Little Millet-Based Products

7.6.1.1 Dosa

Samai dosa is prepared by using one cup of little millet, half cup dhal of black gram, half cup processed rice (puffed), one tablespoon of fenugreek, and a desired amount of salt to taste. Soaked seeds of fenugreek, black gram, and rice are used in the preparation process. All the soaked ingredients are grind properly and kept overnight for fermentation. Mix all the processed ingredients and prepare the dosa.

7.6.1.2 Porridge

Porridge is prepared using the following ingredients; little millet, coconut milk, and salt. Rice is cooked in water and added to coconut milk for 10–15 min, add the desired amount of salt and cool. The porridge is ready to serve.

7.6.1.3 Payasam

Payasam is prepared by using little millet flour, sugar, ghee, milk powder, and milk. The product makes it more healthy and tasty by adding fruits before serving.

7.6.2 Foxtail Millet-Based Products

7.6.2.1 Porridge

Porridge is prepared by using foxtail millet, sugar, powder form of cardamom, and cloves. Foxtail millet is boiled in water, add sugar and further boil and at the final stage use powder of cardamom and cloves. Porridge is served with milk in hot form.

7.6.2.2 Pongal

Pongal is a very tasty and South Indian millet-based product. It is prepared by using foxtail millet, jaggery, coconut, mixture of cloves and cardamom powder. In the presence of millet pongal becomes healthier and can be utilized by any age group.

7.6.2.3 Burfi

This is sweet in nature and very popular after meals in Indian families. Burfi is very healthy, tasty with good texture. It is prepared by using the following ingredients; foxtail millet, grounded nut, dry coconut powder, cardamom, and ghee. Foxtail millet is roast and dry in powder form, make syrup of jaggery and mix all the ingredients. The dried form of final mix ingredients is cooled and made into desired shape.

7.6.2.4 Kabab

Kabab is a very delicious and popular form of millet-based food products in India. The important feature of kabab is its nature of melting in mouth with unique flavor. This is a very familiar food in both vegetarian and nonvegetarian populations. Major ingredients are foxtail millet, potato, raw banana, chilli and cumin powder, oil, salts, and spices.

7.6.3 Finger Millet-Based Products

7.6.3.1 Malt (Ambali)

Malt is a very energetic and good snack for a healthy and active life. Major ingredients used in preparation are finger millet powder, buttermilk, and salt for desired taste. Overnight finger millet is cooked with buttermilk at low temperature and made into a thin porridge form.

7.6.3.2 Idli

Regular use of idli in breakfast is nutritious and healthy for snacks in India. Major ingredients used for idli preparation are whole finger millet grain, parboiled rice, black gram dhal, and fenugreek seed. Finger millet grains are soaked in water for a day and drain the water and develop white sprouts after 3–4 h. Rice and black gram are also soaked for 4–5 h and drain the water. Grind all the soaked ingredients and make light and fluffy materials and add desired amount of salt and kept for overnight. Make the idli using steam with the help of a pressure cooker.

7.6.3.3 Halwa (Pudding)

Finger millet-based halwa is tasty and healthy in use. It is one of the healthy, easy to prepare, and lovely recipes. Major ingredients used for halwa are whole finger millet flour, jaggery, cardamom powder, cashew nuts, and ghee. Finger millet flour is mixed with water and kept for 10–15 min and poured into the pan and add sugar. Cook with ghee and make a thick mixer and in the final mixture add cardamom and cashew nut (fried form).

7.6.4 Kodo Millet Recipes

Kodo millet is popular in tribal regions of India and they are cooked similar to rice and develop several types of products.

7.6.4.1 Papad

The major ingredients used for preparation of papad are kodo millet flour and black gram flour. During preparation of papad both the processed ingredients are used in the same amount and add cumin, sodium bicarbonate, and salt for taste and texture. Finally the dough is rolled and made into circular shape.

7.6.4.2 Vadagam

Vadagam is a traditionally prepared recipe. Usually, it is sundried and whenever required, deep-fried in hot oil. It is prepared by using Kodo millet flour, chilli powder, cumin powder, and desired amount of salt in water. Final product is allowed to sun dry and packed in an airtight container for higher shelf life.

7.6.4.3 Idli and Dosa

Idli and dosa are very common products in south India and are now getting popular in other regions of India. Kodo millet can be used to prepare idli and dosa, using 3:1 ratio with black gram dhal. This product is prepared after soaking and grinding which play crucial roles in preparation.

7.6.4.4 Thatuvadai

This is a unique product and prepared by using kodo millet, Bengal gram dhal, curry leaves, chilli powder, and butter. Airtight packaging improves product quality and shelf life.

Fig. 7.5 Kodo Kheer

7.6.4.5 Kodo Kheer

Kheer is a very popular food in all regions of India. Fermented kodo millet is nutritionally rich with a minimum amount of anti-nutritional compounds (Fig. 7.5). Kodo millet kheer is prepared by using fermented kodo millet, dry fruits, and milk with some flavoring ingredients.

7.6.4.6 Muruku

It is popularly known as chakli in some regions of India. The major ingredients used in preparation of chakli are kodo millet flour, sesame seeds, cumin, chilli, butter, and salt for desired taste. Using coarse kodo (rawa) vadai, upma, kesari Bhat, and cheela (Adai) are also other recipes that can be prepared. Kodo kheer and halwa are other sweet items, which could be prepared using kodo grits.

7.6.4.7 Vadai

It is a soaked kodo millet product. Kodo millet and other ingredients such as Rice and Bengal green dhal are soaked for 4–5 h and properly grind the soaked ingredients to make dense consistency and made into round shape.

7.7 Conclusion

The nutritional and health welfare role of sorghum and millet grains are already established which is comparable to other major crops. The household-level utilization of sorghum and millets are not satisfactory which is urgently required to improve. In the last few years, the importance of sorghum and millets as a staple food has been falling due to a number of reasons such as increasing the income of populations, urbanization, and some important government issues. Due to these reasons, these crops are not mainly used for food purposes but they are becoming alternatives as feed, processed food, and alcohol production in some industries. In some regions of India, the growth of these crops in value-added product development is increased due to health awareness. Development of suitable process technologies for farming, processing, and value-added products from these crops can overcome the limitation of these crops. Crop yield and yield stability in crop yield and improvement programs can play an important role to make these crops economical at farmhouse and end-use stages. Increasing use and demand of some coarse grains in the market for the development of healthy, gluten-free substitute, and value-added products with higher shelf life there is an urgent need for identification and development of processing technologies which will be helpful in development of healthy food with an increased shelf life of sorghum and millet food-based products. Establishment of a proper linkage between consumption of millet with health is urgently required for the promotion of sorghum and millets. It is also necessary to educate the farmers and other families to produce and use these grains for health and nutrition purposes. As per pearl millet which is popularized for health benefits, other millet varieties are also recommended for their health benefits mainly in development of complementary food development for malnourished populations and in mid-day meal programs.

References

Angioloni A, Collar C (2013) Suitability of oat, millet and sorghum in bread making. Food Bioprocess Technol 6:1486–1493

Balasubramanian S (2013) Processing of millets. Paper presented National Seminar on recent advances in processing, utilization and nutritional impact of small millets. Madurai symposium, Thamukkam grounds, Madurai, 13 September 2013

Begum JM (2007) Refined processing and products for commercial use and health benefits from finger millet. In: Krishne Gowda KT, Seetharam A (eds) Food uses of small millets and avenues for further processing and value addition. Project Coordination Cell, All India Coordinated Small Millets Improvement Project. ICAR, UAS, GKVK, Bangalore, India

Cade JE, Berley VJ, Greenwood DC (2007) Dietary fibre and risk of breast cancer in the UK women's cohort study. Int J Epidemiol 36:431–438

Chandrasekhara MR, Swaminathan M (1953) Enzymes of ragi and ragi malt 1. Amylases. J Sci Ind Res 36:191–196

Daelmans B, Saadeh R (2003) Global initiatives to improve complementary feeding. In: Moreira AD (ed) SCN Newsletter: Meeting the challenge to improve complementary feeding. United Nations System Standing Committee on Nutrition. Lavenham Press, UK, pp 10–17

De Wet JMJ, Brink DE, Rao KP, Mengesha MH (1983) Diversity in kodo millet, Paspalum scrobiculatum. Econ Bot 37(2):159–163

Desai AD, Kulkarni SS, Sahu AK, Ranveer RC, Dandge PB (2010) Effect of supplementation of malted ragi flour on the nutritional and sensorial quality characteristics of cake. Adv J Food Sci Technol 2:67–71

Devi PB, Vijayabharathi R, Sathyabama S, Malleshi NG, Priyadarisini VB (2014) Health benefits of finger millet (Eleusine coracana L.) polyphenols and dietary fiber: a review. J Food Sci Technol 5:1021–1040

Dewey KG, Brown KH (2003) Update on technical issues concerning complementary feeding of young children in developing countries and implications for intervention programs. Food Nutr Bull 24:5–28

Eneche EH (1999) Biscuit-making potential of millet/pigeon pea flour blends. Plant Foods Hum Nutr 54:21–27

FAO/WHO/UNICEF (1971) Protein-rich mixtures for complementary foods. Protein Advisory Group of the United Nations, PAG guidelines no. 8. FAO/WHO/UNICEF, New York

Geervani P, Eggum BO (1989) Nutrient composition and protein quality of minor millets. Plant Foods Hum Nutr 39:201–208

Getahun Z, Urga K, Ganebo T, Nigatu A (2001) Review of the status of malnutrition and trends in Ethiopia. Ethiopian J Health Dev 15:2

Hassan AB, Ahmed IAM, Osman NM, Eltayeb MM, Osman GA, Babiker EE (2006) Effect of processing treatments followed by fermentation on protein content and digestibility of pearl millet (Pennisetum typhoideum) cultivars. Pak J Nutr 5(1):86–89

Hegde PS, Chandra TS (2005) ESR spectroscopic study reveals higher free radical quenching potential in kodo millet (Paspalum scrobiculatum) compared to other millets. Food Chem 92 (1):177–182

Kumate J (1983) Relative crispness and oil absorption quality of sandige (extruded dough) from cereal grains. MSc Dissertation, University of Mysore, Mysore

Malleshi NG, Desikachar HS (1981) Studies on the suitability of roller flour mill, hammer mill and plate grinder for obtaining refined flour from malted ragi. J Food Sci Technol 18:37

Malleshi NG, Desikachar HS (1986) Studies on comparative malting characteristics of some tropical cereals and millets. J Inst Brew 92:174

Malleshi NG, Desikachar HS, Rao SV (1986) Protein quality evaluation of a weaning food based on malted ragi and green gram. Plant Foods Hum Nutr 36(3):223–230

Obilana AO (2013) Nutritional, physico-chemical and sensory characteristics of a pearl millet-based instant beverage powder. Doctoral dissertation, Durban University of technology

Parthasarathy Rao P, Birthal PS, Reddy BVS, Rai KN and Ramesh S (2006) Diagnostics of sorghum and pearl millet grain-sbased nutrition in India. International Sorghum and Pearl Millet Newsletter (ISMN) 45–47

Pelto GH, Levitt E, Thairu L (2003) Improving feeding practices: current patterns, common constraints, and the design of interventions. Food Nutr Bull 24(1):45–82

Rao BD, Kalpana K, Srinivas K, Patil JV (2014) Development and standardization of sorghum-rich multigrain flour and assessment of its storage stability with addition of TBHQ. J Food Process Preserv 39:451–457

Rao BD, Vishala AD, Christina GD, Tonapi VA (2016) Millet recipes-a healthy choice. ICAR-Indian Institute of Millets Research, Hyderabad

Ravinder KK, Jain R, Mridula D (2008) Impact of indigenous fiber rich premix supplementation on blood glucose levels in diabetics. Am J Food Technol 3:50–55

Saleh AS, Zhang Q, Chen J, Shen Q (2013) Millet grains: nutritional quality, processing, and potential health benefits. Compr Rev Food Sci Food Saf 12(3):281–295

Samson MA (1993) Mapping of Nigerian staple foods. National Diploma in Nutrition and Dietetics Project, Department of Food Technology, Kaduna Polytechnic, Kaduna, pp 4–5

Senappa M (1988) Sorghum and millets in East Africa with reference to their use in weaning foods. In: Meeting: Improving young child feeding in Eastern and Southern Africa: household level food technology, 12–16 October 1987, Nairobi, KE

Shahidi F, Chandrasekara A (2013) Millet grain phenolics and their role in disease risk reduction and health promotion: a review. J Funct Foods 5(2):570–581

Shreshta KB (1972) Dehusking of varagu (Paspalum scrobiculatum) and its utilization for edible purposes. MSc Dissertation, University of Mysore, Mysore

Singh P, Raghuvanshi RS (2012) Finger millet for food and nutritional security. Afr J Food Sci 6:77–84

Singh P, Singh U, Eggum BO, Kumar KA, Andrews DJ (1987) Nutritional evaluation of high protein genotypes of pearl millet (Pennisetum americanum L.). J Sci Food Agric 38:41–48

Ushakumari SR, Rastogi NK, Malleshi NG (2007) Optimization of process variables for the preparation of expanded finger millet using response surface methodology. J Food Eng 82:35–42

Varma V, Patel S (2012) Value added products from nutria. Cereals: finger millet. Emir J Food Agric 25:169–176

Villapando S (2000) Feeding mode, infections, and anthropometric status in early childhood. Pediatrics 106:1282–1283

8 Seed Storage Proteins and Amino Acids Synthetic Pathways and Their Regulation in Cereals with Reference to Biologically and Nutritionally Important Proteins and Bioactive Peptides in Millets

Anil Kumar, Kavita Gururani, Supriya Gupta, Apoorv Tiwari, Manoj Kumar Tripathi, and Dinesh Pandey

Abstract

Although the food systems in developing countries have changed dramatically since the green revolution, malnutrition still remains a challenge and is now known to include the concurrent dimensions of under-nourishment and micronutrient deficiency as a serious issue in developing as well as developed countries. An average cereal protein value of 10% will give us the total cereal protein production of approximately 17 million tons annually. The accumulation of seed protein is a complicated characteristic and seed storage proteins are proteins that considerably accumulate in developing seed, whose principal role is to behave as the nitrogen, carbon, and sulfur storage reserve. These proteins are mobilized quickly during the germination of seeds and are the principal cause of nitrogen reduction to increasing plantings. In particular, the enzymatic functions of seed storage proteins are not known although proteins are structurally distinct in storage from various crops, they all have certain prevalent features. Plant storage proteins may be categorized into two categories; proteins from seed storage

A. Kumar (✉)
Directorate of Education, Rani Lakshmi Bai Central Agricultural University, Jhansi, Uttar Pradesh, India

K. Gururani · S. Gupta · D. Pandey
Department of Molecular Biology and Genetic Engineering, College of Basic Sciences & Humanities, G.B. Pant University of Agriculture & Technology, Pantnagar, Uttarakhand, India

A. Tiwari
Department of Computational Biology and Bioinformatics, Jacob Institute of Biotechnology and Bio-Engineering, Sam Higginbottom University of Agriculture, Technology and Sciences, Allahabad, Uttar Pradesh, India

M. K. Tripathi
ICAR-Central Institute of Agricultural Engineering, Bhopal, Madhya Pradesh, India

A. Kumar et al. (eds.), *Millets and Millet Technology*,
https://doi.org/10.1007/978-981-16-0676-2_8

(SSPs) and plant storage (VSPs). SSPs are a group of proteins that accumulate in seeds at high concentrations in the late stages of seed development, whereas VSPs are protein accumulation in vegetative tissues, such as roots and tubers, based on plant species. SSPs are depleted during germination, and the subsequent amino acids are used as a food source by the growing seedlings. The most popular proteins in crops are the SSPs and the most commonly consumed plant proteins by human beings are crop proteins. Millets are considered as an enriched source of many essential amino acids derived from many quality proteins. According to World Health Organization, the proteins harboring more than 40% essential amino acids are called quality proteins and upon digestion and hydrolytic cleavage, several bioactive peptides having multiple health attributes are generated. In this chapter, we have briefly described about the proteins and peptides and their role in nutritional improvement present in millet.

Keywords

Seed storage proteins · Amino acids · Metabolic pathway · Cereals · Nutrition · Bioactive peptides and millets

8.1 Introduction

Malnutrition imposes serious issues in developing as well as developed countries which includes concurrent dimensions of undernourishment and micronutrient deficiency. It has been a major challenge for scientists to solve the problem of malnutrition. Half of the hunger is caused by protein malnutrition, according to the World Health Organisation. Protein-energy malnutrition (PEM) is a common disorder and is mainly due to energy, protein, and micronutrient deficiency. PEM and mineral malnutrition are mainly responsible for the high mortality ratio due to the onset of various diseases/illnesses. Cereals are the primary sources of food for the rural population and fulfill over half of the world's population's high dietary protein needs (Mandal and Mandal 2000).

Proteins are one of the main macromolecules present in the system and their deficiency may lead to improper muscle functioning, muscle wasting, stunted growth in children, increased risk of bone fracture, poor immunity, and so on. Amino acids are the building blocks of all proteins. There are 22 amino acids usually present in the proteins which are classified as essential and nonessential amino acids (NEAA), based on their synthesis in human beings. Nonessential amino acids are synthesized within the human system and therefore their deficiencies usually do not occur. On the other hand, essential amino acids (EAA) are not synthesized within the system therefore they are required to take from other sources. The deficiency of EAA may cause severe health issues. Figure 8.1 is showing some important symptoms of EAA deficiency. Cereal grains that are consumed directly in human food and feed, supplying more than 70% of the global calorie intake, also serve as primary reservoir of protein in the human food at global level. The

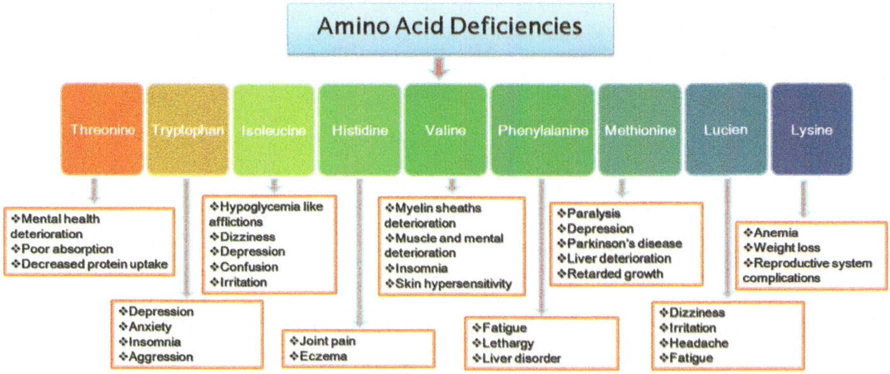

Fig. 8.1 Deficiency symptoms of nine EAA's showing their nutritional importance for human health

evergrowing demand for staple food crops for the increasing population is gradually being met by increasing production of grain seeds with an increased nutrient level (Martínez-Andújar et al. 2012). To meet this challenge, biofortification of cereals, i.e., wheat, rice, barley, and maize varieties and lines with high-quality proteins, high essential nutrients, and high yield is very necessary (Todorovska et al. 2005). In this chapter, we have discussed about the biosynthetic and regulatory pathways involved in the accumulation of seed storage proteins in cereals and also briefly touched on the importance of quality proteins stored in the seed endosperm of Poaceae family crops including millets.

8.2 Storage Proteins of Cereals and Millets

Cereals are the world's leading food crop and their seeds provide a perfect medium for manufacturing value-added food products, as they possess adequate and stable deposition of nutrients (Kawakatsu and Takaiwa 2010). As demand for food is growing worldwide, the attention paid to cereal proteins is increasing. The bulk of required human proteins are produced directly from livestock processing and from grain cereals. The most common protein foods are maize, rice, and wheat because of mainly synthezised and processed proteins in mature endosperm tissue. All of the proteins found in the mature endosperm are also known as the nonenzymatic storage protein for the exclusive purpose of supplying nutrition for germination and for development of a new plant (Doll 1977; Mandal and Mandal 2000). However, the endosperm features a wide variety of proteins different than formation, composition, and function (Doll 1977). Recently, seed proteins are classified into storage, biologically active, and structural proteins (Fukushima 1991). Storage proteins make up 50% of overall protein of mature cereal grains and hold a substantial effect on human and animal food nutritional quality and functionality in the production of food

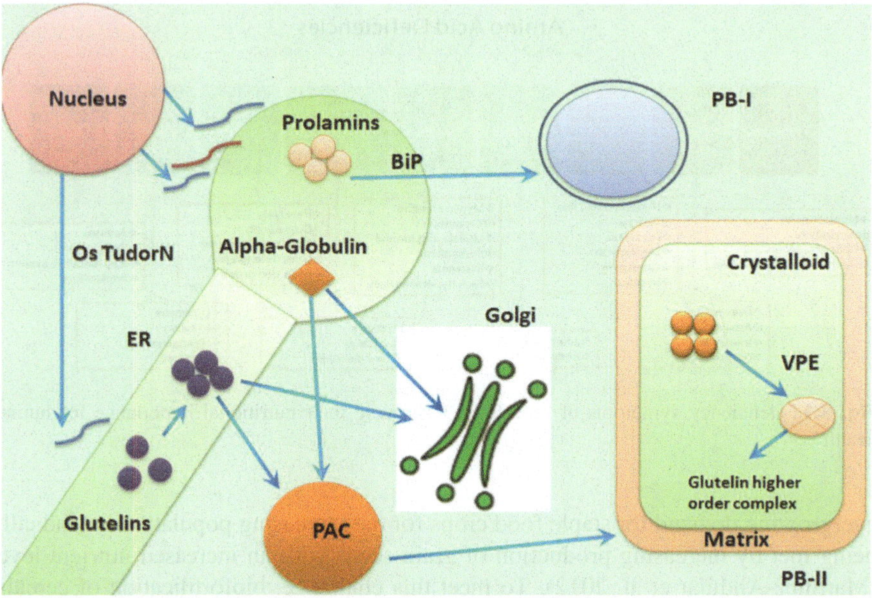

Fig. 8.2 Synthesis, trafficking, and deposition of different SSPs in rice

(Shewry and Halford 2002). These also acted as innovative products and balanced feeds in the processing of conventional local foodstuffs from ancient times. The typical cereal SSP classifications are based on their solubility are albumins (water-soluble), globulins (soluble salt), prolamins (alcohol-soluble) as well as glutelins (diluted acid/base soluble) (Landry et al. 2000; Osborne and Mendel 1914). The SSP structures generally based on DNA and protein sequencing are isolated and discovered by Shewry and Halford (2002). Though all grains commonly contain Albumins, prolamins and glutelins are frequently found in monocotyledonous and dicotyledonous grains, respectively. SSPs are of slightly different molecular masses between 10 and 100 kDa (Shewry et al. 1999). The SSPs are synthesized on the raw endoplasmic reticulum (ER) (Fig. 8.2), but the time the protein biosynthesis is at differing levels of seed processing. Studies of protein synthesis in maturing grains have shown that in the early stages of seed growth the fraction of albumin and globulin protein is synthesized and deposited in aleuronic tissue surfaces in cotyledon, while in later stages of seed ripening prolamin and glutelin are synthesized and located in protein body and matrix deposits. The use of available nitrogen for protein production by the highly productive cereals is usually very successful experiments that have persisted for so many years with the complicated aspects of their packaging, assembly, and partitioning (Doll 1977). The existence of retaining motifs, for example, for the zein motif, the cupin motif for globulin, and the intern series, followed by the hexapeptide repeat for albumin, characterize certain seed storing proteins.

8.2.1 Albumin

The most popular study of albumins in dicot seeds has been found in the Cruciferae and Arabidopsis. They have initially identified as 2s albums based on their sedimentation coefficients (S20wo) (Youle and Huang 1978). They are synthesized as single proteins and protected by the removal of both connection and short peptide from Aminoas well as Carboxyl terminus (Crouch et al. 1983). Both 2s albumins are compact globular proteins, with retained cysteine residues, despite their variations in structure and synthesis. In barley, the sum of albumins in general (3–5% of the total protein) is comparatively poor (Helm et al. 2004).

8.2.2 Globulin

Globulins are the most commonly distributed category of storage proteins, which can be classified into two groups according to their sedimentation coefficiency (S20. w); (a) vicilin-like globulars 7S (Templeman et al. 1987), (b) legumes 11S globulin type. Due to the post-translation process, both classes exhibit significant variance in their form and both types have cysteine and methionine lacking amino acids. In legumes, particularly peas, soybean, broad beans (We/afaba), and french beans were examined in greater detail for globulin storage proteins (*Phaseolus vulgaris*). The embryo and the outer layer of the endosperm contain Globulin proteins which have been more precisely characterized in maize embryo (Kriz 1989, 1999).

8.2.3 Prolamin

Prolamins serve as vital source of dietary protein for both livestock and human, besides being the main storage protein in most cereal seeds. These proteins are known as alcoholic-soluble (Osborne and Mendel 1914) proteins and have been named prolamin because of their high content of proline and glutamine (amid nitrogen). Cereal prolamins are found in the form of monomers or small compounds (Coleman and Larkins 1999) and these are different size and charge heterogeneous protein group (Esen et al. 1985). There are two typical structural features in most prolamins; firstly, there are different regions and domains which follow different structures and origins. Secondly, amino acid chains comprise of blocks reproducing or enriching them into Methionine-like amino acid residues, based on one or shorter peptide motif. These features arise due to increased concentrations of proline, glutamine, prolamine groups and some special amino acids including histidine, phenylalanine, glycine, and methionine (Shewry and Halford 2002). Because of its complexity and unique nomenclature, the nonexpert may have confusion on the prolamine structure and the properties. The existence of full amino acid sequences from all major prolamin groups made it possible for their structural and evolutionary relationship classification to be redefined (Shewry and Tatham 1990). The prolamin family and subfamily are named after each grass species. Table 8.1 provides a set of

Table 8.1 Classification of prolamins present in different species based on their evolution and homology among other crops

Subfamily	Tribe	Species	Gene family	Sub family	Prolamin group	References
Panicoideae	Andropogoneae	Maize (*Zea mays*)	Zein	α, δ	I	Lawton (2002)
				β, γ	II	
		Sorghum (*Sorghum bicolor*)	Kafirin	α, δ	I	Bietz (1982)
				β, γ	II	
	Paniceae	Foxtail millet (*Setaria italica*)	Setarin	α, δ	I	Monterio et al. (1982, 1987), Parameswaran and Thayumanavan (1995)
				β, γ	II	
Ehrhartoideae	Oryzeae	Rice (*Oryza sativa*)	Oryzein	10kd	I	
				13kd, 16kd	II	
Pooideae	Brachypodieae	Brachypodium (*Brachypodium distachyon*)	Brachypodin	Bra1, Bra2	II	
				Bra3	III	
	Triticeae	Wheat (*Triticum aestivum*)	Glutenin, gliadin	S-rich, S-poor	II	Shewry and Tatham (1990), Qi et al. (2012)
				HMW	III	
		Barley (*Hordeum vulgare*)	Hordein	S-rich, S-poor	II	Bietz (1982)
				HMW	III	

examples of prolamin and its names in various crops and even SSPs encoded by various sized multigene families (Feng et al. 2009).

8.3 Transcriptional Regulation of Seed Storage Proteins Mediated by Transcription Factors

Transcription factors are essential gene expression regulators comprising at least four discreet dominants, DNA-binding domain, nuclear localization signals (NLS), and oligomerization sites that function together to control a broad variety of physiological and biochemical processes in order to modulate the rates of transcription of the target genes. Transcription variables are also categorized according to their structural characteristics (a) Helix-turn helix (HTH), (b) Leucine zipper, (c) Zinc finger binds DNA, (d) helicopter-loop-helix (HLH), and (e) high mobility group (HMG) box motifs are the major transcription factors classified. DNA binds helix, leucine-pinching helix as a dimer while zinc finger can bind DNA both as a dimer and as monomers. Of these, the largest family of zinc finger proteins in eukaryotes can bind RNA, DNA, or nither. The term "zinc finger" refers to the sequence motifs in which zinc atom(s) are controlled by cysteines or histidines to form local peptide structures that are necessary for their particular functions. In several transcription factors that play a crucial role in interactions with other molecules involved in gene expression, the sometimes classified Zinc finger motifs dependent on the zinc-binding amino acids are frequently found (Yesudhas et al. 2017).

One of the most important zinc finger TF is Dof TF which highly plants specific and regulates the expression of SSPs. Dof TFs show specific binding with P-box and are identified as PBF (Prolamin Box Binding Factor). These proteins are found to be involved in activation of zein protein (prolamin) expression in maize seeds, in coordination with another basic leucine zipper transcription factor Opaque 2. Indeed the binding sites of these two transcriptional factors are just 20 bp apart on the Zein promoter suggesting their interaction in activating the zein gene expression. Opaque 2 (O2) is a transcriptional regulator for SSP gene expression (Schmidt et al. 1990) that encrypts basic bZIP (leucine zipper) transcription factor analogous to GCN4 of yeast has been shown to include amino acid metabolism. Orthologous genes from different crop plants like (SPA)/putative ESBF-II from wheat, BLZ2 from Barley (a bZIP TF 2) and RISBZ1 from rice (bZIP TF 1) were cloned (Onate et al. 1999). Typically the Opaque 2 resembling TFs bind to the GCN4 motif to trans-active the expression of SSP gene while in the α-globulin gene and the RISBZ2/REB binds with the ACGT motif for its activation (Nakase et al. 1996). GAMYB is another TF that shows interaction with DOF TFs including BPBF and SAD of Barley, RPBF/OsDOF3 of rice (Diaz et al. 2002, 2005). GAMYB expressed coordinately with DOF during seed maturation in both barley and rice. During seed germination GAMYB is coordinately expressed with SAD, while it acts antagonistically with BPBF in barley (Moreno-Risueno et al. 2007). Figure 8.3 is showing the TFs and their binding sites involved directly or indirectly in the regulation of SSP synthesis in cereals.

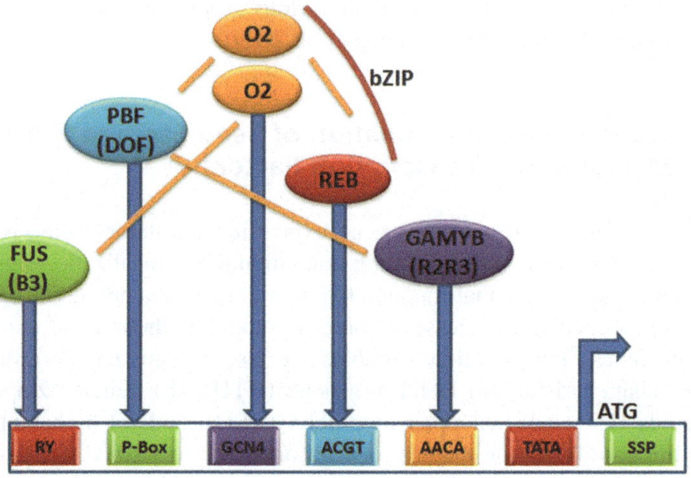

Fig. 8.3 TFs and their cis-regulatory binding sites involved in the synthesis and regulation of SSPs in cereals

8.3.1 DOF: A Zinc Finger Transcription Factor

The expression of plant genes comprises classes of transcription factors that have developed specifically to regulate plant genes and/or mediate a number of plant-specific signals. A typical example for Zinc finger transcription factors is the DOF (DNA link to one finger) family (Yanagisawa 2002, 2004). The DOF family is one of the most well-known transcription factors for plants with various roles (Fig. 8.3). The DOF (One Finger DNA-binding) proteins are defined as DNA-binding proteins, consisting of typically 200–400 amino acids with a strongly conserved Dof domain (Yanagisawa 2002). In maize, there was identified first protein MNB1a with Dof domain, which may connect with the CaMV(35S) promoter (Yanagisawa and Izui 1993). Every identified Dof transcription domain consists of 52 residues of amino acids which are homologous in different domains; however, there has been reported no homology outside these domains in these transcription factors. The lucid molecular description of different Dof domains indicates that it independently functions at structural and functional level (Yanagisawa 1996) (Fig. 8.3).

8.3.2 Dof TFs Involved in C:N Metabolism

Transcriptome studies confirmed that exogenous application of N regulates uptake and assimilation of N, and expression of regulatory genes coordinating C and N metabolism (Kumar et al. 2013). Photosynthesis stimulates the uptake and assimilation of N (Foyer and Noctor 2002) which ensures its correlation with C status. This linkage shows the importance of the study to explore the molecular mechanism and regulatory factors involved in C:N metabolism. There are so many regulatory factors and/or enzymes involved in C and N metabolism among them Dof TFs are the first plant species in its class to be identified as a master regulator to orchestra the C:N metabolism. Dof TFs also affect the NUE, by regulating the expression of multiple genes involved in C and N metabolic pathways and other plant developmental processes that are interrelated and influence the biomass, yield, and quality of a crop. Identification of cis-elements responsible for C and N signaling interactions (Palenchar et al. 2004) provides new avenues for manipulating the pathway to enhance NUE. Thus, alteration in the expression of Dof TFs might be a powerful approach for the generation of crops with enhanced NUE combined with superior agronomic traits and improved photosynthetic performance to achieve nutritionally superior crop with high yield.

The low protein content in most cereal grains with a low nutritional quality stimulated scientific efforts to genetically improve the protein at quantitative as well as qualitative level. Nutritional value for cereal protein has higher possibility for improving the low-prolamine, usually called high-lysine types. However, the observation that most high-lysine type cereals have low grain yield suggests that inhibition of prolamin synthesis also inhibits storing up of carbohydrates in the endosperm (Doll 1977). Although the improvement of quality is very important and necessary for the production of starch should not be compromised as it decreases the production of grain. Therefore, it is important that important nutritional traits of cereals should not be perturbed while improving the protein quality (Doll 1977).

8.4 Regulatory Elements/Genes and Proteins Involved in the Regulation of SSP Synthesis and Accumulation

In endosperm cells, Sulfur and Nitrogen are mainly deposited, in the form of SSPs, during cereal seed growth (Shewry and Tatham 1995). The encoding genes for SSPs are coordinated expressed in the development of endosperm, where spatial and temporal transcription regulation is carried out, by binding of transcription factors containing various cis and trans-acting motifs which bind to promoters of SSP genes. In the conferral of endosperm specificity on cereals, multiple consensus sequences in gene promoters are included (Albani et al. 1997; Mena et al. 1998; Carbajosa et al. 1997). The prolamin box (P-box) in the prolamin gene in barley, wheat, and other members of the Pooideae subfamily (Poaceae) is the most prominent of these cis-motifs. The P-box is based both on the highly conserved nucleotide sequence (5′-TGTAaAG-3′) and on the position (−300 region) in relation to the prolamin genes start translation codon (Fig. 8.4) (Forde et al. 1985).

Studies show the interrelation of these TF's with SSP's synthesis and accumulation and consequence of nitrogen availability on SSP's synthesis and accumulation (Schmidt et al. 1990; Albani et al. 1997; Mena et al. 1998; Yanagisawa and Schmidt 1999; Yanagisawa 2002; Yanagisawa 2004; Diaz et al. 2002; Kumar et al. 2018a, b). The availability of Zn may also affect the expression of these TFs and therefore the synthesis of SSPs (Riechmann et al. 2000). The SSPs are nutritionally more balanced in comparison to other proteins of cereals and millet as they have a high percentage

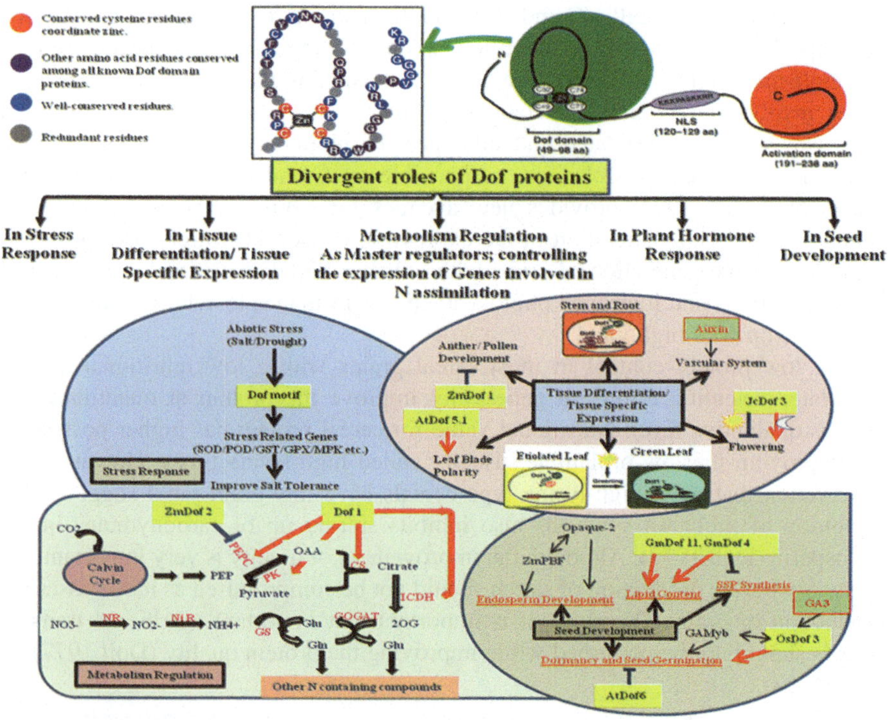

Fig. 8.4 A proposed structure of the Dof domain with N-Terminal amino acid sequence, representing divergent roles of Dof proteins

of essential amino acids which can be further exploited to develop value-added products to solve the problem of malnutrition and health problems.

The conserved P-box motif contains the recognition core for Dof proteins, i.e., (A/T AAAG) (Yanagisawa and Schmidt 1999), which regulates the expression of prolamin protein in coordination with Opaque 2. GCN4 motif provides the binding site to Opaque 2. Mutation analysis of P-box in zein promoter suggested that both PBF and Opaque 2 binding sites are necessary for zein protein accumulation in maize seeds (Carbajosa et al. 1997) and horde in protein accumulation in barley seeds (Diaz et al. 2002). This P-box is found to be conserved in all the prolamin gene promoters among cereals, including maize, rice, wheat, sorghum, barley, oat, and rye activate the transcription of prolamins by binding of PBF. So, the protein content can be improved by over-expression of PBF which in turn will increase the expression of the prolamin gene in cereal seeds and thus we can improve the protein content in cereal seeds (Diaz et al. 2002; Carbajosa et al. 1997).

The three storage proteins of millets are albumin, globulin, and prolamin among which prolamin is predominant. During cereal seed development, the genes encoding these seed storage proteins are simultaneously expressed in developing endosperm (Shewry and Tatham 1995). They are under transcriptional control, involving cis-motifs in their promoters and TFs. Few consensus sequences in gene promoters play an important role in conferring endosperm specificity in cereals (Albani et al. 1997; Mena et al. 1998). In the crops belonging to the Graminae (Poaceae) family, the most well-known cis-motifs is the prolamin box (P-box). The conserved P-box motif contains the recognition core for Dof proteins, i.e., A/T AAAG (Yanagisawa and Schmidt 1999). The Dof protein binds with P-box is identified as PBF (Prolamin Box Binding Factor) and found to activate the expression of prolamin protein in maize and barley (Carbajosa et al. 1997; Diaz et al. 2002; Kumar et al. 2018a, b). These PBF proteins which share a high degree of homology in their protein sequences with each other are specifically expressed during the grain filling stage. Recently 48 Dof genes have been reported in finger millet (Gupta et al. 2018). DOF TFs having DOF (DNA-binding with One Finger) domain is specific to higher plants. These have been shown to interact with both DNA and proteins, and enhance the expression of multiple genes implicated in C4 photosynthetic pathway or nitrogen assimilation (Yanagisawa 2002, 2004). Opaque 2 (O2) is a transcriptional SSP gene regulator that encodes abZIP (leucine zipper) transcription factor close to GCN4 yeast and has also been shown to be involved in amino acid metabolism. These O2-like TFs link to the GCN4 motif and switch SSP genes on expression.

8.5 Synthesis of Essential Amino Acids Determining the Accumulation of Quality Proteins

Nutritional quality of food is a crucial factor in maintaining human health and optimizing human genetic potential. In order to address the issue of deep-rooted food insecurity and starvation, the nutritional content of proteins should be improved (Fanzo 2015). Inherent factors are impacting the value of dietary protein in terms of amino acids affecting the dietary supply and the digestibility of the protein. Protein obtained from plant matter is limited to essential amino acids.

8.5.1 Regulatory Genes Involved in EAA's Metabolism

The crop or genotypes itself have high or low protein content and may be lacking in one or more amino acids. This amino acid deficiency can be resolved by combining two or three protein sources; however, antinutritive factors will still be present, limiting digestibility and bioavailability. Extensive attempts have been made to improve crops with these important amino acids by means of conventional breeding and mutagenesis. Additional attempts were made using genetic engineering methods to improve the production and reduce the metabolism of essential amino acids, as well as the expression of the recombinant proteins enriched with them, which were not sufficient and ineffective. It is therefore important to carry out a thorough analysis of the fundamental biological processes which are associated with the synthesis and aggregation of these amino acids in plants along with different factors linked with the biofortification of crop plants with these essential amino acids (Galili and Amir 2013).

Some regulatory enzymes of the EAA's biosynthetic pathways can alter the concentration of these EAA's by controlling their biosynthesis through a feedback regulatory mechanism. These enzymes are regulated by feedback regulatory mechanism in which the end products (intrinsic) of biosynthetic pathways inhibit the enzyme, e.g., Threonine is the −ve regulator of Homoserine Dehydrogenase (HD) and Aspartate Kinase (AK), which inhibit the biosynthesis of Threonine itself and Lysine, respectively. Besides feedback regulation other regulatory mechanisms are also operative during amino acid biosynthesis (Pratelli and Pilot 2014). It has been proved in different studies that external factors can also modulate the expression of these regulatory enzymes and therefore the synthesis of EAAs (Catala et al. 2007; Devoto et al. 2002). Results from the previous studies also show the possibility that the synthesis of EAA can be improved by altering the expression of these regulatory genes in response to exogenous application of fertilizers.

8.5.2 Millets: Source of Quality Proteins for Combating Malnutrition

Millets being rich in terms of proteins with essential amino acids, minerals like calcium, and phosphorus, vitamins of B group, and antioxidants have higher nutritional value than the common cereals like wheat, rice, and maize and are considered as nutri-cereals. Millets, minor small-seeded annual cereals, hold sixth rank in the world in terms of cereal grain production. They are climate-resilient crops because they grow in tropical and arid climates. Among the millets, Finger millet (FM) is minor millet and an ethnic crop of Uttarakhand which ranks fourth in importance among millets. Commonly known as Madua or Ragi, Finger millet (*Eleusine coracana*) (ragi) is highly nitrogen use efficient crop. It is a high nutritious value crop that has many valuable nutritional characteristics such as high amounts of minerals (Ca, Fe, Zn, etc.), vitamins, polyphenols, fiber, and high-quality grain proteins, which are a good source of essential amino acids. Despite growing under low or minimal nitrogen conditions, the protein content of finger millet grain is as good as that of major cereals such as rice and wheat which consume significant quantities of nitrogen fertilizers. Perhaps the finger millet has evolved special pathways for protein aggregation under the condition of limiting nitrogen conditions as compared to rice and wheat.

8.5.3 Nutritional Quality of Seed Storage Protein

Nitrogen is a crucial factor for the growth and production of plants used by young seedlings during germination and stored in the seed in the form of storage proteins. Roughly 16–50% protein is present in the seeds which include 1/3 of all dietary protein nitrogen (Graham and Vance 2003). Grain crop SSPs fulfill the main dietary protein needs of more than 50% of the world's population. Differences in synthesis and structure are of considerable significance for recognizing the capacity for greater nutritional benefit of SSPs in the endosperm. FM seeds grown under natural conditions produce 7–13% protein content of 100 g dry weight. Oat grain can have a high nutritional value relative to other cereals such as Barley and Wheat due to the high content of globulin storage proteins which implies a significant consideration for using it in livestock feed (Lockhart and Hurt 1986; Cuddeford 1995). Wheat SSPs include gliadins, glutenins, globulins, and albumins. "Legumin" type globulins found in wheat are called triticin, which accounts for around 5% of the total protein in wheat endosperm. It is recognized as nutritionally rich because a special decapeptide repeat pattern motif which is rich in lysine remains present in hyper-variable portion of this gene. Glutelin and globulin and protein fractions of barley possess high lysine content (about 5%) (Shewry et al. 1980). However, the horde in protein fraction of barley is especially rich in glutamine and proline and poor in lysine, i.e., approximately 0.8% (Shewry et al. 1980). Protein profiling of 52 FM genotypes was performed in our lab, which revealed no substantial variations in banding pattern between genotypes, whereas quantitative estimation of SSP fractions using Lowry method showed that glutelin was the highest followed by prolamin, globulin, albumin, and a significant difference was found based on comparative quantitative analysis of total seed protein. The content, yield, and composition determined the nutritional value and usefulness of SSPs (Chen et al. 2018). The essential amino acid content of SSPs of different cereals and millets has also been compared (Table 8.2) which reveals finger millet to be a nutritionally rich crop.

Table 8.2 Composition of essential amino acids in different cereals (mg/100 g)

Nutrients	Finger millet	Barnyard millet	Foxtail millet	Proso millet	Kodo millet	Rice	Wheat
Isoleucine	4.4	8.8	7.6	8.1	3.0	3.3	3.8
Leucine	9.5	16.6	16.7	12.2	6.7	6.7	8.2
Lysine	2.9	2.9	2.2	3.0	3.0	2.8	3.8
Methionine	3.1	1.9	2.8	2.6	1.5	1.5	2.3
Cystine	2.2	2.8	1.6	1.0	2.6	2.2	1.4
Phenylalanine	5.2	2.2	6.7	4.9	6.0	4.5	5.2
Tyrosine	3.6	2.4	2.2	4.0	3.5	3.0	3.9
Threonine	3.8	2.2	2.7	3.0	3.2	2.8	4.1
Tryptophan	1.6	1.0	1.0	0.8	0.8	1.5	1.4
Valine	6.6	3.8	6.9	6.5	3.8	4.4	5.5
Histidine	2.2	1.5	2.1	1.9	1.5	2.3	2.4

Adapted Gopalan et al. (1989)

8.5.4 Factors Affecting the Nutritional Quality of Seed and Proteins

The nutritional content of crops can directly or indirectly affected by several factors. These are soil influences, such as pH, abundance of nutrients, organic matter content, and soil-water relations; environmental and climatic conditions; crops and cultivars; and pesticide applications and cultural practices. Among all, the availability of nutrients and fertilizer applications and management affect the nutritional quality of field crops more directly. Some other factors regulate the nutritional quality of crops at the molecular level. These factors play a key role at transcriptional (promoters, transcription factors), translational, and post-translational levels (modification, refining, and deposition) (Kawakatsu and Takaiwa 2010) and influence the efficiency of seed w.r.t. yield, protein, and nutrient accumulation. Earlier studies into the nutritional status of crops produced using fertilizers are conflicting findings on seed yields and nutritional consistency (Leesawatwong et al. 2005). The supply of nutrients is an environmental variable while modern crop species are optimized for growth in soils with high abundance of nutrients and the genotype portion of G × E interaction is relatively stable, each cultivar has a different degree of efficiency in the acquisition and use of nutrients. It is therefore necessary to maximize the efficiency in which fertilizers are used in the production of crops. Studies also show that the applications of nutrients and fertilizers may alter the expression of genes involved in the acquisition of micro and macronutrients (Hammond et al. 2004).

Bogard et al. (2011) found that the increased absorption of N after synthesis was a significant factor in increasing the protein content of grain (GPC). They have shown that the synchronization of Nitrogen demand and supply in plants and the interaction between N supply and other environmental variables can affect GPC. However, generally, a negative correlation between grain yield and GPC has been observed and that may be an obstacle for improvement in protein accumulation. Since grain protein cannot adequately form without available N; therefore, a supply of N is a prerequisite for high protein yield. It has also been seen that changes to the availability of N to the plant at critical developmental stages can also improve the NUE (Kumar et al. 2013). Grain protein is directly related to the availability of soil nitrogen but the mechanism of Zn accumulation upon nitrogen application is not known. Therefore, the effect of Zn with nitrogen fertilizers/doses on grain protein quality and protein accumulation along with Zn is needed to be studied. Collecting knowledge on how N fertilization influences the aggregation of Zn in the shoot and the grain would further lead to a deeper understanding of the physiological and molecular processes underlying the relation between grain Zn and protein.

Millets are grown throughout the world due to the low farming costs, their biodiversity, such as finger millet (*Eleusine coracana*), foxtail millet (*Setaria italica*), pearl millet (*Pennisetum glaucum*), and proso millet (*Panicum miliaceum*) and the high biological characteristics. Low fat (1.5–5%) is the one character of millet grains but is rich in carbohydrates (60–70%), proteins (7–12%), and fibers (2–7%). They are a good source of magnesium, iron, and calcium. Proteins in millets are rich in essential amino acids that in other cereals and legumes are usually limited. Millet proteins are relatively rich in amino acids that contain sulfur which include methionine and cysteine. Millets are essential ingredients of a nutritionally balanced and healthy diet. The presence of proteins and BPs in millets helps improve the quality of their food protein, adding "functionality" to everyday consumption. It is well-known to be associated with reduction of cancer, cardiovascular disease, obesity, and diabetes and chronic disorders by taking diets rich in full-grain and related products. Because of their hypoglycemic property, their antioxidant activities millets can be used in the management of type 2 diabetes.

8.6 Bioactive Peptides

Peptides may have natural existance or they can be derived from native proteins' cryptic sequences, and are protein molecules smaller than 10 kDa. Bioactive Peptides (BPs) are low molecular weight organic substances composed of amino acids combined with the so-called covalent bonds of amide or peptide bonds. BPs are defined to promote heath with high tissue specificity and effectiveness. Peptides with physiological activity in the body consisted of 2–20 amino acids, which allows them to cross the intestinal barrier to have an effect on the level of tissue.

BPs have been detected in the hydrolysates and fermented milk products of enzymatic protein but gastrointestinal proteins digestion can also be responsible for their release (Gobbetti et al. 2007; Hartmann and Meisel 2007). BPs are mostly found by the digestion of microbial, plant proteolytic enzyme, and also by food processing through hydrolysis (cooking, fermentation, and ripening) BPs are considered as the next generation organic regulatory body because they not only preclude oxidation and microbial degradation in foodstuff but also improve treatment for different diseases and disorders (Lemes et al. 2016). The bioactive peptides display different biological effects of drugs, for example, antiproliferative, antimicrobial, ACE (inhibitory) effects, lowering cholesterol capacity, and antioxidant activities, improved absorption and bioavailability of minerals, and opioid-like activities. There is a dire need to identify various bioactive peptides that are present in various quality proteins of millets. Identification of such bioactive peptides in intact proteins having positive impact on body functions and influence health are given below:

Cardiovascular system: antihypertensive, antithrombotic, antioxidative, hypocholesterolemic activity.
Nervous system: antagonist activity and opioid agonist activity.
Gastrointestinal system: anti-appetizing, mineral binding, antimicrobial activity.
Immune system: immunomodulatory, cytomodulatory, antimicrobial activity.

8.6.1 Millet Proteins and Bioactive Peptides

Albumin, Globulin, cross-linking of prolamin, b-prolamin, and glutein are important constitutional fractions of Millet seed protein. However, these proteins relatively vary in their levels in different millets (Virupaksha et al. 1975). Millet protein amino acid analysis has found the presence of high amounts of amino acids which contain sulfur such as cysteine and methionine and various essential amino acids (Geervani and Eggum 1989). In our lab, efforts were made to focus on understanding the molecular basis of accumulation of quality proteins and accordingly several nutritionally important genes/proteins and promoters were identified. Besides, we are also interested to identify the EF-hand calcium-binding proteins for enhancing calcium accumulation and uptake in human gut.

Recently in our lab, few seed storage proteins have been characterized from finger millet that are nutritionally important as they contain elevated levels of essential amino acids like *Fima 1* lysine-rich gene (upto 9.0%), *Fimp2* methionine rich (upto 15.90%) besides two other Leu-Lys-Met-Pro and Leu-Lys-Val rich proteins having more than 40% EAAs (two patents filed). In order to understand the mechanism by which finger millet stores high calcium in seed, SEM-EDX analysis carried out in different finger millet genotypes revealed that maximum Ca is present in the aleurone layer, followed by seed coat and embryo (Goyal et al. 2005; Nath et al. 2013). It was therefore thought that chelators such as phytate, oxalate, pectate, and calcium-binding proteins (CBPs) may associate with calcium in these components during grain filling process (Goel et al. 2012). Although, phytates and oxalates are reported to be antinutrients, consuming finger millet has never been reported to cause negative health effects. Hence, research endeavors were mainly focused to identify genes encoding calcium-binding proteins involved in calcium sensing, transportation, and sequestration into cellular organelles. Subsequently, using conserved primer approaches, 3′ and 5′ RACE technique and developing seed transcriptome data, several genes encoding calcium-binding proteins such as calcium exchangers, calcium channel, calmodulin, Ca^{2+} ATPase and CaM dependent protein kinase (CDPK), calreticulin and calcinurin were identified in finger millet and their expressions studied in contrasting finger millet genotypes (Mirza et al. 2014; Singh et al. 2014, 2015; Chinchole et al. 2017).

These traditional functional genomics approaches have mainly investigated the changes in mRNA abundance, which due to transcriptional regulation and complex post-translational modifications of proteins, do not provide a true indication of protein expression and activity levels, respectively (Hittalmani et al. 2017). Only a few proteomics studies have so far been attempted to decipher the molecular basis of higher calcium build up in seeds of finger millet. This includes the identification of a 48 kDa Calreticulin as a CBP from finger millet using nano liquid chromatography–tandem mass spectrometry (Nano LC-MS) (Hirschi 2004). With influx of abundant genomic information and improvements in analytical technologies, proteomics has emerged as a powerful tool for studying different aspects of plant properties and growth regulations. Hence, the exploration of the seed proteome of finger millet not only appears to be essential to compliment the transcriptomics studies but also for the identification of new protein(s) with desirable functional characteristics such as nutraceuticals. Alternatively, the characteristic genetic traits of high nutritional value and stress tolerance make finger millet a particularly attractive system for proteomics study to analyze the proteins associated with nutritional quality and stress tolerance. In recent studies on seed storage protein in our lab, about 18 sequences of SSPs were recognized in the transcriptome of finger millet developing spikes namely LTP_DIR1, Germin-like_protein_1–3, Storage_protein, 11S_Globulin_seed_storage_protein, Glycine-rich_Protein-5, NSLTP_A, NSLTP_GPI-anchored_1, LTP_VAS, 19_kDa_globulin, NSLTP-3_like, Alpha-amylase_inhibitor-5, Vicilin-like_SSP, Glutelin_type-D-1, 10_kDa_prolamin, NSLTP-2, NSLTP-1, NSLTP-3, and NSLTP-4.

8.6.2 Characterization of Proteins Specifically Rich in Lysine, Arginine, and Methionine

BPs derived from millet seeds such as finger millet, Sorghum bicolor, buckwheat foxtail, pearl millet, quinoa and chia are described in Table 8.3 (Majid and Priyadarshini 2019).

8.6.3 Antimicrobial Peptides

A wide range of bacteria, fungus, and viruses is affected directly by peptides. The lengths (12–50 amino acids), compositions, charge, and presence of disulfide bonds vary from antimicrobial peptides. Antimicrobial peptides can kill bacteria directly by making pores through the membrane of the bacterial cord or by interacting in the microbial cells with macromolecules. There have been extensive studies on cereal and millet antimicrobial activity.

Camargo et al. (2008) reported antiviral peptides (2 kDa) which strongly inhibited replication of BHV (bovine herpes virus), HSV-1 (herpes simplex virus type 1), and weakly acted against polio virus. A 4 kDa buckwheat peptide is inhibiting the reverse HIV-1 transcriptase activity of *Mycosphaerella arachidicola* and *Fusarium oxysporum* in vitro as well. The antifungal activity in these peptides was caused by glycine and cysteine-rich amino acids (Egorov et al. 2005; Fujimura et al. 2003; Leung and Ng 2007). Fa-AMP1 and Fa-AMP2 are two antibiotical peptides of Buckwheat which have shown a broad range of antimicrobial activity against various gram-positive, gram-negative bacteria, and plant pathogens (Fujimura et al. 2003).

Peptides derived from foxtail millet had a different range of antimicrobial and antifungal activity (Xu et al. 2011). FFMp4, FFMp6, and FFMp10 three peptides derived from foxtail millet were strong antibacterial in *E. coli* ATCC 8099. Also peptides, as isolated from barnyard millet, finger millet, and proso millet reveal activity against *Pseudomonas areuginosa* (Bisht et al. 2016). Millet-derived BPs have high antivirus, bacterial, and fungal activity.

Table 8.3 Biologically active peptides in different millets and cereals crops

Source	Peptide sequence/ hydrolysates	Precursor protein	Bioactivity/against	References
Jower	Protein hydrolysates	Acid-soluble protein	Antiviral activity/herpes simplex virus type 1 (HSV-1), bovine herpes virus (BHV), Poliovirustype 1	Camargo et al. (2008)
Jower	VAITLTMK and VSKSVLVK	Total protein	Antioxidant activity	Agrawal et al. (2017)
Buckwheat	Protein hydrolysates	Total protein	Antifungal activity/ *Mycosphaerella arachidicola* and *Fusarium oxysporum*	Leung and Ng (2007)
Buckwheat	Protein hydrolysates	Total protein	Antiviral activity/HIV-1 reverse transcriptase	Leung and Ng (2007)
Foxtail	Protein hydrolysates	Total protein	Antifungal activity/ *Alternaria alternate*	Xu et al. (2011)
Finger millet	TSSSLNM VRGGLTR and STTVGLGISMRSASVR	Total protein	Antioxidant	Agrawal et al. (2019)
Buckwheat	Protein hydrolysates	Total protein	Antimicrobial activity/ *Agrobacterium rhizogenes* MAFF 210265, *Agrobacterium radiobacter* MAFF 520028, *Clavibacter michiganensis* subsp. *michiganensis* MAFF 301044, *Fusarium oxysporum* IFO 6384, *Geotrichum candidum* and *Curtobacterium flaccumfaciens* pv. *oortii* MAFF 301203	Fujimura et al. (2003)
Foxtail	Protein hydrolysates	Total protein	Antioxidant and antibacterial activity/ *E. coli* ATCC 8099	Mohamed et al. (2012), Amadou et al. (2013)
Pearl millet	SDRDLLGPNNQYLPK	Total protein	Antioxidant effect	Agrawal et al. (2016)
Foxtail	Protein hydrolysates	Total protein	Antihypertensive effect	Chen et al. (2017)
Buckwheat	WPL, VPW, VFPW	Total protein	Antioxidant effect	Ma et al. (2010)
Buckwheat	DVWY, FQ, VVG, VAE, GPP, DTPF, and WTFR	Total protein	Antihypertensive effect	Ma et al. (2006), Koyama et al. (2013)

(continued)

Table 8.3 (continued)

Source	Peptide sequence/ hydrolysates	Precursor protein	Bioactivity/against	References
Finger millet	Protein hydrolysates	Total protein	Antibacterial effect/ *Pseudomonas aeruginosa* (MTCC 424), and *Salmonella entrica* (MTCC 739)	Bisht et al. (2016)
Quinoa	Protein hydrolysates	Total protein	ACE-inhibitory activity and radical scavenging activity	Aluko and Monu (2003)
Buckwheat	Protein hydrolysates	Total protein	Anticancer/ antiproliferative against Hep G2 (hepatoma) cells, L1210 (leukemia) cells, breast cancer (MCF-7) cells, liver embryonic WRL 68 cells	Leung and Ng (2007)
Chia	Protein hydrolysates	Total protein	Antihypertensive effect	Segura Campos et al. (2013)
Quinoa	IQAEGGLT, DKDYPK, GEHGSDGNV	Globulin B	Antidiabetic effect	Vilcacundo et al. (2017)
Buckwheat	LQAFEPLR and EFLLAGN	Globulin protein	Antidiabetic effect	Wang et al. (2015)

8.6.4 Antioxidant Peptides

Oxidative stress causes the occurrence and onset spread of several diseases. Reactive oxygen species (ROS) are widely known to damage all macromolecules as well as lipids, protein, and DNA. Protein hydrolysis and digestion were used to produce antioxidant peptides from millets. Antioxidant properties are found in certain amino acids including Trptophan, Histidine, Lysine, and Tyrosine. Millets contribute significantly to the delivery of antioxidants such as tannins and phenolic acids which counteract in vitro and in vivo oxidative stress (Subba Rao and Muralikrishna 2002; Sudhakar et al. 2015; Han et al. 2018). An excellent antioxidant effect of a 14-Mer peptide (SDRLLPNNQYLPK) was demonstrated which was isolated from pearl millet (Agrawal et al. 2016). The antioxidant effect of 28 Leu-Leu-Pro-His-His synthetic peptides has been reported in soybean protein digestion. Among these peptides maximum antioxidant activity was reported in P-H-H tripeptide. Sorghum-isolating seven peptides, BPs, F2A, F2B, F2D, F2E, F3A, and F3B, of which 74.19% and 78.27% of free radical inhibition were F2B (VAITKTMK) and F3A (VSKSVLVC) higher in antioxidants compared with other peptides (Agrawal et al. 2017). In addition, antioxidant activity of finger millet BPs has also been reported by the same group. Studies related to molecular docking of the peptides (TSSLNMAVRGGLTR and STTVGLGISMRSASVR) showed that they interact with free radicals through the serine and threonine residue (Agrawal et al. 2019). Also, bioactive peptides derived from foxtail millet are enriched with Tyrosin/ Leucine and di, tri, and tetrameric peptides of buckwheat containing proline and tryptophane, and quinoa protein hydrolysates showed radical scavenging activities (Aluko and Monu 2003; Amadou et al. 2013). Together these studies show the application, and may assist in development of functional nutrients or functional foods, of millet-derived peptides to reduce oxidation stress and associated diseases.

8.6.5 Anticancer Peptides

Uncontrolled proliferation of cells is an important characteristic for cancer growth and development. Several BPs extracted from different sources demonstrated activity in anticancer (Wu et al. 2014; Wang et al. 2007). The proliferation of breast cancer (MCF-7), Hep G2 (hepatoma), and WRL 68 cells were inhibited by 4 kDa peptide derivative buckwheat with IC_{50} of 33 mM, 25 mM, and 37 mM respectively (Leung and Ng 2007). In addition, peptides such as LWREGM (F-1), DKDYPK (F-2), RELGEWGI (F-3), DVYSPEAG, and IFQEYI extracted from Quinoa have been suggested as potential anticancer agents due to their antioxidant and chemical protective activities. More studies are needed to identify potential BPs in millets and food products.

8.6.6 Antidiabetic Peptides

Blood sugar caused by inadequate insulin secretion is a metabolic disease. Type 1 insulin-dependent and type 2 noninsulin-dependent diabetes are the two types of diabetes. Millets have low rates of starch and protein digestibility in comparison to other cereals. BPs are isolated from plant source hydrolysates and have a key role to play in energy balance, insulin signaling, and resistance with a significant potential for development of suitable diabetes therapies (Chakrabarti et al. 2018). The hypoglycemic property of Millets has been attributed to the diverse factors, including polyphenols, fiber, interactions among starch, protein, and lipid and the intrinsic structural properties of starch (Kumar et al. 2018a, b).

The inhibitory activities of DPP4 showed peptides identified by oat, buckwheat, and barley proteins have varying IC (50) values. In addition to hydrolysates derived from millet and BPs, DPP4 displayed inhibitory effect in vitro (Vilcacundo et al. 2017). Millets-derived BP are therefore a significant reservoir of diabetes-targeting inhibitory DPP4 peptides.

8.6.7 Antihypertensive Peptides

Prolonged systolic blood pressure is a definite feature of high blood pressure. The renin-angiotensin pathway and kallikrein-kinnine system (KKS) have a vital role to play in regulation of blood pressure and maintenance of sodium homeostasis. The renin and ACE (angiotensin-converting enzymes) are the main enzymes which are involved in the renin-angiotensin system (RAS). ACE also degrades bradykinin, the central molecule of the KKS system, which is generated through the cleavage of kininogen kallikrein. Liberated BK stimulates nitric oxide (NO) and superoxide formation that leads to vasodilation (Erdos et al. 2010; Ceravolo et al. 2014). Different organic peptides, such as wheat, oat, barley, and rice, have been extracted from different cereals (Majumder and Wu 2014; Fan et al. 2019; Ganguly et al. 2019). Six BPs, viz., FQ, VAE, VVG, DVWY, WTFR, and FDART have decreased blood pressure from the lactic buckwheat sprouts (Koyama et al. 2013). The Quinoa-derived BPs and Chia derived from controlled protein hydrolysis significantly inhibited activity of ACE (Aluko and Monu 2003; Segura Campos et al. 2013). These findings demonstrate that RAS as well as KKS pathways are affected by ACE-inhibitory activity of millet BP and that it can be used to treat hypertension and related diseases.

8.6.8 Production and Processing of Bioactive Peptides

The BPs are dormant in the parent protein sequence and become active when released. The release of peptides can occur through digestive enzymes, microorganisms with proteolytic action and microbe or plant-derived enzymes derived in various ways (Wang et al. 2018). The functionality, efficiency of BPs and their absorption through the small intestine, and bioavailability in target tissues depend mainly on their amino acid composition, size, sequence and other parameters including charge, hydrophobicity, and hydrolysis rate (Shen and Matsui 2017; Ganguly et al. 2019; Wang et al. 2019). Therefore, their production is an important step, and research and attention need to be enhanced.

Fermentation using *Lactobacillus* spp. is one of the commonly used methods to release BPs from millet protein hydrolysate. The advantage of this method is that the main amino acid and environmental stability are not lost (Hajfathalian et al. 2018; Raveschot et al. 2018). Millet BPs was derived using enzymes (trypsin and pepsin) to imitate normal human digestion. The production of BPs from millets is challenging, as millets contain lower amount of protein than that from other sources (Agrawal et al. 2016; Bisht et al. 2016).

The sonication and hydrostatic pressure treatment have been shown to enhance protein hydrolysis and release of effective BPs (Li and Aluko 2010). After production, these bioactive peptides have high yield along with greater efficacy and are therefore important for downstream processing. Ultrafiltration techniques are also used to produce small molecular BPs in millets by using varying cutoff membranes of low molecular weight and chromatography with size exclusion. Reverse-phase HPLC is used to fractionate peptides based on hydrophobic properties of peptides (Pownall et al. 2010). A combination of different methods including conventional electrical dialysis, electrophoresis, and an ultrafiltration membrane with separation helps to achieve high efficiency in terms of concentration and desalination of peptides. This procedure was used to divide low molecular weight BPs in the range from 300 to 700 Da from flaxseed hydrolysate protein (Doyen et al. 2013; Firdaous et al. 2009).

Moreover, some affinity adsorbents, such as activated carbon, are also utilized to produce peptide fractions high in amino acids but low in aromatic amino acids (Adachi et al. 1992; Stone and Kozlov 2014; Dias et al. 2016). Pulse electric field was used to improve the ability of antioxidant properties in pine nut (KCHQP) and soybean peptides (SHCMN). The conditions in which BPs are processed to maintain their bioactivity and bioavailability are therefore important to optimize.

8.6.9 In Silico Approach to Identify Bioactive Peptides

Food-based peptides are potential functional food-ingredients for the incorporation of health-promoting diets targeted in the prevention of many chronic diseases. Conventional in vivo methods are tedious and time-consuming to produce, identifying and validating BPs. Due to the focused approach, lower time consumption, rapid results acquisition rate, and the resultant cost-effectiveness, bioinformatics/in silico analysis could be a powerful means for discovering BPs (Dziuba and Dziuba 2010). In order to identify potential BPs in food protectants, bioinformatics tools offer a great opportunity. Elucidating protein sequences is an essential step from BPs and molecular docking in the observation of protein–ligand interactions to determine the therapeutical potentials of novel BPs derived from foods. Many databases such as PepBank, PeptideDB, BIOPEP with antimicrobial peptide, APD, CAMP, and FeptideDB recorded the ample amount of bioactive peptides case already exist. The Web application includes tools to generate all possible peptides cleaved by different available enzymes from the input protein. FeptideDB is ultimately a computer aid for the evaluation of peptide bioactivities is freely available at http://www4g.biotec.or.th/FeptideDB/ (Panyayai et al. 2019).

8.7 Conclusion

Grains are the real basic food and frequently the main source of protein in several developing nations of Africa, Latin America, and Asia. Grains typically supply half of the people with dietary protein and can provide 70% of people's food in developing countries with protein intake. The seed protein ratios switch from 10% (in grains) of dry weight to ~40% (in certain plants and oilseeds) for an exceptional origin of dietary protein. The lack of protein contributes to protein exhaustion and a number of other issues impacting typical organic ability. Among minor grains, finger millet is recognized for its healthy quality (*Eleusine coracana* L. Gaertn). Finger millet calcium is 5–30 times higher than different oats. Finger millet has second main component in the form of grain protein content (GPC), which is 7% natural and ranges from 4.88% to 15.5%, supplying 44.7% of the essential amino acids. It also has various medical advantages including hypoglycemia, hypocholesterol, and ulceration. The yield is adapted to a wide range of situations, can resist critical saltiness, dry spell is moderately impermeable to waterlogging, and has a few true effects. Due to its high dietary benefit and storage features, the finger millet is developed primarily by subsistence farms and is filled as a food safety plant. In yield changes at molecular level, the enhancement of grain quality for food and feed is an important goal. Transcriptional and translational control has been shown to take a significant role in regulating seed development articulation (Curtis and Halford 2014).

At transcriptional level, genetic differences can be seen in the amalgamation of power proteins. Opaque 2 (o2) is an important translation factor (TF) of the leucine zipper (bZIP), binding to and regulating GCN4 value, cytosolic pyruvate cypd kinase (cyPDK) protein b32, and ribosomal inactivating properties. Several systems for DNA markers are actually used for respectable crop analysis. RFLP, AFLP, and Microsatellites or Single Sequence Repeat (SSRs) are the most widely employed marker structures (Gupta et al. 2017). The advantages of finger millet, such as hypoglycemia, hypocholesterol, and ulcerative behavior, are many. Grain is used as flour for cookies, bread, and other baked goods and is packed to provide good baby food (Ramakrishnan et al. 2016). In the coming very long period, the late released finger millet WGS will be helpful for all these assets to be examined for characterization of seed storage proteins, their amino acid composition, and also bioactive peptides present in the nutritionally and biologically important proteins can also be identified.

References

Adachi S, Yamanaka T, Hayashi S, Kimura Y, Matsuno R, Yokogoshi H (1992) Preparation of peptide mixture with high Fischer ratio from protein hydrolysate by adsorption on activated carbon. Bioseparation 3(4):227–232

Agrawal H, Joshi R, Gupta M (2016) Isolation, purification and characterization of antioxidative peptide of pearl millet (Pennisetum glaucum) protein hydrolysate. Food Chem 204:365–372. https://doi.org/10.1016/j.foodchem.2016.02.127

Agrawal H, Joshi R, Gupta M (2017) Isolation and characterisation of enzymatic hydrolysed peptides with antioxidant activities from green tender sorghum. LWT 84:608–616. https://doi.org/10.1016/j.lwt.2017.06.036

Agrawal H, Joshi R, Gupta M (2019) Purification, identification and characterization of two novel antioxidant peptides from finger millet (Eleusine coracana) protein hydrolysate. Food Res Int 120:697–707. https://doi.org/10.1016/j.foodres.2018.11.028

Albani D, Hammond-Kosack MC, Smith C, Conlan C, Colot C, Holdsworth V, Bevan MW (1997) The wheat transcriptional activator SPA: a seed-specific bZIP protein that recognizes the GCN4-like motif in the bifactorial endosperm box of prolamin genes. Plant Cell 9:171–184

Aluko RE, Monu E (2003) Functional and bioactive properties of quinoa seed protein hydrolysates. J Food Sci 68(4):1254–1258. https://doi.org/10.1111/j.1365-2621.2003.tb09635.x

Amadou I, Le GW, Amza T, Sun J, Shi YH (2013) Purification and characterization of foxtail millet-derived peptides with antioxidant and antimicrobial activities. Food Res Int 51 (1):422–428. https://doi.org/10.1016/j.foodres.2012.12.045

Bietz JA (1982) Cereal prolamin evolution and homology revealed by sequenc analysis. Biochem Genet 20:1039–1053

Bisht A, Thapliyal M, Singh A (2016) Screening and isolation of antibacterial proteins/peptides from seeds of millets. Int J Curr Pharm Res 8(3):96–99. https://doi.org/10.22159/ijcpr.2016v8i4.15271

Bogard M, Jourdan M, Allard V, Martre P, Perretant MR, Ravel C et al (2011) Anthesis date mainly explained correlations between post-anthesis leaf senescence, grain yield, and grain protein concentration in a winter wheat population segregating for flowering time QTLs. J Exp Bot 62:3621–3636

Camargo F, Cortez DAG, Ueda-Nakamura T, Nakamura CV, Dias Filho BP (2008) Antiviral activity and mode of action of a peptide isolated from Sorghum bicolor. Phytomedicine 15 (3):202–208. https://doi.org/10.1016/j.phymed.2007.07.059

Carbajosa J, Moose V, Parsons RL, Schmidt RJ (1997) A maize zinc-finger protein binds the prolamin box in zein promoters and interacts with the basic leucine zipper transcriptional activator Qpaque2. Proc Natl Acad Sci 94(14):7685–7690

Catala R, Ouyang J, Abreu IA, Hu Y, Seo H, Zhang X, Chua N (2007) The Arabidopsis E3 SUMO ligase SIZ1 regulates plant growth and drought responses. Plant Cell 19:2952–2966

Ceravolo GS, Montezano AC, Jordão MT, Akamine EH, Costa TJ, Takano AP, Fernandes DC, Barreto-Chaves ML, Laurindo FR, Tostes RC et al (2014) An interaction of renin-angiotensin and kallikrein-kinin systems contributes to vascular hypertrophy in angiotensin II-induced hypertension: in vivo and in vitro studies. PLoS One 9(11):e111117. https://doi.org/10.1371/journal.pone.0111117

Chakrabarti S, Jahandideh F, Davidge ST, Wu J (2018) Milk-derived tripeptides IPP (Ile-Pro-Pro) and VPP (Val-Pro-Pro) enhance insulin sensitivity and prevent insulin resistance in 3T3-F442A preadipocytes. J Agric Food Chem 66(39):10179–10187. https://doi.org/10.1021/acs.jafc.8b02051

Chen J, Duan W, Ren X, Wang C, Pan Z, Diao X, Shen Q (2017) Effect of foxtail millet protein hydrolysates on lowering blood pressure in spontaneously hypertensive rats. Eur J Nutr 56 (6):2129–2138. https://doi.org/10.1007/s00394-016-1252-7

Chen P, Shen Z, Ming L, Li Y, Dan W, Lou G, Peng B, Wu B, Li Y, Zhao D, Gao G, Zhang Q, Xiao J, Li X, Wang G, He Y (2018) Genetic basis of variation in rice seed storage protein (albumin, globulin, prolamin, and glutelin) content revealed by genome-wide association analysis. Front Plant Sci 9:612

Chinchole M, Pathak RK, Singh UM, Kumar A (2017) Molecular characterization of EcCIPK24 gene of finger millet (Eleusine coracana) for investigating its regulatory role in calcium transport. 3 Biotech 7:267. https://doi.org/10.1007/s13205-017-0874-7

Coleman CE, Larkins BA (1999) The prolamins of maize. In: Shewry PR, Casey R (eds) Seed proteins. Kluwer Academic, Dordrecht, pp 109–139

Crouch ML, Tenbarge KM, Simon AE, Ferl R (1983) cDNA clones for Brassica napus seed storage proteins: evidence from nucleotide sequence analysis that both subunits of napin are cleaved from a precursor polypeptide. J Mol Appl Genet 2:273–283

Cuddeford D (1995) Oats for animal feed. In: Welch RW (ed) The oat crop: production and utilization. Chapman & Hall, London, pp 321–368

Curtis T, Halford NG (2014) Food security: the challenge of increasing wheat yield and the importance of not compromising food safety. Ann Appl Biol 164(3):354–372. https://doi.org/10.1111/aab.12108

Devoto A, Nieto-Rostro M, Xie D, Ellis C, Harmston R, Patrick E, Davis J, Sherratt L, Coleman M, Turner JG (2002) COI1 links jasmonate signalling and fertility to the SCF ubiquitin–ligase complex in Arabidopsis. Plant J 32:457–466

Dias AMGC, dos Santos R, Iranzo O, Roque ACA (2016) Affinity adsorbents for proline-rich peptide sequences: a new role for WW domains. RSC Adv 6(73):68979–68988. https://doi.org/10.1039/C6RA10900D

Diaz I, Vicente-Carbajosa J, Abraham Z, Martinez M, Isabel-La Moneda I, Carbonero P (2002) The GAMYB protein from barley interacts with the DOF transcription factor BPBF and activates endosperm-specific genes during seed development. Plant J 29:453–464

Diaz I, Martinez M, Isabel-LaMoneda I, Rubio-Somoza I, Carbonero P (2005) The DOF protein, SAD, interacts with GAMYB in plant nuclei and activates transcription of endosperm-specific genes during barley seed development. Plant J 42:652–662

Doll H (1977) Genetic diversity in plants. In: Muhammed A et al (eds) 8. 978-1-4684-2888-9

Doyen A, Husson E, Bazinet L (2013) Use of an electrodialytic reactor for the simultaneous β-lactoglobulin enzymatic hydrolysis and fractionation of generated bioactive peptides. Food Chem 136:1193–1202

Dziuba M, Dziuba B (2010) In silico analysis of bioactive peptides. In: Mine Y, Li-Chan E, Jiang B (eds) Bioactive proteins and peptides as functional foods and nutraceuticals. Wiley-Blackwell, Oxford, UK, pp 325–340

Egorov TA, Odintsova TI, Pukhalsky VA, Grishin EV (2005) Diversity of wheat anti-microbial peptides. Peptides 26(11):2064–2073

Erdos EG, Tan F, Skidgel RA (2010) Angiotensin I-converting enzyme inhibitors are allosteric enhancers of kinin B1 and B2 receptor function. Hypertension 55(2):214–220. https://doi.org/10.1161/HYPERTENSIONAHA.109.144600

Esen A, Bietz JA, Pauls JW, Wall JS (1985) Isolation and characterization of a methionine-rich protein from maize endosperm. Cereal Sci 3:143–152

Fan H, Liao W, Wu J (2019) Molecular interactions, bioavailability, and cellular mechanisms of angiotensin-converting enzyme inhibitory peptides. J Food Biochem 43(1):e12572. https://doi.org/10.1111/jfbc.12572

Fanzo J (2015) Ethical issues for human nutrition in the context of global food security and sustainable development. Glob Food Secur 7(2015):15–23

Feng LN, Lu DQ, Bei JX, Chen JL, Liu Y, Zhang Y, Liu XC, Meng ZN, Wang L, Lin HR (2009) Molecular cloning and functional analysis of polymeric immunoglobulin receptor gene in orange-spotted grouper (Epinephelus coioides) Comp Biochem. Physiol B Biochem Mol Biol 154:282–289

Firdaous L, Dhulster P, Amiot J, Gaudreau A, Lecouturier D, Kapel R, Lutin F, Vezina LP, Bazinet L (2009) Concentration and selective separation of bioactive peptides from an alfalfa white protein hydrolysate by electrodialysis with ultrafiltration membranes. J Membr Sci 329 (1–2):60–67. https://doi.org/10.1016/j.mem-sci.2008.12.012

Forde BG, Heyworth A, Pywell J, Kreis M (1985) Nucleotide sequence of a B1 hordein gene and the identification of possible upstream regulatory elements in endosperm storage protein genes from barley, wheat and maize. Nucleic Acids Res 13:7327–7339

Foyer CH, Noctor G (2002) Photosynthetic nitrogen assimilation: inter-pathway control and signaling. In: Foyer CH, Noctor G (eds) Photosynthetic nitrogen assimilation and associated carbon and respiratory metabolism, Advances in photosynthesis and respiration, vol 12. Springer, Dordrecht. https://doi.org/10.1007/0-306-48138-3_1

Fujimura M, Minami Y, Watanabe K, Tadera K (2003) Purification, characterization, and sequencing of a novel type of antimicrobial peptides, Fa-AMP1 and Fa-AMP2, from seeds of buckwheat (Fagopyrum esculentum Moench.). Biosci Biotechnol Biochem 67(8):1636–1642. https://doi.org/10.1271/bbb.67.1636

Fukushima D (1991) Recent progress of soybean protein foods: chemistry, technology and nutrition. Food Rev Int 7:323–351

Galili G, Amir R (2013) Fortifying plants with the essential amino acids lysine and methionine to improve nutritional quality. Plant Biotechnol J 11:211–222. https://doi.org/10.1111/pbi.12025

Ganguly A, Sharma K, Majumder K (2019) Chapter 4: Food-derived bioactive peptides and their role in ameliorating hypertension and associated cardiovascular diseases. In: Toldra F (ed) Advances in food and nutrition research, vol 89. Academic, New York, pp 165–207

Geervani P, Eggum BO (1989) Nutrient composition and protein quality of minor millets. Plant Foods Hum Nutr 39(2):201–208. https://doi.org/10.1007/BF01091900

Gobbetti M, Minervini F, Rizzello CG (2007) Bioactive peptides in dairy products. In: Hui YH (ed) Handbook of food products manufacturing. Wiley, Hoboken, NJ, pp 489–517

Goel A, Gaur VS, Arora S, Gupta S, Kumar A (2012) In silico analysis of expression data for identification of genes involved in spatial accumulation of calcium in developing seeds of rice. OMICS J Integrat Biol 16(7–8):402–413

Gopalan C, Rama Sastri BV, Balasubramanian SC (1989) Nutritive value of Indian foods. Book; Government Publication.

Goyal K, Walton LJ, Tunnacliffe A (2005) LEA proteins prevent protein aggregation due to water stress. Biochem J 388:151–157. https://doi.org/10.1042/BJ20041931

Graham PH, Vance CP (2003) Legumes: importance and constraints to greater use. Plant Physiol 131(3):872–877

Gupta SM, Arora S, Mirza N, Pande A, Lata C, Puranik S et al (2017) Finger millet: a "certain" crop for an "uncertain" future and a solution to food insecurity and hidden hunger under stressful environments. Front Plant Sci 8:643. https://doi.org/10.3389/fpls.2017.00643

Gupta S, Pathak RK, Gupta SM, Gaur VS, Singh NK, Kumar A (2018) Identification and molecular characterization of Dof transcription factor gene family preferentially expressed in developing spikes of Eleusine coracana L. 3 Biotech 8(2):82

Hajfathalian M, Ghelichi S, Garcıa-Moreno PJ, Moltke Sørensen AD, Jacobsen C (2018) Peptides: production, bioactivity, functionality, and applications. Crit Rev Food Sci Nutr 58 (18):3097–3129. https://doi.org/10.1080/10408398.2017.1352564

Hammond JP, Broadley MR, White PJ (2004) Genetic responses to phosphorus deficiency. Ann Bot 94(3):323–332. https://doi.org/10.1093/aob/mch156

Hartmann R, Meisel H (2007) Food-derived peptides with biological activity: from research to food applications. Curr Opin Biotechnol 18(2):163–169

Han Y, Wu M, Hao L, Yi H (2018) Sulfur dioxide derivatives alleviate cadmium toxicity by enhancing antioxidant defence and reducing Cd(2þ) uptake and translocation in foxtail millet seedlings. Ecotoxicol Environ Saf 157:207–215. https://doi.org/10.1016/j.ecoenv.2018.03.084

Helm CV, DeFrancisco A, Gaziola SA, Fornazier RF, Pompeu GB, Azevedo RA (2004) Hull-less barley varieties: storage proteins and amino acid distribution in relation to nutritional quality. Food Biotechnol 18:327–341

Hirschi KD (2004) The calcium conundrum. Both versatile nutrient and specific signal. Plant Physiol 136:2438–2442. https://doi.org/10.1104/pp.104.046490

Hittalmani S, Mahesh HB, Shirke MD et al (2017) Genome and transcriptome sequence of finger millet (Eleusine coracana (L.) Gaertn.) provides insights into drought tolerance and nutraceutical properties. BMC Genomics. https://doi.org/10.1186/s12864-017-3850-z

Kawakatsu T, Takaiwa F (2010) Cereal seed storage protein synthesis: fundamental processes for recombinant protein production in cereal grains. Plant Biotechnol J 8:939–953

Koyama M, Naramoto K, Nakajima T, Aoyama T, Watanabe M, Nakamura K (2013) Purification and identification of antihypertensive peptides from fermented buckwheat sprouts. J Agric Food Chem 61(12):3013–3021. https://doi.org/10.1021/jf305157y

Kriz AL (1989) Characterization of embryo globulins encoded by the maize Glb genes. Biochem Genet 27:239–251

Kriz AL (1999) 7S globulins of cereals. In: Shewry PR, Casey R (eds) Seed proteins. Kluwer Academic, Dordrecht, pp 477–498

Kumar A, Kanwal P, Gupta AK, Singh BR, Gaur VS (2013) A full-length Dof1 transcription factor of finger millet and its response to a circadian cycle. Plant Mol Biol Rep

Kumar A, Jaiswal JP, Sharma N, Gupta S, Kumar A (2018a) Understanding the molecular basis of differential grain protein accumulation in wheat (Triticum aestivum L.) through expression profiling of transcription factors related to seed nutrients storage. 3 Biotech 8(2):112

Kumar A, Tomer V, Kaur A et al (2018b) Millets: a solution to agrarian and nutritional challenges. Agric Food Secur 7:31. https://doi.org/10.1186/s40066-018-0183-3

Landry J et al (2000) The silencing protein SIR2 and its homologs are NAD-dependent protein deacetylases. Proc Natl Acad Sci 97(11):5807–5811

Lawton JW (2002) Zein: a history of processing and use. Cereal Chem 79:1–18

Leesawatwong M, Jamjod S, Kuo J, Dell B, Rerkasem B (2005) Nitrogen fertilizer increases seed protein and milling quality of rice. Cereal Chem 82(5):588–593

Lemes AC, Sala L, Ores JDC, Braga ARC, Egea MB, Fernandes KF (2016) A review of the latest advances in encrypted bioactive peptides from protein-rich waste. Int J Mol Sci 17:950

Leung EH, Ng TB (2007) A relatively stable antifungal peptide from buckwheat seeds with antiproliferative activity toward cancer cells. J Pept Sci 13(11):762–767. https://doi.org/10.1002/psc.891

Li H, Aluko RE (2010) Identification and inhibitory properties of multifunctional peptides from pea protein hydrolysate. J Agric Food Chem 58(21):11471–11476. https://doi.org/10.1021/jf102538g

Lockhart HB, Hurt HD (1986) Nutrition of oats. In: Webster FH (ed) Oats: chemistry and technology. American Association of Cereal Chemists, Inc., St Paul, MN, pp 297–308

Ma MS, Bae IY, Lee HG, Yang CB (2006) Purification and identification of angiotensin I-converting enzyme inhibitory peptide from buckwheat (Fagopyrum esculentum Moench). Food Chem 96(1):36–42. https://doi.org/10.1016/j.foodchem.2005.01.052

Ma Y, Xiong YL, Zhai J, Zhu H, Dziubla T (2010) Fractionation and evaluation of radical-scavenging peptides from in vitro digests of buckwheat protein. Food Chem 118(3):582–588. https://doi.org/10.1016/j.foodchem.2009.05.024

Majid A, Priyadarshini PCG (2019) Millet derived bioactive peptides: a review on their functional properties and health benefits. Crit Rev Food Sci Nutr. https://doi.org/10.1080/10408398.2019.1686342

Majumder K, Wu J (2014) Molecular targets of antihypertensive peptides: understanding the mechanisms of action based on the pathophysiology of hypertension. Int J Mol Sci 16 (1):256–283. https://doi.org/10.3390/ijms16010256

Mandal S, Mandal R (2000) Seed storage proteins and approaches for improvement of their nutritional quality by genetic engineering. Curr Sci 79(5):576–589

Martínez-Andújar C, Pluskota WE, Bassel GW, Asahina M, Pupel P, Nguyen TT, Takeda-Kamiya-N, Toubiana D, Bai B, Górecki RJ, Fait A, Yamaguchi S, Nonogaki H (2012) Mechanisms of hormonal regulation of endosperm cap-specific gene expression in tomato seeds. Plant J 71:575–586. https://doi.org/10.1111/j.1365-313X.2012.05010.x

Mena M, Vicente-Carbajosa J, Schmidt RJ, Carbonero P (1998) An endosperm-specific DOF protein from barley, highly conserved in wheat, binds to and activates transcription from the prolamin-box of a native B-hordein promoter in barley endosperm. Plant J 16:53–62

Mirza N, Taj G, Arora S, Kumar A (2014) Transcriptional expression analysis of genes involved in regulation of calcium translocation and storage in finger millet (Eleusine coracana L. Gartn.). Gene 550:171–179

Mohamed TK, Amadou I, Zhou HM (2012) Antioxidant activity of fractionated foxtail millet protein hydrolysate. Int Food Res J 19(1):207–213

Monteiro PV, Sudharshana L, Ramachandra G (1987) Japanese barnyard millet (Echinochloa frumentacea): protein content, quality and SDS-PAGE of protein fractions. J Sci Food Agric 43:17–25

Monterio PV, Virnpaksha TK, Rao DR (1982) Proteins of Italian millet: amino acid composition, solubility fractionation and electrophoresis of protein fractions. J Sci Food Agric 33:1072–1079

Moreno-Risueno MÁ, Díaz I, Carrillo L, Fuentes R, Carbonero P (2007) The HvDOF19 transcription factor mediates the abscisic acid-dependent repression of hydrolase genes in germinating barley aleurone. Plant J 51:352–365. https://doi.org/10.1111/j.1365-313X.2007.03146.x

Nakase M, Hotta H, Adachi T, Alvarez AM, Aoki N, Nakamura R, Masumura T, Tanaka K, Matsuda T (1996) Cloning of the rice seed alpha-globulin-encoding gene: sequence similarity of the 5′-flanking region to those of the genes encoding wheat high-molecular-weight glutenin and barley D hordein. Gene 170:223–226

Nath M, Roy P, Shukla A, Kumar A (2013) Spatial distribution and accumulation of calcium in different tissues, developing spikes and seeds of finger millet genotypes. J Plant Nutr 36:539–550. https://doi.org/10.1080/01904167.2012.748072

Onate L, Vicente Carbajosa J, Lara P, Diaz I, Carbonero P (1999) Barley BLZ2, a seed-specific bZIP protein that interacts with BLZ1 in vivo and activates transcription from the GCN4-like motif of B-hordein promoters in barley endosperm. J Biol Chem 274:9175–9182
Osborne TB, Mendel LB (1914) Nutritive properties of the maize kernel. J Biol Chem 18:1–16
Palenchar PM, Kouranov A, Lejay LV et al (2004) Genome-wide patterns of carbon and nitrogen regulation of gene expression validate the combined carbon and nitrogen (CN)-signaling hypothesis in plants. Genome Biol 5:R91. https://doi.org/10.1186/gb-2004-5-11-r91
Panyayai T, Ngamphiw C, Tongsima S, Mhuantong W, Limsripraphan W, Choowongkomon K, Sawatdichaikul O (2019) FeptideDB: a web application for new bioactive peptides from food protein. Heliyon 5(7):e02076. https://doi.org/10.1016/j.heliyon.2019.e02076
Parameswaran KP, Thayumanavan B (1995) Homologies between prolamins of different minor millets. Plant Food Hum Nutr 48:119–126
Pownall TL, Udenigwe CC, Aluko RE (2010) Amino acid composition and antioxidant properties of pea seed (Pisum sativum L.) enzymatic protein hydrolysate fractions. J Agric Food Chem 58 (8):4712–4718. https://doi.org/10.1021/jf904456r
Pratelli R, Pilot G (2014) Regulation of amino acid metabolic enzymes and transporters in plants. J Exp Bot 65(19):5535–5556. https://doi.org/10.1093/jxb/eru320
Qi WZ, Liu HH, Liu P, Dong ST, Zhao BQ, So HB et al (2012) Morphological and physiological characteristics of corn (Zea mays L.) roots from cultivars with different yield potentials. Eur J Agron 38:54–63
Ramakrishnan M, Ceasar SA, Duraipandiyan V, Al-Dhabi NA, Ignacimuthu S (2016) Using molecular markers to assess the genetic diversity and population structure of finger millet (Eleusine coracana (L.) Gaertn.) from various geographical regions. Genetic Resour Crop Evol 63(2):361–376
Raveschot C, Cudennec B, Coutte F, Flahaut C, Fremont M, Drider D, Dhulster P (2018) Production of bioactive peptides by Lactobacillus species: from gene to application. Front Microbiol 9:2354
Riechmann JL, Heard J, Martin G, Reuber L, Jiang CZ et al (2000) Arabidopsis transcription factors: genome-wide comparative analysis among eukaryotes. Science 290:2105–2110
Schmidt RJ, Burr FA, Aukerman MJ, Burr B (1990) Maize regulatory gene opaque-2 encodes a protein with a "leucine-zipper" motif that binds to zein DNA. Proc Natl Acad Sci U S A 87:46–50
Segura Campos MR, Peralta Gonzalez F, Chel Guerrero L, Betancur Ancona D (2013) Angiotensin I-converting enzyme inhibitory peptides of chia (Salvia hispanica) produced by enzymatic hydrolysis. Int J Food Sci 2013:1. https://doi.org/10.1155/2013/158482
Shen W, Matsui T (2017) Current knowledge of intestinal absorption of bioactive peptides. Food Funct 8(12):4306–4314. https://doi.org/10.1039/C7FO01185G
Shewry PR, Halford NG (2002) Cereal seed storage proteins: structures, properties and role in grain utilization. J Exp Bot 370:947–958
Shewry PR, Tatham AS (1990) The prolamin storage proteins of cereal seeds: structure and evolution. Biochem J 267:1–12
Shewry PR, Tatham AS (1995) Seed storage proteins: structure and biosynthesis. Plant Cell 7:945–956
Shewry PR, Field JM, Kirkman MA, Faults AJ, Miflin BJ (1980) The extraction, solubility, and characterization of two groups of barley storage polypeptides. J Exp Bot 31:393–407
Shewry PR, Tatham AS, Halford NG (1999) The prolamins of the Triticeae. In: Shewry PR, Casey R (eds) Seed proteins. Kluwer Academic, Dordrecht, pp 35–78
Singh UM, Chandra M, Shankhdhar SC, Kumar A (2014) Transcriptome wide identification and validation of calcium sensor gene family in the developing spikes of finger millet genotypes for elucidating its role in grain calcium accumulation. PLoS One. https://doi.org/10.1371/journal.pone.0103963
Singh UM, Metwal M, Singh M et al (2015) Identification and characterization of calcium transporter gene family in finger millet in relation to grain calcium content. Gene 566:37–46. https://doi.org/10.1016/j.gene.2015.04.021

Stone MT, Kozlov M (2014) Separating proteins with activated carbon. Langmuir 30 (27):8046–8055. https://doi.org/10.1021/la501005s

Subba Rao MV, Muralikrishna G (2002) Evaluation of the antioxidant properties of free and bound phenolic acids from native and malted finger millet (ragi, Eleusine coracana Indaf-15). J Agric Food Chem 50(4):889–892. https://doi.org/10.1021/jf011210d

Sudhakar C, Veeranagamallaiah G, Nareshkumar A, Sudhakarbabu O, Sivakumar M, Pandurangaiah M, Kiranmai K, Lokesh U (2015) Polyamine metabolism influences antioxidant defense mechanism in foxtail millet (Setaria italica L.) cultivars with different salinity tolerance. Plant Cell Rep 34(1):141–156. https://doi.org/10.1007/s00299-014-1695-3

Templeman TS, Demaggio AE, Stetler DA (1987) Biochemistry of fern spore germination: globulin storage proteins in Matteuccia struthiopteris L. Plant Physiol 85:343–349

Todorovska E, Abumhadi N, Kamenarova K, Zheleva D, Kostova A, Christov N, Alexandrova N, Jacquemin JM, Anzai H, Nakamura C, Atanassov A (2005) Biotechnological approaches for cereal crops. Improv Biotechnol Biotechnol Equip 19(Suppl 3):91–104

Vilcacundo R, Martınez-Villaluenga C, Hernandez-Ledesma B (2017) Release of dipeptidyl peptidase IV, a-amylase and a-glucosidase inhibitory peptides from quinoa (Chenopodium quinoa Willd.) during in vitro simulated gastrointestinal digestion. J Funct Foods 35:531–539. https://doi.org/10.1016/j.jff.2017.06.024

Virupaksha TK, Geeta R, Dasasetty N (1975) Seed proteins of finger millet and their amino acid composition. J Sci Food Agric 26(8):1237–1246. https://doi.org/10.1002/jsfa.2740260823

Wang J, Zhao M, Zhao Q, Jiang Y (2007) Antioxidant properties of papain hydrolysates of wheat gluten in different oxidation systems. Food Chem 101(72):1658–1663

Wang F, Yu G, Zhang Y, Zhang B, Fan J (2015) Dipeptidyl peptidase IV inhibitory peptides derived from oat (Avena sativa L.), buckwheat (Fagopyrum esculentum), and highland barley (Hordeum vulgare trifurcatum (L.) Trofim) proteins. J Agric Food Chem 63(43):9543–9549. https://doi.org/10.1021/acs.jafc.5b04016

Wang YL, Huang Q, Kong D, Xu P (2018) Production and functionality of food-derived bioactive peptides: a review. Mini Rev Med Chem 18(18):1524–1535. https://doi.org/10.2174/1389557518666180424110754

Wang B, Xie N, Li B (2019) Influence of peptide characteristics on their stability, intestinal transport, and in vitro bioavailability: a review. J Food Biochem 43(1):e12571. https://doi.org/10.1111/jfbc.12571

Wu D, Gao Y, Qi Y, Chen L, Ma Y, Li Y (2014) Peptide-based cancer therapy: opportunity and challenge. Cancer Lett 351(1):13–22. https://doi.org/10.1016/j.canlet.2014.05.002

Xu W, Wei L, Qu W, Liang Z, Wang J, Peng X, Zhang Y, Huang K (2011) A novel antifungal peptide from foxtail millet seeds. J Sci Food Agric 91(9):1630–1637. https://doi.org/10.1002/jsfa.4359

Yanagisawa S (1996) Dof DNA binding proteins contain a novel zinc finger motif. Trends Plant Sci 1:213–214

Yanagisawa S (2002) The DOF family of plant transcription factors. Trends Plant Sci 7 (12):555–560

Yanagisawa S (2004) DOF domain proteins: plant-specific transcription factors associated with diverse phenomena unique to plants. Plant Cell Physiol 45:386–391

Yanagisawa S, Izui K (1993) Molecular cloning of two DNA-binding proteins of maize that are structurally different but interact with the same sequence motif. J Biol Chem 268:16028–16036

Yanagisawa S, Schmidt RJ (1999) Diversity and similarity among recognition sequences of Dof transcription factors. Plant J 17(2):209–214

Yesudhas D, Batool M, Anwar MA, Panneerselvam S, Choi S (2017) Proteins recognizing DNA: structural uniqueness and versatility of DNA-binding domains in stem cell transcription factors. Genes 8(8):192. https://doi.org/10.3390/genes8080192

Youle RJ, Huang AHC (1978) Albumin storage proteins in the protein bodies of castor bean. Plant Physiol 61:13–16

Millets, Phytochemicals, and Their Health Attributes

9

Pallavi Shah, Anil Kumar, Vishnu Kumar, and Manoj Kumar Tripathi

Abstract

Robust small-grained crops are consumed as staple food in developing Asian and African countries, millets are highly rich in health-promoting bioactive compounds. Not only are millets resistant to a number of abiotic stresses they are also an extremely rich source of micronutrients and phytochemicals that have been reported to possess antioxidant, anticancerous, antidiabetic, antiaging, antihypertensive, cardioprotective, and many other health attributes. Their hardy nature coupled with exceptional nutritional value makes them highly valuable crops of modern agroecosystems. This chapter provides basic information on the various kinds of phytochemicals present in millets and their beneficial effects on human health. This chapter will also provide a comprehensive account of synthesis, structural and nutraceutical aspects, and bioavailability of different types of phytochemicals with special emphasis on polyphenols, flavonoids, phytosterols, and bioactive peptides.

Keywords

Millets · Phytochemicals · Bioactive peptides · Bioavailability · Health attributes

P. Shah
Department of Molecular Biology and Genetic Engineering, G.B. Pant University of Agriculture and Technology, Pantnagar, Uttarakhand, India

A. Kumar (✉)
Directorate of Education, Rani Lakshmi Bai Central Agricultural University, Jhansi, Uttar Pradesh, India

V. Kumar
Department of Genetics & Plant Breeding, Rani Lakshmi Bai Central Agricultural University, Jhansi, Uttar Pradesh, India

M. K. Tripathi
ICAR-Central Institute of Agricultural Engineering, Bhopal, Madhya Pradesh, India

A. Kumar et al. (eds.), *Millets and Millet Technology*,
https://doi.org/10.1007/978-981-16-0676-2_9

9.1 Introduction

Despite their high nutritional and nutraceutical potential, millets constitute only a minor portion of cultivated crops at the global level. These are grown mainly by marginal and tribal farmers in developing countries and have not been studied as extensively as the major cereal crops. Millets are nutritionally important because of their high content of calcium (0.38%), dietary fiber (18%), and phenolic compounds (0.3–3%) (Chauhan et al. 2018, Sood et al. 2019). Millet proteins are a rich source of essential amino acids such as tryptophan, threonine, and sulfur containing amino acids excluding lysine and threonine. Millets have a high fat content in comparison to other cereals especially unsaturated fatty acids (Zhang et al. 2015). Abundance of phytochemicals and micronutrients of immense therapeutic potential signifies the importance of all varieties of millets in human nutrition (Mal et al. 2010; Singh and Raghuvanshi 2012). Millets are robust crops with great adaptability to environmental stresses such as drought, high temperature, and poor cultivation practices (Duodu and Awika 2019). Their hardy nature coupled with exceptional nutritional value makes them highly valuable crops of modern agroecosystems. In recent years, there have been major emphases on millets research and promotion that will render the acceptance of millets to modern cropping systems and make them both ecologically and economically viable future crops. In this chapter, we review the different types of phytochemicals present in various millet varieties, their biosynthesis, bioavailability, nutraceutical potential, and numerous health benefits.

9.2 Phytochemicals of Millets: An Overview

Apart from possessing high content of micronutrients, millet grains possess ample amounts of other phytochemicals/nutrients which include bound phenolic acids, free phenolic acids, flavonoids, lignin, insulin-resistant starch, and β-glucan (Liu 2007; Taylor and Duodu 2015). Generally, these phytochemicals belong to the category of secondary metabolites or plant cell wall components with molecular weights ranging from 150 to 30,000 Da (Chethan and Malleshi 2007). Together these phytochemicals impart resistances to millet grains against various diseases and storage pests. The phytochemicals possess a wide range of bioactive properties and are responsible for major health benefits often associated with whole-grain consumption of millets (Duodu and Awika 2019). The main flavonoids present in millet grains are apigenin, catechin, quercetin, and taxifolin (Shahidi and Chandrasekara 2013). Among the carotenoids, lutein, zeaxanthin, and β-carotene are the major ones and their concentration varies among different millet varieties and species (Shen et al. 2015). Similarly, β-sitosterol, campesterol, and stigmasterol are the major phytosterols reported in millet grains (Avato et al. 1990). Similar to certain sorghum varieties, few dark colored finger millet varieties have been reported to possess condensed tannins as well (Dykes and Rooney 2006). In the recent past, these phytochemicals have gained attention of health professionals and consumers for their potent health benefits chiefly recognized for antioxidant, antidiabetic, antimicrobial, anticarcinogenic, anti-atherosclerogenic, and antiaging properties (Taylor 2017; Table 9.1).

9.2.1 Polyphenols of Millets: Types and Biosynthesis

Polyphenols are naturally occurring secondary metabolites of plants and generally constitute of multiple structural phenol units. Recently, dietary polyphenols obtained from plant sources have received tremendous attention across the globe for their potent antioxidant potential and health benefits like protection against cardiovascular and neurodegenerative disorders, anticancer, antidiabetic, and antiaging potential (Pandey and Rizvi 2009). Myriad of polyphenols have been detected from different plant species and majority of them are derivatives of phenylalanine or another precursor-like shikimic acid. Polyphenols are generally divided into classes based on the number of phenol rings and on the basis of how these rings are bound to each other. Polyphenols are abundant in millets and are being largely used as dietary supplements around the world (Chandrasekara and Shahidi 2010). Polyphenols present in millets are classified into flavonoids (tannins and anthocyanins) and phenolic acids (benzoic and cinnamic acid derivatives) (Fig. 9.1) (Awika and Rooney 2004). In millets, the phenolic acids are largely present in the bran layers. Various processing methods can thus reduce the beneficial phenolic content over all by inducing bran separation from the grain endosperm.

The basic structure of flavonoids constitutes of a 15-carbon backbone having A and B benzene rings connected by a heterocycle (C). Malonyl CoA gives rise to aromatic ring A whereas *p*-coumaric acid gives rise to Ring B. Chalcone is formed by the combination of three malonyl CoA molecules with *p*-coumaroyl CoA which gives rise to flavonoid ring after several cyclization reactions (Moreno and Peinado 2012).

Polyphenol biosynthesis is mainly catalyzed by the enzyme phenylalanine ammonia-lyase (Stuper-Szablewska and Perkowski 2019). Figure 9.2 represents a general scheme for the synthesis of polyphenols in plants. Fertility and availability of water determine whether phenylalanine ammonia-lyase will participate in either protein or polyphenol synthesis. Out of the many pathways the precursor shikimic acid is of utmost importance for the synthesis of polyphenols. The simplest phenols like cinnamic acid are synthesized from phenylalanine by the activity of phenylalanine ammonia-lyase. Further, hydroxylation steps lead to the generation of *p*-coumaric acid like phenols. Side chain cleavage of cinnamic acid followed by loss of a water molecule generates the aldehyde form. These aldehydes on oxidation lead to the formation of benzoic acids.

9.2.2 Phenolic Acids

Millet phenolics have been studied extensively for their antioxidant potential and many important therapeutic properties. Different species of millet contain both free

Table 9.1 Major classes of phytochemicals present in millets

S. No.	Phytochemicals	Source	Examples	Properties	References
1	Polyphenols				
1a	Phenolic acids	Seed coat/bran layers	Hydroxycinnamic acids—protocatechuric acid and vanillic acid Hydroxybenzoic acids—ferulicacid, caffeic, coumaric acid, and sinapic acid	Antioxidant, anti-fibrotic, antiviral, anti-atherosclerotic, antibacterial, antiaging, anticancer	Shahidi and Chandrasekara (2013)
1b	Flavonoids	Seed/seed coat	Quercetin, Catechin and its derivatives (Gallocatechin, Epicatechin, and Epigallocatechin), vitexin, tricin, luteolin, myricetin	Antioxidant, anti-inflammatory, antimutagenic effects, antiaging as well as cardiovascular disease-preventive properties	Shahidi and Chandrasekara (2013)
1c	Condensed tannins	Seed coat of brown finger millets	Polymers of flavan-3-ols	Anticancer, anti-ulcer, gastroprotective, and hypocholesteraemic properties, avoiding UTI	Dykes and Rooney (2006)
2	Carotenoids	Seed/seed coat	β-Carotene, lutein, zeaxanthine	Antioxidant	Shen et al. (2015)
3	Phytosterols	Bran layers	β-Sitosterol, campesterol, stigmasterol	Cardiovascular disease-preventive properties	Ryan et al. (2007)
4	Bioactive peptides	Total protein	Specific peptide sequences	Antioxidant, antihypersensitive, immunomodulatory, antidiabetic, anticarcinogenic	Majid and Priyadarshini (2019)

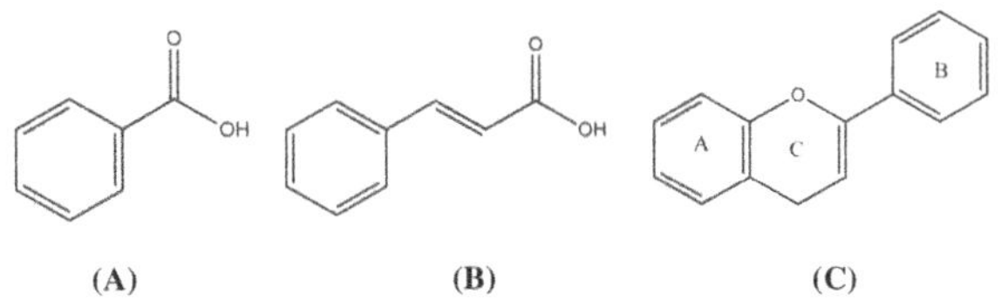

Fig. 9.1 Basic structures of polyphenols present in millets (**a**) Benzoic acid (**b**) Cinnamic acid (**c**) Flavan; the backbone of flavonoid

and bound phenolic acids that are either derived from hydroxycinnamic acid or hydroxybenzoic acid. A major portion of these cinnamic acid derivatives are insoluble bound phenolic acids attached to cell wall polysaccharides by esterification. The soluble phenolic acids include free, nonconjugated forms and phenolic compounds esterified or etherified to soluble carbohydrate moieties (Shahidi and Chandrasekara 2013). Cinnamic acid derivatives like ferulic, caffeic, coumaric, and sinapic acid are dominant bound phenolic acids reported in millet varieties, whereas the free unbound phenolic acids consist of benzoic acid derivatives like vanillic, gallic, protocatechuic, syringic, and gentisic acids (Duodu and Awika 2019). The content and distribution of different phenolic acids vary depending on different millet species. The large diversity of phenolic acids present in millet varieties have been reviewed in detail but comparison of quantitative data from different resources is very difficult because of the variations in methods used for extraction and characterization of several etherified, esterified, free, and bound phenolic acids (Chandrasekara and Shahidi 2011a, b).

Table 9.2 represents the contents of bound, conjugated, and free phenolic acids in some millet varieties. Bound phenolic acids constitute the majority almost 70% of the total phenols present in millet grains and are basically different hydroxycinnamic acid derivatives associated with cell wall polysaccharides (Hassan and Burton 2018). These bound phenolic acids mostly act as crosslinkers of cell wall polysaccharides and lignin by either esterification or etherification. Bound phenolic acids are non-extractable by normal solvent-based methods and thus require alkaline hydrolysis for extraction. The free extractable forms are often present as glycerides or glycosides esterified to monosaccharides (Yang et al. 2012). Apart from monomeric bound hydroxycinnamic acid derivatives several dimers and trimers of ferulic acid have also been reported in different millet varieties. Several reviewers have reported the potent antioxidant and therapeutic potential of phenolic acids, even more so, for their bound dimeric forms. Insoluble forms are the most dominant polyphenols in majority of millets except finger millet where soluble phenolic acids are more in quantity (Chandrasekara and Shahidi 2010).

Phenolic acids that can be extracted are metabolized by microbes and hence contribute to gut microbiota function and associated consequences to gastrointestinal (GI) health. The future prospects for this work include unraveling of the mechanisms on how gut microbiota metabolizes bound phenolics and how they influence GI health (Selma et al. 2009). Table 9.3 states amounts of some important phenolic acids in different millet varieties.

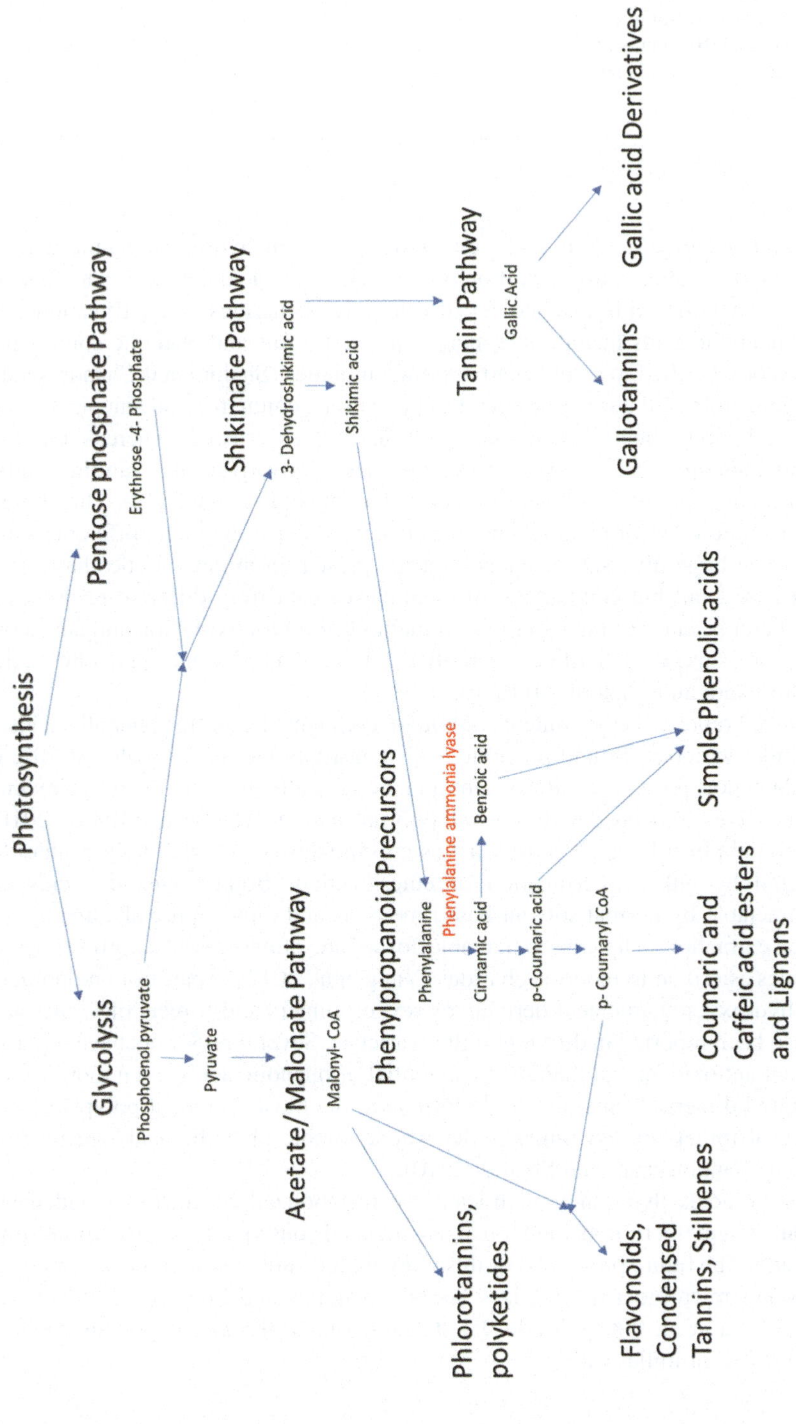

Fig. 9.2 General overview of polyphenol synthesis in plants

Table 9.2 Phenolic acid content in grains of various millet varieties given as μg/g of defatted meal

Millets	Free soluble Phenolic acids		Esterified + etherified soluble phenolic acids		Insoluble bound phenolic acids	
	Hydroxy benzoic acid and derivatives	Hydroxycinnamic acid and derivatives	Hydroxybenzoic acid and derivatives	Hydroxycinnamic acid and derivatives	Hydroxybenzoic acid and derivatives	Hydroxycinnamic acid and derivatives
Pearl	9	51	112	410	32	971
Finger	62	12	60	33	55	468
Proso	28	89	144	237	215	1867
Foxtail	56	171	94	307	38	1769
Kodo	50	74	219	608	101	3687
Little	38	173	146	329	119	1242

Content in defatted meal: Source Shahidi and Chandrasekara (2013), Pradeep and Sreerama 2017, Duodu and Awika (2019)

Table 9.3 Total phenols (mg/100 g) and profiles for some important phenolic acids in different millet varieties

Millets	Pearl	Finger	Proso	Foxtail	Fonio	Teff	Kodo	Barnyard	Little
Total phenolics	269–420 mg GAE/100 g DW	172–302 mg FAE/100 g DW	0.48–27 mg phenolic acid/100 g DW	28.2–59 mg RE/100 g db	3.36 mg FAE/g of flour	326–448 mg GAE/100 g db	55 mg GAE/100 g db	43 mg phenolic acids/100 g defatted meal	24 mM FAE/100 g DW
Hydroxycinnamic acids (μg/g)									
Ferulic acid	29–812	41–405	166–445	225–857	25	411	2210	212	146–171
p-Coumaric acid	55–118	19–41	12–1235	45–943	6	66	802	27	40–84
Caffeic acid	8–30	10–16	16–339	15–38	5	–	324	8	26–83
Hydroxybenzoic acids (μg/g)									
p-Hydroxybenzoic acid	48	4–6	34–126	8–22	–	–	31	11	3–12
Gallic acid	5–39	5–11	4–6	5–7	7	–	2	4	1.7–8.2
Protocatechuic acid	2–18	15–120	72	12	9	47	71	–	–

Source: Nambiar et al. (2012), N'Dri et al. (2013), Zhang et al. (2014), Pradeep and Sreerama (2015), Sharma et al. (2015), Goudar and Sathisha (2016), Shumoy and Raes (2016), Sharma and Saxena (2016), Pradeep and Sreerama (2018), Xiang et al. (2019), Serna-Saldivar and Espinosa-Ramírez (2019)
RE rutin equivalents, *GAE* gallic acid equivalents, *FAE* ferulic acid equivalents, *DW* dry weight, *db* dry basis, – no data available

9.2.3 Flavonoids

Flavonoids are an important class of polyphenols that are extensively studied for their numerous health-related attributes which include cardioprotective, gastroprotective, antioxidant, anti-inflammatory, and anticancer potential (Serna-Saldivar and Espinosa-Ramírez 2019). Significant concentrations of flavonoids have been identified in pericarp and germ of cereal grains (Dykes and Rooney 2006). Previous works have shown that flavonoids play a significant role in protecting cereal crops against biotic and abiotic stresses by providing pest resistance, grain pigmentation, UV protection, and play an important part in physiological developmental functions such as seed dormancy, germination, and auxin regulation (Panche et al. 2016). Hence, they play a major role in plant growth, reproduction, and function.

Flavonoids are present as both bound and free forms that can exist as monomeric or polymeric forms. Monomeric forms are relatively less abundant and hence not studied extensively as compared to polymeric forms such as condensed tannins. Finger millet is the only known millet to be rich in condensed tannins (Dykes and Rooney 2006). Unlike most cereal grains like wheat, maize, rice, and oat, where bound flavonoids are dominant, majority of millet flavonoids are free forms with an exception of teff (Atanasova-Penichon et al. 2016; Shumoy and Raes 2016).

Depending on the degree of hydroxylation and modification in the heterocyclic ring the flavonoids can be divided into many classes the important ones being represented by flavones, flavonols, flavanols, isoflavones, anthocyanidins or anthocyanins, and flavanones (Fig. 9.3). The major flavonoids present in millet varieties have been discussed further. Table 9.4 depicts the total flavonoid content (TFC) and profiles for some important ones reported in different millet varieties.

9.2.4 Flavones

Flavones are an important group of flavonoids with no substitution at the C3 position, oxidized at the C4 position, and having a double bond between C2 and C3 in the basic flavonoid structure (Fig. 9.3). Flavones have important activities in plants like providing pigmentation alongside flavonols and anthocyanins and also act as UV protectants (Hostetler et al. 2017). Apart from being reported chiefly from plants such as celery, parsley, red peppers, and mint they are also found abundantly in cereal grains. Cereal grains chiefly consist of O- or C-glycosides of luteolin and apigenin. Several free forms are also present such as aglycones in some grains, such as fonio and sorghum (Chandrasekara and Shahidi 2011a; Yang et al. 2012, 2015; Panche et al. 2016). In majority of millet varieties except finger millet flavones are the dominant or only flavonoids present. Pearl millet has been reported to contain the maximum amounts of C-glycosylated derivatives of flavones (Table 9.5). Flavones such as orientin, vitexin, luteolin, apigenin, violanthin, tricin, and their glycosides have been reported in different millet varieties (Shahidi and Chandrasekara 2013).

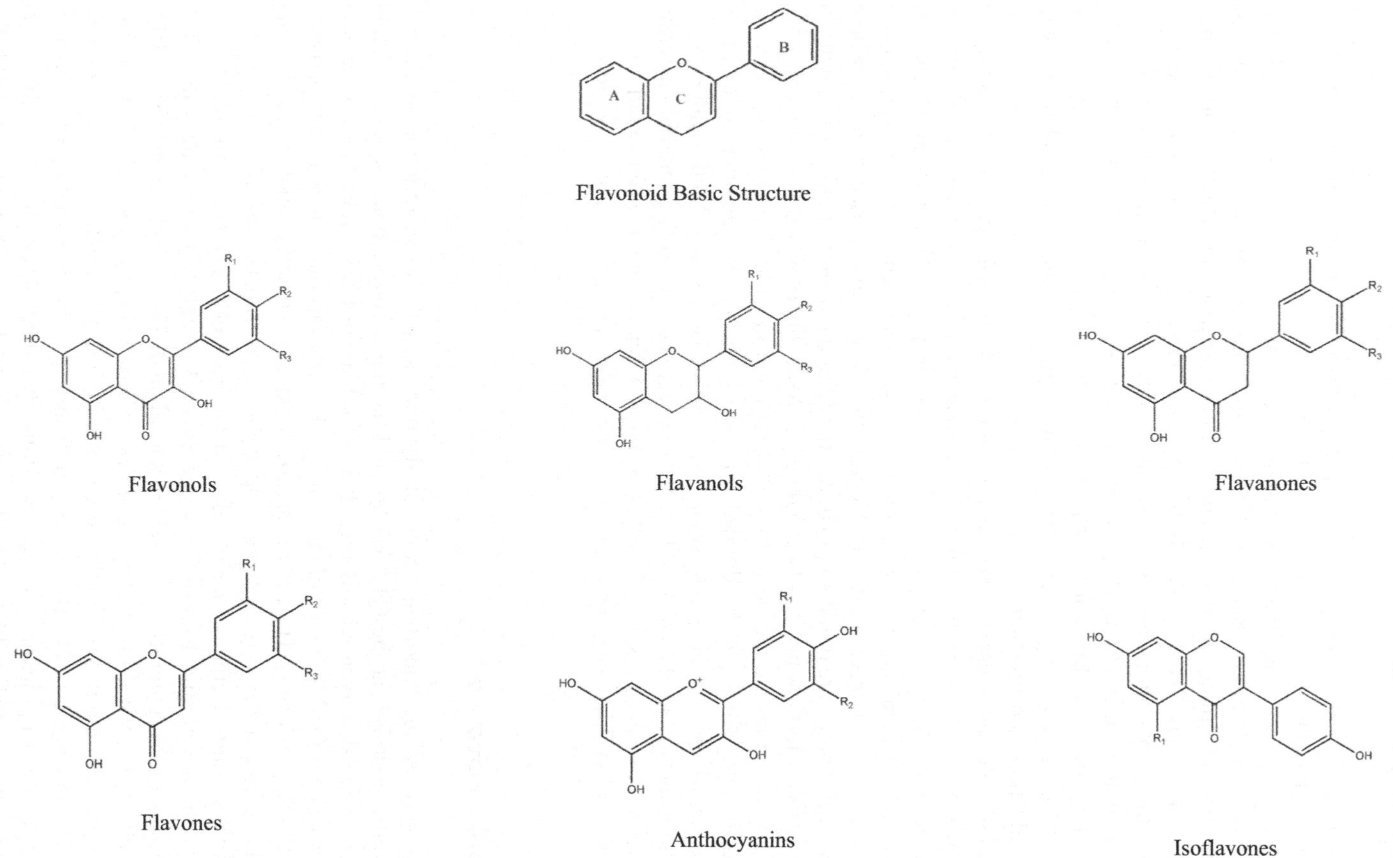

Fig. 9.3 General structures of monomeric flavonoids based on the type of heterocycle and group substitutions in the flavonoid basic structure. Variations within the group may arise as a result of differences in the number and position of OH group and differences in their degree of alkylation or glycosylation

Table 9.4 Total flavonoids (mg/100 g) and profiles for some important flavonoids reported in different millets

Millets	Pearl	Finger	Proso	Foxtail	Fonio	Teff	Kodo	Barnyard	Little
Total flavonoids (mg/100 g)	7	211	5–8	25–88	–	62–116	51	29–58	87–335
Important flavonoids (μg/g)									
Catechin	–	1533	–	7	–	111	–	–	15–16
Apigenin	4	–	2–16	102–160	–	–	2	23	14–25
Myricetin	10	11	7–16	7	–	–	–	15	29–37

Source: Shahidi and Chandrasekara (2013), Duodu and Awika (2019), Serna-Saldivar and Espinosa-Ramírez (2019)
–*no data available*

Table 9.5 Different flavones reported in various millet varieties

Type	Structural backbone	Millet variety								
		Pearl	Finger	Proso	Foxtail	Fonio	Teff	Kodo	Barnyard	Little
Flavones	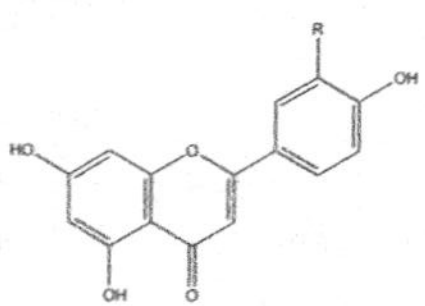 R group substitutions can be -OH and -OCH_3	Mostly C-glycosides of vitexin and oreintin	Minimum amount of flavones accumulates flavan-3-ols preferentially	Mixture of C and O-glycosides	Mostly aglycones and C-glycosides of apigenein	Mixtures of O-glycosides of luteolin and apigenin and aglycones	C-glycosides of chrysoeriol, luteolin, and apigenin	Mixture of C and O-glycosides	C-glycosides of luteolin and tricin	Mostly C-glycosides

Source: Siwela et al. (2007), Shahidi and Chandrasekara (2013), Ravisankar et al. (2018), Duodu and Awika (2019)

9.2.5 Flavanols

Flavanols are an important type of flavonoids often reported in rich quantities in green tea, red grapes, cocoa, red wine, berries, and apples. They have been associated with strong antioxidant properties and potential health benefits. Unlike other flavonoids that exist as glucosides, they exist in aglycone form as monomers, oligomers, or esterified with gallic acid. Epicatechin and catechin are the major monomeric forms which oligomerize to form procyanidins. They can further polymerize and form condensed tannins (proanthocyanidins) (Hackman et al. 2008). Barley, sorghum, and finger millet possess flavanols, along with polymeric condensed tannins in extractable and measurable amounts (Duodu and Awika 2019). In millets, finger millet is the only variety with condensed tannins which accumulate intesta, found in between the endocarp and the aleurone layer parts of the endosperm. The red and white colored finger millet varieties have poor quantity, i.e., 0.04–0.06% catechin equivalents of tannins but the brown varieties showed higher amounts of condensed tannins ranging from 600 to 2100 mg/100 g, i.e., 0.12–3.47% (Siwela et al. 2007; Shahidi and Chandrasekara 2013).

9.2.6 Carotenoids

Carotenoids are colored pigments produced by plants and some microbes and play an important role in photosynthesis and photo-protection. These are fat-soluble red, yellow, or orange colored pigments and possess many health attributes. These can be broadly divided into carotenoids such as β-carotene and xanthophylls such as lutein. Often, most of them are tetraterpenoids consisting of linked 8 units of isoprenoids. Modifications in the basic structure occur by reactions such as cyclization, hydrogenation, dehydrogenation, and oxidation (Mezzomo and Ferreira 2016).

Carotenoid biosynthesis begins with the precursor molecule GGPP (geranylgeranyl pyrophosphate) formed in the plastids by the MEP (methylerythritol 4-phosphate) pathway (Fig. 9.4). Carotenoid synthesis begins with the condensation of GGPP catalyzed by the enzyme phytoene synthase (PSY in plants and CRTB in bacteria) to produce C40 15-*cis*-phytoene. Desaturation reactions catalyzed by PDS (phytoene desaturase) and ZDS (ζ-carotene desaturase) convert phytoene to lycopene and further undergo isomerization to convert poly *cis* to all trans forms in presence of CRTISO (carotenoid isomerase). On the other hand, a single enzyme CRTI (bacterial phytoene desaturase/isomerase) catalyzes both desaturation and isomerization steps in bacteria. Cyclization of lycopene to incorporate ε- and β-rings to produce α- and β-carotene is further carried out by LCYB (lycopene β-cyclase), LCYE (lycopene ε-cyclase) or CRTI in bacteria. Hydroxylation of ε-ring by CYP97C (cytochrome P450 carotene ε-ring hydroxylases) and β-ring by HYDB (alternatively called as BCH β-carotene hydroxylase) or CYP97A (cytochrome P450 carotene β-ring hydroxylase) in plants or CRTZ (bacterial β-carotene hydroxylase) in bacteria lead to formation of xanthophylls (Trono 2019).

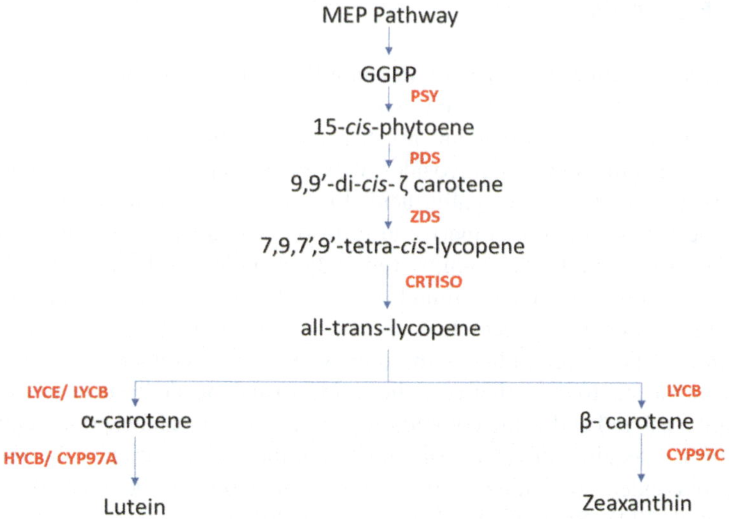

Fig. 9.4 Biosynthesis of carotenoids in plants. *MEP* methylerythritol 4-phosphate, *GGPP* geranylgeranyl pyrophosphate, *PSY* phytoene synthase, *PDS* phytoene desaturase, *ZDS* ζ-carotene desaturase, *CRTISO* carotenoid isomerase, *LYCB* lycopene β-cyclase, *LYCE* lycopene ε-cyclase, *BCH/HYDB* β-carotene hydroxylase, *CYP97A and CYP97C* cytochrome P450 carotene β- and ε-ring hydroxylases

Different millet varieties show a range of carotenoid content. Most of the millet varieties contain lutein and zeaxanthin as dominant xanthophylls. Shen et al. (2015) determined the carotenoid content in foxtail millet cultivars and found all trans-lutein content to vary from approximately 6 to 14 mg/kg of dry weight (DW) whereas all trans-zeaxanthin content varied from approximately 1 to 2 mg/kg of dry weight. The total yellow pigment content was reported to be around 12–23 mg/kg of dry weight. Zhang et al. (2014) reported approximately 0.5– 2 mg/100 g of dry weight of xanthophylls and 1.60–1.68 mg/100 g of dry weight of zeaxanthin in various proso millet cultivars. Khangura et al. (1980) reported that in pearl millet the β-carotene content varies from 36 to 38 mg/100 g. Asharani et al. (2010) have reported varying amounts of carotenoids in proso, finger, and foxtail millets, i.e., 249–518 mg/100 g, 78–316 mg/100 g, and 126–191 mg/100 g, respectively.

9.2.7 Phytosterols

Phytosterols are a class of sterols mainly found in plant cell walls and membranes as structural components. These are cholesterol like lipophilic molecules, mostly abundant in cereal grain bran often extracted along with bran oil waxes. Dietary phytosterols are hypocholesterolemic as they reduce intestinal cholesterol absorption and hence have gained importance for their cardioprotective potential (Jiang and Wang 2005; Duodu and Awika 2019).

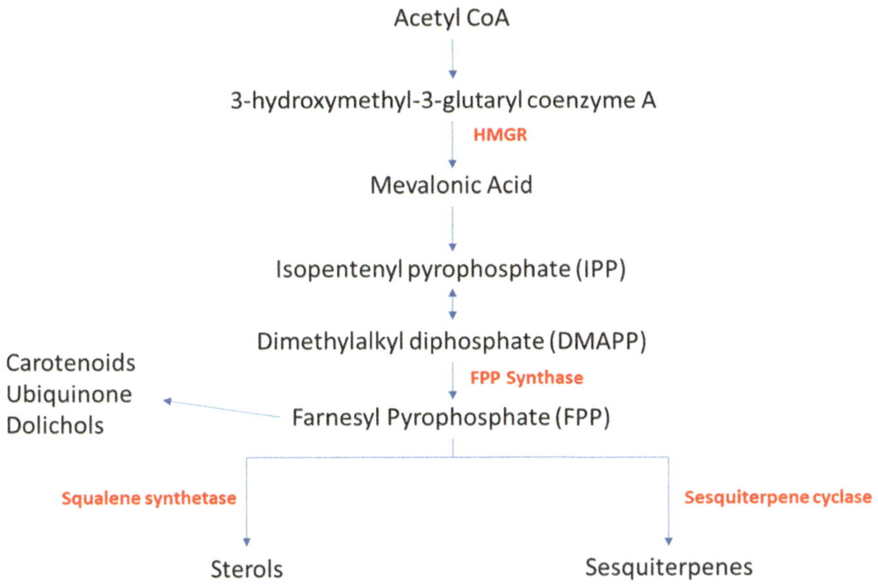

Fig. 9.5 General Scheme for sterol synthesis in photosynthetic plants. *HMGR* 3-hydroxymethyl-3-glutaryl coenzyme A reductase, *IPP* isopentenyl pyrophosphate, *DMAPP* dimethylalkyl diphosphate, *FPP synthase* farnesyl pyrophosphate synthase

The basic plant phytosterols are synthesized by the isoprenoid biosynthetic pathway with IPP (Isopentenyl Pyrophosphate) as the major building block for all the terpenoids (Fig. 9.5). The IPP is derived from the mevalonic acid pathway by the action of the enzyme HMGR, 3-hydroxymethyl-3-glutaryl coenzyme A reductase. IPP forms DMAPP, dimethylalkyl diphosphate which by the action of FPP synthase converts the metabolite to FPP, farnesyl pyrophosphate. From this point forward the pathway separates into synthesis of sesquiterpenes and triterpenes. Squalene synthetase which mainly occurs in the cytoplasm leads to synthesis of sterols whereas sesquiterpene cyclase leads to synthesis of sesquiterpenes. The sterol biosynthesis involves several enzymes ($n \geq 30$) which are associated with the plant membranes. The photosynthetic plants convert squalene oxide into cycloartenol with the help of these enzymes and further processes like methylation (alkylation) lead to formation of alkylated sterol end products (Piironen et al. 2000).

Cereals contain phytosterols present in free forms as well as conjugated forms. The free forms exist as esters of fatty acids such as linoleic acid, oleic acid, and palmitic acid or cinnamic acids (typically ferulate) whereas conjugated forms are associated with sugars mainly glucose (Sridhar and Lakshminarayana 1994). Like other cereals, millet varieties are found rich in stigmasterol, β-sitosterol, and campesterol. β-sitosterol constitutes the largest fraction (more than 70%) of unbound sterols and about 54% of bound sterols in all millet varieties (Avato et al. 1990). Pearl millet consists of 58 mg/100 g of phytosterols (Ryan et al. 2007). Phytosterols of 44–57 mg/100 g and 19–26 mg/100 g were reported in foxtail and proso millet, respectively (Bhandari and Lee 2013; Esche et al. 2013).

Another form of phytosterols known as stanols are also present in millets but at lower concentrations in comparison to other sterols. With no double bond at position 5, they are formed by the hydrogenation of sterols. They have been reported to possess similar health benefits when compared to sterols (Pang et al. 2014). Proso millet possesses meager amount of stanols, constituting less than 2% of the total phytosterols. On the other hand, Foxtail millet is abundant in stanols with majority being sitostanol.

9.2.8 Bioactive Peptides

Bioactive peptides are specific peptide sequences encrypted in the parent protein that influence health by effecting body functions and conditions (Sharma et al. 2011; Sánchez and Vázquez 2017). The amino acid sequence and composition are responsible for the different activities of peptides once they are released from the precursor protein by enzymatic proteolysis or during food processing. These activities range from antimicrobial, anti-oxidative, immunomodulatory, antithrombotic, anticancerous to antihypertensive (Daliri et al. 2016; Waseem et al. 2018). A number of bioactive peptides with detailed sequences have been reported from different cereal crops such as wheat, rice, oat, and barley; however, millet peptides are yet to be studied in detail. Most of the available reports are that pertaining to hydrolysates or mixtures of peptides. Figure 9.6 represents a general scheme for the synthesis of millet-based bioactive peptides.

Bisht et al. (2016) have reported several antimicrobial proteins/peptides extracted from finger millet, barnyard millet, and proso millet. Similarly, FFMp4, FFMp6, and FFMp10 peptides derived from foxtail millets showed significant antimicrobial activity against *E. coli* ATCC 8099 (Amadou et al. 2013a, b). Agrawal et al. (2016) reported a peptide, SDRLLGPNNQYLPK, isolated from the pearl millet to possess significant anti-oxidative potential. Further, Agrawal et al. (2019) reported antioxidant properties of two peptides (TSSSLNMAVRGGLTR and STTVGLGISMRSASVR) derived from finger millet. Foxtail millet derived biopeptides, rich in Tyrosine/Leucine have potent antioxidant properties (Majid and Priyadarshini 2019). Chen et al. (2017) reported the antihypertensive activities of protein hydrolysates from foxtail millet.

Antioxidant activity of peptides may be due to the presence of various amino acids such as alanine, cysteine, histidine, lysine, leucine, methionine, proline, valine, tryptophan, and tyrosine. The valine-alanine-proline epitope has been reported to possess strong ACE (Angiotensin-I Converting Enzyme) inhibitory effect. Millet and fermented millet foods have been found rich in such amino acids and thus have immense antioxidant and antihypertensive potential (Duodu and Awika 2019).

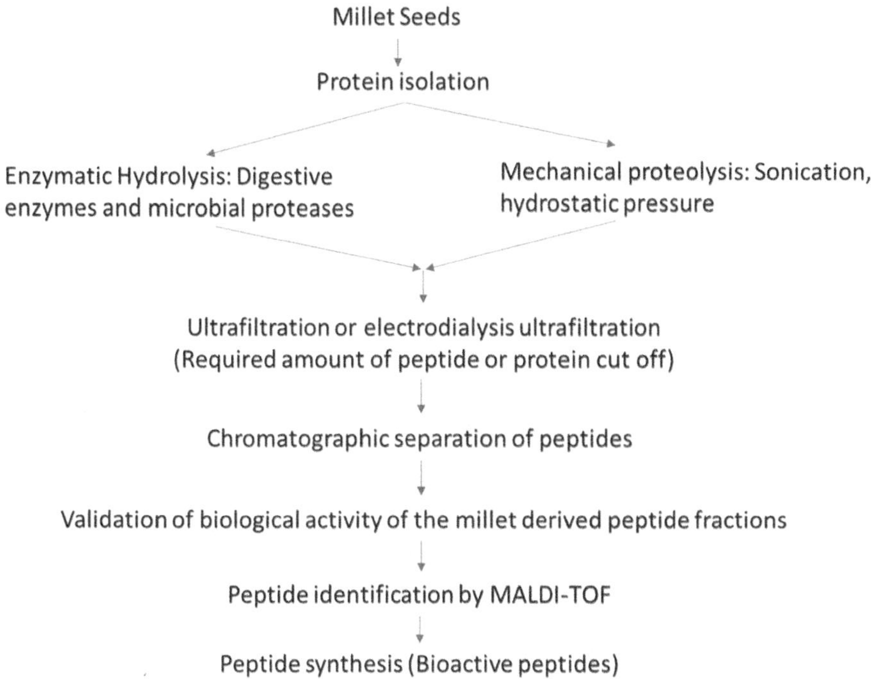

Fig. 9.6 Synthesis of millet-based bioactive peptides

9.2.9 Antinutrients

Millets are known to contain antinutrients like oxalates, phytates, and condensed tannins which reduce nutrient bioavailability. Phytates and oxalates chelate divalent mineral ions and interfere with their uptake in the body. Aleurone layer is abundant in phytates whereas oxalates are abundant in the seed coat (Serna-Saldivar and Rooney 1995).

Phytates are present as myoinositol phosphates with the major function of providing phosphorous and inositol during the course of seed germination. Being present in the aleurone layer, processes like milling and decortication significantly reduce their amount. Previous works have reported phytates in finger millet (529 mg/100 g), pearl millet (500–700 mg/100 g dw), proso millet (170–470 mg/100 g), fonio (514 mg/100 g), and teff (680–1370 mg/100 g) with significant reduction in content on dehulling (Lorenz 1983; Krishnan et al. 2012; Koreissi-Dembele et al. 2013; Baye 2014; Suma and Urooj 2014).

Oxalates chelate dietary calcium by forming calcium oxalate crystals thus adversely affecting the uptake of calcium in the body. Poor calcium uptake is responsible for disorders ranging from rickets in children to osteomalacia in adults. Finger millet, pearl millet, and proso millet have been reported to contain different amounts of oxalates, i.e., 11.3 mg/100 g, 21 mg/100 g, and 33 mg/100 g, respectively (Serna-Saldivar and Espinosa-Ramírez 2019).

In all the millets, only few finger millet varieties are known to possess condensed tannins. Previous works have reported that condensed tannins are associated with poor protein digestibility (Kumar et al. 2016). Being predominantly present in testa, processes like decortication lead to significant reduction in their content.

Vitexin, glucosylvitexin, and glucosylorientin found in pearl millet have been associated with the incidence of endemic goiter in certain communities of Africa. These compounds found mainly in bran layer often inhibit the formation of triiodothyronine from thyroxine (Taylor 2017).

9.3 Bioaccessibility and Bioavailability of Phenolic Compounds from Millets

Bioavailability is a major step in determining the bioefficacy of bioactive phytochemicals present in different foods such as millet and millet products. It involves complex steps such as liberation, absorption, distribution, metabolism, and bioactivity (Rein et al. 2013). Release of these compounds from the food and modification in the digestive system plays an important role in their bioavailability. This liberation step, in other words, bioaccessibility is thus the first step of bioavailability. Bioavailability in actual terms is the final amount of compound that reaches the systemic circulation after absorption.

Several in vitro gastrointestinal digestion models have been used for the study of bioaccessibility of phenolic compounds present in different foods. These models utilize appropriate enzymes and digestive fluids to simulate various steps of gastrointestinal digestion from mouth, stomach, small intestine to large intestine along with use of colonic microorganisms to further mimic the natural process. Bioaccessibilty of phenolics from millets have been associated with the type of phenolic group, presence of phenolic binding compounds, and different methods of processing used such as dehulling, decortication, and cooking. The bioaccessibilty of phenolics from cooked finger, kodo, pearl, proso, and foxtail millet grains increases their release from these grains after in vitro digestion and microbial fermentation in comparison to uncooked grains (Chandrasekara and Shahidi 2012). Rao and Muralikrishna (2002) reported that malting of finger millet showed an increase in the content of free phenolics such as gallic acids whereas there was a decrease in protocatechuic acid content. Majority of polyphenols are concentrated in the seed coat (Akanbi et al. 2019). Mohamed et al. (2010) observed 51% reduction in polyphenol content of pearl millet after dehulling and decortication.

Bioavailability of polyphenols has gained increased attention because of their several health benefits and future prospects of being utilized as therapeutics for the treatment of diseases ranging from diabetes to cardiovascular disorders. Present-day studies on bioavailability utilize in vivo animal and human models. Significant detailed reviews are available for the bioavailability of polyphenols from different food sources emphasizing on their stepwise digestion, release, absorption, metabolism, and elimination (Karas et al. 2017; Teng and Chen 2019). However, the bioavailability of polyphenols from millets is still an unexplored area of great potential with much work to be done. No recent works on animal models are present for millets though much work has been done with bioavailability of sorghum bioactives.

9.4 Health Attributes of Millet Phytochemicals

Millets are considered to be nutraceutical and functional food being a rich source of proteins, vitamins, minerals, dietary fibers, and antioxidant activities (Saleh et al. 2013). In addition to these essential minerals, vitamins, and fibers, the other bioactive compounds such as resistant starch, lipids, phenolic acid, avenanthramides, flavonoids, lignans, phytosterols, phytic acid, and tannins present in millets have increasing health effects (Manach et al. 2005; Scalbert et al. 2005; Chandrasekara and Shahidi 2012). Epidemiological evidences have suggested that inclusion of millets in regular diet has protective and preventive effects against age-related ailments like diabetes, high blood pressure, cardiovascular diseases, cancer, Parkinson's disease, and other metabolic syndromes (Saleh et al. 2013; Rao et al. 2017; Ofosu et al. 2020). Besides, soothing alkaline nature of millets is helpful in maintaining pH balance in the body (Sarita and Singh 2016). Millets are gluten-free hence serve as an alternative food to wheat, barley, oat, and other gluten-containing cereals for people allergic to gluten and suffering from Celiac disease (Amadou et al. 2013a, b).

9.4.1 Antioxidant Potential

Millets, including kodo, barnyard, proso, finger, foxtail, and little millet, are rich source of both soluble and insoluble bound polyphenols, which have rich antioxidant potential with metal chelating and reducing powers (Saleh et al. 2013). A number of antioxidants present in millets have beneficial effect by neutralizing free radicals (reactive oxygen species) and boost overall immunity and human health. The free radicals can cause oxidative stress, which in turn is associated with ailments like cancer, arthritis, respiratory, and compromised immunity. Thus, consumption of millets is good for health and in scavenging free radicals from liver and kidney. Chandrasekara and Shahidi (2012) studied soluble and insoluble bound phenolics in various millets using trolox equivalent antioxidant capacity (TEAC) and reducing power. They reported that kodo millet had the highest total phenolic content, while proso millet possessed the minimum content. Further, in HPLC analysis they found that bound forms had high ferulic acid and *p*-coumaric acid than the soluble forms. Sreeramulu et al. (2009) conducted antioxidant studies in cereals, millets, and pulses and detailed that finger millet showed the highest ferric reducing antioxidant power (FRAP) and DPPH (2,2-Diphenyl-1-picryl hydrazyl) radical scavenging assay and total phenolic content (TPC) as 471.71/100 g, 1.73/100 g, and 373/100 g, respectively, among cereals. Saleh et al. (2013) summarized that presently over 50 phenolic compounds: phenolic acids, their derivatives like dehydrodiferulates and dehydrotriferulates, flavones, flavonols, and flavanonols have been identified in millets using HPLC and HPLC mass spectrometry (MS). The polyphenolic compounds vary among the millets, as foxtail millet has 47 mg polyphenolics/100 g, while proso millet possesses 29 mg polyphenolics/100 g (Saleh et al. 2013). Chandrasekara and Shahidi (2012) demonstrated that dehulling and hydrothermal treatment have an adverse effect on total phenolic content in millets and found that the phenolic extract values were higher in hull followed by whole grain, dehulled grain, and cooked dehulled. Colored varieties of sorghum, finger, and foxtail millets have higher quenching than white seeded varieties indicating role of phenolics in seed coat color for higher antioxidant activities.

9.4.2 Antidiabetic Properties

Diabetes mellitus causes different body disorders and characterized by high blood sugar levels with imbalance in protein, carbohydrate, and lipid metabolism. It can be categorized into four major types; insulin-dependent, noninsulin-dependent, gestational, and secondary caused by other conditions. Type 1 diabetes is most common childhood disease recognized by loss of insulin production as a result of autoimmune destruction of pancreatic cells. Type 2 diabetes is the most common in adults above the age of 40 and characterized by resistance for insulin and relative insulin deficiency. Albeit, artificial inhibitors of alpha-glucosidase and pancreatic amylase are important in the management of postprandial hyperglycemia but the inclusion of natural inhibitors is safer. It has been reported that consumption of whole grains is helpful and clinically proven in lowering the diabetic effects (Shobana et al. 2009; Kim et al. 2011; Saleh et al. 2013). Rao et al. (2017) reported that regular intake of millets is helpful in preventing diabetes due to the higher level of magnesium in these small grains, which in turn enhances the efficiency of insulin and glucose receptors. Kumari and Sumathi (2002) described that finger millet-based diet has a lower glycemic index due to the high fiber content and α-amylase inhibition properties, which are helpful in reducing starch digestibility and absorption. Further, millet-based diets have been reported for their antioxidant potential, nerve growth factor production, and wound healing properties. Ugare et al. (2011) documented the role of dehulled and heat-treated barnyard in controlling type 2 diabetes. Finger millet phenolic compounds like gallic, protocatechuic, *p*-coumaric, *p*-hydroxybenzoic, ferulic, syringic, vanillic, trans-cinnamic acids and quercetin inhibited aldose reductase activity and cataract eye lens effectively (Saleh et al. 2013). This aldose reductase inhibition was due to the hydroxyl group at the fourth position in phenolics and a neighboring *o*-methyl group. Rao et al. (2017) summarized that slowly digestible starch, high fiber, magnesium, vitamin E, phenolics, and tannins present in sorghum were helpful in lowering diabetic effects and hyperlipidemia. Proso millets and foxtail millets also have anti hyperglycemic and hyperlipidemic effects (Park et al. 2008).

9.4.3 Millets for Cardiovascular Diseases

Physical inactivity, stress, unhealthy diet, smoking, and many other reasons may cause strokes and cardiovascular diseases. Regular physical activity, exercise, and intake of healthy diet may curb cardiovascular diseases. Millets are fully loaded with magnesium, potassium, lignans, antioxidants, and fibers and have the ability to reduce the risk of heart strokes, high blood pressure, low density cholesterol, and atherosclerosis. Rao et al. (2017) reported that the inclusion of barnyard, proso, and finger millets in diet enhanced plasma level of adiponectin, high density lipoprotein (HDL), and reduced triglycerides in hyperlipidemia. The study conducted for cholesterol absorption, hyperlipidemia, high blood pressure, and non-HDL cholesterol concentrations indicated the beneficial role of grain sorghum lipid extract and policosanol on lowering non-HDL and inhibiting endogenous cholesterol synthesis. Saleh et al. (2013) summarized that phenolic extracts of little, pearl, proso, kodo, foxtail, and finger millets had inhibitory effects on lipid peroxidation in vitro copper-mediated human LDL cholesterol oxidation resulting in inhibition by 1–41%, where kodo millet showed superior inhibition. Therefore, millets, including proso, finger, barnyard, foxtail, and pearl millet have significant reducing effects on hyperlipidemia, hyperglycemia, triglycerides, and non-HDL and curtailing risk of cardiovascular diseases.

9.4.4 Anticancer Properties

Cancer is characterized by changes at cellular level due to uncontrolled cell growth and cell division. In general, millets have high phenolic compounds, tannins, fibers, and other phytonutrients than other cereals, which improve health status and lower the risks of colon, breast, and esophageal cancers (Graf and Eaton 1990; Chandrasekara and Shahidi 2011b; Saleh et al. 2013, Anantharaju et al. 2016). It is recommended that intake of 30 g of fiber everyday may reduce the risk of breast cancer chances by 50% (Rao et al. 2017). Sorghum polyphenols and tannins possess anti-mutagenic and anticarcinogenic activities showing potent activity against human melanoma cells (Grimmer et al. 1992; Rao et al. 2017). Shan et al. (2014) extracted a 35 kDa protein from foxtail bran (FBMP), which suppressed colon cancer cell growth by arresting G_1 phase and caused a loss of mitochondrial transmembrane potential.

9.4.5 Millets for Gluten Sensitivity

Gluten is a complex storage protein in wheat, rye, oat, and barley composed of different distinct proteins including gliadins and glutenin in wheat, hordeins in barley, and avenins in oat. Celiac disease and gluten sensitivity are triggered by gluten in genetically susceptible organisms. The disease affects 0.6–1% population globally (Fasano and Catassi 2012). The haplotype HLA-DQ2 is expressed in majority of the patients with celiac disease, while in 5% of cases the haplotype HLA-DQ8 is expressed. Among the gluten-free diets, including amaranth, buckwheat, quinoa, corn, and sorghum, millets are considered as gluten-free functional and nutraceutical food. Millet shave a lot of potential in making gluten-free food and beverages targeting the population suffering from gluten sensitivity and celiac disease (Taylor and Emmambux 2008; Chandrasekara and Shahidi 2011b; Saleh et al. 2013).

9.4.6 Antiaging Properties

Aging is referred to the biological process of growing older and efficiency loss of biochemical and physiological processes. Glycation is a natural nonenzymatic reaction of aldehyde or carbonyl group of reducing sugars with free amino groups of proteins, lipids, and DNA to form Amadori products (Kim et al. 2017). These Amadori products undergo various irreversible dehydration, oxidation, polymerization, cross-linking, and reconformation processes and ultimately give rise to advanced glycation end products. The process of aging is further accelerated with the adverse effects of hyperglycemia and tissue oxidative stress (Kim et al. 2017). Millets, including kodo, proso, finger, foxtail, little millet, are nutri-rich functional food enriched with high levels of micronutrients, vitamins, minerals, phenolic compounds, fibers, and antioxidants therefore helpful in inhibiting glycation and cross-linking of collagen (Saleh et al. 2013).

9.4.7 Antihypertensive Properties

Previous works have reported the antihypertensive potential of bioactive peptides obtained from different millets. The bioactive peptides obtained from millet proteins are able to inhibit angiotensin-I converting enzyme (ACE) and thus interfere with the conversion of angiotensin-I to angiotensin-II. Angiotensin-II constricts the blood vessels and thus leads to hypertension. Further, ACE participates in the degradation of bradykinin which is a vasodilator. The inhibition of ACE hence plays a significant role in reducing hypertension and signifies the antihypertensive potential of millet-derived bioactive peptides. In a study conducted by Chen et al. (2017), it was found that foxtail millet hydrolysates were able to control blood pressure in rats with hypertension.

9.5 Conclusion

In recent times, millets have gained immense importance as robust and highly nutritious crops possessing numerous health benefits. Being a rich source of phytochemicals of great nutraceutical potential the present times require more detailed and collaborative studies for utilizing the true potential of these neglected and under-researched cereals. Most of the present studies have been focused on the whole grain and extracts obtained from millets. Future work on these crops should emphasize on better methods of processing and their effects on the bioavailability and content of their phytochemicals. More detailed studies are required for isolation and identification of important phytochemicals followed by the validation of their nutraceutical potential in vitro and in vivo in different cell lines, animals, and humans. More concrete efforts are further required for establishing cost-effective management systems for extensive farming and distribution of millet species around the globe.

References

Agrawal H, Joshi R, Gupta M (2016) Isolation, purification and characterization of antioxidative peptide of pearl millet (*Pennisetum glaucum*) protein hydrolysate. Food Chem 204:365–372. https://doi.org/10.1016/j.foodchem.2016.02.127

Agrawal H, Joshi R, Gupta M (2019) Purification, identification and characterization of two novel antioxidant peptides from finger millet (*Eleusine coracana*) protein hydrolysate. Food Res Int 120:697–707. https://doi.org/10.1016/j.foodres.2018.11.028

Akanbi TO, Timilsena Y, Dhital S (2019) Chapter 10: Bioactives from millet: properties and effects of processing on bioavailability. In: Bioactive factors and processing technology for cereal foods. Springer, Singapore, p 171

Amadou I, Gounga ME, Le GW (2013a) Millets: nutritional composition, some health benefits and processing-a review. Emir J Food Agric 25(7):501–508

Amadou I, Le GW, Amza T, Sun J, Shi YH (2013b) Purification and characterization of foxtail millet-derived peptides with antioxidant and antimicrobial activities. Food Res Int 51 (1):422–428. https://doi.org/10.1016/j.foodres.2012.12.045

Anantharaju PG, Gowda PC, Vimalambike MG, Madhunapantula SV (2016) An overview on the role of dietary phenolics for the treatment of cancers. Nutr J 15(1):1–16. https://doi.org/10.1186/s12937-016-0217-2

Asharani VT, Jayadeep A, Malleshi NG (2010) Natural antioxidants in edible flours of selected small millets. Int J Food Prop 13(1):41–50. https://doi.org/10.1080/10942910802163105

Atanasova-Penichon V, Barreau C, Richard-Forget F (2016) Antioxidant secondary metabolites in cereals: potential involvement in resistance to Fusarium and Mycotoxin accumulation. Front Microbiol 7:566. https://doi.org/10.3389/fmicb.2016.00566

Avato P, Bianchi G, Murelli C (1990) Aliphatic and cyclic lipid components of sorghum plant organs. Phytochemistry 29:1073–1078

Awika JM, Rooney LW (2004) Sorghum phytochemicals and their potential impact on human health. Phytochemistry 65:1199–1221

Baye K (2014) Teff: nutrient composition and health benefits. Working paper 67 Ethiopia Strategy Support Program. Addis Ababa, Ethiopia. http://ebrary.ifpri.org/cdm/ref/collection/p15738coll2/id/128334

Bhandari SR, Lee YS (2013) The contents of phytosterols, squalene, and vitamin E and the composition of fatty acids of Korean landrace Setaria italica and Sorghum bicolor seeds. Korean J Plant Res 26:663–672

Bisht A, Thapliyal M, Singh A (2016) Screening and isolation of antibacterial proteins/peptides from seeds of millets. Int J Curr Pharm Res 8(3):96–99. https://doi.org/10.22159/ijcpr.2016v8i4.15271

Chandrasekara A, Shahidi F (2010) Content of insoluble bound phenolics in millets and their contribution to antioxidant capacity. J Agric Food Chem 58(11):6706–6714. https://doi.org/10.1021/jf100868b

Chandrasekara A, Shahidi F (2011a) Determination of antioxidant activity in free and hydrolyzed fractions of millet grains and characterization of their phenolic profiles by HPLC-DAD-ESI-MSn. J Funct Foods 3(3):144–158. https://doi.org/10.1016/j.jff.2011.03.007
Chandrasekara A, Shahidi F (2011b) Antioxidant phenolics of millet control lipid peroxidation in human LDL cholesterol and food systems. J Am Oil Chem Soc. https://doi.org/10.1007/s11746-011-1918-5
Chandrasekara A, Shahidi F (2012) Bio-accessibility and antioxidant potential of millet grain phenolics as affected by simulated in vitro digestion and microbial fermentation. J Funct Foods 4(1):226–237
Chauhan M, Sonawane SK, Arya SS (2018) Nutritional and nutraceutical properties of millets: a review. Clin J Nutr Dietetics 1(April):513–518. Retrieved from https://asclepiusopen.com/clinical-journal-of-nutrition-and-dietetics/volume-1-issue-1/7.pdf
Chen J, Duan W, Ren X, Wang C, Pan Z, Diao X, Shen Q (2017) Effect of foxtail millet protein hydrolysates on lowering blood pressure in spontaneously hypertensive rats. Eur J Nutr 56 (6):2129–2138. https://doi.org/10.1007/s00394-016-1252-7
Chethan S, Malleshi NG (2007) Finger millet polyphenols: characterization and their nutraceutical potential. Am J Food Technol 2:582–592
Daliri EBM, Lee BH, Oh DH (2016) Current perspectives on antihypertensive probiotics. Probiotics Antimicrob Proteins 9:91–101. https://doi.org/10.1007/s12602-016-9241-y
Duodu KG, Awika JM (2019) Phytochemical-related health promoting attributes of sorghum and millets. In: Taylor JRN, Duodu KG (eds) Sorghum and millets, chemistry, technology and nutritional attributes, 2nd edn. AACC International, Eagan, pp 225–258
Dykes L, Rooney LW (2006) Sorghum and millet phenols and antioxidants. J Cereal Sci 44 (3):236–251. https://doi.org/10.1016/j.jcs.2006.06.007
Esche R, Scholz B, Engel KH (2013) Online LCeGC analysis of free sterols/stanols and intact steryl/stanyl esters in cereals. J Agric Food Chem 61:10932–10939
Fasano A, Catassi C (2012) Celiac disease. N Engl J Med 367(25):2419–2426
Goudar G, Sathisha GJ (2016) Effect of processing on ferulic acid content in foxtail millet (Setaria italica) grain cultivars evaluated by HPTLC. Orient J Chem 32(4):2251–2258. https://doi.org/10.13005/ojc/320458
Graf E, Eaton JW (1990) Antioxidant functions of phytic acid. Free Radic Biol Med 8(1):61–69
Grimmer HR, Parbhoo V, McGrath RM (1992) Anti-mutagenicity of polyphenol-rich fractions from Sorghum bicolor grain. J Sci Food Agric 59(2):251–256
Hackman RM, Polagruto JA, Zhu QY, Sun B, Fujii H, Keen CL (2008) Flavanols: digestion, absorption and bioactivity. Phytochem Rev 7(1):195–208. https://doi.org/10.1007/s11101-007-9070-4
Hassan AS, Burton RA (2018) Role, importance and biosynthesis of cell wall-bound phenolic acids in cereals. Annu Plant Rev 1:737–766. https://doi.org/10.1002/9781119312994.apr0629
Hostetler GL, Ralston RA, Schwartz SJ (2017) Flavones: food sources, bioavailability. Adv Nutr 8:423–435. https://doi.org/10.3945/an.116.012948.FIGURE
Jiang Y, Wang T (2005) Phytosterols in cereal by-products. J Am Oil Chem Soc 82(6):439–444. https://doi.org/10.1007/s11746-005-1090-5
Karas M, Jakubczyk A, Szymanowska U, Złotek U, Zielinska E (2017) Digestion and bioavailability of bioactive phytochemicals. Int J Food Sci Technol 52:291–305
Khangura BS, Gill KS, Phul PS (1980) Combining ability analysis of beta-carotene, total carotenoids and other grain characteristics in pearl millet. Theor Appl Genet 56(1–2):91–96. https://doi.org/10.1007/BF00264433
Kim JS, Hyun TK, Kim MJ (2011) The inhibitory effects of ethanol extracts from sorghum, foxtail millet and proso millet on α-glucosidase and a-amylase activities. Food Chem 124:1647–1651
Kim CS, Park S, Kim J (2017) The role of glycation in the pathogenesis of aging and its prevention through herbal products and physical exercise. J Exerc Nutr Biochem 21(3):55

Koreissi-Dembele Y, Fanou-Fogny N, Hulshof PJ, Brouwer ID (2013) Fonio (*Digitaria exilis*) landraces in Mali: nutrient and phytate content, genetic diversity and effect of processing. J Food Compos Anal 29(2):134–143

Krishnan R, Dharmaraj U, Malleshi NG (2012) Influence of decortication, popping and malting on bioaccessibility of calcium, iron and zinc in finger millet. LWT Food Sci Technol (Lebensmittel-Wissenschaft-Technol) 48:169–174

Kumar SI, Babu CG, Reddy VC, Swathi B (2016) Anti-nutritional factors in finger millet. J Nutr Food Sci 06(03):5–6. https://doi.org/10.4172/2155-9600.1000491

Kumari PL, Sumathi S (2002) Effect of consumption of finger millet on hyperglycemia in non-insulin dependent diabetes mellitus (NIDDM) subjects. Plant Foods Hum Nutr 57:205–213

Liu RH (2007) Whole grain phytochemicals and health. J Cereal Sci 46(3):207–219. https://doi.org/10.1016/j.jcs.2007.06.010

Lorenz K (1983) Tannins and phytate content in proso millet (*Panicum miliaceum*). Cereal Chem 60:424–426

Majid A, Priyadarshini CG (2019) Millet derived bioactive peptides: a review on their functional properties and health benefits. Crit Rev Food Sci Nutr:1–10. https://doi.org/10.1080/10408398.2019.1686342

Mal B, Padulosi S, Ravi SB (2010) Minor millets in South Asia: learnings from IFADNUS Project in India and Nepal. Bioversity Intl and Chennai India, Maccarese, Rome, Italy and M.S. Swaminathan Research Foundation. pp 1–185

Manach C, Williamson G, Morand C, Scalbert A, Remesy C (2005) Bioavailability and bioefficacy of polyphenols in humans. 1. Review of 97 bioavailability studies. Am J Clin Nutr 81:230S–242S

Mezzomo N, Ferreira SRS (2016) Carotenoids functionality, sources, and processing by supercritical technology: a review. J Chem. https://doi.org/10.1155/2016/3164312

Mohamed EA, Ali NA, Ahmed SH, Ahmed IAM, Babiker EE (2010) Effect of refrigeration process on antinutrients and HCl extractability of calcium, phosphorus and iron during processing and storage of two millet cultivars. Radiat Phys Chem 79:791–796

Moreno J, Peinado R (2012) Non-flavonoid phenols. In: Enological chemistry. Elsevier Inc., San Diego, CA, pp 53–76. https://doi.org/10.1016/B978-0-12-388438-1.00005-4

N'Dri D, Mazzeo T, Zaupa M, Ferracane R, Fogliano V, Pellegrini N (2013) Effect of cooking on the total antioxidant capacity and phenolic profile of some whole-meal African cereals. J Sci Food Agric 93(1):29–36. https://doi.org/10.1002/jsfa.5837

Nambiar VS, Sareen N, Daniel M, Gallego EB (2012) Flavonoids and phenolic acids from pearl millet (Pennisetum glaucum) based foods and their functional implications. Functional Foods Health Dis 2(7):251. https://doi.org/10.31989/ffhd.v2i7.85

Ofosu FK, Elahi F, Daliri EB, Chelliah R, Ham HJ, Kim J, Oh D (2020) Phenolic profile, antioxidant, and antidiabetic potential exerted by millet grain varieties. Antioxidants (Basel) 9 (3):254. https://doi.org/10.3390/antiox9030254

Panche AN, Diwan AD, Chandra SR (2016) Flavonoids: an overview. J Nutr Sci 5. https://doi.org/10.1017/jns.2016.41

Pandey KB, Rizvi SI (2009) Plant polyphenols as dietary antioxidants in human health and disease. Oxidative Med Cell Longev 2(5):270–278. https://doi.org/10.4161/oxim.2.5.9498

Pang M, He S, Wang L, Cao X, Cao L, Jiang S (2014) Physicochemical properties, antioxidant activities and protective effect against acute ethanol-induced hepatic injury in mice of foxtail millet (*Setaria italica*) bran oil. Food Funct 5:1763–1770

Park KO, Ito Y, Nagasawa T, Choi MR, Nishizawa N (2008) Effects of dietary Korean proso-millet protein on plasma adiponectin, HDL cholesterol, insulin levels, and gene expression in obese type 2 diabetic mice. Biosci Biotechnol Biochem 72(11):2918–2925

Piironen V, Lindsay DG, Miettinen TA, Toivo J, Lampi AM (2000) Plant sterols: biosynthesis, biological function and their importance to human nutrition. J Sci Food Agric 80(7):939–966. https://doi.org/10.1002/(SICI)1097-0010(20000515)80:7<939:AID-JSFA644>3.0.CO;2-C

Pradeep PM, Sreerama YN (2015) Impact of processing on the phenolic profiles of small millets: evaluation of their antioxidant and enzyme inhibitory properties associated with hyperglycemia. Food Chem 169:455–463. https://doi.org/10.1016/j.foodchem.2014.08.010

Pradeep PM, Sreerama YN (2017) Soluble and bound phenolics of two different millet genera and their milled fractions: comparative evaluation of antioxidant properties and inhibitory effects on starch hydrolysing enzyme activities. J Funct Foods 35:682–693. https://doi.org/10.1016/j.jff.2017.06.033

Pradeep PM, Sreerama YN (2018) Phenolic antioxidants of foxtail and little millet cultivars and their inhibitory effects on α-amylase and α-glucosidase activities. Food Chem 247:46–55. https://doi.org/10.1016/j.foodchem.2017.11.103

Rao MVSSTS, Muralikrishna G (2002) Evaluation of the antioxidant properties of free and bound phenolic acids from native and malted finger millet (ragi, *Eleusine coracana* Indaf-15). J Agric Food Chem 50:889–892

Rao BD, Bhaskarachary K, Christina GDA, Devi GS, Tonapi VA (2017) Nutritional and health benefits of millets. ICAR Indian Institute of Millets Research (IIMR), Rajendranagar, Hyderabad, p 112

Ravisankar S, Abegaz K, Awika JM (2018) Structural profile of soluble and bound phenolic compounds in teff (Eragrostis tef) reveals abundance of distinctly different flavones in white and brown varieties. Food Chem 263:265–274. https://doi.org/10.1016/j.foodchem.2018.05.002

Rein MJ, Renouf M, Cruz-Hernandez C, Actis-Goretta L, Thakkar SK, da Silva Pinto M (2013) Bioavailability of bioactive food compounds: a challenging journey to bioefficacy. Br J Clin Pharmacol 75(3):588–602. https://doi.org/10.1111/j.1365-2125.2012.04425.x

Ryan E, Galvin K, O'Connor TP, Maguire AR, O'Brien NM (2007) Phytosterol, squalene, tocopherol content and fatty acid profile of selected seeds, grains, and legumes. Plant Foods Hum Nutr 62:85–91

Saleh ASM, Zhang Q, Chen J, Shen Q (2013) Millet grains: nutritional quality, processing, and potential health benefits. Compr Rev Food Sci Food Saf 12(3):281–295. https://doi.org/10.1111/1541-4337.12012

Sánchez A, Vázquez A (2017) Bioactive peptides: a review. Food Qual Saf 1:29–46. https://doi.org/10.1093/fqsafe/fyx006

Sarita ES, Singh E (2016) Potential of millets: nutrients composition and health benefits. Journal of Scientific and Innovative Research 5(2):46–50

Scalbert A, Manach C, Morand C, Remesy C, Jimenez L (2005) Dietary polyphenols and the prevention of diseases. Crit Rev Food SciNutr 45:287–306

Selma MV, Espín JC, Tomás-Barberán FA (2009) Interaction between phenolics and gut microbiota: role in human health. J Agric Food Chem 57(15):6485–6501. https://doi.org/10.1021/jf902107d

Serna-Saldivar SO, Espinosa-Ramírez J (2019) Grain structure and grain chemical composition. In: Sorghum and millets: chemistry, technology, and nutritional attributes. Woodhead Publishing, Duxford, UK. https://doi.org/10.1016/B978-0-12-811527-5.00005-8

Serna-Saldivar S, Rooney LW (1995) Structure and chemistry of sorghum and millets. In: Dendy DAV (ed) Sorghum and millets: chemistry and technology. American Association of Cereal Chemists, St. Paul, MN, pp 69–124

Shahidi F, Chandrasekara A (2013) Millet grain phenolics and their role in disease risk reduction and health promotion: a review. J Funct Foods 5(2):570–581. https://doi.org/10.1016/j.jff.2013.02.004

Shan S, Li Z, Newton IP, Zhao C, Li Z, Guo M (2014) A novel protein extracted from foxtail millet bran displays anti-carcinogenic effects in human colon cancer cells. Toxicol Lett 227(2):129–138

Sharma S, Saxena J (2016) Phytochemical screening and quantitative estimation of total phenolic content and total flavonoid content of grains of Paspalum scrobiculatum. Asian J Pharm Clin Res 9(6):73–76. https://doi.org/10.22159/ajpcr.2016.v9i6.13552

Sharma S, Singh R, Rana S (2011) Bioactive peptides: a review. Int J Bioautomat 15(4):223–250. https://doi.org/10.1093/fqs/fyx006

Sharma S, Saxena DC, Riar CS (2015) Antioxidant activity, total phenolics, flavonoids and antinutritional characteristics of germinated foxtail millet (Setaria italica). Cogent Food Agric 1(1). https://doi.org/10.1080/23311932.2015.1081728

Shen R, Yang S, Zhao G, Shen Q, Diao X (2015) Identification of carotenoids in foxtail millet (Setaria italica) and the effects of cooking methods on carotenoid content. J Cereal Sci 61:86–93. https://doi.org/10.1016/j.jcs.2014.10.009

Shobana S, Sreerama YN, Malleshi NG (2009) Composition and enzyme inhibitory properties of finger millet (Eleusine coracana L.) seed coat phenolics: mode of inhibition of a-glucosidase and pancreatic amylase. Food Chem 115(4):1268–1273

Shumoy H, Raes K (2016) Antioxidant potentials and phenolic composition of tef varieties: an indigenous Ethiopian cereal. Cereal Chem 93(5):465–470. https://doi.org/10.1094/CCHEM-10-15-0210-R

Singh P, Raghuvanshi RS (2012) Finger millet for food and nutrition security. Afr J Food Sci 6:77–84

Siwela M, Taylor JRN, de Milliano WAJ, Duodu KG (2007) Occurrence and location of tannins in finger millet grain and antioxidant activity of different grain types. Cereal Chem 84:169–174

Sood P, Singh RK, Manoj P (2019) Millets genetic engineering: the progress made and prospects for the future. Plant Cell Tissue Organ Cult 137:421–439

Sreeramulu D, Reddy CVK, Raghunath M (2009) Antioxidant activity of commonly consumed cereals, millets, pulses and legumes in India. Indian J Biochem Biophys 46(1):112–115

Sridhar R, Lakshminarayana G (1994) Contents of total lipids and lipids classes and composition and fatty acids in small millets: foxtail (Setaria italica), proso (Panicumm iliaceum) and finger (Eleusine coracana). Cereal Chem 71:355–359

Stuper-Szablewska K, Perkowski J (2019) Phenolic acids in cereal grain: occurrence, biosynthesis, metabolism and role in living organisms. Crit Rev Food Sci Nutr 59(4):664–675. https://doi.org/10.1080/10408398.2017.1387096

Suma PF, Urooj A (2014) Nutrients, antinutrients and bioaccessible mineral content (in vitro) of pearl millet as influenced by milling. J Food Sci Technol 51(4):756–761

Taylor JRN (2017) Millets: their unique nutritional and health-promoting attributes. In: Gluten-free ancient grains: cereals, pseudocereals, and legumes: sustainable, nutritious, and health-promoting foods for the 21st century. Elsevier, Amsterdam. https://doi.org/10.1016/B978-0-08-100866-9.00004-2

Taylor JR, Duodu KG (2015) Effects of processing sorghum and millets on their phenolic phytochemicals and the implications of this to the health-enhancing properties of sorghum and millet food and beverage products. J Sci Food Agric 95(2):225–237. https://doi.org/10.1002/jsfa.6713

Taylor JR, Emmambux MN (2008) Gluten-free foods and beverages from millets. In: Gluten-free cereal products and beverages. Academic, Cambridge, MA, p 119-V

Teng H, Chen L (2019) Polyphenols and bioavailability: an update. Crit Rev Food Sci Nutr 59(13):2040–2051. https://doi.org/10.1080/10408398.2018.1437023

Trono D (2019) Carotenoids in cereal food crops: composition and retention throughout grain storage and food processing. Plants (Basel, Switzerland) 8(12):551. https://doi.org/10.3390/plants8120551

Ugare R, Chimmad B, Naik R, Bharati P, Itagi S (2011) Glycemic index and significance of barnyard millet (Echinochloa frumentacae) in type II diabetics. J Food Sci Technol. https://doi.org/10.1007/s13197-011-0516-8

Waseem M, Kumar S, Kumar A (2018) Bioactive peptides. In: Secondary metabolite and functional food components: role in health and disease. Nova Science, New York, pp 259–287. https://doi.org/10.3390/foods6050032

Xiang J, Apea-Bah FB, Ndolo VU, Katundu MC, Beta T (2019) Profile of phenolic compounds and antioxidant activity of finger millet varieties. Food Chem 275:361–368. https://doi.org/10.1016/j.foodchem.2018.09.120

Yang L, Allred KF, Geera B, Allred CD, Awika JM (2012) Sorghum phenolics demonstrate estrogenic action and induce apoptosis in nonmalignant colonocytes. Nutr Cancer 64:419–427

Yang L, Allred K, Dykes L, Allred C, Awika J (2015) Enhanced action of apigenin and naringenin combination on estrogen receptor activation in non-malignant colonocytes: implications on sorghum-derived phytoestrogens. Food Funct 6:749–755

Zhang L, Liu R, Niu W (2014) Phytochemical and antiproliferative activity of proso millet. PLoS One 9(8). https://doi.org/10.1371/journal.pone.0104058

Zhang A, Liu X, Wang G, Wang H, Liu J, Zhao W, Zhang Y (2015) Crude fat content and fatty acid profile and their correlations in foxtail millet. Cereal Chem 92(5):455–459

Science-Led Innovation for Searching and Creating Values in Natural Gene Pool of Millets for Agri-Food Nutrition and Health

10

Anil Kumar, Divya Sharma, Rajesh Kumar Pathak, Ram Sewak Singh Tomar, Aparna Agrawal, and Manoj Kumar Tripathi

Abstract

The green revolution has increased the quantum of major cereals (wheat and rice), which otherwise would have been substantially low across the developing countries; a significant contribution that one can never undervalue. However, focused monoculture of either crop was realized particularly with respect to loss of agro-biodiversity and sustenance of nutri-rich minor crops resulting in poor food grain diversity. Albeit, these crops provide sufficient calories, they do not make a complete diet resulting in malnutrition of over 2000 million people worldwide. Millets are versatile grains valued for their exceptional nutritional profile. Being the reservoir of essential micronutrients and trace elements they are often termed as nature's nutraceutical basket. Furthermore, their climate resilient nature and adaptation to low input agriculture makes them "harbingers for evergreen revolution." The latest advancements in genomics and automate

Anil Kumar and Divya Sharma contributed equally with all other contributors.

A. Kumar (✉)
Directorate of Education, Rani Lakshmi Bai Central Agricultural University, Jhansi, Uttar Pradesh, India

D. Sharma · A. Agrawal
G.B. Pant University of Agriculture and Technology, Pantnagar, Uttarakhand, India

R. K. Pathak
Govind Ballabh Pant Institute of Engineering & Technology, Pauri Garhwal, Uttarakhand, India

R. S. S. Tomar
Department of Biotechnology & Crop Improvement, ICAR-Central Institute of Agricultural Engineering, Bhopal, Madhya Pradesh, India

M. K. Tripathi
ICAR-Central Institute of Agricultural Engineering, Bhopal, Madhya Pradesh, India

A. Kumar et al. (eds.), *Millets and Millet Technology*,
https://doi.org/10.1007/978-981-16-0676-2_10

phenotyping techniques for searching genes and metabolites, coupled with high-throughput transformation processes have opened new avenues for product development in millets. Furthermore, various computational biology platforms help us in analyzing big molecular data of crop plants to identify valuable genes hidden in them. In addition, molecular breeding platforms may be utilized to speed up the introgression of value-added genes in high welding and widely adapted genetic backgrounds. This chapter focuses on searching for values in the natural millet gene pool of millets for developing climate smart crops with value-added traits.

Keywords

Minor millets · Nutraceuticals · Genome · Biofortification · Data mining

10.1 Introduction

In nature, plants are a treasure trove of interesting and useful substances because they must collect everything from the place on earth where they are rooted and when challenged they cannot escape; thus, despite these constraints, they have created the most impressive panoply of products to survive in ever-changing environments. Major cereals comprise a major part of the diets of people, but they lack the essential micronutrients resulting in micronutrient malnutrition across the world. Among the cereal crops, millets, which are having the status of neglected and minor crops, are quite important from the point of nutritional and food security.

Millets are the best-kept secrets of our ancestors. Millets were important staple food in human history, especially for poor people of Asia and Africa's semi-arid tropics. As per an old *Kannada* (south Indian language) saying, it is said that "The rice eater is weightless like a bird; the one who eats *Jowar* (sorghum) is strong like a wolf: one who eats *ragi* (finger millet) remains 'nirogi' (free from illness) throughout his life." Sorghum and pearl millet share the major cultivated area under millets and share 95% of total millet output across the world (Yadav and Rai 2013; Nedumaran et al. 2014). Among the six minor millets, foxtail millet is cultivated as a food crop in semi-arid tropics of Asia and Africa and as a forage crop in Europe, North America, Australia, and North Africa (Austin 2006). Finger millet is grown as a major food crop in Central and East Africa and Asia (Vijayakumari et al. 2003). Millets are hardy in nature, thrive well in drylands under low soil fertility, and humid environments. Some of them have a short life cycle as from seed planting to harvest it takes only 65–70 days. When properly stored and preserved, whole millets can be used as seed for 5 years or more. Therefore, improvement in millet production and value addition through harnessing the genes available in natural gene pool is one of the sustainable approach for achieving food and nutritional security.

10.2 Minor Millets: Neglected Crop Goes Mainstream

Kids, teenagers, pregnant women, and lactating mothers are most vulnerable parts of the population suffering from malnutrition (Arlappa et al. 2010). For instance, the national family health survey of India revealed that approximately 46% of the children under 5 years of age are severely underweight, 38% are stunted, and 19% are wasted (Kanjilal et al. 2010). The overdependence on modern monocultural agricultural systems of major crops (rice, wheat, and potato) for dietary requirements have resulted in the loss of diversity of many nutritionally rich plant species from agricultural production systems. One such group of highly promising crop plant is that of minor millets. Once regarded as major player in household food security, they have been replaced by rice and wheat in the last few decades. However, in the context of the ongoing weather extremes and rising insecurities for food supplies, there has been a revived interest in cultivation and consumption of minor millets (Ravi et al. 2010).

10.3 Natural Gene Pool of Millets and Its Ex Situ Conservation

Mainstream and wild gene pool of millets harbor genes for agronomic and nutritional traits. Modest efforts have been made for collection and conservation of millet genetic resources across the world (Table 10.1). International Crop Research Institute for Semi-arid Tropics (ICRISAT) gene bank is the global repository for millets with a total of 24.7% accessions belonging to millets (Upadhyaya et al. 2015). The Indian germplasm collection of millets is the second largest collection in the world (Table 10.1). The millet diversity spots are well represented in these collections. However, major share belongs to the landraces and genetic stocks of

Table 10.1 Major ex situ gene pools of millets

Crop	Total germplasm holding at global collection of ICRISAT gene bank	Total germplasm holding at national gene bank NBPGR, New Delhi	References
Sorghum	37,491	–	Upadhyaya et al. (2014)
Pearl millet	22,888	7268	Yadav et al. (2017)
Finger millet	6804	10,507	Joshi et al. (2020)
Foxtail millet	1535	4330	Joshi et al. (2020)
Barnyard millet	985	1718	Joshi et al. (2020)
Kodo millet	650	2100	Joshi et al. (2020)
Proso millet	842	–	Joshi et al. (2020)

the cultivated gene pool. Therefore, systematic and exhaustive efforts are required for enriching the collections with wild and unadapted gene pool of millets.

10.4 Searching Nutraceuticals Properties Hidden in Millets

10.4.1 Harnessing Nutraceutical Potential of the Natural Gene Pool

Diverse germplasm collections of millets assembled in gene banks made a modest impact on cultivar development (Upadhyaya et al. 2015). These large collections need rigorous phenotyping for nutraceutical traits in diverse environmental conditions for identification of genes and molecular mechanism governing the traits. However, extremely large size of these collections is the major hindrance in large-scale characterization for searching valuable genes. Concept of core and minicore collection to retain the total genetic diversity within a fraction of the entire germplasm collection with 95% certainty has been proposed for the enhanced utilization of germplasm in breeding programs (Brown 1989; Upadhyaya et al. 2015). Core and minicore collections with drastically reduced size, capturing maximum diversity, and minimum repetitiveness have been developed in millets and served as an ideal gene pool in extracting genes regulating nutraceutical traits.

Promising trait-specific accessions containing high concentration of micronutrients (Fe, Zn, and Ca) and protein in grains have been identified through rigorous multi-locational evaluation of core and minicore collections in sorghum), finger millet (Upadhyaya et al. 2011a), and foxtail millet (Upadhyaya et al. 2011b). In addition to conventional core and minicore subsets, global composite germplasm collections (GCGC), and genotyping-based reference sets have also been developed in millets (Upadhyaya et al. 2015). These collections comprised of wild and weedy relatives, advanced breeding lines, and accessions selected from phenotypic and genotypic evaluation of core and minicore collections.

Genome-wide analysis of the core and diversity panels and its integration with physiochemical characterization can be used to select parental combinations that can maximize genetic polymorphism for traits required for biofortification. Additionally, characterization of these collections through high fidelity next-generation molecular markers will facilitate analysis of population structure and association genetics to identify the genomic regions (QTLs) governing nutraceutical traits in millets.

10.4.2 Genomic Resources for Exploring Nutraceutical Potential of Millets

Deciphering molecular mechanism controlling nutraceutical traits is essential for Designing strategies for crop biofortification program. However, dissection of the nutritional quality traits through conventional methods is cumbersome, owing to their complex genetic control and high G $\times$ E interaction. In recent years, phenomenal development of high-throughput genomic tools facilitated comprehensive genome analysis of nutritional traits in coarse cereals (Singh et al. 2016). Application of such tools has helped in the identification and characterization of genes for genetic biofortification of crops (Velu et al. 2016). Despite well documented genetic variation for grain physiochemical properties in millets, little information is available with respect to genetics and genomics of these traits. Well assembled reference genome sequence is one of the most important requirements for gene identification and trait improvement through molecular breeding. Among the millets, genome sequences are available for sorghum (Paterson et al. 2009), pearl millet (Varshney et al. 2010), foxtail millet (Zhang et al. 2012), and finger millet (Hittalmani et al. 2017). Utilizing the genome sequencing information, a large set of simple sequence repeat (SSR) markers and a physical map has been developed in foxtail millet (Muthamilarasan et al. 2014). Thus, genome sequencing of these four species will provide a thorough insight into the gene structure and genetic network system governing the nutrient trait pathways in millets. However, no genome sequencing information is available for other lesser explored millets. In the absence of reference genome sequence, mutagenesis-based high-throughput genomic approaches such as TILLING and EcoTILLING may be exploited for the identification of genetic determinants governing nutraceutical properties in these neglected small millets. Among the millets, exploitation of genomic resources has just begun for characterizing QTLs, genes, and gene families governing valuable traits (Mudge et al. 2016). For instance, genome-wide association analysis (GWAS) followed by candidate gene-based approach identified genomic regions for ten grain quality traits in community resource sorghum diversity panel comprising 300 lines (Sukumaran et al. 2012). Likewise, QTLs and candidate genes controlling grain iron and zinc content in pearl millet were identified (Anuradha et al. 2017) in GWAS diversity panel. There have been no reports on application of GWAS analysis for genomic regions deciphering nutritional traits in other millets until recent times, though GWAS for agro-morphological traits and stress tolerance has been conducted in foxtail millet and finger millet (Jia et al. 2013).

In addition to QTLs, some specific genes or mechanisms associated with nutritional traits have been identified in millets. In sorghum, GWAS followed by candidate gene identification revealed that polymorphism in five genes (*Sh2*, *Bt*, *Sssl*, *Ae*1, and *Wx*) regulating starch biosynthesis and one gene governing protein content (*O2*) was associated with grain quality traits such as grain hardiness, endosperm texture, and protein content. In another study, polymorphism in coding regions of two alleles of *Tannin-1* gene controlling tannin biosynthesis in sorghum grains revealed that *tan-1a* has a 1-bp deletion and the *tan-1b* has a 10-bp insertion in the coding region (Wu et al. 2012). The polymorphism observed in these genes can be exploited for the development of SNPs to improve protein digestibility of sorghum grain. Haplotype trait association analysis revealed considerable allelic variation in starch metabolism gene pullulanase (*SbPUL*) of sorghum and demonstrated that low-frequency haplotype of *SbPUL* gene enhances the starch digestibility, a critical factor for poor acceptability of sorghum grain (Gilding et al. 2013).

In recent years, transcriptome sequencing has emerged as a cost-effective and high-throughput approach to characterize the genes regulating nutraceutical

Table 10.2 Nutritional genes/transporters responsible for production of nutritionally and health-related molecules in millets

Nutritional genes/ transporters	Characteristic features	Target millet crops	References
Wx gene	Encodes a key enzyme, granule-bound starch synthase 1 (GBSS1), that catalyzes amylase formation	Foxtail millet, barnyard millet, and proso millet	-
Setarin gene	Encoding 16 prolamins which are essential for the accumulation of quality protein in seed	Foxtail millet	Muthamilarasan and Prasad (2015)
NAS 1, *NAS 3*, and *YSL 2*	Genes/transporters associated with high grain Fe contents	Barnyard millet, little millet, and kodo millet	Patel et al. (2015)
NAAT	Genes/transporters associated with high grain Zn contents	Barnyard millet, little millet, and kodo millet	Patel et al. (2015)
FER 1, *FER 2*, *IRT 1*, *NAS 2*, *ZIP 1*, *ZIP 5*, *ZIP 7*, *NAC 5*, *NRAMP 5*, and *NRAMP 7*	Genes/transporters associated with both high Fe and Zn contents in the grain	Barnyard millet, little millet, and kodo millet	Patel et al. (2015)
CAX1, *TPC1*, *CaMK1*, *CaMK2*, *CIPK2*, *CIPK9*, and *CIPK11*	Ca sensor genes responsible for positive influence on seed calcium content	Finger millet	Mirza et al. (2014), Vinoth and Ravindhran (2017)

properties in millets. Recently, finger millet seed transcriptome has emerged as a model to decipher the complex mechanism involved in higher calcium accumulation in grains (Kumar et al. 2014; Kumar et al. 2015a, b; Singh et al. 2014; Mirza et al. 2014, Table 10.2). Singh et al. (2014) generated a unique transcriptome assembly based on Illumina sequencing platform comprising 120, 130 Transcript Assembly Contigs (TACs) through RNA-seq analysis of two finger millet genotypes with contrasting grain calcium content. Analysis of TACs identified 82 calcium sensors and transporter genes classified into eight gene families, namely one calmodulin gene (CaM), one two-pore channel (TPC) protein, four calcium interacting protein kinases (CIPKs), and two calcium dependent protein kinases (CDPKs). These interesting findings in finger millet provide valuable functional markers for improving calcium content in crop breeding programs and encourage development of comprehensive transcriptome assemblies in all millets for functional characterization of genes and gene families governing nutritional traits.

Studies using transcriptomics to understand grain quality traits have been limited in other millets. Nevertheless, studies analyzed transcriptomes of various tissues to provide molecular understanding of other essential traits. For instance, several informative differentially expressed sequence tags (ESTs) during salinity and moisture stress have been generated through cDNA libraries (Lata et al. 2010) and suppression subtraction hybridization (SSH) (Puranik et al. 2011) in foxtail millet. However, major disadvantage of these traditional transcriptomes is that it is expensive, time-consuming, and labor intensive. Keeping this in view, whole transcriptome of foxtail millet through RNA-seq analysis of various tissues based on Illumina GA II platform have been generated (Zhang et al. 2012). In sorghum, a large transcriptome database "Morokoshi" comprising high-quality ESTs and TACs generated from cDNA libraries and RNA-seq analysis of various tissues and 23 previous studies have been developed (Makita et al. 2014). The foxtail and sorghum transcriptome data generated in these recent studies would enable identification of functional markers and candidate genes associated with different traits of interest including nutraceutical properties. Therefore, efforts need to be directed towards the development of transcriptome databases in other neglected millets for getting insight into the molecular mechanisms of nutraceutical traits and ultimately help to develop biofortified millet cultivars.

RNAi-induced gene silencing has been well exploited in deciphering the molecular mechanism underlying poor digestibility of sorghum protein and stable transformants producing grains with improved digestibility were developed (Grootboom et al. 2014). This approach could potentially be used in many other millets as well mainly for silencing the genes governing anti-nutritional factors reducing digestibility and bioavailability of various nutritional compounds including starch, protein, and micronutrients. Availability of an efficient in vitro regeneration protocol is a prerequisite for utilizing RNAi-mediated gene silencing. Therefore, immediate attention need to be paid towards developing efficient regeneration and transformation protocols in many of the small millets for harnessing the potential of functional gene validation through RNAi approach.

Proteomics involves large-scale study of the complete set of proteins and detailed analysis of their expression, structure, and function. In millets, proteomics pipelines can play a driving role in deciphering molecular events involved in nutrient trait pathways by characterizing bioactive proteins and peptides to address the question of nutritional bio-efficacy through proteome mapping, comparative proteomics, and protein–protein interactions. Furthermore, integration of experimental biology and computational modeling with proteomics will greatly facilitate characterization of versatile properties of proteins governing nutraceutical properties. So far, proteomics studies for nutritional traits in millets have been carried out only in finger millet (Kumar et al. 2014, 2016a, b) and sorghum (Cremer et al. 2014). Two calcium-binding proteins, calmodulin (Kumar et al. 2014) and calreticulin (Kumar et al. 2016a, b) responsible for higher calcium accumulation during grain filling stage in finger millet were identified through peptide mass fingerprinting. In sorghum, seed proteome analysis of different sorghum genotypes revealed significant genetic variation in concentration of β, γ, and δ kafirin proteins (Cremer et al. 2014). Further proteome investigation revealed that many of these proteins were involved in formation of kafirin–starch complex, which is a major reason for poor digestibility of sorghum protein. Allelic variation for these proteins may be utilized for improving protein digestibility of sorghum cultivars. Comparative proteomics of small millets and coarse cereals could be exploited for extracting the common functional and regulatory proteins involved in nutrition biosynthesis pathways. Although lesser information is available for nutraceutical traits currently, proteomics enriched datasets for the minor millets will be enhanced with the increasing availability of millet reference genome sequences (Hittalmani et al. 2017).

In recent years, metabolomics has emerged as a potential tool for identification and quantification of essential metabolites produced in specific tissues at a particular developmental stage (Kumar et al. 2016a, b). In addition to structural and functional properties of metabolites, metabolomics can also be used to analyze the metabolism pattern of nutrients in the gastrointestinal tract and its effect on growth and health of the organism (Fig. 10.1).

Almost all the analytical techniques utilized for metabolomics profiling are based on the principle of mass spectroscopy (MS) and nuclear magnetic resonance (Pandey et al. 2016). Metabolomics profiling of finger millet grain through the combination of analytical techniques such as high performance liquid chromatography (HPLC), electrospray ionization mass spectrometry (ESI-MS), and NMR revealed that a wide range of secondary metabolites (polyphenols) are associated with its high nutritive value (Chandrasekara and Shahidi 2010; Banerjee et al. 2012). The integration of this metabolic information with transcriptome datasets and reverse genetics approaches will facilitate the development of biomarkers for effective metabolomics profiling of large breeding populations for deciphering the novel metabolic QTLs in millets.

10.4.3 Bioinformatics and Computational Biology Tools

The increasing availability of big molecular data about the gene pool of crop plants resulting from enhancements in the precision of molecular biology results and in systems biology-based integrated simulation technologies provides us a unique opportunity to analyze and integrate these data through bioinformatics for novel discovery (Kumar et al. 2015a, b; Pathak et al. 2017). Data-driven research has made data integration strategies which are crucial for the advancements and novel insight. Technically, data integration methods have been developed and utilized in both corporate and academic sectors. When it arrives to biological research, there are different types of interpretations and levels of data integration ranging from genomics to proteomics. Data integration and annotation is necessary to ensure the reproducibility of the experimental findings through systems biology and bioinformatics (Lapatas et al. 2015; Kumar et al. 2015a, b).

10.4.3.1 Integration and Visualization of Big Molecular Data

In the post genomic era, big data has really become the most popular term. It refers to the huge quantity of structured, semi-structured, and unstructured data (Rodríguez-Mazahua et al. 2016). These data are generated through high-throughput platforms by using molecular sample of organisms obtained via specialized molecular biology protocol so far it is said to be big molecular data in the molecular biology and

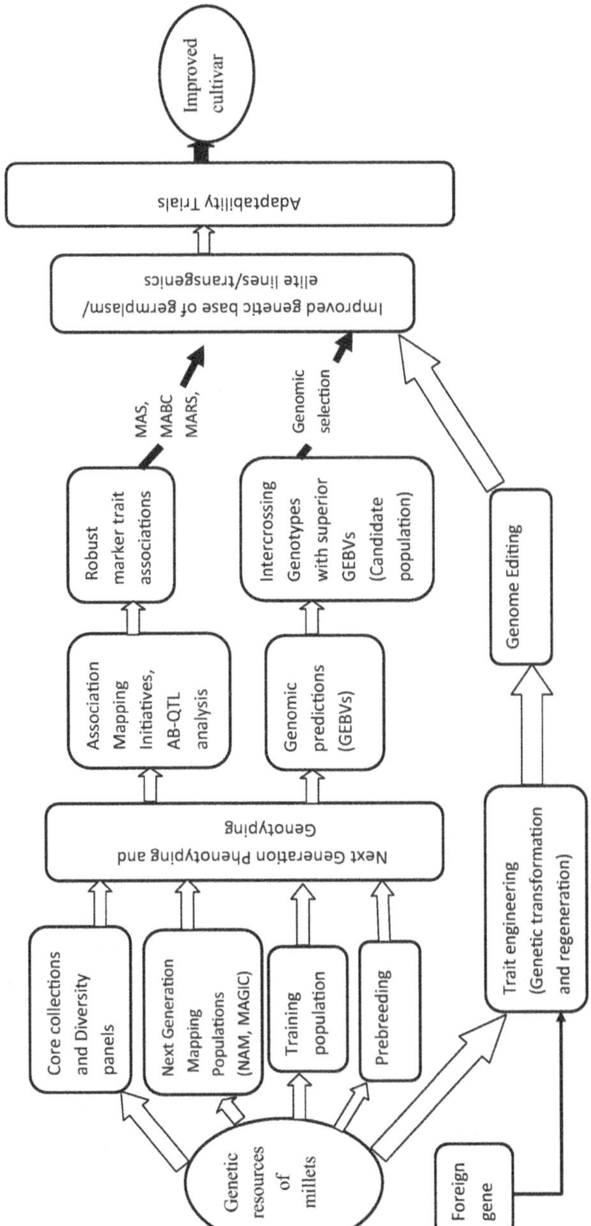

Fig. 10.1 An overview of an integrative genomics and breeding approach for achieving higher genetic gain in millets biofortification program

biotechnology domain. During the last 10–12 years, technical advances in the field of plant molecular biology and biotechnology have made the opportunity to produce enormous amounts of data, heretofore it was impossible to generate such type of data in minimum time and cost. Advances in DNA sequencing technology have shortened the time for the gathering of the raw sequencing data from years to days, and the costs have also reduced 1000-fold from millions to few thousands of dollars. Besides, a similar reduction was also seen in the amount of required material for generation of such data (Frelinger 2015).

In recent years rapid developments in the field of computational systems biology and bioinformatics have made tremendous contributions to decoding the intricacy of such big data. The application of computational methods in the analysis of big molecular data for detection of important genes/proteins, which are directly linked to human health and nutrition have gained a lot of attention. From that aspect, these computational approaches are very crucial for novel insight (Demir 2015). Therefore, harnessing the potential of such computational approaches for integration and analysis of the gene pool data has significant opportunity to identify and visualize the key genes and alleles that can be further utilized to enhance the performance of major crop species for ensuring food and nutritional security (Henry 2011).

10.4.3.2 Data Mining for Computational Systems Biology and Bioinformatics

In recent years, rapid developments in omics platforms have generated big molecular data about crop plants. Decoding of these data for identification of novel genes/ proteins which are essential for biofortification program can be utilized for agri-food nutrition and health requires sophisticated computational analyses (Kumar et al. 2015a, b). Due to availability of huge molecular data in public domain, the development of data mining tools and techniques to solve the complexity of biological problems recently received a lot of attention. Data Mining is the finding of new interesting knowledge from big data. It is also known as Knowledge Discovery in Databases. It requires next-generation technologies and the motivation to investigate the hidden knowledge that resides in the data. It has several tasks to uncover the complexity of big molecular data for extraction of meaningful informations such as classification, estimation, prediction, association rule, clustering, and visualization (Raza 2012). It was suggested that data mining approaches should be applied to overcome the drawbacks of traditional methods. Accordingly, future work should be focussed on the development of integrated databases and user-friendly software for routine use to analyze the natural gene pool data of crop plants (Kumar et al. 2015a, b). This is challenging because gene pools are highly complex and several things are defined poorly. Nonetheless, data mining should facilitate researchers to build earlier and critical decisions in the searching of useful information that can further be utilized in development of the functional food for human health and nutrition (Lee et al. 2008).

Data-driven research has made data integration strategies which are crucial for the advancements and novel insight. Technically, data integration methods have been developed and utilized in both corporate and academic sectors. When it arrives to biological research, there are different types of interpretations and levels of data integration ranging from genomics to proteomics. Data integration and annotation is necessary to ensure the reproducibility of the experimental findings through computational systems biology and bioinformatics (Lapatas et al. 2015; Kumar et al. 2015a, b).

It is mainly based on three approaches called as top-down approach, intermediate approach, and bottom-up approach. In case of top-down, big molecular data of natural gene pool of plants generated through wet-lab experimentation were used to develop integrative models that can be used to predict dynamic behavior of systems at specific scale. This is also known as integrative systems biology. Whereas, bottom-up approach deals with the molecular mechanism and function of cell components via formulating the hypothesis based on perturbation results that can be further utilized in validation through experimental approaches. To determine the intricacy of systems at different levels such as intercommunicating system of systems that cell components, interaction of many molecules in any pathway, and altered states to build profiling data that can be used for filling the gaps between top-down and bottom-up approaches. Intermediate approaches are utilized for the integration of various biomolecules in a controlled way at molecular to systems level that can assess the complexity of biological systems for filling the unknown information (Pathak et al. 2013, 2017; Kumar et al. 2015a, b; Gupta and Misra 2016). Therefore, systems biology and bioinformatics have immense potential to decode the natural gene pool of crops for agri-food nutrition and health.

10.4.3.3 Data Mining Approaches in Gene Discovery

Converting the biological data repository into an informative electronic format is the key function of system biology tools widely used in genomics studies. These large genomic data repositories help in understanding the sequences and expression pattern of the genes regulating complex biological possesses such as biosynthesis and accumulation of nutrients (Zweiger 1999; Kumar et al. 2015a, b). Classification, estimation, prediction, clustering, description, and visualization are the common data mining approaches used in defining big genomic data sets. Based on these available information multiple algorithms are developed for each task. All of these algorithms attempt to fit either a descriptive or predictive model to the data (Dunham 2002). A descriptive model deciphers patterns or relationships in data, whereas a predictive model makes a prediction about data based on already available examples (Sindhu and Sindhu 2017). The key tasks well suited for mining of meaningful new patterns from the big data that information will be further utilized for sustainable agriculture.

10.5 Searching of Genes Responsible for Production of Nutritionally and Health-Related Molecules in Millets with Higher Contents for Biofortification

Various epidemiological reports have suggested a reverse connection between the daily intake of polyphenols-rich diet like millets and threat of human sickness (Scalbert et al. 2005; Arts and Hollman 2005). Millets are a family of cereals, very rich source of nutritionally and health-related molecules like phenolic acids, carotenoids, and various minerals in comparison to wheat and rice (Saleh et al. 2013; Hegde et al. 2005). These small molecules act as therapeutic agents in severe diseases such as diabetes, cancer, and cardiac disorders. These small molecules are synthesized in various pathways as secondary metabolites with numerous metabolic roles in plants. Apart from these poly-phenolic molecules some millet proteins also have the capability to work against chronic human diseases.

Diabetes mellitus is a chronic metabolic disease recognized by hyperglycemia either due to insufficient insulin production or secretion. It was very well known that all millets have some phytochemical which is present in the outer membrane of their seeds have antioxidant properties, and help in lowering glycemic response (Hegde et al. 2005, Table 10.2). Quercetin, a particular polyphenol extract from finger millet seed coats showed effective inhibition against aldolase enzyme (Shobana et al. 2009) and may act as potential alternatives to clinically tested artificially synthesized AR inhibitors (Lee and Kim 2001). Quercetin lowers the hyperglycemic condition by various pathways one is by modulating insulin-dependent pathways such as akt, another one by stimulating GLUT4 translocation which is associated with AMPK activation (Coskun et al. 2005; Stewart et al. 2009; Eid et al. 2015). In another study, proso millet protein concentrate (PMP) upon feeding revamped the glycemic response in severe type-2 diabetic mice by enhancing the plasma levels of adiponectin, HDL cholesterol, and insulin. Some previous studies have shown that intake of calcium and magnesium minerals in daily diet can reduce the effect of type-2 diabetes. Enrichment of finger millet with these calcium, magnesium, iron, and potassium could be directly correlated with its role in reducing the type-2 diabetes risk. Perhaps the presence of some anti-nutritional molecules in millets especially in finger millet like tannins, phenolics, and phytates may also reduce the glucose concentration in cells due to reduction in starch digestibility and absorption. Oxidation stress to cells or ROS generation is an interdependent process which is responsible for regulating several pathological conditions and aging (Chandrasekara and shahidi 2011c). In millet grains, insoluble bound forms of phenolics are present in the majority of free form of phenolics, and main contributor as an antioxidant as well as anti-aging. But apart from the above statement, both free and bound phenolic form present in proso millet were shown to possess very good antioxidation property (Zhang et al. 2014). Use of flavonoids, including quercetin protects cell damage from oxidative stress by reducing the activity of induced nitric oxide synthase (Shoskes 1998).

Millets have varieties of phytochemicals that may protect human population from different types of cancers by suppressing cellular oxidation. Millet's phenolics, tannins, and phytates are known to possess a variety of anticarcinogenic properties (Graf and Eaton 1990; Chandrasekara and Shahidi 2011a, b). For example, ferulic acid, a major bound phenolic acid in millets is known to have anticarcionogenic effect on tongue, colon, and breast cancer cells (Mori et al. 1999; Kawabata et al. 2000; Choi and Park 2015). Similarly, extracts of edible part of proso millet, rich in ferulic acid and chlorogenic acid inhibited the growth of human cancer cell lines, providing a better supplement as an anticancerous agent (Zhang et al. 2014). Phenolics of finger and pearl millet have shown anticancerous activity in HepG2 hepatic cancer cell lines (Singh et al. 2015). Studies conducted in Africa and China reported lower incidence of oesophageal cancer on consumption of sorghum and millet (van Rensburg 1981; Chen et al. 1993). Not only the phenolics and phytates but also the protein of millets is known to possess anticancerous properties. Recently, novel 35 kDa protein homologus to peroxidize extracted from foxtail millet seeds is reported to inhibit colon cancer.

In recent years, various reports have shown a strong relationship between uptake of polyphenols rich diet with decrease risk of coronary heart diseases (Renaud and de Lorgeril 1992; Nardini et al. 2007). Atherosclerosis, major cause of myocardial infarction or cardiac death is caused by oxidation of LDL followed by deposition in narrow arteries (Aviram et al. 2000; Vita 2005). Polyphenols play anti-coronary diseases by function as anti LDL oxidation, antioxidant, anti-platelet, anti-inflammatory as well as by increasing HDL, and improving endothelial function 6 g/kg of pearl millet bran have showed similar antihyperlipidemic effect to with 0.6 mg/kg Simvastatin, a synthetic cholesterol reducing drug in albino rats through stimulating fecal bile acid secretion and by lowering all lipid profile. In one study, streptozotocin-induced diabetes rats showed lower lipid levels after treatment with Nigerian finger millet seed coat matter, suggested high phenolic content in finger millet seed coat. Biosynthesis of all these secondary metabolites initiate by formation of phenylalanine with the help of intermediate products of cellular metabolism through shikimate pathway which further proceed to phenylpropanoid pathway and flavonoid branch pathway.

Expression of genes involved in biosynthesis pathway regulated by different transcription factors, which influenced the generation of specific secondary metabolite in the particular plant in an environment conditions. Myeloblastosis viral oncogene homolog (MYB) represents a large class of transcription factors which is the main regulator of synthesis of plant phenolics. 1R-MYB, 2R-MYB, and 3R-MYB of foxtail millet have been characterized as phenylpropanoid pathway regulator (Muthamilarasan et al. 2014). Expression level of flavonoid biosynthesis genes encoding enzymes were affected in overexpressed soyabean R2R3-MYB in transgenic Arabidopsis plants. In *Arabidopsis thaliana*, *AtDOF4*;2 have shown environmental and tissue-specific regulation on phenylpropanoid metabolism (Skirycz et al. 2007). Searching and identification of genes involved in the biosynthesis, degradation pathways, and in regulation of these metabolites and proteins in millets through various biological approaches will help us to create a plant with desired metabolites as use of nutraceuticals and as a therapeutic.

10.6 Genomic Selection

In recent years, genomic selection (GS) has emerged as a robust tool for enhancing genetic gain of complex traits with low heritability. In GS, potential of a genotype for a trait of interest is estimated based on its genomic estimated breeding value (GEBV). Based on the GEBVs, derived from numerous genome-wide distributed SNPs, training population (TP) and candidate population (CP) are formed for deducing genetics of the trait. GS has been successfully utilized for deducing genetics and enhancing genetic gain for grain micronutrient content (Velu et al. 2016). So far, genomic selection has not been tested in millets. However, with the emerging genome sequence information in many of the minor millets, it will be possible to use GS for enhancing genetic gains for nutritional traits in the biofortification program.

10.7 Genome Editing

In recent years, genome editing technology opened exciting new avenues for trait improvement. Unlike MAS, genome editing inserts desirable alleles or QTLs into a promising genotype to obtain desired phenotypes without following a lengthy backcross cycle. Different genome editing methods modify plant genomes through different sequence-specific nucleases (SSNs) that can be engineered to target genomic sites. The traditional nuclease-based gene editing approaches requiring complex gene constructs are now replaced with more robust and easy to handle CRISPR/Cas9 mediated gene editing. To date, genome editing has not been used in any of the millets. However, advances in establishing regeneration and transformation protocols present an opportunity to address the major production constraints of millets through genome editing technology in near future. One further application of CRISPR/Cas9 that is likely to expand in the future is the molecular stacking of genes governing nutraceutical properties and potential medical compounds in millets.

10.8 Metabolic Pathway Engineering

The metabolic engineering of plants with desirable quality traits involves both conventional plant breeding and biotechnology techniques. Metabolic engineering involves redirecting cellular behavior by modifying the cell's enzymatic, transport, and regulatory functions. In recent years, metabolic engineering has revealed the molecular mechanism of import pathways and key regulatory genes have been cloned. With the rapid advances in omics, bioinformatics, and system biology tools it is possible to identify and clone the valuable genes governing important nutrient pathways in millets through metabolic engineering.

10.9 Conclusion

The emerging tools for next-generation sequencing favors rapid millet genome sequencing. The various omics platforms have great potential in enhancing the use of millets to cope up with malnutrition by micronutrients and foster millets as model systems for sustainable crop biofortification. Identifying rate-limiting synthesis steps may provide objectives for the genetic manipulation of important nutrient accumulation pathways to enhance the production of nutraceutical compounds. Targeted gene expression and gene knockdown through genome editing may help to minimize or remove anti-nutritional factors. However, millet value enhancement is constrained by poor knowledge of regulatory mechanism of complex metabolic pathways. With the advancements in "omics" and "informatics" millet researchers have the ability to search valuable genes in the gene pool and simultaneously define their role in value addition and biofortification. To accomplish this, interdisciplinary endeavor ranging from plant science to human physiology and clinical science are needed.

References

Anuradha N, Satyavathi CT, Bharadwaj C, Nepolean T, Sankar SM, Singh SP, Srivastava RK (2017) Deciphering genomic regions for high grain iron and zinc content using association mapping in pearl millet. Front Plant Sci 8:412

Arlappa N, Laxmaiah A, Balakrishna N, Brahmam GNV (2010) Consumption pattern of pulses, vegetables and nutrients among rural population in India. Afr J Food Sci 4(10):668–675

Arts ICW, Hollman PCH (2005) Polyphenols and disease risk in epidemiologic studies. Am J Clin Nutr 81:317–325

Austin D (2006) Fox-tail millets (Setaria: Poaceae): abandoned food in two hemispheres. Econ Bot 60:143–158

Aviram M, Dornfeld L, Rosenblat M, Volkova N, Kaplan M, Coleman R, Hayek T, Presser D, Fuhrman B (2000) Pomegranate juice consumption reduces oxidative stress, atherogenic modifications to LDL, and platelet aggregation: studies in humans and in atherosclerotic apolipoprotein E-deficient mice. Am J Clin Nutr 71:1062–1076

Banerjee S, Sanjay KR, Chethan S, Malleshi NG (2012) Finger millet (Eleusine coracana) polyphenols: investigation of their antioxidant capacity and antimicrobial activity. Afr J Food Sci 6:362–374

Brown AHD (1989) Core collections: a practical approach to genetic resources management. Genome 31(2):818–824

Chandrasekara A, Shahidi F (2010) Content of insoluble bound phenolics in millets and their contribution to antioxidant capacity. J Agric Food Chem 58:6706–6714. https://doi.org/10.1021/jf100868b

Chandrasekara A, Shahidi F (2011a) Antiproliferative potential and DNA scission inhibitory activity of phenolics from whole millet grains. J Funct Foods 3:159–170

Chandrasekara A, Shahidi F (2011b) Inhibitory activities of soluble and bound millet seed phenolics on free radicals and reactive oxygen species. J Agric Food Chem 59:428–436

Chandrasekara A, Shahidi F (2011c) Determination of antioxidant activity in free and hydrolysed fractions of millet grains and characterization of their phenolic profiles by HPLC-DAD-ESI-MSn. J Funct Foods 3:144–148

Chen F, Cole P, Mi Z, Zing LY (1993) Corn and wheat-flour consumption and mortality from esophageal cancer in Shanxi, China. Int J Cancer 53:902–906

Choi YE, Park E (2015) Ferulic acid in combination with PARP inhibitor sensitizes breast cancer cells as chemotherapeutic strategy. Biochem Biophys Res Commun 458:520–524

Coskun O, Kanter M, Korkmaz A, Oter S (2005) Quercetin, a flavonoid antioxidant, prevents and protects streptozotocin-induced oxidative stress and beta-cell damage in rat pancreas. Pharmacol Res 51:117–123

Cremer JE, Bean SR, Tilley MM, Ioerger BP, Ohm JB, Kaufman RC, Godwin ID (2014) Grain sorghum proteomics: integrated approach toward characterization of endosperm storage proteins in kafirin allelic variants. J Agric Food Chem 62(40):9819–9831

Demir A (2015) Possible effect of biotechnology on plant gene pools in Turkey. Biotechnol Biotechnol Equip 29(1):1–9

Dunham MH (2002) Data mining: introductory and advanced topics. Prentice Hall, Upper Saddle River, NJ

Eid HM, Nachar A, Thong F, Sweeney G, Haddad PS (2015) The molecular basis of the antidiabetic action of quercetin in cultured skeletal muscle cells and hepatocytes. Pharmacogn Mag 11:74–81

Frelinger JA (2015) Big data, big opportunities, and big challenges. J Invest Dermatol Symp Proc 17(2):33–35

Gilding EK, Frere CH, Cruickshank A, Rada AK, Prentis PJ, Mudge AM, Godwin ID (2013) Allelic variation at a single gene increases food value in a drought-tolerant staple cereal. Nat Commun 4:1483

Graf E, Eaton JW (1990) Antioxidant functions of phytic acid. Free Radic Biol Med 8:61–69

Grootboom AW, Mkhonza NL, Mbambo Z, O'Kennedy MM, Da Silva LS, Taylor J, Mehlo L (2014) Co-suppression of synthesis of major α-kafirin sub-class together with γ-kafirin-1 and γ-kafirin-2 required for substantially improved protein digestibility in transgenic sorghum. Plant Cell Rep 33(3):521–537

Gupta MK, Misra K (2016) A holistic approach for integration of biological systems and usage in drug discovery. Netw Model Anal Health Informatics Bioinformatics 5(1):4

Hegde PS, Rajasekaran NS, Chandra TS (2005) Effects of the antioxidant properties of millet species on oxidative stress and glycemic status in alloxan induced rats. Nutr Res 25:1109–1120

Henry RJ (2011) Next-generation sequencing for understanding and accelerating crop domestication. Brief Funct Genomics 11(1):51–56

Hittalmani S, Mahesh HB, Shirke MD, Biradar H, Uday G, Aruna YR, Mohanrao A (2017) Genome and transcriptome sequence of finger millet (Eleusine coracana (L) Gaertn) provides insights into drought tolerance and nutraceutical properties. BMC Genomics 18(1):465

Jia G, Huang X, Zhi H, Zhao Y, Zhao Q, Li W, Zhu C (2013) A haplotype map of genomic variations and genome-wide association studies of agronomic traits in foxtail millet (Setaria italica). Nat Genet 45(8):957

Joshi DC, Meena RP, Chandora R (2020) Genetic resources: collection, characterization, conservation and documentation. In: Singh M, Sood S (eds) Millets and pseudocereals genetic resources and breeding advancements. Wood Head Publishing, Cambridge, UK, pp 19–28

Kanjilal B, Mazumdar PG, Mukherjee M, Rahman MH (2010) Nutritional status of children in India: household socio-economic condition as the contextual determinant. Int J Equity Health 9 (1):19

Kawabata K, Yamamoto T, Hara A, Shimizu M, Yamada Y, Matsunaga K, Tanaka T, Mori H (2000) Modifying effects of ferulic acid on azoxymethane-induced colon carcinogenesis in F344 rats. Cancer Lett 157:15–21

Kumar A, Mirza N, Charan T, Sharma N, Gaur VS (2014) Isolation, characterization and immunolocalization of a seed dominant CaM from finger millet (Eleusine coracana L Gartn) for studying its functional role in differential accumulation of calcium in developing grains. Appl Biochem Biotechnol 172:2955–2973

Kumar A, Gaur VS, Goel A, Gupta AK (2015a) De novo assembly and characterization of developing spikes transcriptome of finger millet (Eleusine coracana): a minor crop having nutraceutical properties. Plant Mol Biol Report 33(4):905–922

Kumar A, Pathak RK, Gupta SM, Gaur VS, Pandey D (2015b) Systems biology for smart crops and agricultural innovation: filling the gaps between genotype and phenotype for complex traits linked with robust agricultural productivity and sustainability. Omics 19(10):581–601

Kumar A, Metwal M, Kaur S, Gupta AK, Puranik S, Singh S, Yadav R (2016a) Nutraceutical value of finger millet [Eleusine coracana (L) Gaertn], and their improvement using omics approaches. Front Plant Sci 7:934

Kumar A, Sharma D, Tiwari A, Jaiswal JP, Singh NK, Sood S (2016b) Genotyping-by-sequencing analysis for determining population structure of finger millet germplasm of diverse origins. Plant Genome 9. https://doi.org/10.3835/plantgenome2015070058

Lapatas V, Stefanidakis M, Jimenez RC, Via A, Schneider MV (2015) Data integration in biological research: an overview. J Biol Res-Thessalon 22(1):9–9

Lata C, Sahu PP, Prasad M (2010) Comparative transcriptome analysis of differentially expressed genes in foxtail millet (Setaria italica L) during dehydration stress. Biochem Biophys Res Commun 393(4):720–727

Lee MS, Kim MK (2001) Rat intestinal α-glucosidase and lens aldose reductase inhibitory activities of grain extracts. Food Sci Biotechnol 10:172–177

Lee JK, Williams PD, Cheon S (2008) Data mining in genomics. Clin Lab Med 28(1):145–166

Makita Y, Shimada S, Kawashima M, Kondou-Kuriyama T, Toyoda T, Matsui M (2014) MOROKOSHI: transcriptome database in Sorghum bicolor. Plant Cell Physiol 56(1):e6

Mirza N, Taj G, Arora S, Kumar A (2014) Transcriptional expression analysis of genes involved in regulation of calcium translocation and storage in finger millet [Eleusine coracana (L) Gartn]. Gene 550:171–179. https://doi.org/10.1016/jgene201408005

Mori H, Kawabata K, Yoshimi N, Tanaka T, Murakami T, Okada T, Murai H (1999) Chemopreventive effects of ferulic acid on oral and rice germ on large bowel carcinogenesis. Anticancer Res 19:3775–3778

Mudge SR, Campbell BC, Mustapha NB, Godwin ID (2016) Genomic approaches for improving grain quality of Sorghum. In: The Sorghum genome. Springer, Cham, pp 189–205

Muthamilarasan M, Prasad M (2015) Advances in Setaria genomics for genetic improvement of cereals and bioenergy grasses. Theor Appl Genet 128:1–14

Muthamilarasan M, Khandelwal R, Yadav CB, Bonthala VS, Khan Y, Prasad M (2014) Identification and molecular characterization of MYB transcription factor superfamily in C-4 model plant foxtail millet (Setaria italica L). PLoS One 9:e109920

Nardini M, Natella F, Scaccini C (2007) Role of dietary polyphenols in platelet aggregation a review of the supplementation studies. Platelets 18:224–224

Nedumaran S, Bantilan MCS, Gupta SK, Irshad A, Davis JS (2014) Potential welfare benefit of millets improvement research at ICRISAT: multi country-economic surplus model approach. Socioeconomics Discussion Paper Series Number 15. ICRISAT, Hyderabad

Pandey MK, Roorkiwal M, Singh VK, Ramalingam A, Kudapa H, Thudi M, Varshney RK (2016) Emerging genomic tools for legume breeding: current status and future prospects. Front Plant Sci 7:455

Patel GS, Dubey M, Chandel G (2015) Characterization of metal homeostasis related rice gene orthologs in nutri-rich minor millets. Int J Plant Anim Environ Sci 5(4):14–23

Paterson AH, Bowers JE, Bruggmann R, Dubchak I, Grimwood J, Gundlach H, Schmutz J (2009) The Sorghum bicolor genome and the diversification of grasses. Nature 457(7229):551

Pathak RK, Taj G, Pandey D, Arora S, Kumar A (2013) Modeling of the MAPK machinery activation in response to various abiotic and biotic stresses in plants by a system biology approach. Bioinformation 9(9):443

Pathak RK, Baunthiyal M, Pandey N, Pandey D, Kumar A (2017) Modeling of the jasmonate signaling pathway in Arabidopsis thaliana with respect to pathophysiology of Alternaria blight in Brassica. Sci Rep 7:16790

Puranik S, Jha S, Srivastava PS, Sreenivasulu N, Prasad M (2011) Comparative transcriptome analysis of contrasting foxtail millet cultivars in response to short-term salinity stress. J Plant Physiol 168:280–287

Ravi SB, Hrideek TK, Kumar AK, Prabhakaran TR, Mal B, Padulosi S (2010) Mobilizing neglected and underutilized crops to strengthen food security and alleviate poverty in India. Indian J Plant Genet Resour 23(1):110–116

Raza K (2012) Application of data mining in bioinformatics. arXiv preprint arXiv:12051125

Renaud S, de Lorgeril M (1992) Wine, alcohol, platelets, and the French paradox for coronary heart disease. Lancet 339:1523–1526

Rodríguez-Mazahua L, Rodríguez-Enríquez CA, Sánchez-Cervantes JL, Cervantes J, García-Alcaraz JL, Alor-Hernández G (2016) A general perspective of Big Data: applications, tools, challenges and trends. J Supercomput 72(8):3073–3113

Saleh AS, Zhang Q, Chen J, Shen Q (2013) Millet grains: nutritional quality, processing, and potential health benefits. Compr Rev Food Sci 12(3):281–295

Scalbert A, Manach C, Morand C, Remesy C (2005) Dietary polyphenols and the prevention of diseases. Crit Rev Food Sci Nutr 45:287–306

Shobana S, Sreerama YN, Malleshi NG (2009) Composition and enzyme inhibitory properties of finger millet (Eleusine coracana L) seed coat phenolics: mode of inhibition of a-glucosidase and pancreatic amylase. Food Chem 115:1268–1273

Shoskes DA (1998) Effect of bioflavonoids quercetin and curcumin on ischemic renal injury: a new class of renoprotective agents. Transplantation 66:147–152

Sindhu S, Sindhu D (2017) Data mining and gene expression analysis in bioinformatics. Int J Comput Sci Mob Comput 6(5):72–83

Singh UM, Chandra M, Shankhdhar SC, Kumar A (2014) Transcriptome wide identification and validation of calcium sensor gene family in developing spikes of finger millet genotypes for elucidating their role in grain calcium accumulation. PLoS One. https://doi.org/10.1371/0103963

Singh N, Meenu G, Sekhar A, Abraham J (2015) Evaluation of antimicrobial and anti-cancer properties of finger millet (Eleusine coracana) and pearl millet (Pennisetum glaucum) extracts. Pharma Innov J 3:82–86

Singh AK, Singh R, Subramani R, Kumar R, Wankhede DP (2016) Molecular approaches to understand nutritional potential of coarse cereals. Curr Genomics 17(3):177–192

Skirycz A, Jozefczuk S, Stobiecki M, Muth D, Zanor MI, Witt I, Roeber BM (2007) Transcription factor AtDOF4;2 affects phenylpropanoid metabolism in Arabidopsis thaliana. New Phytol 175:425–438

Stewart LK, Wang Z, Ribnicky D, Soileau JL, Cefalu WT, Gettys TW (2009) Failure of dietary quercetin to alter the temporal progression of insulin resistance among tissues of C57BL/6 J mice during the development of diet-induced obesity. Diabetologia 52:514–523

Sukumaran S, Xiang W, Bean SR, Pedersen JF, Kresovich S, Tuinstra MR, Yu J (2012) Association mapping for grain quality in a diverse sorghum collection. Plant Genome 5(3):126–135

Upadhyaya HD, Ramesh S, Sharma S, Singh SK, Varshney RK, Sarma NDRK, Ravishankar CR, Narasimhudu Y, Reddy VG, Sahrawat KL et al (2011a) Genetic diversity for grain nutrients contents in a core collection of finger millet (Eleusine coracana (L) Gaertn) germplasm. Field Crops Res 121:42–52. https://doi.org/10.1016/jfcr201011017

Upadhyaya HD, Ravishankar CR, Narasimhudu Y, Sarma NDRK, Singh SK, Varshney SK, Nadaf HL (2011b) Identification of trait-specific germplasm and developing a mini core collection for efficient use of foxtail millet genetic resources in crop improvement. Field Crop Res 124 (3):459–467

Upadhyaya HD, Sharma S, Dwivedi SL, Singh SK (2014) Sorghum genetic resources: conservation and diversity assessment for enhanced utilization in Sorghum improvement. In: Genetics, genomics and breeding of Sorghum, Genetics, genomics and breeding of crop plants. CRC Press (Taylor & Francis), Boca Raton, FL, pp 28–55

Upadhyaya HD, Wang YH, Sastry DV, Dwivedi SL, Prasad PV, Burrell AM, Klein PE (2015) Association mapping of germinability and seedling vigor in sorghum under controlled low-temperature conditions. Genome 59(2):137–145

Van Rensburg SJ (1981) Epidemiological and dietary evidence for a specific nutritional disposition to esophageal cancer. J Natl Cancer Inst 67:243–241

Varshney RK, Glaszmann JC, Leung H, Ribaut JM (2010) More genomic resources for less-studied crops. Trends Biotechnol 28(9):452–460

Velu G, Crossa J, Singh RP, Hao Y, Dreisigacker S, Perez-Rodriguez P, Tiwari C (2016) Genomic prediction for grain zinc and iron concentrations in spring wheat. Theor Appl Genet 129 (8):1595–1605

Vijayakumari J, Mushtari Begum J, Begum S, Gokavi S (2003) Sensory attributes of ethnic foods from finger millet (Eleusine coracana). In: Proc of the National seminar on Processing and utilization of millet for nutrition security: recent trends in millet processing and utilization. CCSHAV, Hisar, pp 7–12

Vinoth A, Ravindhran R (2017) Biofortification in millets: a sustainable approach for nutritional security. Front Plant Sci 8:29

Vita JA (2005) Polyphenols and cardiovascular disease: effects on endothelial and platelet function. Am J Clin Nutr 81:292–297

Wu Y, Li X, Xiang W, Zhu C, Lin Z, Wu Y, Li J et al (2012) Presence of tannins in sorghum grains is conditioned by different natural alleles of Tannin1. Proc Natl Acad Sci U S A 109:10281–10286. https://doi.org/10.1073/pnas1201700109

Yadav OP, Rai KN (2013) Genetic improvement of pearl millet in India. Agric Res 2:275–292

Yadav OP, Upadhyaya HD, Reddy KN, Jukanti AK, Pandey S, Tyagi RK (2017) Genetic resources of pearl millet: status and utilization. Indian J Plant Genet Res 30:31–47

Zhang G, Liu X, Quan Z, Cheng S, Xu X, Pan S, Tao Y (2012) Genome sequence of foxtail millet (Setaria italica) provides insights into grass evolution and biofuel potential. Nat Biotechnol 30 (6):549

Zhang L, Liu R, Niu W (2014) Phytochemical and antiproliferative activity of proso millet. PLoS One 9(8):e104058. https://doi.org/10.1371/journalpone0104058

Zweiger G (1999) Knowledge discovery in gene-expression-microarray data: mining the information output of the genome. Trends Biotechnol 17(11):429–436

Processing Technology for Value Addition in Millets

11

Adinath Kate and Anupama Singh

Abstract

Millets are the harbinger of nutrition required for human health. Besides the diverse essential nutritional constituents like minerals, vitamins, micronutrients, etc., millet grains also contain a considerable amount of anti-nutritional constituents. The removal of anti-nutritional compounds from the particular type of millet is a mandatory requirement prior to its consumption otherwise it creates serious health hazards. Different techniques and treatments have been given to the millet grains at household level from ancient times to make them suitable for human consumption. These operations include soaking, heating, roasting, fermentation, cooking, etc. Now, days various unit operations have been standardized and optimized independently for each type of millet. All these unit operations like soaking, dehulling, grinding, roasting, puffing, fermentation, malting, etc., with appropriately designed processing equipment and machines are well-established in industries for commercial-scale processing of millets. The selection criteria for appropriate processing operation, equipment, and production scale are based on the targeted output.

Keywords

Millet processing · Processing equipments · Value addition · Millet products

A. Kate (✉)
ICAR-Central Institute of Agricultural Engineering, Bhopal, Madhya Pradesh, India

A. Singh
National Institute of Food Technology and Entrepreneurship Management, Sonipat, Haryana, India

A. Kumar et al. (eds.), *Millets and Millet Technology*,
https://doi.org/10.1007/978-981-16-0676-2_11

11.1 Introduction

Millets are one of the important crop for world's food and nutrition economy and are recognized as one of the necessary food grain for balanced nutrition of humans. Millets are also playing a role as cash crop for small farmers and can adapt themselves to marginal soil for nearly one-third of the world's population (Singh et al. 2010). Millets are also playing a vital role in food security due to its sustainability in adverse and variety of agro-climatic conditions (Ushakumari et al. 2004). They specifically include major (Pearl millet and sorghum) and minor millets (*Finger millet, Foxtail millet, proso millet, little millet, kodo millet*, etc.). The crops belonging to this category have huge potential in broadening the genetic diversity and hence potential to key contribution in nutrition security (Mal et al. 2010). These millets are not only rich in nutrition but at the same time offer various health benefits (Veena 2003). In recent years, millets have received attention due to their unparalleled nutritional profile and it can easily be consumed in any suitable form. Because of excellent nutritional and functional qualities millets can be transformed into various value-added products. Besides the rich quality nutrients, the utilization of millets as a staple food is restricted to traditional consumers only, then on-availability of consumer-friendly, ready-to-eat food products is the major reason for it (Hulse et al. 1980). Poor quality attribute of the millets and their traditional preliminary form of consumption like flours such as dark and dull color, coarse and gritty texture, astringent taste, high fiber content, larger cooking time, anti-nutritional constituents, low shelf life have limited their utilization. Most of these limitations can be overcome through the better utilization of appropriate processing technologies. Appropriate techniques of millet processing lead to the promising and successful utilization of millets in various traditional and convenience health foods. Different processed products from millets have been developed in various categories like popped, flaked, puffed, extruded, roller dried, fermented, malted and composite flours, weaning foods, etc. Some of the processing technologies have promising effects on the food qualities and its nutrition like extrusion of pearl millet in weaning foods enhances the protein digestibility (Cisse et al. 1998). Similarly, germination and probiotic fermentation catalyzed in protein profile improvement and mineral availability (Arora et al. 2011).

In this chapter, the focus has been made on detailed description of various processing technologies and development of various value-added products from millets including both major and minor millets.

11.2 Processing Technologies for Millets

The processing and utilization of millet crops are mainly carried out at home scale and hence it is largely confined to production catmint and its valuable nutrients remain unavailable to wide population. In India, various millets are consumed in the form of different products like *porridge*, *chapati*, *laddu*, *machula*, *dubkee*, *churkani*, and various geographical location-specific dishes. Anti-nutritional constituent present in the millet grain is also one of the common reason for its limited utilization. But due to deficiency in proper processing knowledge, the consumption of these crops in desired consumption rate especially in nontraditional area is limited. Different age-old techniques have been carried out to eliminate or reduce such anti-nutritional factors present in different millets and converted into edible form. Home practices carried out using household utensils significantly improve the functional properties with considerable reduction in anti-nutritional constituents. Various traditional operations are followed for the processing of millets prior to its consumption. Most of these techniques are manual, laborious, tedious, crude, and unhygienic without any standardization. Therefore, the uniformities are not maintained in the processed products/dishes. Though there is valuable nutritional importance of millets, they are neglected for commercial production of processed and value-added products as the traditional practices lack with a systematic scientific approach. Keeping in view of these limitations the upgradation with proper scientific study is very much essential. The development and introduction of farmer's friendly equipment and optimization processing protocol is one of the possible appropriate approach for this. Also, the large involvement of women in household traditional processing need to be taken into account while design and introducing machinery for different unit operations.

Different processes mentioned in Table 11.1 have been traditionally carried out from ancient times with the aim of removal of the anti-nutritional factors and alter the taste in different types of millets. Different operations such as soaking, dehulling, and cooking have been carried out from ancient times prior to its consumption at the household level. Development of simple technological inputs in the form of appropriate standardization process parameters, operating conditions, and development of suitable equipment geared the millets utilization from household level to commercial scale industrial level. Various advances based on the improvement of traditional practices studied by the researchers are presented in Table 11.2.

The unit operation wise processing operations with their aim and advancement over traditional home scale practice are as follows:

11.2.1 Soaking

Soaking is one of the popular technique for grains for improvement of mineral bioavailability and reduction of anti-nutritional compounds like phytic acid. Degradation and leaching of compounds like phytates, changes in phytase activity, and concentration of minerals like iron, zinc, etc., get influenced by soaking. Hence, their quantities and concentrations are monitored after soaking different forms of millets utilized for household consumption. The loss due to such leaching was reported quite severe in case of soaking of dehulled and milled grains. At the same time cooking grains by soaking in water has not found any significant increase in phytate degradation (Lestienne et al. 2007) but a considerable amount of leaching of nutritional constitute was observed. Similarly, about 25% loss of iron with endogenous phytates degradation was reported by Eyzaguirre et al. (2006) during soaking

Table 11.1 Traditional processing techniques used for effective utilization of millets and other underutilized crops

S. No.	Unit operation	Objective	Technique used	Limitations
1	Threshing	Separation of grains, husks, and panicle	Beating method	High labor and time, lower output efficiency
2	Winnowing	Separation of foreign materials like chaff, straw, husk, etc., from desired fraction	Manual air blowing and winnowing	Low output capacity, weather dependent
3	Dehulling	Aleuron layer separation as hull from kernel	Wooden pounding	Required huge labor and higher time
4	Soaking	Maintain the desired softness by addition of moisture	Water pounding	Weather-dependent, risk of unwanted contamination, lower output capacity
5	Germination	Alteration of physiological form of grains to achieve nutritional change in grain	Using moist cloth	Lower output capacity and efficiency
6	Blanching	Inactivation of undesirable enzymatic activities	Hot water or steam exposure	Difficult to achieve accurate temperature time combination, batch process
7	Roasting	Flavor and taste improvement	Open type of roasting pan or *Kadhai*	No control on burning of grains
8	Popping	Flavor and taste improvement with defines texture	Popping pans/ *kadhai*	Difficult to maintain accurate process conditions, risk of unwanted contamination, lower output

of pearl millet grains. Minerals like iron, calcium, and phosphorus present in pearl millet were found to be decreased during its soaking in acid medium.

11.2.2 Dehulling/Dehusking/Decortication/Pearling

Traditionally, the dehusking of millet grains (dry, moistened, or wet) is normally done through the method of pounding with a pestle in a stone or wooden mortar. Proper moisture condition in the grains prior to its dehulling is a key requirement of the desired output. Accordingly, about 10% moisture is required for proper detachment of fibrous bran and separation of germ and endosperm. The dehulling of millets removed nonedible part of grain and improves the bioavailability of present nutrients, reduction in the anti-nutritional constituents, and consumer acceptability. Various types of dehullers, decorticators, and pearlers are available for different

Table 11.2 Advanced processing technologies for millets and other underutilized crops

S. No.	Grain processing methods	Process description	Advantages of process	Equipment/ technology available
1	Decortication/ Dehusking	The outer layer of husk removed from kernel of grain	Significantly reduces anti-nutritional factors	Decorticator and Dehusking machines
2	Pearling	In this operation aleurone layer from dehusked grain is peeled/scratched off	Reduces non-digestible and off taste compounds	Mechanical pearlers and millet mills
3	Grinding	Size reduction of the millet grains in different forms based on the targeted processed product	Grain converted into suitable form required for recipe	Millet mill, attrition mill
3	Cooking	Cooking is a process in which desirable changes taking place like starch gelatinization which makes them suitable for digestion and intestinal absorption	Required physicochemical changes taking place to make it suitable for digestion	Open vessel and pressure cooker
5	Roasting	In the roasting operation, the grains are exposed to intensive high heat for short time	Enhance the sensory qualities and decreased anti-nutritional constituents	Open pan, specially designed roasters
6	Puffing	In puffing whole unhusked or decorticated grains with defined moisture content is mixed with hot sand (250 °C, about 15–60 s)	Reduces antinutrients and enhances the taste and flavor. The operation also deactivate the bacteria and hence improves storage quality	In salt using open hot pan
7	Sprouting	Soaking of whole undamaged grains for 2–24 h, and kept in humid space with desired humidity up to 24–48 h	Increases availability of micronutrients, improve digestibility, and reduces antinutrients	Incubators, humidity chambers, germinator
8	Fermentation	In the fermentation process, growing specific strain of microorganisms at controlled conditions over the raw material as identified medium	Enhance the sensory qualities with improvement in nutritional value and digestibility. Also a considerable reduction in anti-nutritional constituents	Fermenters
9	Malting	It involves the combined process of	Resulted in the better digestibility of starch	Malting units

(continued)

Table 11.2 (continued)

S. No.	Grain processing methods	Process description	Advantages of process	Equipment/ technology available
		steeping, germination, drying, toasting, grinding, and sieving	and improve availability of minerals	

Fig. 11.1 Typical roller mill

types of millets. These motorized mechanical machines have very high output over the conventional laborious, tedious manual operation.

Abrasive type: This type of decorticator are most common for the removal of surface husk or bran from grain surfaces. During the operation, a main functional component, i.e., abrasive disk or roller creates abrasion and friction force on the surface of grain due to which fibrous hull or bran present in pericarp get removed in the form of fine powder or course flaky form. Because of the nonuniformity in the grains, individual grain is abraded at varying extent at different surface points and hence sometime more loss of endosperm occurrs. Such instances are quite common in case of damaged grains even at very low degree of abrasion. According to the suitability of different millet grains, some of the common machines recommended for millets includes:

Roller mills: In these types of machines, generally pearling is done by passing the grains through a set of two steel rollers rotating inside the cylindrical chamber with equal or differential speed in cocurrent or countercurrent manner (Fig. 11.1). Properly tempered grains when comes in contact with rotating

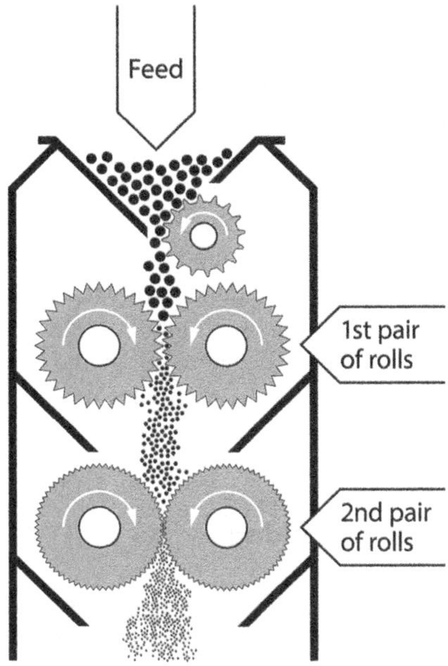

Fig. 11.2 CIAE millet mill

rollers, the pericarp is rubbed with adjacent grain and shearing with the surface of the roller. The degree of pearling and particle size in the flour is maintained by adjusting the spacing between the rollers. The temperature and moisture content of the grain is vital in this process to avoid loss of edible material as well as the machinery.

CIAE-Millet mill: ICAR-CIAE Bhopal has designed and developed CIAE-Millet Mill aiming to process millets with eliminating drudgery involved in the traditional method (Fig. 11.2). The capacity of the machine is 100 kg/h of millet grains with 10–12% of moisture content. The mill operates with 1 Hp electric motor as a power source. The mill is flexible for varying quantities of millets for its milling likewise even a kilogram of grains too be milled in a single pass with the desired degree of dehusking. The machine has a provision of detachment as well as separation and collection of husk simultaneously during dehusking operation. The mill is suitable for dehusking most common millets like kodo millet, proso millet, foxtail millet, little millet, barnyard millet, etc. The provision of adjustment of clearance between dehusking surfaces makes the machine suitable for different types of millets. The dehusking efficiency of this machine is about 95%. The compact processing zone attached with cyclone separator does not allow spillage of any dehusking fraction into direct environment. Similarly, the air and noise pollution during operation of machine is under control. Also, the machine is women-friendly, simple in operation, and does not require any qualified labor and hence machine is suitable to operate in production catchment as well as domestic level.

11.2.3 Grinding

Grinding is one of the important unit operation that deals with the secondary processing of millets. Actual size reduction of the millet grains takes place during the grinding. Particle size and fineness modulus are the representative indices of the grinding process. Different types of particle sizes are selected based on the targeted product. Various types of grinding mills, viz., attrition mills, hammer mills, burr mills are recommended for the millets (Fig. 11.3). Every grinding mill have a specific type of adjustment for maintaining the final size of output particles in the ground sample, i.e., flour. The shelf life of millet grains significantly decreases after its grinding. The increase in the free fatty acid (FFA) is the major concern for reduced shelf life and creates rancidity in millet flour (Yawatkar et al. 2010).

11.2.4 Cooking

Cooking is the essential unit operation that needs to be performed prior to the human consumption of any type of millet grains. The complex changes in physicochemical and functional properties in terms of starch gelatinization, protein denaturation, release of bound phenolic and antioxidants have been taking place at molecular level (Khamgaonkar 2012). At the same time, a considerable level of reduction in anti-nutritional constituents is also found. The time of cooking and temperature is very crucial for maintaining the nutritional constituents especially minerals and phenolic contents present in millets. The FFA content in the decorticated pearl millet grains and flour was found to be significantly reduced due to heat provided during cooking and hence it enhances the shelf life (Meera et al. 2003). There are various types of equipment used for cooking based on the dish to be prepared. The capacity, design, size, etc., of the equipment is based on its application at household level to commercial industrial-scale application. Different size and capacity extruders are the most common type of equipment used in industries for cooking millets.

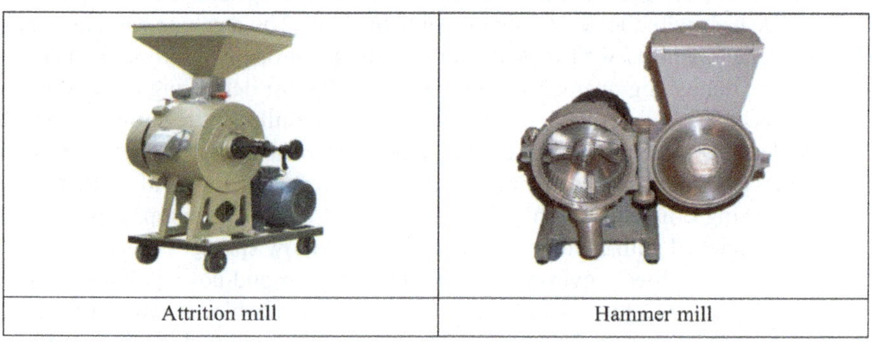

Fig. 11.3 Grinding mills for millet grains

11.2.5 Roasting

It is also one of the age-old technique in which millet grains with appropriate amount of moisture content are exposed to intense, high temperature (160–200 °C) conductive heat medium for varying periods of time. The required moisture in grain is maintained through pre-soaking followed by shade drying. During the roasting operation, the cotyledons of grain exposed to heat get shrink more than the outer husk and form a gap in-between which facilitates the loosening of husk from the cotyledon. This loosening of husk further facilitates husk removal during dehusking or pearling. This heat treatment also effectively creates a unique flavor and deactivate unwanted anti-nutritional and toxic constituents present in millets (Siegel and Fawcett 1976). It was found that some of the physicochemical and functional properties of roasted grains get changed. Similarly, roasting enhances the water absorption and water solubility index and due to this considerable changes were observed in their pasting and some of the thermal properties of flours (Sharma et al. 2011). Variety of the different roasters specifically designed for millets are available based on their mode of heat transfer operation. The roasting equipment working on the principle of conduction heating have better performance in terms of the degree of roasting and quality of the roasted product. Nowadays, equipments equipped with appropriate instrumentation for precise control of operating parameters (temperature, time) are quite popular due to the uniform quality of output and flexibility to use different types of commodities.

11.2.6 Puffing

It is the common processing technique mainly carried out for preparation of ready-to-eat expanded snacks and other products from any type of grains. In the process of popping/expansion, the millets were presoaked at desired moisture content and exposed to hot sand in the ratio of 1:6 at high-temperature (230–250 °C) and short-time (20–30 s). Popping of the decorticated finger millet was quite common among the millets where due to the high-temperature short-time heat treatment raw millet grains expanded significantly gets converted into desired expanded form (Fig. 11.4). To obtain the maximum expansion ratio, raw grains need to be flattened to desired shape and maintain desired moisture content in it prior to exposure to

Fig. 11.4 Popped millet grains (pearl millet, finger millet, and sorghum)

heating environment (Saleh et al. 2013). This process of popping of grains not only improves the expansion properties in its physical form but also significantly improves its functional properties. Also a considerable amount of different anti-nutritional constituents present in the millet grains gets reduced (Sreerama et al. 2008). The physical and textural properties of the millet grains also completely get changed due to its puffing. The common millets like sorghum, pearl millet, and finger millet are generally used for puffing based on its expansion and puffing ability (Fig. 11.4). Besides millet, puffing is quite common for grains like maize, chickpea, horse gram, etc., in Asian countries. The equipment/machine required for puffing includes a traditional individual house level system to a large cottage/industrial level system with varying capacities. Nowadays popping gun operating with hot air are used popularly due to its small size, lightweight, robust construction, and desired output for household consumption.

11.2.7 Sprouting

Sprouting is the germination of grains under controlled conditions. During sprouting of the millet grains, the dormant embryo present in the grain gets activated under specified moisture content and environmental condition (temperature and humidity). For the millets, this germination period varies from 48 to 72 h at an ambient temperature of 25–30 °C. The availability of the minerals and vitamins present in millets get significantly increased due to sprouting, whereas the anti-nutritional components like phytic acid and tannin were found to be significantly reduced. Sprouting for more than 48 h resulted in the loss in dry weight of source grain without much improvement in nutrition (Mbithi-Mwikya et al. 2000). Therefore, the process temperature and time need to be precisely maintained during sprouting to maintain desired nutrition in sprouted millets. Traditionally, sprouting of millets have been carried out in environmental chambers but nowadays specialized grain sprouters with the provision of adjustment of environmental conditions including the source of light, wavelength, illumination, etc., to get a quality product with enough safety and shelf life are available. Figure 11.5 represents the sprouted form of some of the common millets.

Fig. 11.5 Sprouted millet grains (barnyard millet, finger millet, pearl millet, and sorghum)

11.2.8 Malting

The malting of the millet grains at appropriate processing conditions improves the sensory attributes, nutritional quality, and digestibility of the grains with a considerable amount of reduction in antinutrients. It is a three-step process carried out in sequence like (1) steeping (soaking of grains in water), (2) germination (sprout development and enhance enzymatic activity), and (3) kilning (grain drying and stopping the enzymatic activity). All three steps may take place in separate equipment or in integrated single unit. An integrated single unit is quite popular nowadays due to ease in operation, time, and processing cost. In the integrated single malting unit, the grains are spread on the grain holding sieves for maintaining the uniform layer of thickness (Kumar et al. 2016). The water sprinkling arrangements made at the top provides the water required for the soaking of grains. This soaking water is replaced 3–4 times by opening the bottom valve during entire period of steeping, which also provides aeration to the soaked grains by the freshwater. After completion of the steeping process, all water is drained out by opening the bottom valve and the grain is left for germination on sieves. After completion of germination process, the side valve is open for the last step of malting, i.e., drying, hot air of the desired temperature comes from the side and cross the sieves so that the hot air spread equally over all sieves and the grain is dried. After the removal of sieves from the unit, we get malt.

11.2.9 Fermentation

Fermentation is one of the age-old preservation technique used for a variety of food products throughout the world. There are a variety of traditional dishes prepared from millets in different parts of the globe. Fermentation process significantly alter the physiochemical attributes in the millet grains and final fermented product. Fermentation process has key benefit due to the decrease in antinutrients, increase in protein availability, protein digestibility, and overall nutrition profile. In the pearl millet fermentation, a considerable amount of improvement in the IVPD was found with traction in unwanted antinutrients. During the fermentation of pearl millet grains, it was observed that significant reduction in anti-nutritional constituents present in the grain (Hassan et al. 2006). At the same time, because of this process, the availability of protein and starch for digestion significantly increased. In addition to amino acids, some of the vitamins were also found increased, while some specific flavaonoids and past behavior gets decreased during fermentation of some millet grain (Akingbala et al. 2002). A simultaneous increase in some of the nutrients was observed while decrease in some of them was also reported due to specific fermentation conditions for specific type of millet. The quantifiable changes in the macronutrients like fiber, protein, fat, etc., and micronutrients like K, Mg, Na, Fe, Cu, Mn, and Zn were observed in the fermentation of pearl millet grain. The enhancement in crude protein and crude fiber was also observed during fermentation of some of the millets especially after 16 h. The reduction in phytic acid and increase in the Zinc availability during the 24 h fermentation of finger millet was also observed (Murali and Kapoor 2003).

The selected microbial strains for fermentation of different types of millets may be chosen based on the properties of raw material and targeted output of the fermentation process (Khetarpaul and Chauhan 1991). Literature suggests that the fermentation with individual or in combination with some other processes and possible to get millet-based final product with rich nutrition value. But the application of these techniques appropriately at commercial scale and development of the innovative millet products are limited and most of them are utilized at household level for the development of traditional food dishes. Hence, the development of modern equipment and proper optimization of processing parameters are vital and crucial things for industrial scale application and commercial production of such products. At the same time, quality and safety need to be taken care for the final product.

11.3 Effect of Different Processing Techniques on Different Bioactive Compounds of Millets

Though the different processing techniques improves the nutrition status of the millets, sometimes they may effect the bioactive constituents. The benefit as well as adverse effects on bioactive active compounds always needs to take into account during optimization of the process parameters. The processing methods like soaking, decortication, malting, and cooking were significantly influenced by the total antioxidant content in the millet grains and their activity. During malting of finger millet, the antioxidant capacity contacting free phenolic acids was found to be increased especially after 96 h while at the same time bound phenolic content gets decreased (Rao and Muralikrishna 2002). Similarly, during the roasting of finger millets and cooking of kodo millet antioxidant value was observed to decrease. The separation of husk and endosperm of kodo millet during the milling or pearling also decreases DPPH activity and phytochemicals (Hegde and Chandra 2005). Similar effect was also observed for little millets when subjected to germination, steaming, and roasting where TPC, TFC, and tannin get increased and affected the nutraceutical and antioxidant value of raw grain (Pradeep and Guha 2011). A physical process like dehulling of whole millet grains was also found to exhibit a considerable reduction in phenolic and antioxidant capacity. Various processing treatments have a significant effect on the change in phenolics and antioxidants of different types of millets (Kalam Azad et al. 2019). Most of these reductions especially in antioxidant and phenolic during different processes might be due to oxidation reactions happened at particular condition. While the reduction due to simple physical operation like dehulling is due to the removal of the outermost protective layer from the grains. Hence, millet processing in terms of grains, flours, and other value-added products needs to be carried out at properly optimized conditions of operating parameters to protect their original quality.

11.4 Millet-Based Food Products

11.4.1 Composite Flour

Consumption of millet in the form of flour is one of the popular traditional form for human consumption. It may be consumed as an individual type of millet flour or by blending them with other common types of flours at appropriate proportion (Singh and Kaur 2005). The blending of different flours is carried out especially to get desired physicochemical, nutritional, and functional characteristic in the flour. Accordingly, in case of wheat flour this substitution was possible in the range of 10–20% level like barnyard millet, finger millet, and proso millet with 20%, 10%, and 15%, respectively. The increase in the quantity of millet flour in the blend increases the ash content and decreases the gluten value, dough loaf volume, percentage of damaged starch, and protein. The blending of flour some may be prepared by more complex way to improve the nutritional status keeping in view of all types of constituents likewise Khamgaonkar (2012) formulated millet-legume-wheat based composite flour by mixing 5% malted finger millet, 5% heat-processed black soybean, 5% popped horse gram, and 85% wheat flour under optimized conditions. This composite flour was found to be nutritionally rich and was used for preparing bread.

11.4.2 Bakery Products

Baked products are foods mainly based on cereal flours which are blended with other ingredients. Popular bakery products such as biscuits, cakes, and cookies were developed using different types of millets. This type of millet-based bakery cum confectionery food products are getting popular day by day due to its low gluten content and rich in dietary fibers. Because of low gluten content in most of the millets, it needs to be incorporated in different variations from 10% to 50% levels to standardize bread (20%), cake (30%), cookies (50%), soup sticks (20%), and khari (40%) with refined wheat flour. The features of bakery products are affordable cost, variation in taste and texture, packaging and better shelf life made them popular all over the world. Furthermore, addition of millets in baked products made them more superior in nutritional value, fiber content, and various micronutrients (Verma and Patel 2013).

11.4.3 Extruded Products

It is a high-temperature short-time cooking process in which feed material (grains) in ground form is fed into the extrusion device called extruder. The appropriate setting of feed rate, temperature, residence time, and pressure is maintained in the equipment. Variety of different types and different properties of millet-based extruded products can be developed with varying grain compositions, feed rate, temperature, pressure, and residence time during the cooking process. The millet-based products in the category of pasta, noodles, and others get popular day by day. An extruded snack was also prepared by Deshpande and Poshadri (2011) mixing flour of foxtail millet, rice, amaranths, cowpea, and Bengal gram in a ratio of 60:05:05:10:20, respectively. In some of the millet blended extruded products, expansion ratio was observed to decrease especially in case of high level of foxtail millet flour. Noodles, one of the popular product among all age groups and having higher shelf life was most commonly prepared through cold extrusion with the addition of different millet flours in varying proportions. The glycaemic index value of such products is quite low.

11.4.4 Flaked and Popped Products

Popping of the grains is one of the age-old traditional practice for value addition. It is generally used for the preparation of snacks or breakfast cereals which are either plain or incorporated with spices, salts, and sweeteners. At present, cereal popped products and flakes prepared from corn are popular breakfast foods. But after the suitable processing at optimized process conditions it is possible to prepare popped and flacked products from millets. Nowadays, the demand for such millet-based ready-to-eat products is increasing because of its properties like better nutrition value, crisp and friable texture, relative smaller size, quick hydration, etc. This technology can be applied successfully for the preparation of RTE from foxtail millet using high-temperature short-time (HTST) process with processing at 230 ± 5 °C temperature (Ushakumari et al. 2004). Similar category products can also be prepared from finger millet using the same technology.

11.4.5 Fermented Products

Different types of fermented millet products were used from ancient times in India. The recipes of these types of products are different at different geographical locations and socioeconomic cultures. Besides the unique taste, fermented products are rich in various major macro and micronutrients and lower level of antinutrients as compared to raw grains or grains processed by other process technique (Maha et al. 2003; Verma and Patel 2013). Millet-based products in this category also have significantly higher protein digestibility. Fermented finger millet flour showed lower antinutrients with an increase in mineral availability like zinc, iron, calcium, etc., as compared to raw nonfermented flour (Antony and Chandra 1998). Some of the popular fermented millet-based recipes are cutlets, weaning mixtures, vermicelli, and biscuits. The special types of fermented weaning foods and beverages developed from the different combinations of millets and specially identified microbial cultures are quite popular nowadays.

11.5 Conclusion

The adoption of appropriate type of processing technology with suitable type of equipment surely decreases the anti-nutritional constituents, off taste, and off flavors from the millet. Because of this, day by day the consumers for millet are continuously increasing. Also, the commercial scale production of various value-added products of millets are boosted due to the availability of suitable type of processing machineries and equipment. The advancement in the processing technologies of millets opens a new horizon and it will help to raise millets at competitive level of staple food.

References

Akingbala JO, Uzo-Peters PI, Jaiyeoba CN, Baccus-Taylor GSHP (2002) Changes in the physical and biochemical properties of pearl millet (*Pennisetum americanum*) on conversion to ogi. J Sci Food Agric 82:1458–1464

Antony U, Chandra TS (1998) Antinutrient reduction and enhancement in protein, starch, and mineral availability in fermented flour of finger millet (*Eleusine coracana*). J Agric Food Chem 46(7):2578–2582

Arora S, Jood S, Khetarpaul N (2011) Effect of germination and probiotic fermentation on nutrient profile of pearl millet based food blends. Br Food J113(4):470–481

Cisse B, Wilson JP, Hess DE, Hanna WW, Youm O (1998) Striga hermonthica infection of wild Pennisetum germplasm is related to time of flowering and downy mildew incidence. Int Sorghum Millets Newsl 3(9):149–150

Deshpande HW, Poshadri A (2011) Physical and sensory characteristics of extruded snacks prepared from foxtail millet based composite flours. Int Food Res J18(2):34–41

Eyzaguirre RZ, Nienaltowska K, De Jong LE, Hasenack BB, Nout MR (2006) Effect of food processing of pearl millet (Pennisetum glaucum) IKMP-5 on the level of phenolics, phytate, iron and zinc. J Sci Food Agric 86(9):1391–1398

Hassan AB, Ahmed IAM, Osman NM, Eltayeb MM, Osman GA, Babiker EE (2006) Effect of processing treatments followed by fermentation on protein content and digestibility of pearl millet (*Pennisetum typhoideum*) cultivars. Pak J Nutr 5(1):86–89

Hegde PS, Chandra TS (2005) ESR spectroscopic study reveals higher free radical quenching potential in kodo millet (Paspalum scrobiculatum) compared to other millets. Food Chem 92:177–182

Hulse JH, Laing EM, Pearson OE (1980) Sorghum and the millets: their composition and nutritive value. Academic press

Kalam Azad MO, Jeong DI, Adnan M, Salitxay T, Heo JW, Naznin MT, Park CH (2019) Effect of different processing methods on the accumulation of the phenolic compounds and antioxidant profile of broomcorn millet (*Panicum miliaceum* L.) flour. Foods 8(7):230

Khamgaonkar SG (2012) Optimization of processing parameters for development of underutilized cereals and legumes based composite flour. Unpublished Thesis, Govind Ballabh Pant University of Agriculture and Technology

Khetarpaul N, Chauhan BM (1991) Biological utilisation of pearl millet flour fermented with yeasts and lactobacilli. Plant Foods Hum Nutr 41:309–319

Kumar S, Singh A, Kohli D, Mishra R (2016) Fabrication of integrated malting unit for production of malts. Int J Eng Res Appl 6(8):33–36

Lestienne I, Buisson M, Lullien-Pellerin V, Picq C, Treche S (2007) Losses of nutrients and anti-nutritional factors during abrasive decortication of two pearl millet cultivars (Pennisetum glaucum). Food Chem 100(4):1316–1323

Maha AMA, Tinay E, Abdalla AH (2003) Effect of fermentation on the in vitro protein digestibility of pearl millet. Food Chem 80(1):51–54

Mal B, Padulosi S, Bala RS (2010) Minor millets in South Asia: learnings from IFAD-NUS Project in India and Nepal. Bioversity International, Maccarese, Rome, Italy and the MS Swaminathan Research Foundation, Chennai, India

Mbithi-Mwikya S, Van Camp J, Yiru Y, Huyghebaert A (2000) Nutrient and antinutrient changes in finger millet (Eleusine coracan) during sprouting. LWT-Food Sci Technol 33(1):9–14

Meera MS, Bhashyam MK, Ali SZ (2003) Effect of heat processing of pearl millet grains on shelf life and functional properties of flour and quality of final products. Paper presented at the 5th International food convention, Mysore, 5–8 December. p 124

Murali A, Kapoor R (2003) Effect of natural and pure culture fermentation of finger millet on zinc availability as predicted from HCI extractability and molar ratios. J Food Sci Technol 40 (1):112–114

Pradeep SR, Guha G (2011) Effect of processing methods on the nutraceutical and antioxidant properties of little millet (*Panicum sumatrense*) extracts. Food Chem 126:1643–1647

Rao MVSSTS, Muralikrishna G (2002) Evaluation of the antioxidant properties of free and bound phenolic acids from native and malted finger millet (ragi, Eleusine coracana Indaf-15). J Agric Food Chem 50(4):889–892

Saleh AS, Zhang Q, Chen J, Shen Q (2013) Millet grains: nutritional quality, processing, and potential health benefits. Compr Rev Food Sci Food Saf 12(3):281–295

Sharma P, Singh H, Cristina G, Rosell M (2011) Effects of roasting on barley β-glucan, thermal, textural and pasting properties. J Cereal Sci 53:25–30

Siegel A, Fawcett B (1976) Food legume processing and utilization (with special emphasis on application in developing countries). International Development Research Centre, Ottawa Canada, vol 4

Singh N, Kaur M (2005) Studies on functional, thermal and pasting properties of flours from different chickpea (*Cicer arietinum* L.) cultivars. Food Chem 91(3):403–411

Singh TP, Rao BD, Patil JV, Ansari A (2010) Overview of millets cultivation in India. In: "Research and development in millets: present status and future strategies", National seminar on millets. pp 1–11

Sreerama YN, Sasikala VB, Pratape VM (2008) Nutritional implications and flour functionality of popped/expanded horse gram. Food Chem 108:891–899

Ushakumari SR, Latha S, Malleshi NG (2004) The functional properties of popped, flaked, extruded and roller-dried foxtail millet (*Setaria italica*). Int J Food Sci Technol 39(9):907–915

Veena B (2003) Nutritional, functional and utilization studies on barnyard millet. MSc Thesis, University of Agricultural Sciences, Dharwad (Karnataka), India

Verma V, Patel S (2013) Value added products from nutri-cereals: finger millet (*Eleusine coracana*). Emir J Food Agric 25(3):169

Yawatkar AP, Unde PA, Patil AP (2010) Effect of grinding mills on quality of bajra flour and its products. Int J Agric Eng 3(1):144–146

12 Fermented Millet Technology and Products

Jyoti Semwal, Mohammad Hassan Kamani, and M. S. Meera

Abstract

Millets are small pack of grains with several health benefits and dense nutrients including protein, essential fatty acids, dietary fiber, B-vitamins, and minerals. However, its utilization is still limited to the local consumers due to the lack of convenience food products. Processing technology, such as fermentation, is being used to prepare traditional millet products in Asian and African countries. For millet fermentation, specifically two types of fermentation are used, i.e., lactic acid bacteria fermentation and yeast fermentation that can be spontaneous or nonspontaneous. Fermentation technology is known to improve the nutritional quality of food products though it has some limitations such as contamination with toxic microorganisms, etc. Therefore, the intervention of innovative approaches with the fermentation can make it a safer and better technology. This chapter discusses the effect of the fermentation technology on nutritional composition of the millets, current developments in fermented millet-based products, and the technological advancements explored in the millet fermentation technology.

Keywords

Millets · Fermentation technology · Lactic acid bacteria · Yeast · Product

J. Semwal · M. H. Kamani · M. S. Meera (✉)
Department of Grain Science and Technology, CSIR-Central Food Technological Research Institute, Mysore, Karnataka, India
e-mail: meera@cftri.res.in

A. Kumar et al. (eds.), *Millets and Millet Technology*,
https://doi.org/10.1007/978-981-16-0676-2_12

12.1 Introduction

Millets are small-seeded grains mainly produced in the arid and semi-arid regions of Africa and Asia. The global production of millets has increased from 28.20 to 31.01 M tonnes from 2015 to 2018 (FAOSTAT 2018). Millets are drought-tolerant crops and hence are of great importance in the present era of climate change, water scarcity, population growth and inflation. The various types of millets that are grown and consumed worldwide include pearl millet (*Pennisetum glaucum*), finger millet (*Eleusine coracana*), kodo millet (*Paspalum setaceum*), proso millet (*Penicum iliaceum*), foxtail millet (*Setaria italic*), little millet (*Panicum umatrense*), barnyard millet (*Echinochloa esculenta*), and sorghum (*Sorghum bicolor*).

Millets are not only beneficial from the agriculture and economic perspectives but also a boon for human health. These small seeds are a dense nutrient source and can keep several health disorders at bay. They have better nutritional and health-promoting properties than commonly consumed grains such as wheat and milled rice (Parameswaran and Sadasivam 1994), and hence are referred as nutri-cereal. Millets are known to contain substantial amounts of protein, essential fatty acids, dietary fiber, B-vitamins, and minerals such as calcium, iron, zinc, potassium, and magnesium (Rao et al. 2017). They are also rich in health-promoting factors like polyphenols, lignans, phytosterols, phytoestrogens, and phytocyanins. These factors act as antioxidants, immune modulators, detoxifying agents, etc., and hence protect against age-related degenerative diseases like cardiovascular diseases (CVD), diabetes, cancer, metabolic syndrome, Parkinson's disease, etc. (Manach et al. 2005; Scalbert et al. 2005; Chandrasekara and Shahidi 2012). Millets are non-glutinous, and hence are safe for people suffering from gluten allergy and celiac disease. They are non-acid forming, easy to digest, and non-allergenic (Saleh et al. 2013).

There are several millet processing methods, such as thermal processing, mechanical processing, soaking, fermentation, and germination/malting. Among these, fermentation is one of the oldest and widely used methods as millets are known for their prebiotic potential (Amadou et al. 2013; Di Stefano et al. 2017). It is a process dependent on the biological activity of certain microorganisms to release energy from a carbohydrate source to produce a wide range of metabolites such as alcohol, acetic acid, and lactic acid (Ross et al. 2002). Fermentation can be classified based on microorganisms, end product, aeration, phase, starter culture, and substrate feeding method (Fig. 12.1). Fermentation of the millet is commonly done using lactic acid bacteria (LAB) and yeast.

In addition to a convenient processing technique, it is also known to be one of the most popular and crucial processes that considerably lowers the anti-nutrient factors (ANFs) (Tsafrakidou et al. 2020), and hence enhances the overall nutritive value. Several studies have shown that the fermentation of millets might reduce the ANFs present in millets, thereby improving the bioavailability of the nutrients (Table 12.1 and Fig. 12.2). This reduction in ANFs can happen in two ways; either the microbial metabolic activity led to the phytase production or the alteration in the pH during fermentation provides optimum pH for the action of an endogenous enzyme (Samtiya et al. 2020). The breakdown of phytic acid releases the minerals in their

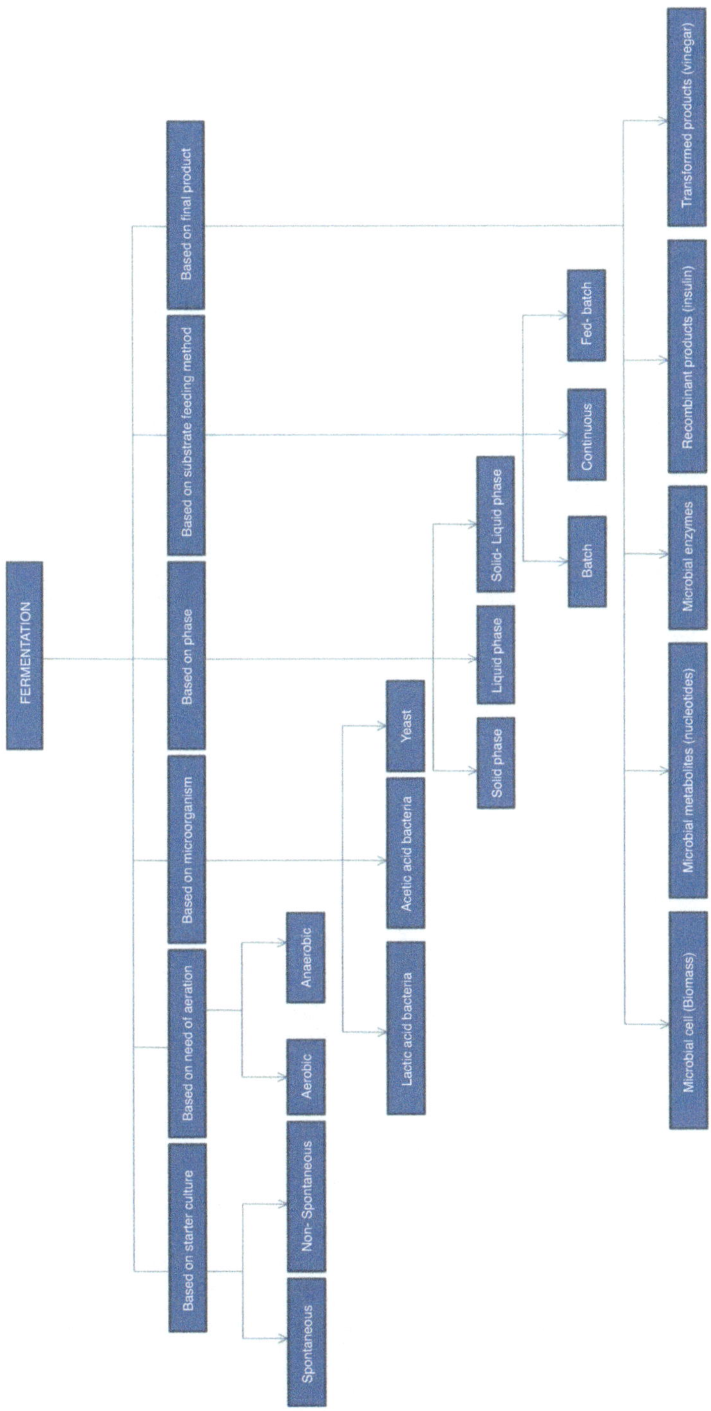

Fig. 12.1 Classification of the fermentation process

Table 12.1 The effect of the fermentation technology on the nutritional profile of millets

Raw material	Fermentation conditions	Effect on nutrients	References
Red sorghum, white sorghum, and pearl millet flour	Native microflora at 25 °C for 48 h	Protein digestibility increased by 97.4–98.3%. Phytic acid reduced by 20–21%.	Onyango et al. (2013)
Pearl millet flour	Native microflora at room temperature for 72 h	Protein content increased by 117.96%. Tannin reduced by 31.53%. Phytate reduced by 48.78%.	Ojokoh and Bello (2014)
Finger millet grains	*Lactobacillus salivarius* subsp. *salivarius* (LMG 9477T) at 30 °C for 48 h	Tryptophan increased by 17.8%, lysine increased by 7.1%, and phenylalanine decreased by 3.3%.	Mbithi-Mwikya et al. (2000)
Pearl millet flour	Native microflora at 30 °C for 24 h	Reduction in phytic acid by 51.9%.	Osman (2011)
Pearl millet	*Lactobacillus plantarum* at 37 °C for 96 h	Increase in protein content from 8.73% to 20.21% and reduction in fat content from 10.39% to 0.64%. Reduction in phytates from 1.78% to 0.09% and tannins from 2.80% to 1.40%.	Chinenye et al. (2017)
Pearl millet flour	Native microflora at 37 °C for 12 or 24 h	Improved protein digestibility to 93.6%.	Hassan et al. (2006)
Pearl millet slurry	*Lactobacillus acidophilus* (NCDC-16) at 37.8 °C for 12 h	Increase in thiamine from 0.46 to 1.02 mg/100 g, niacin from 1.25 to 2.52 mg/100 g, and total lysine from 2.53 to 6.30 g/100 g. Increase in albumin from 0.45 to 0.72 g/100 g, globulin from 3.75 to 4.34 g/100 g, and glutelin from 1.8 to 2.06 g/100 g. Prolamin decreased from 4.45 to 3.85 g/100 g. Increase in calcium availability from 40.54% to 80.57%, iron from 20.62 to 68.41, and zinc from 36.01% to 77.97%.	Arora et al. (2011)
Dehulled pearl millet flour (standard cultivar and Ugandi variety)	Native microflora at 30 °C for 14 h	Increase in in-vitro protein digestibility by 82% and 84%. Decrease in polyphenols by 59.5% and 30.45% in standard and Ugandi variety, respectively. Decrease in phytic acid by 59.5% and 46.1% in standard and Ugandi variety, respectively.	El Hag et al. (2002)

Sorghum	Baker's yeast fermentation by *Saccharomyces cerevisiae* at 37 °C for 48 h Amylolytic yeast fermentation by *Lipomyces kononenkoae* NRRL Y-11553 at 37 °C for 48 h	Baker's yeast fermentation led to the higher level of increase in protein digestibility. No significant difference in amino acid profile. Amylolytic yeast fermentation reduced phytate content to a higher extent.	Day and Morawicki (2018)
Little millet	Probiotic yeast (*S. boulardii*) for 5 days	Slight increase in contents of protein and phosphorus. Slight decrease in fat and total carbohydrates content. Marked decreased in phytic acid (from 188.95 to 167.56 mg/100 g). No significant change in calcium, magnesium, iron, and zinc.	Pampangouda et al. (2015)
Millet-wheat composite bread	Baker's yeast (*S. cerevisiae*) for 1 h	Three types of breads enriched with millets (composite bread) prepared, namely LAB fermented composite bread, yeast fermented composite bread, and commercially available white bread. The highest vitamin B group and calcium found in the yeast fermented composite bread.	Mythrayee and Pavithra (2017)
Pearl millet flour	*S. cerevisiae* and *S. diastaticus* for 72 h at 27 °C	Decreased protein content and bulk density, least gelation concentration, tannin, and total phenol. Increased fat content, thiamine, total soluble sugar, reducing sugar, water, and oil-absorption capacity.	Rathore and Singh (2018)
Pearl millet fermented gruel	*P. kudriavzeviiin* combination with LAB strains for up to 24 h	Significant increase in the concentration of folate after 2, 4, and 24 h of treatment due to co-fermentation with yeast.	Greppi et al. (2017)
Sorghum-based fermented *Kisra*	*S. cerevisiae* in combination with LAB strains for different intervals (up to 19 h)	The presence of *S. cerevisiae* reduced the fermentation time.	Ali and Mustafa (2008)

(continued)

Table 12.1 (continued)

Raw material	Fermentation conditions	Effect on nutrients	References
Fermented broom corn millet sour porridge	*S. cerevisiae* versus *L. brevis* and *Acetobacter aceti* for 24 h at 30 °C	*S. cerevisiae* showed the minimum titratable acidity and sensory scores compared with *L. brevis* and *A. aceti*. Mixed-strains fermentation (1:1:1, v/v/v) was found to be the best combination of strains starter.	Wang et al. (2019)
Fermented millet-based *Kunu*	*S. cerevisiae* and *Lactobacillus* species for 48 h at 35 °C versus naturally fermented *Kunu*	*S. cerevisiae* increased the contents of nitrogen (from 0.5 to 0.62 mg/100 ml), calcium (from 1.13 to 7.35 mg/100 ml), magnesium (from 22.99 to 43 mg/100 ml), and potassium (from 50.45 to 60.01 mg/100 ml), but it reduced the contents of sodium (from 30.1 to 21 mg/100 ml) and anti-nutrients (tannin, oxalate, and phytate) compared to natural fermentation. The combination of *S. cerevisiae* and *Lactobacillus* species maximized the contents of nitrogen, calcium, magnesium, potassium and minimized the content of tannin and oxalate content among all samples.	Agboola and Ojo (2018)
Fermented sorghum-Irish potato based gruel (*Ogwo*)	*S. cerevisiae*, *G. candidum*, *Lactobacillus* strains and natural fermentation at 28 °C for 48 h	*S. cerevisiae* significantly reduced the contents of oxalate, phytic acid, tannin, saponin, and flavonoids, but it increased the contents of potassium, magnesium, and phosphorus when compared to the unfermented and naturally fermented sample. *Lactobacillus* strains minimized the content of anti-nutrients (tannin and phytic acid) and maximized the concentration of magnesium, calcium, and manganese compared to other fermented samples.	Adegbehingbe (2015)

Pearl millet	*S. diastaticus* and *S. cerevisiae* at 30 °C for 72 h	Increased starch digestibility (from 17.8% to 38.3% and 35.6% for *S. diastaticus* and *S. cerevisiae*, respectively). Increased protein digestibility (from 51% to 67.6% and 81.6% for *S. diastaticus* and *S. cerevisiae*, respectively).	Khetarpaul and Chauhan (1990a)
Pearl millet	*S. cerevisiae* and *S. diastaticus* at 30 °C for 72 h	Increased the total contents of soluble sugars (from 1.7 to 3.09 and 2.82 g/100 g for *S. diastaticus* and *S. cerevisiae*, respectively) and reducing sugar (from 0.36 to 2.01 and 0.66 g/100 g for *S. diastaticus* and *S. cerevisiae*, respectively). Decrease in the starch content (from 68.5 to 50.4 and 50.2 g/100 g for *S. diastaticus* and *S. cerevisiae*, respectively).	Khetarpaul and Chauhan (1990b)
Pearl millet	*S. diastaticus*, *S. cerevisiae* for 30 °C for 72 h	Significant reduction in phytic acid (from 990 to 585 and 570 mg/100 g for *S. diastaticus* and *S. cerevisiae*, respectively). Significant reduction in polyphenols (from 761 to 641 and 719 mg/100 g for *S. diastaticus* and *S. cerevisiae*, respectively).	Khetarpaul and Chauhan (1989a)
Pearl millet	*S. diastaticus*, *S. cerevisiae* for 20–30 °C for 72 h	Increased thiamine content (from 1.97 to 5.76 and 840 µg/100 g for *S. diastaticus* and *S. cerevisiae*, respectively).	Khetarpaul and Chauhan (1989b)

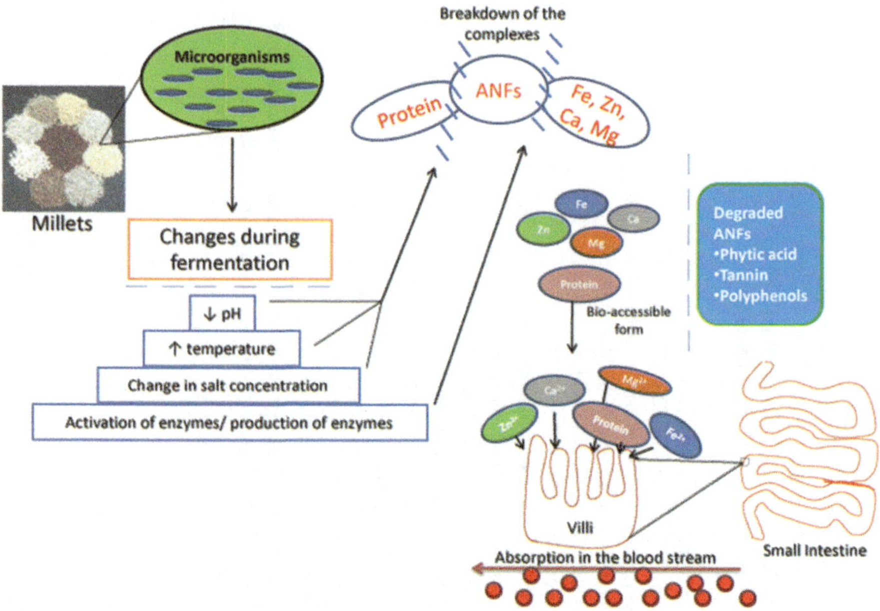

Fig. 12.2 Changes during the fermentation process

readily absorbable form, thereby increasing their bioavailability (Samtiya et al. 2020). The breakdown of phytic acid increases protein digestibility as well (Melini et al. 2019). Fermentation conditions like pH, temperature, and ionic strength can directly or indirectly influence protein digestibility (Joye 2019). Lactic acid fermentation may also induce favorable conditions for the interaction of starch and protein, thereby decreasing glycemic index (Melini et al. 2019). Furthermore, the microbial metabolism during fermentation produces more essential nutrients, such as thiamine, folate, riboflavin, vitamin C, and vitamin E, which are of great importance to human health (Kohajdova 2017).

Fermentation technology may be used for the development of probiotic or synbiotic fermented foods from millet. Such products may have some beneficial effects on some of the pathological conditions as diarrhea (Lei et al. 2006).

12.2 Lactic Acid Fermentation of Millets

Lactic acid bacteria (LAB) represent a ubiquitous and heterogeneous species with typical characteristics of lactic acid production by metabolizing sugar and creating acidic conditions with a pH of 3.5 (Fig. 12.3) (Charlier et al. 2009; Ghaffar et al. 2014). LAB fermentation ability is mainly associated with dairy products, but nowadays, its potential in cereal fermentation is being attempted. However, the

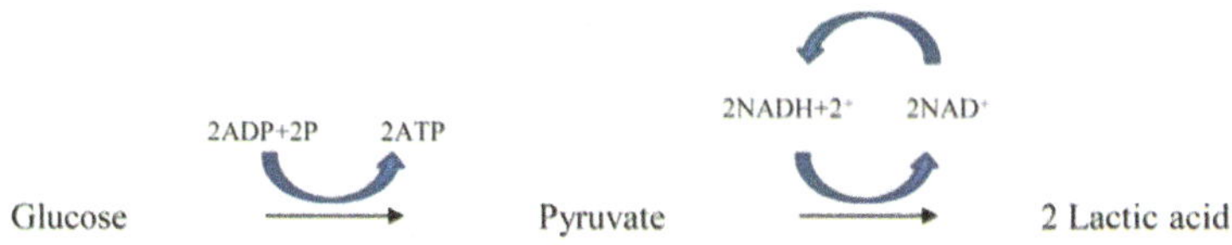

Fig. 12.3 Breakdown of sugar during LAB fermentation (Ghaffar et al. 2014)

Fig. 12.4 Fermented millet-based products: (**a**) *Dosa* (**b**) *Idli* (**c**) Curd

nutritionally fastidious nature of LAB for specific amino acids, B-vitamins, and other growth factors has made it an appropriate choice for millet fermentation.

LAB fermentation of the millets has been shown to enhance vitamin B and K, lysine, folate, and micronutrients in the fermented products (Tamene et al. 2019). One of the advantages of LAB fermentation is higher mineral bio-accessibility due to the reduction in pH. The use of LAB for the fermentation of millets has a long tradition that can be observed in some of the traditional fermented food such as *Idli*, *Dosa*, and *Ambli* in the Asian subcontinent (Rawat et al. 2018) and *Ben saalga*, *Uji*, *Mangisi*, etc., in African subcontinents (Amadou et al. 2011). These traditional products are the result of spontaneous or natural fermentation where native microorganisms metabolize the substrate (Fig. 12.4).

At present, LAB fermentation technique is being explored for the development of products, such as curd, yogurt, *Rabadi*, fermented milk, and other fermented beverages, from millets (Table 12.2). For the development of fermented millet products, processes like germination, fermentation, malting, etc., are used individually or in combination. For example, the millets are first germinated to extract milk, which is further fermented to develop curd. Sheela et al. (2018) demonstrated that the germination process directly influences the millet milk yield and subsequently, the curd yield. Currently, millets such as foxtail, proso, kodo, barnyard, and little millet have been explored to produce fermented milk. Among the millets, foxtail and proso millets had the highest extraction yield (Sheela et al. 2018).

A millet yogurt with acceptable organoleptic characteristics has been developed with the combination of millet milk (15%) and dairy milk with 6 h fermenting (Miaomiao 2007). Fermented millet sprout beverage has been developed using a combination of pearl millet (23.9%), finger millet (30%), and sorghum (21.1%) with

Table 12.2 LAB fermentation for the development of the millet products

Product	Fermentation conditions	Product characteristics	References
Fermented millet sprout milk beverage from sorghum, pearl millet, and finger millet	Millets were germinated for 48 h followed by grinding and filtration. The milk was fermented by 2% LAB culture for 12 h at 37 °C	• Sedimentation rate—0.7 ml • Viscosity—156.5 cP • Whey off—0.014% • Acidity (LA %)—0.63% • Sensory score—7.1/10	Sudha et al. (2016)
Non-dairy gluten-free millet-based fermented beverages	*Lactobacillus delbrueckii* subsp. *bulgaricus*, and *Streptococcus thermophilus* at 45 °C for 5 h	• pH —4.2–4.4 • Product had a thick consistency, velvety texture, light clotting density, and taste similar to cow's yogurt	Ziarno et al. (2019)
Fermented millet-based curd from foxtail millet, little millet, kodo millet, proso millet, barnyard millet	Soaking for 16 and 24 h, germination for 24 h, and milk extraction The extracted milk was fermented by using commercial curd culture NCDC 261 for 6 h at 30 °C	• Acidity %—0.74–1.2% • pH —3.5–4.5 • Sensory score—8/10	Sheela et al. (2018)
Rabadi from pearl millet	Germinated grain slurry was heated at 90 °C and cooled to 37 °C, fermented by mesophilic mixed-strain curd culture NCDC-167 for 12 h and finally cooled to 5 °C	• Fat—0.65% • Total solids—8.7% • Protein—2.2% • Ash—1.3% • Sensory score—7.4/10	Modha and Pal (2011)
Millet yogurt	*Lactobacillus bulgaricus* and *Streptococcus thermophilus* (ratio 1:1) culture 3%, sugar 8%, and fermentation at 42 °C for 6 h	• Acceptable sensory score	Miaomiao (2007)

25% of skim milk that took 12 h soaking and 48 h germination time (Sudha et al. 2016). The duration of fermentation may depend on the type of millet and its grain structure, such as the presence/absence of seed coat, size, and hardness. LAB fermentation of millets can be an appropriate technology to develop milk-like products for people with lactose intolerance.

LAB fermentation process can be improved by regulating the growth and survival rates of the microorganisms. For instance, the addition of calcium ion during LAB fermentation can enhance the heat resistance of the LAB, increase the survival rate, and shorten the regrowth lag times of the bacterial cell (Huang and Chen 2013). Furthermore, at low inoculum concentration of LAB (*Lactobacillus bulgaricus*), if the yogurt dispersion is agitated at the speed of 200 rpm for 5 min, it can increase the specific growth rate and final microbial count of LAB hence shortening the lag period (Aguirre-Ezkauriatza et al. 2008). Therefore, the technological intervention during LAB fermentation can be beneficial from economic perspective.

12.3 Yeast Fermentation of Millets

Yeast fermentation is another type of fermentation that is used for millet fermentation to produce alcoholic products such as wine, beer, and bread. The characteristic features of this fermentation is that yeast utilizes sugar and produces alcohol (ethanol) and carbon dioxide (Fig. 12.5) (Alba-Lois and Segal-Kischinevzky 2010).

Yeast fermentation produces several metabolites other than alcohol, such as esters, acids, terpenes, and lactones that impart a peculiar flavor to the final product and influence the organoleptic characteristics of the product (Geetha 2013). Similar to LAB fermentation, it can improve the nutritive value of millet products by reducing the level of ANFs such as tannin, phytic acid, and saponin (Adegbehingbe 2015; Agboola and Ojo 2018), that might improve the bio-availability of minerals (e.g., calcium, magnesium, potassium, and phosphorus) (Adegbehingbe 2015; Agboola and Ojo 2018) and vitamins (Mythrayee and Pavithra 2017; Khetarpaul and Chauhan 1989b; Rathore and Singh 2018). It may also contribute to improved starch and protein digestibility (Khetarpaul and Chauhan 1990a). Rathore and Singh (2018) pointed out that the fermentation by yeast may substantially improve key functional properties of millet flour such as water-absorption capacity, oil-absorption capacity, and least gelation concentration, which is useful from technological standpoint (Rathore and Singh 2018).

Yeast fermentation is employed to prepare several fermented products such as *Ogi*, *Kolo*, *Kenkey*, *Enjara*, *Jandh*, *Fura*, *Kodokojaanr*, and *Oti-oka* (Karovicova and Kohajdova 2007; Tamang 2012; Amadou 2019). These products can be prepared either by spontaneous yeast fermentation or by directly inoculating yeast under controlled conditions. Different strains of yeasts such as *S. cerevisiae*, and *S. diastaticus* can be used for the fermented millets (Mugula et al. 2003; Amadou et al. 2011; Geetha 2013; Adegbehingbe 2015; Rathore and Singh 2018). Among them, *S. cerevisiae* species is the most well-known and commercially significant (Zarnkow et al. 2010). In this context, several studies exhibited the effectiveness and positive impacts of *S. cerevisiae* in the fermentation of various types of millet-based products such as *Ogwo* (Adegbehingbe 2015), *Kunu* (Agboola and Ojo 2018), *Kisra* (Ali and Mustafa 2008), and broomcorn millet porridge (Wang et al. 2019).

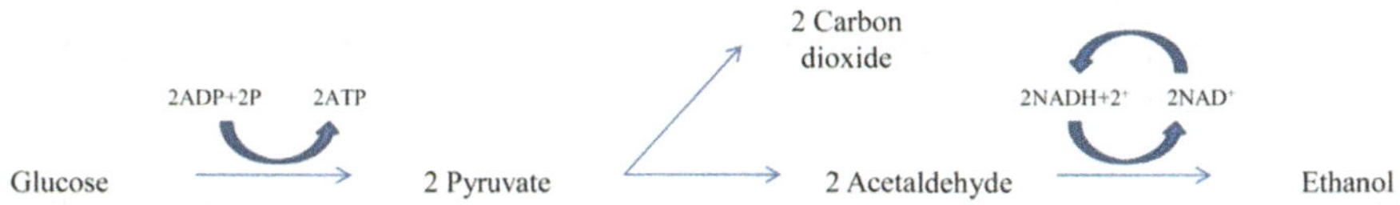

Fig. 12.5 Breakdown of sugar during yeast fermentation (Alba-Lois and Segal-Kischinevzky 2010)

Yeasts can either be used in a single form or combination with LAB for fermentation of the millet (Mugula et al. 2003; Taylor and Duodu 2015). A combination of yeasts with LAB might offer additional advantages in the quality of final products. For instance, Agboola and Ojo (2018) reported that the combination of *S. cerevisiae* and *Lactobacillus* species resulted in a higher content of calcium, magnesium, and potassium than single inoculation of each strain in fermented *Kunu*. They also observed that this combination is more effective in reducing tannin and oxalate contents than individual strain. Similarly, a recent report by Wang et al. (2019) also indicated the mixed-strains (*S. cerevisiae*, *Lactobacillus brevis*, and *Acetobacter aceti*) as the best combination starter for the fermentation of broomcorn millet sour porridge. Therefore, it can be interpreted that the use of yeast and LAB strains as mixed-strain cultures might be an effective technique to improve the final quality of fermented millet products. Table 12.1 indicates some of the studies related to the use of yeast (either single or in combination with LAB) in different fermented millet products.

Yeast fermentation technology is also known for its efficiency in the alcohol brewing industry. For this purpose, *S. cerevisiae*, cultured yeast, is commonly used in brewery industries as it provides control over the fermentation process, unlike other wild yeasts (Soden 1998). The utilization of yeast fermentation for the development of alcoholic drinks goes back to the time of Antoine Lavoiser when he used yeast paste or "ferment" to determine the chemical reaction of yeast fermentation (Alba-Lois and Segal-Kischinevzky 2010).

Some Asian countries like China, Korea, and Taiwan are known for their rice wines that are now shifting towards the millet wine development because of their better nutritional profile, unique medicinal values, and typical healthcare functions. A recent study indicated that millet yellow wine contains biologically active peptides, oligosaccharides, phenolic compounds, gamma aminobutyric acid, and other functional compounds of high nutritional values (Shang et al. 2011).

Among millets, foxtail millet (Kim and Koh 2004; Yang et al. 2006) and proso millet (Liu et al. 2018) have been explored for the production of millet wine. However, the millet wine's colloidal stability is very poor, which is associated with the prolamin of 14 and 18 kDa molecular weight (Yang et al. 2006). Moreover, a strong correlation between chill haze and prolamin proteins in millet wine was found (Passaghe 2014). Therefore, the presence of the prolamin protein can be considered as a critical indicator to determine the colloidal instability of millet wine. However, further researches are needed to minimize certain processing drawbacks in millet wine.

12.4 Technological Advancements in the Millet Fermentation Technology

The action of microorganisms in fermentation gives desirable biochemical changes and significant modification of food quality. It enriches the diet through the development of a diverse flavor, aroma, and texture in food substrate. The fermentation of the food material often prolongs its shelf-life due to the production of the metabolites such as acids and alcohols. It also enriches the food biologically with protein, essential amino acids, and vitamins (Haard et al. 1999). Although fermentation is an essential traditional technology for food processing, it has some limitations and drawbacks. The major concern with fermentation is the risk of foreign microflora. The contamination of the fermented products with toxic microorganisms can cause food-borne infection and intoxication due to microbial metabolites such as mycotoxins, ethyl carbamate, and biogenic amines (Nout 1994). Major risk factors include the use of contaminated raw materials, poorly controlled fermentation conditions, etc. These limitations of fermentation process can be eliminated by reducing the processing time either by exercitation of microorganisms or by altering the conditions during fermentation. Therefore, in order to make fermentation a safe and better technology, the intervention of advanced technologies might be useful.

One of the advancements in millet fermentation technology is mixed-culture fermentation. The use of a single strain of microorganism has been a norm for the production of fermented products. However, the utilization of the mixed culture of the microorganisms for the fermentation process seems to be an advantageous and better alternative as it can increase the final output yield with better organoleptic properties. Additionally, it is a new method to produce different metabolites such as enzymes and antimicrobial compounds in one process. All this is possible through the use of the synergistic utilization of different metabolic pathways. However, the growth of one microbe can enhance or hamper another microorganism's growth hence increasing or decreasing the final production rate. Nonetheless, the mixed-culture technology still has potential to increase the acidification, fermentation rate, and improvement of the functionality of the final fermented product (Adebo et al. 2018). For instance, the traditional preparation of *Ting*, a traditional fermented sorghum product of Botswana, requires 2–3 days to attain a pH below 4 to initiate the fermentation process. However, the use of mixed culture for the *Ting's* preparation reduced the fermentation time to 8 h (Sekwati-Monang and Ganzle 2011).

Another advanced fermentation technique is "Very High Gravity" (VHG) technology. This technology is defined as "the use of mash containing more than 27 g or more solids dissolved in 100 g of mash" (Puligundla et al. 2011). In case of the finger millet fermentation, the VHG fermentation technology using *S. bayanus* with additional nutritional supplements increased final ethanol production by 15% that can be of great economic advantage at the industrial scale (Puligundla et al. 2010).

Other technologies such as extrusion have also been used in combination with millet fermentation technology. These technologies have the potential to develop a novel product as well as may simplify the cumbersome process of millet fermentation. Sangeetha and Devi (2012) explored the dehydrated finger millet milk powder for the development of the pasta using the extrusion technique. The extrusion technology has also been explored for the development of ready-to-eat *Uji*, a fermented East African food prepared from maize, millet, sorghum, or cassava. Onyango et al. (2004) showed that the application of extrusion technology in the development of *Uji* from maize and finger millet (1:1) composite flour could produce self-preserving low moisture *Uji* that readily reconstitutes in warm water.

Although, there are different advanced technologies that have emerged in the fermentation technology field, the application of which has not been extended to millet fermentation. These novel technologies include non-thermal processes such as high-pressure processing, ultrasound, gamma irradiation, microwave irradiation, and pulsed electric field; and thermal processes such as Ohmic heating, radiofrequency, and microwave heating. The non-thermal methods are usually aimed at acceleration of the rate of chemical reactions and also used for monitoring of fermentation process. In contrast, thermal methods are usually employed to inactivate pathogens, improve metabolic activities, and shorten the fermentation process (Adebo et al. 2018). The use of these modern techniques combined with the existing millet fermentation technologies seems promising to make fermentation much safer and efficient. These technological advancements can aid in expanding the spectra of fermented millet products in the market.

12.5 Conclusion

Millets are small-seeded, drought-tolerant crops with several health-promoting nutrients including protein, essential fatty acids, dietary fiber, B-vitamins, and minerals. In spite of that, its utilization is limited to the local consumers due to the lack of suitable processing technology. In Asian and African countries, millets are fermented to prepare traditional dishes, such as *Kisra*, *Ogwa*, *Kunu*, *Ogi*, *Idli*, *Dosa*, and *Rabadi*. Although fermentation is an old millet processing technology, it demands technological advancements to develop convenience food products.

Several attempts are being made to improve the efficiency of millet fermentation technology and a number of novel fermented millet products, such as milk, milk powder, yogurt, and curd, are being explored. Technologies such as mixed-culture fermentation, very high gravity fermentation, as well as the combination of extrusion technology with fermentation have been shown to improve the product yield and quality. However, various novel thermal and non-thermal technologies have emerged in the field of fermentation technology, the application of which is still unexplored in millet fermentation. The intervention of these advanced technologies in millet fermentation can promote and expand the spectra of fermented millet products in the market.

References

Adebo O, Njobeh PB, Adeboye A, Adebiyi J, Sobowale S, Ogundele OM, Kayitesi E (2018) Advances in fermentation technology for novel food products. In: Panda SK, Shetty PH (eds) Innovations in technologies for fermented food and beverage industries. Springer, Cham, pp 71–87. ISBN: 978-3-319-74820-7

Adegbehingbe KT (2015) Effect of starter cultures on the anti-nutrient contents, minerals and viscosity of ogwo, a fermented sorghum-Irish potato gruel. Int Food Res J 22(3):1247–1252

Agboola SA, Ojo OC (2018) Effect of *Lactobacillus Species* and *Saccharomyces cerevisiae* on the mineral and anti-nutrient composition of *kunu*—a fermented millet based food. Asian Food Sci J 4(2):1–8

Aguirre-Ezkauriatza EJ, Galarza-Gonzalez MG, Uribe-Bujanda AI, Ríos-Licea M, Lopez-Pacheco-F, Hernandez-Brenes CM, Alvarez MM (2008) Effect of mixing during fermentation in yogurt manufacturing. J Dairy Sci 91(12):4454–4465

Alba-Lois L, Segal-Kischinevzky C (2010) Beer & wine makers. Nat Educ 3(9):17

Ali AA, Mustafa MM (2008) Use of starter cultures of lactic acid bacteria and yeasts in the preparation of *kisra*, a Sudanese fermented food. Pak J Nutr 8(9):1349–1353

Amadou I (2019) Millet based fermented beverages processing. In: Alexandru MG, Alina MH (eds) Fermented beverages. Woodhead Publishing, Cambridge, pp 433–472. ISBN: 9780128157039

Amadou I, Gbadamosi OS, Le GW (2011) Millet-based traditional processed foods and beverages—a review. Cereal Foods World 56(3):115–121

Amadou I, Mahamadou EG, Guo-Wei L (2013) Millets: nutritional composition, some health benefits and processing—a review. Emir J Food Agric 25(7):501–508

Arora S, Jood S, Khetarpaul N (2011) Effect of germination and probiotic fermentation on nutrient profile of pearl millet based food blends. Br Food J 113(4):470–481

Chandrasekara A, Shahidi F (2012) Bioaccessibility and antioxidant potential of millet grain phenolics as affected by simulated in vitro digestion and microbial fermentation. J Funct Foods 4(1):226–237

Charlier C, Cretenet M, Even S, Le LY (2009) Interactions between *Staphylococcus aureus* and lactic acid bacteria: an old story with new perspectives. Int J Food Microbiol 131(1):30–39

Chinenye OE, Ayodeji OA, Baba AJ (2017) Effect of fermentation (natural and starter) on the physicochemical, anti nutritional and proximate composition of pearl millet used for flour production. Am J Biosci Bioeng 10(5):12–16

Day CN, Morawicki RO (2018) Effects of fermentation by yeast and amylolytic lactic acid bacteria on grain sorghum protein content and digestibility. J Food Q 2018:1–8

Di Stefano E, White J, Seney S, Hekmat S, McDowell T, Sumarah M, Reid G (2017) A novel millet-based probiotic fermented food for the developing world. Nutrients 9(5):529

El Hag ME, El TAH, Yousif NE (2002) Effect of fermentation and dehulling on starch, total polyphenols, phytic acid content and in vitro protein digestibility of pearl millet. Food Chem 77 (2):193–196

FAOSTAT: Agriculture Organization of the United Nations Statistics Division (2018) Data–Crops-Production. http://www.fao.org/faostat/en/#data/QC/visualize. Accessed 17 Aug 2020

Geetha T (2013) Microbial and biochemical characterization of native millet fermentation and process standardization. PhD Thesis, VIT University, India. https://shodhganga.inflibnet.ac.in

Ghaffar T, Irshad M, Anwar Z, Aqil T, Zulifqar Z, Tariq A, Kamran M, Ehsan N, Mehmood S (2014) Recent trends in lactic acid biotechnology: a brief review on production to purification. J Radiat Res Appl Sci 7(2):222–229

Greppi A, Saubade F, Botta C, Humblot C, Guyot JP, Cocolin L (2017) Potential probiotic *Pichia kudriavzevii* strains and their ability to enhance folate content of traditional cereal-based African fermented food. Food Microbiol 62:169–177

Haard NF, Odunfa S, Lee CH, Quintero-Ramírez R, Lorence-Quinones A, Wacher-Rodarte C (1999) Cereals: rationale for fermentation. In: di Caracalla VT (ed) Fermented cereals: a global perspective. FAO, Rome, pp 1–27

Hassan AB, Ahmed IA, Osman NM, Eltayeb MM, Osman G, Babiker EE (2006) Effect of processing treatments followed by fermentation on protein content and digestibility of pearl millet (*Pennisetum typhoideum*) cultivars. Pak J Nutr 5(1):86–89

Huang S, Chen XD (2013) Significant effect of Ca^{2+} on improving the heat resistance of lactic acid bacteria. FEMS Microbiol Lett 344(1):31–38

Joye I (2019) Protein digestibility of cereal products. Foods 8(6):199

Karovicova ZK, Kohajdova J (2007) Fermentation of cereals for specific purpose. J Food Nutr Res 46(2):51–57

Khetarpaul N, Chauhan BM (1989a) Effect of fermentation by pure cultures of yeasts and *lactobacilli* on phytic acid and polyphenol content of pearl millet. J Food Sci 54(3):780–781
Khetarpaul N, Chauhan BM (1989b) Effect of fermentation on protein, fat, minerals and thiamine content of pearl millet. Plant Foods Hum Nutr 39(2):169–177
Khetarpaul N, Chauhan BM (1990a) Effect of fermentation by pure cultures of yeasts and lactobacilli on the available carbohydrate content of pearl millet. Food Chem 36(4):287–293
Khetarpaul N, Chauhan BM (1990b) Fermentation of pearl millet flour with yeasts and lactobacilli: in vitro digestibility and utilisation of fermented flour for weaning mixtures. Plant Foods Hum Nutr 40(3):167–173
Kim JY, Koh JS (2004) Fermentation characteristics of Jeju foxtail millet-wine by isolated alcoholic yeast and saccharifying mold. Appl Biol Chem 47(1):85–91
Kohajdova Z (2017) Fermented cereal products. In: Pandey A, Du G, Sanromán M, Soccol CR, Dussap C-G (eds) Current developments in biotechnology and bioengineering. Elsevier, Amsterdam, pp 91–117. ISBN: 9780444636775
Lei V, Friis H, Michaelsen KF (2006) Spontaneously fermented millet product as a natural probiotic treatment for diarrhoea in young children: an intervention study in northern Ghana. Int J Food Microbiol 110(3):246–253
Liu J, Zhao W, Li S, Zhang A, Zhang Y, Liu S (2018) Characterization of the key aroma compounds in proso millet wine using headspace solid-phase microextraction and gas chromatography-mass spectrometry. Molecules 23(2):462
Manach C, Mazur A, Scalbert A (2005) Polyphenols and prevention of cardiovascular diseases. Curr Opin Lipidol 16(1):77–84
Mbithi-Mwikya S, Ooghe W, Van CJ, Ngundi D, Huyghebaert A (2000) Amino acid profiles after sprouting, autoclaving, and lactic acid fermentation of finger millet (*Eleusine coracan*) and kidney beans (*Phaseolus vulgaris* L.). J Agric Food Chem 48(8):3081–3085
Melini F, Melini V, Luziatelli F, Ficca AG, Ruzzi M (2019) Health-promoting components in fermented foods: an up-to-date systematic review. Nutrients 11(5):1189
Miaomiao GH (2007) Millet yogurt processing technology. J Chin Cereals Oils Assoc 2007:2
Modha H, Pal D (2011) Optimization of *Rabadi*-like fermented milk beverage using pearl millet. J Food Sci Technol 48(2):190–196
Mugula JK, Narvhus JA, Sorhaug T (2003) Use of starter cultures of lactic acid bacteria and yeasts in the preparation of *togwa*, a Tanzanian fermented food. Int J Food Microbiol 83(3):307–318
Mythrayee R, Pavithra A (2017) Comparative study on nutritive content of finger millet-wheat composite bread fermented with lactic acid bacilli and yeast. IOSR J Biotechnol Biochem 3 (3):15–21
Nout MJ (1994) Fermented foods and food safety. Food Res Int 27(3):291–298
Ojokoh A, Bello B (2014) Effect of fermentation on nutrient and anti-nutrient composition of millet (*Pennisetum glaucum*) and soyabean (*Glycine max*) blend flours. J Life Sci 8(8):668–675
Onyango C, Noetzold H, Bley T, Henle T (2004) Proximate composition and digestibility of fermented and extruded *uji* from maize–finger millet blend. LWT-Food Sci Technol 37 (8):827–832
Onyango CA, Ochanda S, Mwasaru M, Ochieng JK, Mathooko FM, Kinyuru J (2013) Effects of malting and fermentation on anti-nutrient reduction and protein digestibility of red sorghum, white sorghum and pearl millet. J Food Res 2(1):41
Osman MA (2011) Effect of traditional fermentation process on the nutrient and antinutrient contents of pearl millet during preparation of *Lohoh*. J Saudi Soc Agric Sci 10(1):1–6
Pampangouda P, Munishamanna KB, Gurumurthy H (2015) Effect of *Saccharomyces boulardii* and *Lactobacillus acidophilus* fermentation on little millet (*Panicum sumatrense*). J Appl Nat Sci 7(1):260–264
Parameswaran KP, Sadasivam S (1994) Changes in the carbohydrates and nitrogenous components during germination of proso millet, *Panicum miliaceum*. Plant Foods Hum Nutr 45(2):97–102

Passaghe P (2014) The colloidal stability of craft beers obtained with gluten-free adjuncts: an assessment of aspects related to technology, composition and analysis. PhD Thesis, University of Udine, Italy. https://silo.tips/download

Puligundla P, Goud GKK, Reddy OV (2010) Optimization of very high gravity (VHG) finger millet (ragi) medium for ethanolic fermentation by yeast. Chiang Mai J Sci 37(1):116–123

Puligundla P, Smogrovicova D, Reddy OV, Ko S (2011) Very high gravity (VHG) ethanolic brewing and fermentation: a research update. J Ind Microbiol Biotechnol 38(9):1133–1144

Rao DB, Bhaskarachary K, Arlene CGD, Sudha DG, Vilas AT (2017) Nutritional and health benefits of millets. ICAR Indian Institute of Millets Research (IIMR), Hyderabad

Rathore S, Singh K (2018) Analysis of the effects of natural and pure culture fermentation for the qualitative enhancement of pearl millet flour. Nutrafoods 17:145–153

Rawat K, Kumari A, Kumar S, Kumar R, Gehlot R (2018) Traditional fermented products of India. Int J Curr Microbiol Appl Sci 7(4):1873–1883

Ross RP, Morgan S, Hill C (2002) Preservation and fermentation: past, present and future. Int J Food Microbiol 79(1–2):3–16

Saleh ASM, Zhang Q, Chen J, Shen Q (2013) Millet grains: nutritional quality, processing, and potential health benefits. Compr Rev Food Sci Food Saf 2(3):281–295

Samtiya M, Aluko RE, Dhewa T (2020) Plant food anti-nutritional factors and their reduction strategies: an overview. Food Product Process Nutr 2(1):1–4

Sangeetha N, Devi PM (2012) Effect of dehydration on the quality characteristics of extruded pasta using millet milk powder. Nutr Food Sci 2:10

Scalbert A, Manach C, Morand C, Remesy C, Jimenez L (2005) Dietary polyphenols and the prevention of diseases. Crit Rev Food Sci Nutr 45(4):287–306

Sekwati-Monang B, Ganzle MG (2011) Microbiological and chemical characterisation of *ting*, a sorghum-based sourdough product from Botswana. Int J Food Microbiol 150(2–3):115–121

Shang XL, Hui M, Tian Q (2011) Present status of the production of millet yellow wine & research progress in its functional components. Liquor-Making Sci Technol 2011:1

Sheela P, Umamaheswari T, Kanchana S, Kamalasundari S, Hemalatha G (2018) Development and evaluation of fermented millet milk based curd. J Pharmacogn Phytochem 7:714–717

Soden A (1998) The fermentation properties of non-*Saccharomyces* wine yeast and their interaction with *Saccharomyces cerevisiae*. PhD Thesis, The Australian Wine Research Institute, Australia. digital.library.adelaide.edu.au

Sudha A, Devi KSP, Sangeetha V, Sangeetha A (2016) Development of fermented millet sprout milk beverage based on physicochemical property studies and consumer acceptability data. J Sci Ind Res 5:239–243

Tamang JP (2012) Plant-based fermented foods and beverages of Asia. CRC Press, Boca Raton, FL

Tamene A, Baye K, Kariluoto S, Edelmann M, Bationo F, Leconte N, Humblot C (2019) *Lactobacillus plantarum* P2R3FA isolated from traditional cereal-based fermented food increase folate status in deficient rats. Nutrients 11(11):2819

Taylor JR, Duodu KG (2015) Effects of processing sorghum and millets on their phenolic phytochemicals and the implications of this to the health-enhancing properties of sorghum and millet food and beverage products. J Sci Food Agric 95(2):225–237

Tsafrakidou P, Michaelidou AM, Biliaderis CG (2020) Fermented cereal-based products: nutritional aspects, possible impact on gut microbiota and health implications. Foods 9(6):734

Wang Q, Liu C, Jing Y, Fan S, Cai J (2019) Evaluation of fermentation conditions to improve the sensory quality of broomcorn millet sour porridge. LWT 104:165–172

Yang J, Lee Y, Siebert KJ (2006) Study of colloidal instability of millet wine. J Am Soc Brew Chem 64(2):86–93

Zarnkow M, Faltermaier A, Back W, Gastl M, Arendt EK (2010) Evaluation of different yeast strains on the quality of beer produced from malted proso millet (*Panicum miliaceum* L.). Eur Food Res Technol 231(2):287–295

Ziarno M, Zaręba D, Henn E, Margas E, Nowak M (2019) Properties of non-dairy gluten-free millet-based fermented beverages developed with yoghurt cultures. J Food Nutr Res 58 (1):21–30

Processing-Mediated Changes in the Antinutritional, Phenolic, and Antioxidant Contents of Millet

13

M. I. Bhat, Rajeev Kapila, and Suman Kapila

Abstract

Millet grains are the treasure trove of essential micronutrients, resistant starch, and gluten-free proteins making them ideal food for people suffering from hidden hunger, chronic health disorders such as diabetes, obesity, and celiac allergy. Further, the presence of rich quantities of various antioxidant compounds vitamin E, proteins and peptides, carotenoids, flavonoids, and others across various varieties of millets enhance their utility as functional food. However, millets also contain certain phenolic substances such as phytates, tannins along with some protease inhibitors that can severely hamper the availability of these nutrients by acting as anti-nutritional factors. Fortunately, the presence of such anti-nutritional compounds can be reduced by some conventional and modern processing methods that make them more edible with better nutritional and sensory properties. For instance, fermentation mediated by microorganisms including yeasts, lactic, and acetic acid bacteria can modulate the anti-nutritional factors along with enrichment in health-promoting compounds that increase their demand from the nutrition point of view. Therefore, this chapter is aimed to highlight the various intentional and nonintentional food processing techniques that could modulate the nutritional aspects of millets with a special focus on antioxidant compounds, anti-nutritional factors as well as phenolic substances in millet-based food products.

Keywords

Anti-nutritional factors · Antioxidants · Fermentation · Millets · Processing methods

M. I. Bhat · R. Kapila (✉) · S. Kapila
Animal Biochemistry Division, ICAR-National Dairy Research Institute, Karnal, Haryana, India

A. Kumar et al. (eds.), *Millets and Millet Technology*,
https://doi.org/10.1007/978-981-16-0676-2_13

13.1 Introduction

Millets are small-seeded grasses that occur in variable forms and are mainly cultivated as grain crop. These are members of the *Poaceae* family and are known as super grains from the very beginning. With their better capability to adapt to a wide range of temperatures and moisture-regimes, millets are emerging as a natural food choice for the common people particularly to the people of developing nations. Even in some industrialized countries, various millet-based food products are used as primal material for the large-scale manufacturing of nutrient-rich potable alcohols along with starch production. These features have forced the researchers across the globe to contemplate on the utility of millet grains over the other cereals. Scientific evidences have further shown that the nutritional aspects of millets are even better than the other cereals and the presence of bioactive factors can be further enriched by applying the common food processing techniques. For instance, studies have reported and validated the nutraceutical aspects of fermented millet products and are contemplating to substitute other cereals with millets with the latter being as enriched as others with ample beneficial effects (Balli et al. 2020). Therefore, this chapter is to highlight the nutritional attributes of millet and millet-based food products along with special mention of various traditional and modern food processing methods that induce the desirable changes including the reduction in the quantities of some anti-nutritional constituents that in turn boost the digestibility and functionality of millet-based food products.

13.2 Nutritive Value of Millets

Millets are often cultivated as small-grained grasses across the globe primarily as human food as well as fodder and bird feed (Sarita and Singh 2016). Millets are nutrient-enriched crops that contain about 60–70% of carbohydrates, 7–11% proteins, 1.5–5% of lipids, 2–7% of crude fiber along with ample amounts of vitamins and minerals. Interestingly, the protein levels of millets are as rich as are found in notable cereals with ample amounts of essential amino acids with the exception of lysine and threonine (Chauhan et al. 2018). Millets are also known for rich quantities of methionine and cysteine amino acids. The critical nutritional attribute of millets in relation to proteins is that they are gluten-free and hence are suitable for the section of population that are usually gluten allergic. As revealed through prior studies that gluten-induced inflammation damages the small intestine's lining and causes variable symptoms that range from bloating, wind, fatigue, low blood count (anemia) to osteoporosis and other medical complications (Caio et al. 2019; Krupa-Kozak 2014). Presence of gluten in food is also closely linked with the mal-absorption of some of the other nutrients that subsequently lead to shortage of energy demands of the cell. Therefore, millet-based food products without gluten could be one possible way to avoid these complications that in turn can promote better gut health that is very critical for overall well-being of an individual.

Apart from these constituents, millets are also well known for their ample contents of essential fatty acid molecules such as linoleic, oleic, and palmitic acids that are present either in free form or bound form as monogalactosul, diacylglycerols, digalactosyl diacylglycerols, phosphatidyl serine, phosphatidylcholine, and phosphatidylethanolamine (Adu-Gyamfi et al. 2019). Millet-based oils are also a rich source of linoleic acid along with tocopherols (Moreno Amador et al. 2014). Interestingly, some common varieties of millet grains also possess high fat content as compared to the other available cereals. Millets are considered as a good source of vitamin B complex (Niacin, folacin, riboflavin, and thiamine) along with important minerals such as magnesium, manganese, phosphorus, and iron. Even the quantities of macro and micronutrient of millets are as good as are the other major cereals that are often consumed. Finger millet, for example, is known for the rich presence of minerals especially calcium. The presence of biologically active compounds like phenolic phytochemicals (phenolic acids, tannins, and flavonoids) lignans, β-glucan, inulin, resistant starch, phytates, sterols, tocopherol, and carotenoids also renders a special nutritive attribute to millets (Sarita and Singh 2016). These phytochemicals are well-known for their antioxidant and antimicrobial properties and hence could be critical for better human health along with the protective action against the diseases like cancer and cardiovascular diseases that are linked with oxidative stress related complications (Liang and Liang 2019). In addition, millet-based grains could serve as good prebiotic substances that can stimulate the growth and activity of specific gut microbiota population that could subsequently improve the gut functions (Li et al. 2019). Millet food products fermented with specific microbes containing rich bioactive factors could be a natural way of decreasing and preventing the common health complications that are very much prevalent within modern society (Kumar et al. 2016).

13.3 Emerging Insights on Therapeutic Effects Associated with Millet Consumption

The presence of sufficient and wide varieties of nutrients and nutraceutical compounds has made millet as a natural choice to combat the modern-day health complications that have appeared due to changed lifestyle, food habits, and environmental exposure (Niro et al. 2019). Presence of variable biologically active compounds in millets imparts many health benefits including the ability to combat different medical conditions such as diabetes, hypertension, cancer, cardiovascular diseases, tumors, time span of gastric emptying in addition to supplementation of rich quantities of roughage to gastrointestinal track (Sarita and Singh 2016). Since, millet-based foods are alkaline in nature and hence could be an important food that can contribute immensely towards the maintenance of healthy pH balance within the body that is very crucial for proper execution of cellular functions. Millets have a soothing effect on the body due to the presence of a sufficient amount of tryptophan amino acid that is used for the synthesis of serotonin within the body (Saleh et al. 2013). Presence of good amounts of magnesium in millets could be helpful in reducing migraines and heart attacks (Grober et al. 2015). Further, the consumption of millet foods boost the immune system functions because of the presence of a rich quantity of antioxidant activity phenolic compounds (Ofosu et al. 2020).

Aqueous extracts of *Setaria italic* (foxtail millet) exhibit anti-hyperglycaemic and hypolipidemic effects in experimental diabetic rat model because of the presence of good quantities of alkaloids or glycosides (Sireesha et al. 2011). Likewise, Shobana et al. (2010) reported the hypoglycemic, hypocholesterolemic, nephroprotective, and anti-cataractogenic effects in the rat diabetic model when supplemented with seed coat matter of millet (20%) in their diet. In another interesting investigation, Murtaza et al. (2014) observed that supplementation of finger millet bran to mice lowered the obesity-induced oxidative stress and inflammation along with a change in gut microbiota composition with increased abundance of beneficial gut microbes *Lactobacillus*, *Lifidobacteria*, and *Roseburia* but diminished counts of *Enterobacter*. Arabinoxylans, a type of non-starch polysaccharides extracted from finger millet bran showed mitogenic activity and macrophage activation property and hence could be used as a potent natural immunomodulation agent (Prashanth et al. 2015). Due to the presence of these biologically active chemical agents, millet-based food products thus might be very useful in combating various immune-related problems including celiac disease, aging, and allergy. Previously, Lee et al. (2010) have shown that consumption of foxtail millet reduced C reactive protein levels that often elevate during progression and development of inflammatory conditions. Similarly, the bran extracts of different varieties of millets from Japan including foxtail millet, proso millet, and barnyard millet suppressed the cellular levels of nitric oxide (NO) as well as inflammation causing cytokine molecules such as tumor necrosis factor-alpha (TNF-α) and interleukin (IL)-6 in lipopolysaccharide (LPS) stimulated mouse macrophage cell line (RAW264.7 cells) suggesting the immunomodulatory role of millet-based foods (Hosoda et al. 2012). In addition, protein hydrolyzates and peptide fractions (Molecular mass < 3.0 kDa) extracted from millets demonstrated inhibitory effect against the activity of angiotensin-converting enzyme (ACE) as well as α-amylase and α-glucosidase that are closely involved in the pathogenesis of metabolic syndrome and hence could be pivotal protein-based antidiabetic and antihypertensive agents (Karas et al. 2019). With these scientific evidences it obvious that millet-based food products could be a natural remedy to different health problems.

13.4 Anti-nutritional Components of Millets

Many food products particularly with plant origin have shown a wide range of anti-nutritional factors in addition to high quantities of macro and micronutrients (Samtiya et al. 2020). These anti-nutritional factors have the ability to combine with nutrients to form insoluble complexes that have enormous impact with respect to nutrient bioavailability for the host. Presence of anti-nutritional factors often results in micronutrient malnutrition and mineral deficiencies especially in relation to iron and zinc minerals (Nkhata et al. 2018). Phenolic substances, phytates, oxalic acid, tannins, and fibers (Table 13.1) are the important anti-nutritional food factors that impair the bioavailability of various host nutrients.

Phytates are considered as one of the strong anti-nutritional factors that are present in grains, pulses, and oleaginous products because of its strong chelating ability with multivalent metal ions especially those of iron, calcium, and zinc (Norhaizan and Nor Faizadatul Ain 2009; Palacios 2008). Phytates can also form insoluble complexes with divalent cations thereby restricting their bioavailability (Weaver and Kannan 2002). Previously, high contents of anti-nutritional factors like phytic acid and tannins have been reported from the pearl millet-based biscuits which made them unfit for infant consumption (Handa et al. 2017). Recently, Akbar et al. (2018) found the presence of the biosynthesis pathway genes for anti-nutritional factor oxalic acid in finger millet. The presence of oxalic acid in finger millet restricted the availability of calcium ions to host cells because of formation of oxalate calcium insoluble complex that subsequently resulted in the accumulation of calcium within the cells.

In addition to phytochemicals mentioned above, different enzyme inhibitors especially those that inhibit trypsin/chymotrypsin and alpha-amylase have been reported that compromise the quality of different varieties of millets. These inhibitors hinder the action of enzymes that subsequently prevent the digestion and absorption of protein and carbohydrates resulting in poor bioavailability of these essential nutrients to the host cells (Fenwick et al. 2019). Further, some non-starch polysaccharides moieties like betaglucans have been reported in millets that also hamper the digestibility of nutrients (Boncompagni et al. 2018). Besides poor nutrient bioavailability, antinutrient factors can be toxic to the host if their levels are beyond a certain level. Therefore, it is very essential to look for the methods that can be employed to reduce the anti-nutritional factors in millet-based food products so as to overcome their toxic effects and associated health complications when consumed (Samtiya et al. 2020). Interestingly, the contents of these anti-nutritional factors can be reduced by applying common door house food processing techniques that include fermentation, germination, and others. With the abundance of so many biologically active compounds, millets could be food for the future and a natural therapy to promote human health especially for those nations where there are too much food scarcity and economical crises. Interestingly, the traditional food processing techniques have shown promising results to further enhance the nutritional value of cereals including millet-based food products that too in an easy and cost-effective manner.

13.5 Various Food Processing Techniques Used for Nutritional Improvement in Millet

The common food processing techniques that are usually employed before the consumption of food include germination, decortications, milling, soaking, cooking, popping, fermentation, and others. These easy food processing techniques have shown enough potential to modulate the nutritional attributes of millet-based food

Table 13.1 Millet based anti-nutritional factors along their structure and inhibitory action

Anti-nutritional factor	Structure	Inhibitory action
Phytic acid (*Myo*-inositol-1,2,3,4,5,6 hexakisphosphate)	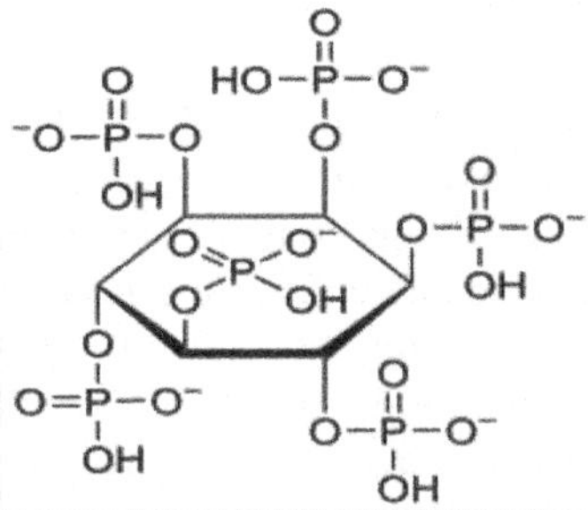	Strong chelating ability with both monovalent and multivalent cations thereby impairing their cellular bioavailability (Norhaizan and Nor Faizadatul Ain 2009)
Condensed tannins (proanthocyanidins or procyanidins)	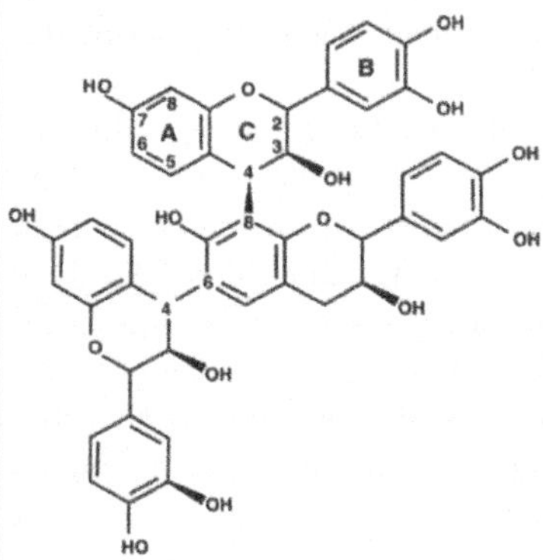	Tannins form irreversible protein complexes that hamper activity of digestive enzymes and cause protein indigestibility (Cirkovic Velickovic and Stanic-Vucinic 2018)
Oxalic acid	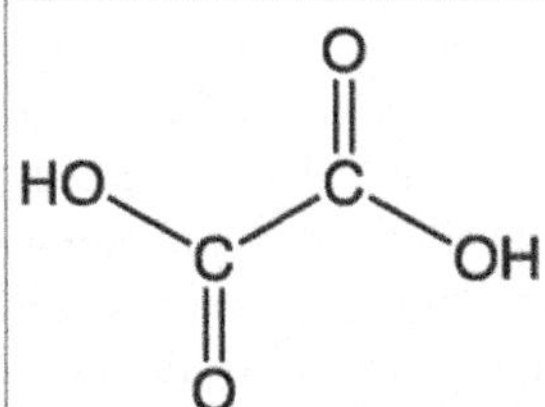	Oxalic acid binds with calcium resulting in formation of insoluble metal complex that impair calcium absorption (Nambiar et al. 2011)
Goitrogenic C-glycosylflavones (Vitexin)	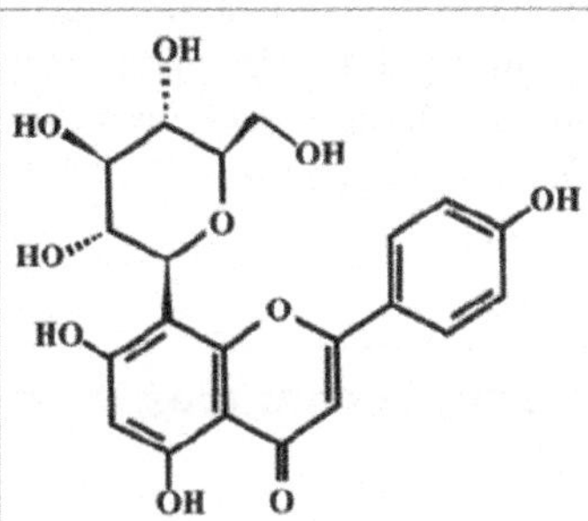	Goitrogen molecules interfere with iodine metabolism in the body and thus impair the functioning of thyroid gland resulting in goiter formation (Boncompagni et al. 2018)

(continued)

Table 13.1 (continued)

Anti-nutritional factor	Structure	Inhibitory action
Alpha-galactosides (raffinose)	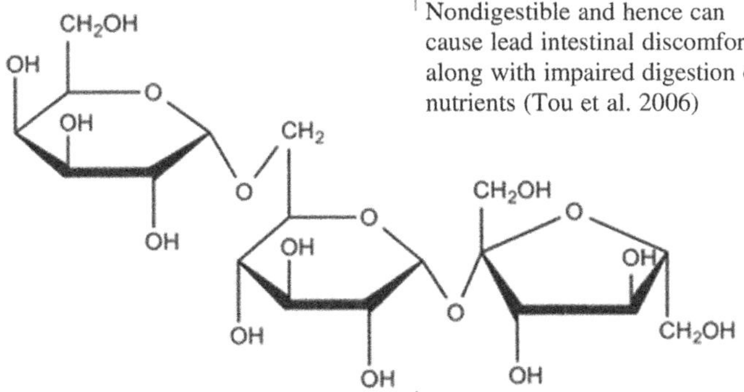	Nondigestible and hence can cause lead intestinal discomfort along with impaired digestion of nutrients (Tou et al. 2006)

products in a desired way (Ramashia et al. 2019). These food processing techniques enhance the nutritional quality as well as improve their digestibility and bioavailability along with the reduction in the antinutrient substances. On the basis of ample scientific evidences, it is now well established that these food processing techniques especially germination and fermentation together or individually modulate the antinutritional components such as oxalic acid, tannins, phytic acid, and trypsin inhibitors that otherwise hamper the utilization of important nutrients like protein, calcium, iron, and zinc present in various millet foods (Inyang and Zakari 2008; Budhwar et al. 2020). Food and Agriculture Organisation (FAO) has also reported and suggested the use of these traditional food processing techniques to improve the nutritional, edible, and sensory properties of millets and other cereals (FAO 2018). In this section, these processing techniques especially the fermentation and germination will be discussed in relation to their role in modulating the anti-nutritional, phenolics, and antioxidant contents of millet-based food products. Fermentation and germination in particular have been focussed here because both these processes in combination or individually modulate the nutrition quality of cereals and legumes that constitutes a major portion of diet plans in developing countries.

13.5.1 Germination-Mediated Changes in Nutritional Attributes of Millets

Germination or malting is one of the common household processing techniques that could be critical in enriching the nutritional value of millet food products (Nkhata et al. 2018). In germination, the seed grains are usually soaked for 2–24 h followed by spreading on a damp cloth for about 24–48 h or can be incubated for 48 h at 30 °C (Ramashia et al. 2019). It is basically a biochemical enrichment tool where the seed undergoes a transition from dormant state to vital active state under the moisture conditions. The term malting is however used for the limited germination of cereals under moist controlled environment. Germination has been found to improve the nutritive aspects of cereals through the reduction of antinutrient factors and also an increase in the optimum level of absorbable nutrients (Nkhata et al. 2018). Previously, it was observed that germination processed finger millet significantly increased the calcium and iron content but decreased the contents of tannin, phytic acid, oxalic acid as well as the activity of the trypsin inhibitor (Budhwar et al. 2020). In another recent revelation, germination along with popping was observed to increase the nutritional profile of finger millet including the total protein content, mineral availability especially the calcium and iron levels (Chauhan 2018a). It was also noticed that these processes significantly reduced the levels of phytic acid and oxalic acid along with trypsin inhibitor activity reflecting that germination along with popping increased the nutritional profile but at the same time diminished the antinutrient contents in the finger millet. Similarly, in barnyard millet (PRJ-1), optimized germination conditions had a tremendous impact on the levels of phenolic compounds along with dietary fibers, minerals, and γ-amino butyric acid that subsequently enhanced the availability of bioactive compounds (Sharma et al. 2016). The germination processed barnyard millet flour also posses better nutraceutical quantities with increased amount of gamma aminobutyric acid, octadecadienoic acid (Z,Z) (14.59%), hexadecanoic acid (5.59%), 9,12 octadecadienoic acid, ethyl ester (11.00%), c-sitosterol (4.00%), campesterol (7.59%), and lupeol (5.18%) in contrast to unprocessed millet flour. Further, it was observed that the levels of dietary fibers and minerals also enhanced whereas phytates decreased significantly which reflects the potential application of germination process in enhancing the better utilization of millet.

Ample scientific investigations have shown the strong antioxidant potential of various millet-based food products. Previously, Sharma et al. (2015) have reported the presence of antioxidant compounds along with and anti-nutritional qualities in foxtail millet (*Setaria italica*). The post-germination analysis revealed that the germination processed foxtail millet flour under optimized conditions induced nutritional enrichment with better amounts of dietary fiber (27.42 g/100 g), protein (14.32 g/100 g), magnesium (107.16 mg/kg), calcium (25.62 mg/kg), iron (54.23 mg/kg) along with sodium (69.45 mg/kg) per kg in contrast to unprocessed millet flour. The germinated millet flour showed higher antioxidant activity, with better levels of total phenolic and flavonoids as compared to control millet flour. They also observed a significant reduction in anti-nutritional factors like tannin and phytate with a significant reduction from 2.803 to 0.983 mg/100 g and 0.341 to 0.102 mol/kg, respectively with increased germination time indicating that processed foxtail millet could be a natural remedy for many potential health benefits particularly in relation to phytochemical bioavailability after germination. In temporal manner, Nefale and Mashau (2018) observed the changes in finger millet properties in relation to physicochemical, functional, and sensory attributes when processed through germination. A strong decline in the pH was observed (6.24–5.72) with a progressive increase in total titratable acidity (0.50–1.34%) in millet germinating seeds after 72 h incubation. In addition, they found a significant ($p < 0.05$) enhancement in the water and oil absorption capacity along with the increased solubility in germination processed millet flour in contrast to significant reduction in the bulk density and swelling power. In a similar type of study, while analyzing the physicochemical characteristics of both malted finger millet (*Eleusine coracana*) as well as pearl millet (*Pennisetum glaucum*) it was revealed that germination leads to decline in the pH (8.50–7.60) of millet varieties with consequent increase in the protein content (7.61–7.81% and 10.57–11.87%) and crude fiber (5.54–8.81% and 1.07–2.55%), respectively (Owheruo et al. 2019). The germination process was also observed to reduce the fat (3.84–2.73%; 7.69–2.30%) contents of both the millets varieties. Mineral levels of germinated millets were also found to be significantly higher especially with respect to the sodium, magnesium, calcium, potassium, and iron contents. During the same analysis, it was further revealed that the phytochemical composition including alkaloid and saponin contents were increased as compared to significant reductions in tannin and phytate quantities suggesting the potential use of germination to enhance the functional attributes of different varieties of millets.

Thus, germination-mediated changes in the nutritional attributes of different millet varieties could make them potential healthy foods that could be used to either inhibit or prevent the incidences of diseases. Recently, it was reported that finger and pearl millet-based foods could be used as a natural therapy to combat celiac disease, especially in economically downtrodden countries. The finger and pearl millets when processed with germination and popping improved the nutrient levels in millets with high protein, carbohydrate, iron and phosphorus, and ash quantities that made them a good alternative for celiac patients (Chauhan and Sarita 2018). Similarly, the addition of germinated varieties of finger (*Eleusine coracana*) and pearl (*Pennisetum glaucum*) millets to baked food products reduced the antinutrient factors such as oxalic acid, phytic acid, and trypsin inhibitor activity (Chauhan 2018b). The incorporation of germinated millet varieties increased the enzymatic activities of amylase, acid, and alkaline phosphatase along with increased percentage of energy, protein, fat, iron, and calcium that rendered the nutraceutical and therapeutical attributes in millet. Such nutritionally enriched millet foodstuffs thus could play a critical role in the prevention of many health complications such as in case of gluten-sensitive patients, diabetes, obesity, allergy, cancer, nephroprotective, aging, and many other health complications.

Likewise, germinated multigrain flour containing millet showed better organoleptic quality along with intact nutritional components including increased carbohydrate, energy, and protein levels reflecting the utility of such suitable health mix products to overcome the common metabolic problem including diabetes (Agrawal et al. 2016). Further, the germinated pearl millet showed considerable enhancement in the bio-accessibility of iron and calcium with the corresponding decrease in antinutritional factors like phytate and oxalate that resulted in the enhanced nutritional benefits of millets. Similarly, Arora et al. (2011) observed the combined effects of germination and probiotic *Lactobacillus acidophilus* mediated fermentation that significantly elevated the protein fractions, sugars thiamine, niacin, total lysine as well as soluble dietary fiber in pearl millets. The investigators also observed the better in vitro bioavailability of essential minerals like Ca, Fe, and Zn as compared to non-germinated food blends reflecting the effectiveness of germination and fermentation processes in improving the better availability and enrichment of essential nutrients for maintenance and promotion of good health conditions. Previously, it was reported that the incidence of anemia development in Tanzanian children was reduced when germinated finger millet-based food recipe was supplemented to infants that subsequently resulted in improved hemoglobin status and growth. The anti-anemic effect of germinated finger millet was due to improved nutrient density as well as better iron bioavailability (Tatala et al. 2007). All these scientific evidences indicate the germination-mediated improvement of nutritional aspects of millet food that is very much important to present world scenario where there is too much food shortage across the globe mostly due to population explosion and climatic extremities. Further being an easy and affordable method, germination can be easily applied to improve the functional aspects of millets. Especially, germination could be a great asset in the lives of people who are economically and technically far behind than the other developed parts of the world.

13.5.2 Fermentation and the Nutrition Attributes of Millet

Fermentation is another common food processing technique in which the nutritional attributes of the food and beverages are modified by using microorganisms. In addition to the improved nutrient contents of food, fermentation also enhances their digestibility. For the fermentation process, the raw material used actually serves as a medium for the growth of the microorganism but during this period the microbes in turn induce some desirable changes in the food material. Importantly, fermentation may occur naturally or can also be intentionally encouraged to change the organoleptic characteristics (viz., texture, flavor, and aroma) so that the food with better nutritional and functional properties could be used that impart the consumers with better health benefits to host (Rezac et al. 2018). However, the term "fermentation" in a broader sense is used to denote the intentional use of microbes such as bacteria, yeast, and fungi to make different food products that are useful to human that includes recombinant products, enzymes, primary and secondary metabolites, product biotransformation and others on a large scale (Teixeira and Vicente 2013).

Millet grains have a natural ability to undergo fermentation and pre-fermenting them is really an easy task. However, the fermentation process can also be induced by external supplementation of specific microbes such as bacteria and yeast specifically to bring desired changes within the nutritional aspects of food products. Probiotic microbes defined as microbes that when consumed in adequate amounts confers health benefit to host are nowadays commonly used to induce fermentation of coarse grains so as to improve their nutritive value (Budhwar et al. 2020). Though probiotic microbes have been in human use from a long time but with the exploration of health benefits associated with their consumption, more focus is given for their regular use in the form of functional food products, fermented drinks, or beverages across the globe that will naturally improve the immunity along with protection against diseases. Ample scientific evidences have shown that probiotic-mediated fermentation could modulate the various attributes of foods including the anti-nutritional, phenolic, and antioxidant characteristics (Peerajan et al. 2016; Marco et al. 2017; Oh et al. 2017). As a result researchers across the globe especially in

Asian countries have extended the utility of fermentation process to improve the nutritional attributes of millets. Recently, it was reported that fermentation of pearl millet (*Pennisetum glaucum*) using surface response methodology increased the overall reducing sugar, iron level, and antioxidant capability with the concurrent decline in amount of tannins optimally at 40 °C temperature, pH 5 during 8 h of incubation (Srivastava et al. 2020). Similarly, it was found that adding probiotic Fiti sachet consortium of *Lactobacillus rhamnosus* GR-1 and *Streptococcus thermophilus* C106 decreased the phytic acid that subsequently improved the bio-availability of specific micronutrients along with better sensory evaluation parameters (Di Stefano et al. 2017). Ranasalva and Visvanathan (2014) also reported that the growth of probiotic bacteria (*Lactobacillus* sp.) in finger millet-based food reduced the anti-nutrient phytic acid (858.4 mg/100 g–380.3 mg/100 g) in contrast to unprocessed millet. Further, they noticed better textural and physical properties of fermented pearl millet flour than what was shown by with nonfermented variety. Now it is thought that the utility of fermented millet products could be an essential component in European markets where the use of such food items is very much limited despite the more nutritive value of millets compared to other cereals (Balli et al. 2020). Millets when fermented with yeast and lactobacilli were having more phenolic contents than the nonfermented varieties. The fermented millets also had increased levels of cinnamic acids and flavonoids approximately by 30%. The higher levels of vitexin and vitexin 2″-*O*-rhamnoside flavonoids in fermented foods could prove valuable for the management of type-2 diabetes because of their ability to hamper the activity of protein tyrosine phosphatase enzyme whose expression is usually impaired during diabetic conditions. Further, the phenolic extracts obtained from fermented millet have shown higher antioxidant activities in human erythrocytes under ex vivo investigation (Balli et al. 2020). Prior investigation by Osman (2011) reported a considerable decline in the activities of amylase and trypsininhibitors along with glycine, lysine, and arginine and phytic acid level in pearl millet flour processed through traditional fermentation for 24 h. However, the fermentation process leads to a significant ($P < 0.05$) enhancement in the tannin and carbohydrate contents in pearl millet flour that is commonly employed for lohoh bread preparation in the South-West region of Jazan, Saudi Arabia.

Probiotic-mediated fermentation in millet has a great potential in developing functional foods with enriched biologically active components and novel properties that could be critical for better health promotion. Recently, Salar et al. (2017) observed the protection against DNA damage from different pearl millet (PUSA-415) based extracts which were prior fermented with *Aspergillus sojae* (MTCC-8779). This protective activity against DNA damage was due to induction of enzymes such as α-amylase, β-glucosidase, and xylanase along with some specific bioactive compounds such as gallic acid, ascorbic acid, and *p*-Coumaric acid due to fermentation with *Aspergillus sojae*. Likewise, when little millet (*Panicum sumatrense*) was fermented with combination of probiotic strains *Lactobacillus acidophilus* and *Saccharomyces boulardii* continuously for 5 days showed significantly elevated contents of protein by 10.95% compared to significant decline in fat (2.61%) and carbohydrate (82.01%) levels (Pampangouda et al. 2015). However, when millets were added to *L. acidophilus* separately, a significant reduction in the fiber (1.05%) and ash (2.30%) contents were observed in contrast to elevated levels of Calcium (41.44 mg), Magnesium (141.08 mg), Phosphorus (238.43 mg), Iron (7.98 mg), and Zinc (4.69 mg). *L. acidophilus* addition also reduced the phytic acid levels reflecting their potential utility for enriching the nutritive characteristics of millets. Previously, it has been found that millet drink fermented by lactic acid bacteria were having a protective effect against antibiotic-associated diarrhea in Ghanaian children that could help in long-term reduction of persistent diarrhea (Lei et al. 2006). In different varieties of millet-based food products, fermentation in combination with germination and cooking has been reported to change phenolic content, anti-nutritional contents thereby increasing the antioxidant potential along with better bioavailability of nutrients (Gabaza et al. 2016; Gadir and Adam 2000; Anand 2017).

In consonance with the above discussion, Shahidi and Chandrasekara (2013) noticed the presence of antioxidant compounds such as hydroxycinnamic acids, hydroxybenzoic acids, and flavonoids in millets when subjected to in vitro enzymatic digestion and microbial fermentation. Further, it was observed that fermentation processed millets were having more soluble phenolic contents compared to their insoluble fiber aggregation prior to fermentation processing. In another intriguing investigation, Palaniswamy and Govindaswamy (2017) reported the presence of enzymes feruloyl esterase and glutamate decarboxylase producing lactic acid bacteria *Lactobacillus fermentum* in pearl millet porridge (kambu koozh) that could be critical for the formulation of next-generation functional foods because of their role to release the cell wall bound ferulic acid which is known for many cellular benefits including antioxidant, antidiabetic, anticancer, and antihypertensive properties. So, enriching of feruloyl esterase producing microbes in millet could be one way to improve the benefits associated with fermented food consumption. Similarly, the presence of ferulic acid esterase enzyme has been confirmed in *Streptomyces* and *Lactobacillus plantarum* that are extensively used while preparing a wide range of fermented dairy products, meat, vegetables, and bakery products and hence could be key future ingredients for functional food applications (Mukherjee et al. 2007; Inmaculada et al. 2013). Previously differential amounts of phenolic compounds including both soluble as well as insoluble fractions were obtained across a wide variety of millet including foxtail, finger, kodo, little, pearl, and proso forms of millet (Chandrasekara and Shahidi 2010). Using high-performance liquid chromatography and mass spectrometry (MS) both the soluble as well as insoluble forms of ferulic and *p*-coumaric acids were present. Among the different millet forms, kodo was having the maximum quantity of total phenolic content compared to proso form which has least amount reflecting the variety-based differential antioxidant activities. Similarly, Palaniswamy and Govindaswamy (2017) reported that the presence of millet bound polyphenolic compounds such as catechin and ferulic acid in large quantities but minor amounts of vanillic acid and resveratrol that were responsible for the inhibition of hydrogen peroxide and peroxyl-AAPH induced peroxidation of protein, DNA, and erythrocytes. They also observed time-delayed response for hemolysis by polyphenols obtained from different millet varieties reflecting their immense cytoprotective potential.

In addition to the changed nutritional attributes of food, the fermenting microbes also produce their own by-products such as acids or antibiotics that in turn augment the health benefits of foods. The presence of microbial products, antibiotics, and acids inhibit the growth of pathogenic microorganisms and thus prevent the host from the ill effects of such harmful microbes (Ranasalva and Visvanathan 2014). The microbial fermentation also increased the shelf-life of food products along with improved amino acid balance and sensory quality in addition to nutritional enrichment of the grains. Previously, Osamwonyi and Wakil (2012) reported the antimicrobial activity of various yeast strains including *Penicillium* sp. (FM1), *Rhizopus* spp. (FM2, FM3, and FM6), *Aspergillus flavus* (FM7), and *Aspergillus niger* (FM8) isolated from fermented pearl millet gruel against the pathogens like *B. lichieniformis*, *Aspergillus flavus* (FM7), *E. coli*, *P. flourescens* and *S. aureus* with *Aspergillus flavus* exhibiting maximum zone of inhibition (24 mm) against *P. flourescens* reflecting that fermenting microbes may also impart antimicrobial activity to foods against various pathogen. Similarly, Lactic acid bacteria (LAB) have been reported to change the sensory characteristics of fermented sausages by producing a mixture of acetic acid, ethanol, acetoin, pyruvic acid, carbon dioxide along with some protein-based aromatic substances (Cocolin and Rantsiou 2012; Fontana et al. 2012). The fermenting microbes also release some important biologically active compounds such as bacteriocins, biogenic amines, bioactive peptides, and others that improve the food quality (Srivastava 2018). Recently, Chaudhary et al. (2020) identified a number of metabolites from various fermented food that are already well-established for their role as defrosting agent, anti-inflammatory, antioxidant, antimicrobial, anti-sleep disorder, and anticancer agents using GC-MS based metabolomic analysis. These all studies reflect the utility of fermenting microbes in changing the nutritional characteristics of foods. Particularly with the evolution of probiotic microbes, fermented foods may serve as useful vehicle for probiotic organisms or their health-promoting metabolites. Thus, the food processing techniques especially germination and fermentation have a massive potential to augment the functional utility of millet products with much relevance nations where people have not much access to healthcare facilities and have very much limited resources.

13.5.3 Other Traditional Processes

Besides fermentation and germination, food products can also be processed by other techniques including milling, decortication/dehulling, soaking, roasting and microwave, and steam treatments that also could prove vital in relation to nutritional enrichment of millet. Previously Chowdhury and Punia (1997) reported a gross change in the chemical composition of millets when subjected to treatment of milling and heat. The combined treatment of milling and heating decreased the polyphenols and phytic acid levels but significantly improved the digestibility of both protein and starch in millet bread products. Similarly, roasting and soaking of barnyard millet-based flour brought a significant change in nutritional as well as sensory qualities of millet with a remarkable decline of anti-nutritional compounds tannins (2.21 $\pm$ 0.01 from 2.96 $\pm$ 0.03 mg/100 g) and phenolic compounds (5.2 $\pm$ 0.01 from 5.31 $\pm$ 0.01 mg/100 g) respectively (Anbalagan and Nazni 2020). In the same study, a significant change was noticed in the viscosity and osmolarity properties of millet flour reflecting the better utility of millet-based food product as diarrheal replacement fluid. In another investigation, roasted whole grain millet samples exhibited higher antioxidant activity compared to nonroasted samples due to the release of bound phenolic compounds by the supplied heat (Pradeep and Sreerama 2015). Hithamani and Srinivasan (2014) also reported the roasting mediated increase in 17% of the total phenolic substances of finger millet grains indicating the utility of roasted millet food products as potential functional foods.

An interesting revelation by Pradeep and Guha (2011) demonstrated that the steam and microwave exposure of barnyard millet diminished the phenolic contents but elevated the phenolic contents both in proso and foxtail millet varieties. Though both treatments diminished the phenolic content of barnyard millet, higher phenolic content was noticed in foxtail and proso millet. With respect to flavonoid contents, all the three millet varieties demonstrated lower flavonoid levels with steam exposure but only foxtail millet showed depleted flavonoids upon microwave treatment as compared to nonsignificant changes in other two millet varieties. In another study, combination of parboiling, grinding, and roasting was found to increase the total phenolic contents in finger millet porridges. When the processed finger millet porridges were fed to human volunteer subjects, improved glycemic response with better antioxidant activities as reflected by increased total plasma antioxidant substances were noticed reflecting the potential application of millets to combat diabetes and other metabolic syndromes. In another study, six varieties of fox millet when exposed to different treatments including dehulling, cooking, and steaming showed differential changes in phenolic compounds of millets. The dehulled grains were having more total phenolic compounds but less total flavonoids as compared to whole millet (Zhang et al. 2017). However, both steamed and cooked millet grains showed decreased phenolic and flavonoid contents as compared to dehulled millet though cinnamic acid level was more in cooked millet than others. Further, it was noticed that cooked millet exhibited better radical scavenging capacity due to high contents of natural antioxidant compounds. Similarly, Kundgol et al. (2013) also observed the decortication mediated changes in the nutritional composition and antioxidant aspects of little millet. Using chemical assays, they found that the level of antioxidant compound tocopherol and other bioactive constituents were more in bran than the decorticated grain and whole millet but the level of phytic acid were significantly reduced in decorticated grain. In similar lines to previous work, processing of foxtail millet through combination of dehulling, soaking, and cooking showed significantly reduced levels of antinutrient factors such as polyphenols and phytate that subsequently improved the cellular bioavailability of minerals such as iron and zinc in addition to improved protein digestibility (Pawar and Machewad 2006). All these scientific evidences reflect the utility of these common house techniques towards the augmentation of nutritional as well as functional aspects of millets that could subsequently prove very critical towards the promotion of better healthy life.

13.6 Conclusion

Millets are one of the richest natural food materials enriched with adequate amounts of proteins, complex carbohydrates, fat as well as minerals, vitamins, and fiber and hence could serve as a natural remedy for the promotion of better health. With short growing season and capability to withstand dry and high-temperature conditions, millets can be utilized in a better way to serve the nations where there is looming threat of starvation, malnutrition, and economic crisis. However, the presence of certain anti-nutritional substances such as phytic acid, polyphenols, and tannins which restrict the nutrient availability, some ways need to be explored so as to diminish these anti-nutritional factors. Interestingly, the common household food processing techniques including germination and fermentation have shown immense potential to not overcome these anti-nutritional compounds but to enhance the nutritional aspects of cereal grains especially with respect to millets. Intentional or unintentional addition of microbes to millet-based food products have been observed to change the antioxidant, anti-nutritional, and phenolic components of millets that have made them food with enormous health benefits. Fermented foods with the unique addition of microbes could be one way to feed the people who are deprived of basic food and are malnourished. Fermented food definitely has given hope to economically downtrodden countries with respect to addressing the problems related to a balanced diet.

References

Adu-Gyamfi C, Tse EK, Axala E, Djidjor EK (2019) Production and consumers' acceptability of meat turnovers produced from wheat and millet flour blends. J Food Technol Food Chem 2:106

Agrawal A, Verma A, Shiekh S (2016) Evaluation of sensory accessibility and nutritive values of multigrain flour mixture products. Int J Health Sci Res 6:459–465

Akbar N, Gupta S, Tiwari A, Singh KP, Kumar (2018) Characterization of metabolic network of oxalic acid biosynthesis through RNA seq data analysis of developing spikes of finger millet (Eleusine coracana): deciphering the role of key genes involved in oxalate formation in relation to grain calcium accumulation. Gene 649:40–49

Anand R (2017) Effect of germination and fermentation on the antioxidant activity of millet koozh prepared using sorghum bicolour (*Monech. L*), pearl millet (*Pennisetum glaucum*) and finger millet (*Eleusine coracana*). Int J Multidiscip Res Dev 4:155–157

Anbalagan S, Nazni P (2020) Comparative study on physico-chemical characteristics of different periods of soaked minor millets flour-based diarrheal replacement fluids. Int J Res Pharm Sci 11:2189–2197

Arora S, Jood S, Khetarpaul N (2011) Effect of germination and probiotic fermentation on nutrient profile of pearl millet based food blends. Br Food J 113:470–481

Balli D, Bellumori M, Pucci L, Gabriele M, Longo V, Paoli P, Melani F, Mulinacci N, Innocenti M (2020) Does fermentation really increase the phenolic content in cereals? A study on millet. Foods 9:303

Boncompagni E, Orozco-Arroyo G, Cominelli E, Gangashetty PI, Grando S, Kwaku Zu TT, Daminati MG, Nielsen E, Sparvoli F (2018) Anti-nutritional factors in pearl millet grains: phytate and goitrogens content variability and molecular characterization of genes involved in their pathways. PLoS One 13:0198394

Budhwar S, Sethi K, Chakraborty M (2020) Efficacy of germination and probiotic fermentation on underutilized cereal and millet grains. Food Prod Process Nutr 2:1–7

Caio G, Volta U, Sapone A, Leffler DA, De Giorgio R, Catassi C, Fasano A (2019) Celiac disease: a comprehensive current review. BMC Med 17:1–20

Chandrasekara A, Shahidi F (2010) Content of insoluble bound phenolics in millets and their contribution to antioxidant capacity. J Agric Food Chem 58:6706–6714

Chaudhary A, Verma K, Saharan BS (2020) A GC-MS based metabolic profiling of probiotic lactic acid bacteria isolated from traditional food products. J Pure Appl Microbiol 14:657–672

Chauhan ES (2018a) Effects of processing (germination and popping) on the nutritional and anti-nutritional properties of finger millet (*Eleusine Coracana*). Curr Res Nutr Food Sci 6:566–572

Chauhan ES (2018b) Exploration of gluten-free baked food products incorporated by germinated finger (*Eleusine coracana*) and pearl (*Pennisetum glaucum*) millets: a therapeutic approach. Int J Health Sci Res 8:232–243

Chauhan ES, Sarita (2018) Development of gluten-free food products incorporated by germinated and popped finger and pearl millets. Indian J Nutr Diet 55:291–307

Chauhan M, Sonawane SK, Arya SS (2018) Nutritional and nutraceutical properties of millets: a review. Clin J Nutr Diet 1:1–10

Chowdhury S, Punia D (1997) Nutrient and antinutrient composition of pearl millet grains as affected by milling and baking. Food Nahrung 41:105–107

Cirkovic Velickovic TD, Stanic-Vucinic DJ (2018) The role of dietary phenolic compounds in protein digestion and processing technologies to improve their antinutritive properties. Compr Rev Food Sci Food Saf 17:82–103

Cocolin L, Rantsiou K (2012) Meat fermentation. In: Handbook of meat and meat processing, 2nd edn. CRC Press, Boca Raton, FL, pp 557–572

Di Stefano E, White J, Seney S, Hekmat S, McDowell T, Sumarah M, Reid G (2017) A novel millet-based probiotic fermented food for the developing world. Nutrients 9:529

FAO (2018) Anti-nutritional factors within feed ingredients. Aquaculture Feed and Fertilizer Resources Information System, Food and Agriculture Organizations of the United Nations, Rome. http://www.fao.org/fishery/affris/feed-resources-database.anti-nutritional-factors-within-feedingredients/en/. Accessed 28 Nov 2018

Fenwick S, Vanga SK, DiNardo A, Wang J, Raghavan V, Singh A (2019) Computational evaluation of the effect of processing on the trypsin and alpha-amylase inhibitor from Ragi (*Eleusine coracana*) seed. Eng Rep 1:12064

Fontana C, Fadda S, Cocconcelli PS, Vignolo G (2012) Lactic acid bacteria in meat fermentations. In: Lactic acid bacteria: microbiological and functional aspects, 4th edn. CRC Press, Taylor & Francis, Boca Raton, FL, pp 247–264

Gabaza M, Shumoy H, Muchuweti M, Vandamme P, Raes K (2016) Effect of fermentation and cooking on soluble and bound phenolic profiles of finger millet sour porridge. J Agric Food Chem 64:7615–7621

Gadir WA, Adam S (2000) Effects of pearl millet (*Pennisetum typhoides*), and fermented and processed fermented millet on Nubian goats. Vet Hum Toxicol 2:133–136

Grober U, Schmidt J, Kisters K (2015) Magnesium in prevention and therapy. Nutrients 7:8199–8226

Handa V, Kumar V, Panghal A, Suri S, Kaur J (2017) Effect of soaking and germination on physicochemical and functional attributes of horsegram flour. J Food Sci Technol 54:4229–4239

Hithamani G, Srinivasan K (2014) Effect of domestic processing on the polyphenol content and bioaccessibility in finger millet (Eleusine coracana) and pearl millet (Pennisetum glaucum). Food Chem 164:55–62

Hosoda A, Okai Y, Kasahara E, Inoue M, Shimizu M, Usui Y, Sekiyama A, Higashi-Okai K (2012) Potent immunomodulating effects of bran extracts of traditional Japanese millets on nitric oxide and cytokine production of macrophages (RAW264.7) induced by lipopolysaccharide. J UOEH 34:285–296

Inmaculada NG, Álvaro SF, Francisco GC (2013) Overexpression, purification, and biochemical characterization of the esterase Est0796 from *Lactobacillus plantarum* WCFS1. Mol Biotechnol 54:651–660

Inyang CU, Zakari UM (2008) Effect of germination and fermentation of pearl millet on proximate, chemical and sensory properties of instant "Fura"-a Nigerian cereal food. Pak J Nutr 7:9–12

Karas M, Jakubczyk A, Szymanowska U, Jęderka K, Lewicki S, Złotek U (2019) Different temperature treatments of millet grains affect the biological activity of protein hydrolyzates and peptide fractions. Nutrients 11:550

Krupa-Kozak U (2014) Pathologic bone alterations in celiac disease: etiology, epidemiology, and treatment. Nutrition 30:16–24

Kumar A, Metwal M, Kaur S, Gupta AK, Puranik S, Singh S, Singh M, Gupta S, Babu BK, Sood S, Yadav R (2016) Nutraceutical value of finger millet [*Eleusine coracana* (L.) Gaertn.], and their improvement using omics approaches. Front Plant Sci 7:934

Kundgol NG, Kasturiba B, Math KK, Kamatar MY, Usha M (2013) Impact of decortication on chemical composition, antioxidant content and antioxidant activity of little millet landraces. Int J Adv Res Technol 2:1705–1720

Lee SH, Chung IM, Cha YS, Park Y (2010) Millet consumption decreased serum concentration of triglyceride and C-reactive protein but not oxidative status in hyperlipidemic rats. Nutr Res 30:290–296

Lei V, Friis H, Michaelsen KF (2006) Spontaneously fermented millet product as a natural probiotic treatment for diarrhoea in young children: an intervention study in northern Ghana. Int J Food Microbiol 110:246–253

Li X, Hui Y, Ren J, Song Y, Liu S, Che L, Peng X, Dai X (2019) Millet-based supplement restored microbiota diversity of acute undernourished pigs. BioRxiv. https://doi.org/10.1101/2019.12.13.875013

Liang S, Liang K (2019) Millet grain as a candidate antioxidant food resource: a review. Int J Food Prop 22:1652–1661

Marco ML, Heeney D, Binda S, Cifelli CJ, Cotter PD, Foligne B, Gsanzle M, Kort R, Pasin G, Pihlanto A, Smid EJ (2017) Health benefits of fermented foods: microbiota and beyond. Curr Opin Biotechnol 44:94–102

Moreno Amador MD, Comino Montilla IM, Sousa Martin C (2014) Alternative grains as potential raw material for gluten-free food development in the diet of celiac and gluten-sensitive patients. Austin J Nutr Food Sci 2:9

Mukherjee G, Singh RK, Mitra A, Sen SK (2007) Ferulic acid esterase production by *Streptomyces* sp. Bioresour Technol 98:211–213

Murtaza N, Baboota RK, Jagtap S, Singh DP, Khare P, Sarma SM, Podili K, Alagesan S, Chandra TS, Bhutani KK, Boparai RK (2014) Finger millet bran supplementation alleviates obesity-induced oxidative stress, inflammation and gut microbial derangements in high-fat diet-fed mice. Br J Nutr 112:1447–1458

Nambiar VS, Dhaduk JJ, Sareen N, Shahu T, Desai R (2011) Potential functional implications of pearl millet (*Pennisetum glaucum*) in health and disease. J Appl Pharm 1:62

Nefale FE, Mashau ME (2018) Effect of germination period on the physicochemical, functional and sensory properties of finger millet flour and porridge. Asian J Appl Sci Eng 6:360–367

Niro S, D'Agostino A, Fratianni A, Cinquanta L, Panfili G (2019) Gluten-free alternative grains: nutritional evaluation and bioactive compounds. Foods 8:208

Nkhata SG, Ayua E, Kamau EH, Shingiro JB (2018) Fermentation and germination improve nutritional value of cereals and legumes through activation of endogenous enzymes. Food Sci Nutr 6:2446–2458

Norhaizan ME, Nor Faizadatul Ain AW (2009) Determination of phytate, iron, zinc, calcium contents and their molar ratios in commonly consumed raw and prepared food in Malaysia. Malays J Nutr 15:213–222

Ofosu FK, Elahi F, Daliri EB, Chelliah R, Ham HJ, Kim JH, Han SI, Hur JH, Oh DH (2020) Phenolic profile, antioxidant, and antidiabetic potential exerted by millet grain varieties. Antioxidants 9:254

Oh BT, Jeong SY, Velmurugan P, Park JH, Jeong DY (2017) Probiotic-mediated blueberry (*Vaccinium corymbosum* L.) fruit fermentation to yield functionalized products for augmented antibacterial and antioxidant activity. J Biosci Bioeng 124:542–550

Osamwonyi UO, Wakil SM (2012) Isolation of fungal species from fermentating pearl millet gruel and determination of their antagonistic activities against indicator bacterial species. Niger Food J 30:35–42

Osman MA (2011) Effect of traditional fermentation process on the nutrient and antinutrient contents of pearl millet during preparation of Lohoh. J Saudi Soc Agric Sci 10:1–6

Owheruo JO, Ifesan BO, Kolawole AO (2019) Physicochemical properties of malted finger millet (*Eleusine coracana*) and pearl millet (*Pennisetum glaucum*). Food Sci Nutr 7:476–482

Palacios JJ (2008) Multinational corporations and the emerging network economy in Asia and the Pacific. Routledge, Oxon

Palaniswamy SK, Govindaswamy V (2017) Inhibition of metal catalyzed H_2O_2 and peroxyl-AAPH mediated protein, DNA and human erythrocytes lipid oxidation using millet phenolics. J Plant Biochem Biotechnol 26:406–414

Pampangouda P, Munishamanna KB, Gurumurthy H (2015) Effect of *Saccharomyces boulardii* and *Lactobacillus acidophilus* fermentation on little millet (*Panicum sumatrense*). J Appl Nat Sci 7:260–264

Pawar VD, Machewad GM (2006) Processing of foxtail millet for improved nutrient availability. J Food Process Preserv 30:269–279

Peerajan S, Chaiyasut C, Sirilun S, Chaiyasut K, Kesika P, Sivamaruthi BS (2016) Enrichment of nutritional value of Phyllanthus emblica fruit juice using the probiotic bacterium, *Lactobacillus paracasei* HII01 mediated fermentation. Food Sci Technol 36:116–123

Pradeep SR, Guha M (2011) Effect of processing methods on the nutraceutical and antioxidant properties of little millet (*Panicum sumatrense*) extracts. Food Chem 126:1643–1647

Pradeep PM, Sreerama YN (2015) Impact of processing on the phenolic profiles of small millets: evaluation of their antioxidant and enzyme inhibitory properties associated with hyperglycemia. Food Chem 169:455–463

Prashanth MS, Shruthi RR, Muralikrishna G (2015) Immunomodulatory activity of purified arabinoxylans from finger millet (*Eleusine coracana* v. Indaf 15) bran. J Food Sci Technol 52:6049–6054

Ramashia SE, Anyasi TA, Gwata ET, Meddows-Taylor S, Jideani AI (2019) Processing, nutritional composition and health benefits of finger millet in sub-saharan Africa. Food Sci Technol 39:253–266

Ranasalva N, Visvanathan R (2014) Development of bread from fermented pearl millet flour. J Food Process Technol 5:1–5

Rezac S, Kok CR, Heermann M, Hutkins R (2018) Fermented foods as a dietary source of live organisms. Front Microbiol 9:1785

Salar RK, Purewal SS, Sandhu KS (2017) Fermented pearl millet (Pennisetum glaucum) with *in vitro* DNA damage protection activity, bioactive compounds and antioxidant potential. Food Res Int 100:204–210

Saleh AS, Zhang Q, Chen J, Shen Q (2013) Millet grains: nutritional quality, processing, and potential health benefits. Compr Rev Food Sci Food Saf 12:281–295

Samtiya M, Aluko RE, Dhewa T (2020) Plant food anti-nutritional factors and their reduction strategies: an overview. Food Prod Process Nutr 2:1–4

Sarita ES, Singh E (2016) Potential of millets: nutrients composition and health benefits. J Sci Innov Res 5:46–50

Shahidi F, Chandrasekara A (2013) Millet grain phenolics and their role in disease risk reduction and health promotion: a review. J Funct Foods 5:570–581

Sharma S, Saxena DC, Riar CS (2015) Antioxidant activity, total phenolics, flavonoids and antinutritional characteristics of germinated foxtail millet (*Setaria italica*). Cogent Food Agric 1:1081728

Sharma S, Saxena DC, Riar CS (2016) Analysing the effect of germination on phenolics, dietary fibres, minerals and γ-amino butyric acid contents of barnyard millet (*Echinochloa frumentaceae*). Food Biosci 13:60–80

Shobana S, Harsha MR, Platel K, Srinivasan K, Malleshi NG (2010) Amelioration of hyperglycaemia and its associated complications by finger millet (Eleusine coracana L.) seed coat matter in streptozotocin-induced diabetic rats. Br J Nutr 104:1787–1795

Sireesha Y, Kasetti RB, Nabi SA, Swapna S, Apparao C (2011) Antihyperglycemic and hypolipidemic activities of Setaria italica seeds in STZ diabetic rats. Pathophysiology 18:159–164

Srivastava RK (2018) Enhanced shelf life with improved food quality from fermentation processes. J Food Technol Preserv 2:8–14

Srivastava U, Saini P, Singh A (2020) Effect of natural fermentation on antioxidant activity of pearl millet (Pennisetum glaucum). Curr Nutr Food Sci 16:306–313

Tatala S, Ndossi G, Ash D, Mamiro P (2007) Effect of germination of finger millet on nutritional value of foods and effect of food supplement on nutrition and anaemia status in Tanzania children. Tanzan J Health Res 9:77–86

Teixeira JA, Vicente AA (eds) (2013) Engineering aspects of food biotechnology. CRC Press, New York

Tou EH, Guyot JP, Mouquet-Rivier C, Rochette I, Counil E, Traoré AS, Trèche S (2006) Study through surveys and fermentation kinetics of the traditional processing of pearl millet (*Pennisetum glaucum*) into ben-saalga, a fermented gruel from Burkina Faso. Int J Food Microbiol 106:52–60

Weaver CM, Kannan S (2002) Phytate and mineral bioavailability. In: Reddy NR, Sathe SK (eds) Food phytates. CRC Press, Boca Raton, FL, pp 211–223

Zhang L, Li J, Han F, Ding Z, Fan L (2017) Effects of different processing methods on the antioxidant activity of 6 cultivars of foxtail millet. J Food Qual. https://doi.org/10.1155/2017/8372854

Technology for Millet Value-Added Products 14

S. D. Deshpande and P. K. Nishad

Abstract

Despite their exceptional nutritional profile, food use of millets is still confined only to traditional consumers and economically deprived sections of the society. Diversification of food resources by incorporating less popular millets is essential for achieving the nutritional security and combat with emerging climatic vagaries and life-threatening diseases. The gluten-free nature of protein, bioactive compounds with medicinal value, and high micronutrient density makes them an ideal candidate for developing several functional and value-added food products. Several value-added products of millets like biscuits, cakes, pasta, and infant foods are available in the market and gaining the attention of economically rich and health concerned masses of the society. This chapter provides a brief account of various processes, traditional and functional including ready-to-eat (RTE) millet value-added food products developed from millets and their characteristic features.

Keywords

Millets · Current scenario · Nutritional values · RTE products

14.1 Introduction

Millets are a group of small-seeded crops and probably the world's earliest food plants domesticated by humans. Their ability to grow in the harshest climatic conditions of the world has provided them a status of "famine crops" in the semi-arid tropics and drylands of Africa and Asia. Sorghum (*Sorghum bicolor*), pearl

S. D. Deshpande (✉) · P. K. Nishad
ICAR-Central Institute of Agricultural Engineering, Bhopal, Madhya Pradesh, India

A. Kumar et al. (eds.), *Millets and Millet Technology*,
https://doi.org/10.1007/978-981-16-0676-2_14

millet (*Pennisetum glaucum*), finger millet (*Eleusine coracana*), foxtail millet (*Setaria italica*), barnyard millet (*Echinochloa frumentacea*), kodo millet (*Paspalum scrobiculatum*), proso millet (*Panicum miliaceum*), and little millet (*Panicum miliare*) are the important millets cultivated in African and Asian countries (Jaybhaye et al. 2014).

Despite rich nutritional profile, some inherent features like hard seed coat, anti-nutritional factors, poor digestibility, and low micronutrient bioavailability are the major hindrances in the processing and cooking of millets. In the absence of primary, secondary, and tertiary processing technologies value-added RTE millet products are not readily available in the market for dietary diversification (Malleshi 2014). The growing awareness about the nutraceutical properties of millets has provided an avenue for their marketing and branding in the form of well-processed value-added products (Sehgal and Kwatra 2003; Nehir El and Simsek 2012). Therefore, this chapter briefly highlights the processing techniques, their effect on nutritional composition, and different traditional and value-added products for realizing the real potential of millets as wonder grains (Hulse et al. 1980).

14.2 Current Production Scenario of Global Millets

The total world production of millets was estimated as 31.02 million tons in 2018 (FAOSTAT 2018). India was the largest millet producer in 2018 with a contribution of 37.40%, followed by Niger and Sudan with contributions of 12.25% and 8.53%, respectively. The top ten millet production countries are presented in Table 14.1 and millet production contribution by continent is presented in Fig. 14.1.

The evolution of world millet area and production from 1999 to 2018 is presented in Fig. 14.2. Africa is the leading continent in millet production followed by Asia (FAO 1991, 1995, 2018). It is understandable that all major millet-producing countries are represented on these two continents. Europe, the Americas, and Oceania were the lowest contributors towards millet production in 2018.

According to FAO statistics (2018), world millet production was 31.02 million tons on an area of 33.56 million hectares. Almost two decades earlier (1999), global millet production in an area of 36.14 million hectares had fallen to 27.31 million tons. Africa was the largest millet producer in 2018 (15.86 MT), followed by Asia (14.38 MT), and India (11.64 MT). Relative to wheat, rice, maize, and barley, sorghum ranks fifth in terms of production and sown area and accounts for 5% of global grain production.

Global millet production has come down from 32.80 million tons in 2010 to 31.02 million tons in 2018. Asia and Africa contribute most to world millet production with a share of over 98%. The share of Asian countries in world millet production has decreased from 53.6% in 1999 to 46.36% in 2018, while the contribution of African countries has increased from 43.2% in 1999 to 51.15% in the year 2018.

Table 14.1 Top ten millet producing country in the world—2018

Rank	Country	Production (tones)
1	India 	1,16,40,000
2	Niger 	38,56,344
3	Sudan 	26,47,000
4	Nigeria 	22,40,744
5	Mali 	18,40,321
6	China, mainland 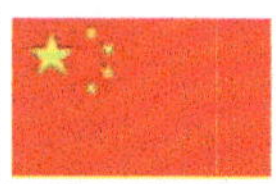	15,65,965
7	Burkina Faso 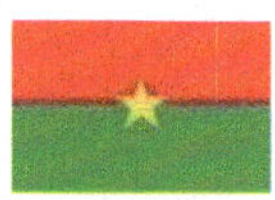	11,89,079
8	Ethiopia 	9,82,958
9	Chad 	7,56,616
10	Russian Federation 	2,17,200
World total		31,019,370

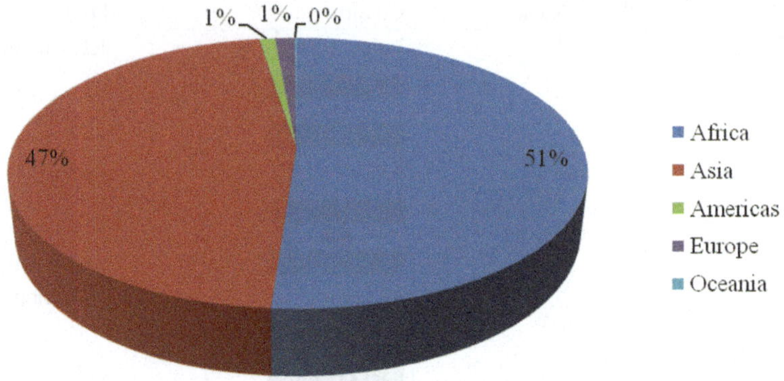

Fig. 14.1 Millet Production contribution by continent (2018)

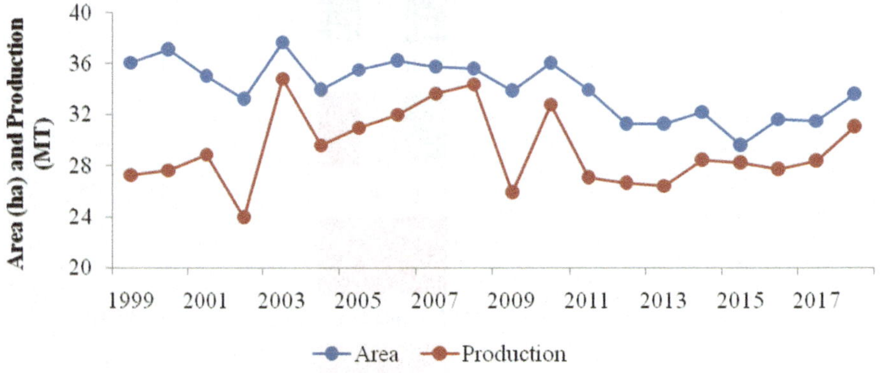

Fig. 14.2 World millet harvesting area and production (1999–2018)

14.3 Processing of Millets

Various techniques which convert the raw millet grains into edible form with enhancement in quality are known as processing. Small millet grains can be converted into various edible forms like rice, flour, sprouts, salted ready-to-eat, flaked, popped, porridge, and fermented products through primary processing techniques. The different processing techniques are described in this section.

14.4 Decortication/Dehulling

By virtue of a very hard seed coat, the first and foremost step in millet processing is dehusking or decortication. Traditionally, millets are decorticated through manual pounding. Traditional millstones (Fig. 14.3) utilized for dehusking and grinding of millets grain are usually comprised of a small stone that is held in the hand and a

Fig. 14.3 Traditional grinding stones for millet dehulling

larger flat stone that is placed on the ground. The friction between these grinding stones results in Dehulling and grinding of millet grains. In general, 4–5 passes are required through this method to get the completely dehusked seeds. The traditional method of Dehulling is laborious and time-consuming with low Dehulling yield (50%) (MP Rural Livelihood Proj, 2011).

The new automated dehulling systems like centrifugal dehulling system (Fig. 14.4) have been introduced to resolve the problems associated with traditional systems. The dehulling yield (72–76%) is much higher than the percentage of dehulled grains (50%) obtained through the traditional compounding method (Gupta and Das 1999). The major determinants of dehulling yield obtained through centrifugal system are moisture content of the grain, the peripheral speed of the impeller, and the feed rate (Gupta and Das 1999). Soaking is another cheap method of dehulling utilized in millets. In pearl millet, soaking in a laboratory scarifier for 1–3 min removed 8–15% husk of the grains (Pawar and Parlikar 1990).

Dehulling or decortication is known to reduce the total mineral content of the grains but enhance the bioavailability of calcium, iron, and zinc by 15, 26, and 24 g/100 g, respectively (Krishnan et al. 2012). It also enhances the digestibility of protein by reducing the anti-nutritional factors like phytic acid, polyphenols, and tannins.

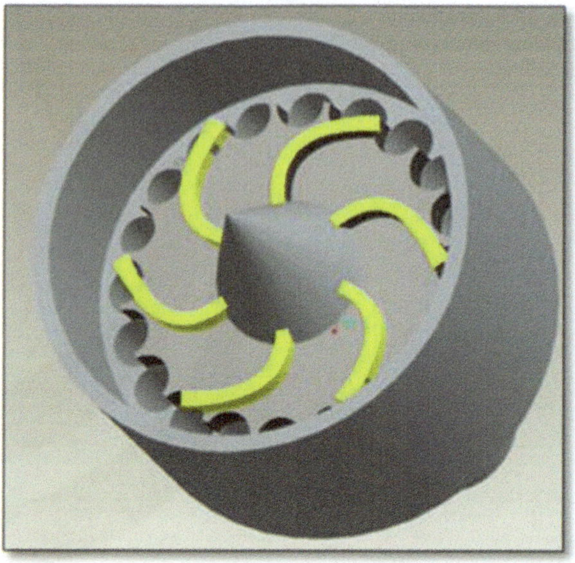

Fig. 14.4 Centrifugal dehulling system

14.5 Milling

The major limiting factors for the utilization of millets as ready-to-use value-added products are the coarse grains, hard seed coat, pigmented seeds, acidic or bitter taste, and poor shelf life of the processed products (Desikachar 1975). Therefore, proper milling technologies are required for the value addition and commercialization of millet-based products. Various processing techniques like pearling, debranning, and chemical treatment of millets are reported to overcome some of these limitations to improve consumer acceptability (Akingbala 1991).

14.6 Technology for Millet Value Addition

14.6.1 Composite Flour

Blending of millet grains with widely utilized cereals like wheat and maize and some nutritious pulses is one possible way to enhance their widespread utilization. Many functional food products of wheat like cakes, pasta, macaroni, vermicelli, noodles, spaghetti, and flakes are widely consumed in developed and developing countries. In general, the major ingredient of these products is refined wheat flour or semolina. Addition of millet flour in a certain ratio will enhance the nutritional and functional properties and change the physicochemical properties of these products (Singh et al. 2005). Blending of millet flour (10–30%) with wheat flour (70–90%) is reported to increase the percentage of protein, fat, and ash but reduced the percentage of total carbohydrates (Singh et al. 2012).

14.6.2 Puffed/Popped and Flaked Millets

Consumption of major cereals in the form of a variety of puffed and popped products is a common practice in developed and developing countries. Starch is the major component of millet and cereal grain. Various processing treatments mainly the thermal treatment of grains induces the physicochemical and structural changes in starch–protein matrix which ultimately leads to the expansion of the grains to produce a puffed product. High treatment short time (HTST) is a common and convenient method to make expanded flakes and other popped products (Jaybhaye et al. 2014). A comparative study of brown (Fig. 14.5) and white grain finger millet cultivars revealed that brown seeded genotypes have a higher percentage of puffed grain. However, the organoleptic properties of white seeded genotypes are better than the brown seeded cultivars (Shukla et al. 1986). Air blowing puffing machines have been designed for the mass production of puffed and popped millet grains (Verma and Patel 2013). The puffed and popped products are dehydrated to lower down the moisture content (3–5%) which enhances their shelf life.

Coarse cereals including finger millet make an integral part of human diet in the North West Himalayan region of India and consumed primarily as *roti* (unleavened flatbread). Unlike commonly consumed wheat *roti*, the finger millet *roti* lacks the puffing ability upon baking due to the absence of gluten in its flour. The hard texture and inability to swell upon baking reduce the consumer preference of these traditional finger millet *rotis*. In the hills of Uttarakhand, a traditional method is used for imparting softness and puffability of finger millet flour. The method involves the use of dried bark of a tree (*Boehmeria rugulosa*), vernacularly known as "*gethi*." About 250 g pieced dried bark is mixed with about 10 kg finger millet grain and the two are milled together The dough of *gethi* bark-incorporated flour is not only easy to roll out into circular shape but the *rotis* made from it puff up like wheat *rotis* and retain softness even after becoming cold (Khulbe et al. 2014). Extension of this knowledge to other parts of the country will enhance the consumer acceptability of millets and its products among rural and urban masses.

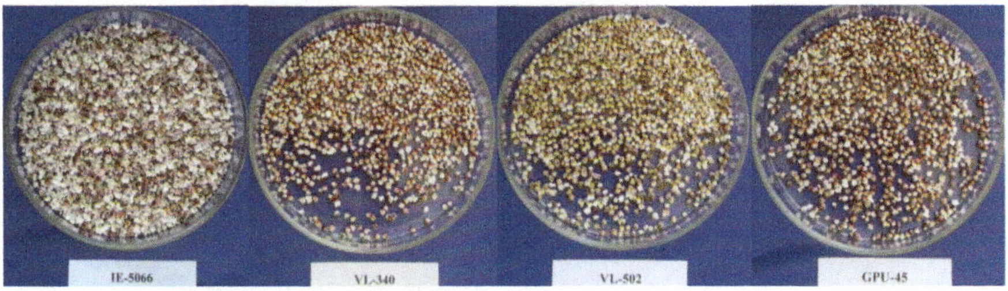

Fig. 14.5 Variation in popping ability of different finger millet genotypes

14.6.3 Pasta, Noodles, and Other Products

Noodles and pasta are the widely consumed food products among all the age groups in both developed and developing countries. Compared to other products noodles and pasta have a longer shelf life and economic significance. A variety of noodles and pasta are prepared from millets. It includes noodles solely made from finger millet flour, noodles made from finger millet and wheat flour in a ratio of 1:1 and finger millet mixed with wheat and soy flour in the ratio of 5:4:1. Fortified small millets noodles are prepared by supplementing with lysine to overcome the deficiency of amino acid on heat treatment (Geerwani and Eggum 1989).

14.6.4 Baked Products

By virtue of their longer shelf life, easy marketing, and handsome packaging baked products are popular among all age groups of people across the world (Patel and Rao 1996). However, the baking industry is mainly occupied by the different value-added products of wheat. Due to high proportion of gluten, these products are not preferred by people suffering from celiac disease especially in developed countries. Millet-based bakery products will not only be superior in terms of nutrition but will also fetch the higher price in the market (Verma and Patel 2013). Flour of foxtail millet and finger millet is preferred for making biscuits and muffins, while barnyard millet is preferred for making cakes (Fig. 14.6). Recently demand has increased for finger millet-based baking products in urban areas. However, the dark brown color is the major hindrance for popularizing it in the baking industry. Among both white

Fig. 14.6 Barnyard millet cake

and brown seeded finger millet, white seeded is more suitable for the baking and processing industry due to high protein, low fiber, low tannins, and higher consumer acceptability (Sharathbabu Sonnad et al. 2008).

14.6.5 Extruded Products

Cooking procedure which is utilized for the post-harvest processing of starchy and protein-rich materials at high temperature for short timing is known as extrusion cooking. Extrusion cooking enhances the protein digestibility, quality, and versatility of the processed food items (Dahlin and Lorenz 1992). It is performed by direct application of heat through a steam injection or indirectly through a jacket by the dissipation of mechanical energy through shearing occurring within the blend. The ready-to-cook extruded products of millets by mixing pearl millet and finger millet flour are reported to have nutrient content, color, texture, and cooking quality and sensory properties in the acceptable range (Devi and Narayanasamy 2013).

14.6.6 Fermented Products

In general, millets are a good source of protein but the protein quality in terms of essential amino acid profile is low (Jaybhaye et al. 2014). Interestingly, probiotic fermentation and germination of millets are known to enhance the protein digestibility and content of lysine, thiamine, niacin, sugars, protein fractions, soluble fibers, and in vitro availability of micronutrients (Arora et al. 2011). Fermented food products (Fig. 14.7) such as *idli* and *dosa* are the popular and common breakfast in many parts of India (Antony et al. 1996; Mbithi-Mwikya et al. 2000).

Fig. 14.7 Fermented millet products

14.6.7 Malting and Weaning Foods

Traditionally, the millet malt is used for infant feeding purposes. Finger millet has good malting properties and its malt is popular in tribal communities. Malting helps to significantly improve the nutrient composition such as minerals, fiber, crude fat, vitamin B, the bioavailability of nutrients, and sensory attributes of the grains (Sangita and Srivastava 2000).

14.6.8 Summary and Outlook

Processing and utilization of millets in new product development offer promising prospects for nutrition, quality, and health benefits and can be an alternative to other grains. Although some of the studies cited here addressed the processes and health benefits of small millets, more studies using advanced techniques and various cooking methods are needed to assess the bioavailability of trace elements including vitamins to realize their nutritional advantage in vivo.

References

Akingbala JO (1991) Effect of processing on flavonoids in millet (*Pennisetum americanum*) flour. Cereal Chem 68:180–183

Antony U, Sripriya G, Chandra TS (1996) Effect of fermentation on the primary nutrients in finger millet (*Eleusine coracana*). J Agric Food Chem 44:2616–2618

Arora S, Jood S, Khetarpaul N (2011) Effect of germination and probiotic fermentation on nutrient profile of pearl millet based food blends. Br Food J 113(4):470–481

Dahlin K, Lorenz K (1992) Protein digestibility of extruded cereal grains. Food Chem 48:13–18

Desikachar HSR (1975) Processing of maize, sorghum and millets for food uses. J Sci Ind Res 43:231–237

Devi MP, Narayanasamy S (2013) Extraction and dehydration of millet milk powder for formulation of extruded product. J Environ Sci Toxicol Food Technol 7:63–70

FAO (1991) Food and Agriculture Organization. Amino acid scoring pattern. In: Protein quality evaluation. FAO/WHO Food and Nutrition Paper, Italy. pp 12–24

FAO (1995) Food and Agriculture Organization. Sorghum and millets in human nutrition. FAO, Rome, Italy

FAO (2018) Food and Agriculture Organization. Economic and Social Department: The Statistical Division.

Geerwani P, Eggum BO (1989) Effect of heating and fortification with lysine on protein quality of minor millets. Plant Foods Hum Nutr 39:349–357

Gupta RK, Das SK (1999) Performance of centrifugal dehulling system for sunflower seeds. J Food Eng 42(4):191–198

Hulse JH, Laing EM, Pearson OE (1980) Sorghum and the millets: their composition and nutritive value. Academic, New York, pp 187–193

Jaybhaye RV, Pardeshi IL, Vengaiah PC, Srivastav PP (2014) Processing and technology for millet based food products: a review. J Ready Eat Food 2:32–48

Khulbe RK, Sood S, Sharma A, Bhatt JC (2014) Value addition and nutritional fortification of finger millet [*Eleusine coracana* (L.) Gaertn.] using bark of Gethi (*Boehmeria regulosa* Wedd.) tree. Indian J Tradit Knowl 13:519–524

Krishnan R, Dharmaraj U, Malleshi NG (2012) Influence of decortication, popping and malting on bioaccessibility of calcium, iron and zinc in finger millet. LWT-Food Sci Technol 48 (2):169–174

Malleshi NG (2014) Post-harvest processing of millets for value addition. http://isites.harvard.edu/fs/docs/icb.topic868074. Cited 21 Jul 2014

Mbithi-Mwikya S, Ooghe W, Van Camp J, Nagundi D, Huyghebaert A (2000) Amino acid profile after sprouting, autoclaving and lactic acid fermentation of finger millet (*Eleusine coracana*) and kidney beans (*Phaseolus vulgaris* L.). J Agric Food Chem 48:3081–3085

Nehir El S, Simsek S (2012) Food technological applications for optimal nutrition: an overview of opportunities for the food industry. Compr Rev Food Sci Saf 11:1–11

Patel MM, Rao V (1996) Influence of untreated, heat treated and germinated black flours on biscuit making quality of wheat flour. J Food Sci Technol 33:53–56

Pawar VD, Parlikar GS (1990) Reducing the polyphenols and phytate and improving the protein quality of pearl millet by dehulling and soaking. J Food Sci Technol 27:140–143

Sangita K, Srivastava S (2000) Nutritive value of malted flours of finger millet genotypes and their use in preparation of Burfi. J Food Sci Technol 37(4):419–422

Sehgal A, Kwatra A (2003) Processing and utilization of pearl millet for nutrition security. Proceeding of national seminar on Recent trend in millet processing and utilization held at CCS HAU, Hissar, India. pp 1–6

Sharathbabu Sonnad SK, Santhakumar G, Salimath PM (2008) Genotype environment interaction effect on seed yield and its component characters in white ragi (*Eleusine coracana* Gaertn). Karnataka J Agric Sci 21:190–193

Shukla S, Gupta O, Sharma Y, Sawarkar N (1986) Puffing quality and characteristics of some ragi cultivars. J Food Sci Technol 23:329–330

Singh P, Singh G, Srivastava S, Agarwal P (2005) Physico-chemical characteristics of wheat flour and millet flour blends. J Food Sci Technol 42:340–343

Singh KP, Mishra HN, Saha S (2012) Changes during accelerated storage in millet–wheat composite flours for bread. Food Bioprocess Technol 5:2003–2011

Verma V, Patel S (2013) Value added products from nutri-cereals: finger millet (*Eleusine coracana*). Emir J Food Agric 25:169–176

Millet-Based Traditional Processed Food Beverages

15

V. Ratnavathi Chamarthy and V. A. Tonapi

Abstract

A variety of cereals are used as substrates for fermented foods and beverages. In Africa, to produce different types of food beverages, cereal grains like sorghum (*Sorghum bicolor* (L.) Moench), millets (pearl and finger millets (*Pennisetum glaucum* (L.) and *Eleusine coracana*)), and maize (*Zea mays* (L.)) are used as raw materials. Beverages are two types, alcoholic and nonalcoholic. Usually, millet beverages were made by fermentation using microorganisms such as yeasts, lactic acid bacteria, and acetic acid bacteria. Dry malt and germinated millet are the main components of these beverages to enhance the amylolytic enzymes for starch degradation. Sorghum-based alcoholic beverages include *Amgba*, *Bili-bili*, *Burukutu*, *Dolo*, *Ikigage*, *Kaffir beer*, *Oti-oka*, *Pito*, *Chakpalo*, *Tchapalo*, *Tchoukoutou*, and nonalcoholic beverages include, *Bushera*, *Gowé*, *Kunun-zak*. The use of malted sorghum grains as a source of fermentable sugars was found to be advantageous in Nigeria. For malting, sorghum varieties, viz., SK 5912, Farafara, and HQSV grains are also used. Of which HQSV variety was found to be superior to other varieties for malting purposes. The by-product of malting like dried shoot and root are being used as organic manures. Apart from malt, unmalted cereals like maize and rice either as flakes or grits are used as adjuncts in the brewery sector. Essentially, adjuncts are starchy materials which have little or no protein content. These adjuncts improve the quality of lager beer. Sorghum malting is a traditional practice in South Africa which is often referred as African '*bantu*' beer. Among the nonalcoholic beverages most traditional and famous beverage is Rabadi. *Rabadi* is a traditional natural cereal like pearl millet based lactic fermented milk beverage commonly used in North-Western semiarid

V. R. Chamarthy (✉) · V. A. Tonapi
ICAR-Indian Institute of Millets Research, Hyderabad, Andhra Pradesh, India
e-mail: ratnavathi@millets.res.in

A. Kumar et al. (eds.), *Millets and Millet Technology*,
https://doi.org/10.1007/978-981-16-0676-2_15

regions of India. Traditionally Rabadi is prepared by fermenting pearl millet (*Pennisetum typhoideum* (L.)) flour with sour buttermilk in different proportions. *Apong* and *Madua Apong* are two popularly used beverages of Arunachal Pradesh, which are produced from rice and millets, respectively. Pearl millet (bajra) is processed into porridge, flour, and a fermented acidic beverage, known as *ontaku* or *oshikundu*. The malt-based millet beverages have nutritional and health benefits.

Keywords

Malt · Sorghum · Beverage · Raw material · Millets · Lager beer

15.1 Introduction

Millets have great potential for utilization as raw material for industrial purposes. Sorghum and millets are traditionally used for alcoholic and nonalcoholic fermented food beverages in African countries. Generally, food beverages are nutritious as they are either malted or fermented. In the brewery sector Sorghum also has great potential of being used as a raw material. Owuyama (1997) explained the potential of sorghum as malt and adjunct. In savanna regions of Africa and Asia, the people prefer common Alcoholic drinks like *burukutu*, *tala*, *pito*, *sorghum wine*, *kafir beer*, and nonalcoholic beverages such as *kunu-zaki*. (Owuyama and Okafor 1990). Nasha is one of the traditional weaning foods (infant porridge) prepared after sorghum flour fermentation (Graham et al. 1986).

In general, barley is used for malt preparation. Barley malt is used for lager beer preparation, which is a fermented beverage. Barley, in particular, malts are prepared after the process of steeping, germination, and kilning the grains. Traditionally for malting and brewing lager beer barley is used. During world war II, for lager beer production Sorghum was first used as a substrate (Hahn 1966) and its usage is continued as a substitute to barley and an adjunct. In Mexico and Nigeria, sorghum is also used for lager beer (Okafor and Aniche 1980).

15.2 Sorghum as Malt

Barley malt import was banned during 1988 in Nigeria and the industry shifted for utilizing maize and sorghum as raw materials in beer production. Sorghum can also be used as raw grains, grits, or malted material. The bulk of the sorghum grain varieties, SK 5912 and farafara were used by the brewers in Nigeria. The grains of sorghum genotypes vary in texture, size, hardness, color, and biochemical parameters (Novellie 1962c; Palmer 1989; Swanston et al. 1994). The selection of a genotype for malt depends on size of the kernel, malting loss, free amino nitrogen, beta glucans, and diastatic activity. The malting quality of the cultivars was also influenced by seasonal effects (Loggerenberg and Pretorius 2004).

In India, sorghum is used as a food grain, while in many African states sorghum is having importance in commercial brewing. The feasibility of using sorghum malt for brewing ager beer has been indicated by recent studies. While production of lager beer from barley malt along with sorghum as a cereal adjunct poses no problem, whereas lager beer brewing from 100% sorghum is confronted with problems of wort recovery, viscosity, fermentability, mashing, lautering, and consumer acceptability. During sorghum brewing, development of nonbiological haze caused by polyphones and insoluble proteins present in sorghum malt and the presence of high lipid content or other unfavorable aspects were observed. There is scope to identify sorghum cultivars with desirable malting qualities through searching for natural variability and barley malt with sorghum malt will provide a better and cheaper indigenous alternative cereal for lager beer brewing in countries with substantial sorghum production.

During germination, Alpha-and beta-amylases and other glucanases regulates starch hydrolysis. In sorghum, Alpha-amylase is the most important and is accounting for about 75% of the saccharifying activity, whereas the beta-amylase content of sorghum malt is very low compared to barley malt. South African sorghum varieties were documented to have beta-amylase properties, which resulted in 18–39% of starch saccharification. During germination, synthesis of all malting enzymes takes place in sorghum. Alpha- and beta-amylase activities have been found to increase gradually in the proportion of 3:1 during germination process till a peak of amylolytic activity is observed. The malting quality of African bird-proof cultivars was better than non-bird-proof types for modification and the production of peptone soluble diastolic activity. The water-extractable diastolic activities in kilts from the bird-proof cultivars were much lower than in those from the non-bird-proof types. This could be due to the interaction between polyphenols and diastase enzymes resulting in insoluble polyphenol–enzyme complexes. The enzymes in the form of insoluble polyphenol–enzyme complexes are active during mashing, as the enzymes and polyphenols are located in separate tissues.

During the malting process of sorghum, amylases and other enzymes production is dependent on moisture, temperature, humidity, as well as on genotypic properties. The optimal temperature regimes and soaking recipes for better enzymatic production have been established in sorghum. Genetic variability in sorghum for all important grain characters except malt ability is well understood. However, the possibility of finding out specific malting conditions suitable for enhancing the diastolic power in particular genotypes are required, studies aimed for evaluating genetic variability for malting ability require malting of different genotypes under optimal conditions that would promote the optimal enzymes in all of them. Moreover, it is underway to set the steeping and germination times, for the better development of α-amylase in the different sorghum genotypes and to establish the association between α-amylase and diastatic power (Ratnavathi and Ravi 1991).

The sorghum cultivars CSH 9, SPV 736, IS 1347, and IS 20503 with different endosperm properties were germinated for 24, 48, 72, 96, and 120 h following 24 h steeping to determine the suitable germination period for maximum alpha-amylase activity. In Table 15.1, the mean alpha-amylase activities at different durations of

Table 15.1 Alpha-amylase activities (g maltose liberated in 3 min/ml extract) in four sorghum cultivars after different periods of germination

S. No.	Cultivar	Germination time (h)					
		24	48	72	96	120	Mean
1	IS 1347	120	443	530	845	230	434
2	IS 20503	218	723	845	980	313	616
3	SPV 736	150	275	288	408	300	284
4	CSH 9	90	390	520	620	463	417
5	Mean	144	458	546	713	326	
6	C.D @0.05						
7	Cultivar mean						55
8	Germination time mean						61
9	Interaction mean						123

Table 15.2 Effect of steeping period on alpha-amylase activity in sorghum

S. No.	Cultivar	Period of steeping (h)				
		12	16	20	24	Mean
1	IS 1347	680	635	880	845	760
2	IS 20503	790	720	814	840	791
3	IS 23860	795	588	830	618	708
4	IS 23930	365	590	815	550	580
5	SPV 736	565	805	765	505	660
6	CSH 9	790	825	940	830	846
7	Mean	664	694	841	698	
8	CD @0.05					
9	Cultivar mean					71
	Steeping time mean					58
10	Interaction mean					143

germination are presented. A steady increase in alpha-amylase activity from 24 to 96 h of germination and a subsequent decline in activity were conspicuous in all cultivars (Table 15.1). Analysis of variance of these results (Table 15.2) indicated that alpha-amylase production was significantly influenced by cultivar, germination time, and their interactions. Cultivars differed significantly in their mean amylase activities. The highest activity was found in the genotype IS 20503, whereas the variety SPV736 showed the lowest activity. At 96 h germination, all cultivars showed highest amylase activities, which were significantly different from the mean activities at other durations of germination (Table 15.2). The cultivars showed varied responses for germination times. The cultivar IS 20503 exhibited significantly higher amylase activity at 24 h of germination than the genotype CSH 9. The amylase activities increased at different time intervals, as germination progressed (Table 15.1). Similarly, all the cultivars showed dipped amylase activity at 120 h, except SPV 736.

Table 15.3 Analysis of lager beer from sorghum

Component	Malt beer	Adjunct beer
Reducing sugars (%)	0.05 ± 0	0.02 ± 0.02
Total sugars (%)	1.54 ± 0.13	0.95 ± 0.07
Alcohol (%)	4.66 ± 0.12	4.5 ± 0.17
Color (EBC)	5.89 ± 0.03	8.58 ± 0.14
Specific gravity	1.03 ± 0.002	0.98 ± 0.03
Bitterness (BU)	6.56 ± 0.08	7.58 ± 0.15
Proteins (%)	2.68 ± 0.07	1.98 ± 0.07
Free α amino nitrogen (mg/L)	74.61 ± 5.18	119.01 ± 2.63

In Table 15.2, the results of the studies on the effect of steeping time on alpha-amylase production in six cultivars possessing different endosperm characteristics are presented. Analysis of variance (Table 15.3) indicated that the production of alpha-amylase activity was influenced by cultivar, steeping time, and their interactions. A fixed period of 96 h was used for these studies. The mean alpha-amylase activity increased steadily in all cultivars with steeping times from 12 to 20 h, and decreased thereafter, the highest activity was observed at 20 h steeping. Alpha-amylase activities at 24, 16, and 12 h steeping were lesser than the peak activity, with the nonsignificant differences. Differences were also discernible among cultivars in their mean alpha-amylase activities across the range of steeping times (Table 15.2). These results presented that steeping for 20 h led to the highest enzyme activities nearly in all the sorghum cultivars.

Alpha-amylase and diastolic activities in 29 cultivars were compared using steeping and germination times at 30 °C for 20 h and 96 h, respectively. The dextrinsing activity of aqueous extracts of malt prepared from these cultivars showed a threefold difference ranging from 427 to 1462 μg maltose released in 3 min/ml extract. The specific and diastatic activities ranged from 456 to 1234 μg maltose/mg protein and 24.2 to 84.3 SDU, respectively. The genotypes CSV 8R, SPV 346, SPV 351, SPV 462, SPV 475, and SPV 504 showed enzymatic activities of more than 75 SDU. Cultivar SPV 504 showed the highest diastatic activity did not have the highest alpha-amylase activity. A significant positive correlation ($r = 0.88$) was noted between alpha-amylase and diastatic activities.

With specific pH and other conditions for the extraction of alpha-amylase, the beta-amylase was also extracted. The extracted beta-amylase was not selectively inactivated prior to alpha-amylase assay, so the alpha-amylase activity may have a small fraction of beta-amylase activity. The beta-amylase activity may also increase the alpha-amylase saccharification activity.

Ratnavathi and Ravi (1991) studied the standardization of conditions for malt preparation. The diastatic activity of sorghum is denoted by Sorghum Diastatic Units (SDU) ranging from 144.5 (IS 14384) to 200 SDU (WS 1297). The malting losses (27–39%) were also reported in these cultivars, which can be minimized with air rest and minimum water supply. Malting causes a decrease in the caryopsis density (Beta et al. 1995), decreases lysine from 0.25% to 0.18% (Okoh et al. 1989), and reduces milling energy (Swanston et al. 1994). Lasekan et al. (1995) have reported that fine milling (0.2 mm particle size) of malt increases hot water extract, diastatic power, and sugar contents.

Sorghum malts with high enzymatic activity had a large albumin-globulin fraction with reduced paste viscosity (Malleshi and Desikachar 1986; Beta et al. 1995). Subramanian et al. (1995) reported that the water extract and water-extractable proteins of the malt increase with the diastatic activity. However, diastatic power of 60–80 Kaffir corn diastatic units (KDU/g) is recommended for commercial malting purposes (Novellie 1962b).

Micro malting technique for rapid screening of diastatic activity in sorghum grains (10 g) was developed for germplasm evaluation in sorghum (Jaya et al. 2001). Sorghum malt has low levels of β-amylase, β-1,3 and β-1,4-glucanase, and β-D glucans, in contrast to barley. Sorghum malts show alpha-amylase activity of 25–183 U/g and β-amylase activity of 11–41 U/g (Beta et al. 1995). During the mashing operations, external heat stable enzymes such as α-amylase, neutral protease, β-glucanase, cellulose, and amyl glucosidase are required (Aisien 1989). Generally, the beers from barley and sorghum differ in their nutritional content, texture, and other sensory properties (Faparusi et al. 1973; Asiedu 1992). The differences could be due to variations in preparation process of worts or addition of adjuncts or spices to worts and organisms associated with fermentation of the wort and post-fermentation treatment.

Pilot scale (1000 L) brewing of grist comprising of unmalted sorghum and malted barley (50:50) was undertaken. The sorghum beers were lower in total alcohol due to lower fermentability of the sorghum worts. Sensory analysis indicated that with regard to aroma, mouth-feel, after-taste, and clarity there are no significant differences existed between the sorghum beer and both the control beer and a commercial malted barley beer (Goode and Arendt 2003).

15.3 Lager Beer from 100% Sorghum

By using exogenous enzymes, Lager beer is possible to be produced from 100% sorghum (Arri 1989). In general, sorghum malt was poor than barley malt however showed relatively higher total soluble sugars and cold water (%) extract (Ogundiwin et al. 1989). The cultivars that have high diastatic power, low gelatinization temperature, low lipid, low polypeptides, and readily soluble protein are good for lager beer preparation. For experiments, beer was prepared using sorghum malt (100%), sorghum adjuncts, and hops in the ratio 96:3:1 (Ratnavathi 2005). Sorghum beer had 4% alcohol and higher free amino nitrogen content than sole sorghum malt beer (Table 15.3). In sorghum, the pentosanase complex does not embody a xylosidase but displays arabinosidase activity as in barley. Polymeric pentosans of barley worts were found to range from 162 mg/100 ml in the unmalted grains to 239 mg/100 ml in the 6 day malts, while ranged from 41 to 79 mg/ml for the same period in sorghum (EtokAkpan 2004a). The diastatic power of the freshly kilned sorghum malt at 68.1 WK had a 29% drop after 6 months of storage. Freshly kilned sorghum malt

Table 15.4 Analysis of sorghum beer

Component	Sorghum malt and sorghum adjunct beer[a]	Barley malt and sorghum adjunct beer[a]
Reducing sugars	0.05 ± 0	0.02 ± 0.02
Total sugars	1.54 ± 0.13	0.95 ± 0.07
Alcohol	4.66 ± 0.12	4.5 ± 0.17
Color (EBO)	5.89 ± 0.03	8.58 ± 0.14
Specific gravity	1.03 ± 0.002	0.98 ± 0.03
Bitterness (BU)	6.56 ± 0.08	7.58 ± 0.15
Proteins	2.68 ± 0.07	1.98 ± 0.07
Free α amino nitrogen	74.61 ± 5.18	119.01 ± 2.63

Source: Final Report, National Agricultural Technology project, RNPS-24, IIMR (NRCS), 2005
[a]All the data expressed are Mean ± S.D.

displayed high wort turbidity (4.9 EBC) which dropped to 0.95 EBC and 1 EBC after 2 and 6 months of storage respectively (EtokAkpan 2004b).

The composition of lager beer made from sorghum malt and sorghum adjunct are shown in Table 15.4. The lager beer made from sorghum malt and adjunct was found superior for increased free alpha amino nitrogen and color units.

Germinated sorghum flour showed lower protein, fat, and carbohydrate but had higher ash content and fiber than non-germinated sorghum flour. Whereas, germinated millet flour was found with higher protein content, moisture, and fiber than the non-germinated flour, which had higher ash content and carbohydrates. Germination led to an increase in the concentration of sugars in both sorghum and millet grains (Muyanja et al. 2003).

In Nigeria, the use of malted sorghum was found to be advantageous as a source of fermentable sugars than the raw kernels. Grains of sorghum genotypes SK 5912, Farafara, and HQSV are being used for malting. HQSV sorghum variety was superior to other varieties for malting purposes. The by-product of malting like dried shoot and root are used as organic manures. The increased use of white grain sorghum by breweries result in competition in the market between grain for food and grain for brewing purpose (Ikediobi 1989).

15.4 Sorghum as Adjunct

In the brewery sector, unmalted cereals like rice, maize, and barley flakes or grits are used as adjuncts. Adjuncts improve the quality of lager beer as they are starchy materials with little or no protein content. As most of the information available from South Africa on sorghum malting, it is referred to as African "bantu" beer (Novellie 1968).

Sorghum is a favorable brewing adjunct due to the presence of amylase in the unmalted sorghum (Owuyama and Okafor 1990). A study showed that the genotypes CSH-5 and SPV 462 with high enzymatic activity were found suitable adjuncts under the Indian situation. Ratnavathi et al. (2000), in India, carried out a systematic approach for the suitability of sorghum as an adjunct. Thirteen sorghum cultivar grain samples with diverse chemical compositions were assessed for their suitability as brewing adjuncts based on the proximate analysis. Besides, grain size, free amino nitrogen (FAN), and malt extract (HWE), hot water-extractable protein (HWEP) were also estimated. The genotypes with high amylose and starch contents coupled with low fat and protein were best suited as adjuncts with their HWE and HWEP yields. The released cultivars, CSH-5, CSV-11, and CSV-13, were found promising to be used as adjuncts with barley malt in preparation of lager beers. Sorghum beer was prepared using two cultivars CSV-15 and CSH-16 grains as adjunct and sorghum malt prepared from elite breeding line (SPV 824) having high diastatic activity. The hybrid CSH 16 with 15% concentration was found suitable for mixing as an adjunct. In Table 15.4, the chemical analysis of the beers prepared from sorghum as malt and as adjunct is presented.

To determine the suitability of nationally released varieties of sorghum as an adjunct, in comparison to broken rice, study was conducted by NRCS in collaboration with Hindustan Breweries Limited; Mumbai. In Table 15.5, adjunct parameters analyzed in the sorghum are presented. All sorghum genotypes tested are equivalent to broken rice in all the parameters. The hybrids CSH-13 and CSH-14 showed high extract values followed by CSH-17, CSV-15, and SPV-462. The hybrids CSV-13, CSH-9, and CSH-17 were found similar to broken rice for color of adjunct (4.5 units). Low saccharification time indicating the property of a good adjunct. Sorghum genotypes CSV-13, SPV-462, and CSH-9 depicted less saccharification time (9 min) than broken rice (10 min) (Ratnavathi et al. 2005).

At NRCS in collaboration with Hindustan Breweries Limited, a pilot experiment with sorghum grain was done by blending with broken rice. No differences were observed for color and taste with the normal beer. Hence, the beer made from sorghum has no problem in marketing (Table 15.5). The hybrids CSH 13, CSH 17, and CSH 14 recorded the highest percent extraction of 61%, 60%, and 61%, respectively (Ratnavathi et al. 2005).

Goode et al. (2002) reported that an increase of the relative proportion of sorghum in the grist resulted in decrease of wort filtrate, color, viscosity, attenuation limit, free amino nitrogen, high molecular weight nitrogen, and a corresponding increase in pH ($p < 0.01$). The addition of small quantities of unmalted sorghum with commercial hydrolytic enzymes to malted barley was found to improve the potential for brewing a high-quality lager beer. Lager beer was prepared using the isolated thermotolerant yeast strain in laboratory fermentor at NRCS. Malt of sorghum variety SPV 824 was ground in a UDY cyclone sample mill (Tecator, Sweden) and unmalted sorghum grain flours of the varieties CSH 9, CSH 16, and CSH 18 were used as adjuncts (Table 15.6).

The maximum alcohol was recovered with adjuncts of CSH 16 and CSH 18 with SPV 824 malt. The highest amount of free alpha amino nitrogen was observed in SPV 824 malt with the genotypes CSH 9, CSH 16, and CSH 18 as adjuncts. Protein content in sole SPV 824 malt was 2.05%, while it was 1.98%, 2.01%, and 2.14% with CSH 9, CSH 16, and CSH 18 adjuncts, respectively (Ratnavathi et al. 2005).

Table 15.5 Analysis of sorghum grain as an adjunct in lager beer preparation

Parameter	CSH 15	CSV 13	CSH 18	SPV 462	SPV 824	CSH 9	CSH 14	CSH 17	CSH 16	CSH 13	Broken rice (check)
PH	5.75	5.70	5.3	5.60	5.40	5.45	5.60	5.68	5.35	5.50	5.60
Gravity	6.80	6.40	6.00	6.80	6.40	6.50	7.00	6.90	6.60	7.10	7.40
Moisture %	5.00	4.70	4.90	4.80	4.50	4.6	4.90	5.20	5.30	5.00	4.50
Saccharification (min)	10.00	9.00	10.00	9.00	10.00	9.00	12.00	11.00	13.00	12.00	10.00
Filtration rate	Good	Good	Good	Good	Good	Good	Good	Good	Good	Good	Good
Clarity	Clear	Clear	Clear	Clear	Clear	Clear	Clear	Clear	Clear	Clear	Clear
Color (EBC units)	4.40	4.50	4.40	4.40	4.40	4.50	4.40	4.50	4.40	4.40	4.50
Extraction %	59.00	55.00	52.00	59.00	55.00	56.00	61.00	60.00	57.00	61.00	64.00

Table 15.6 Analysis of beer produced with different sorghum adjuncts

S. No.	Analysis details	SPV 824 malt without adjunct	SPV 824 malt with CSH 09 as adjunct	SPV 824 malt with CSH 16 as adjunct	SPV 824 malt with CSH 18 as adjunct
1	Reducing sugars	0.05 ± 0.00	0.03 ± 0.01	0.09 ± 0.01	1.8 ± 0.01
2	Total sugars	1.44 ± 0.1	0.95 ± 0.07	2.14 ± 0.06	1.47 ± 0.13
3	Free amino nitrogen	74.61 ± 5.18	119.02 ± 2.6	117.65 ± 0.00	120.33 ± 0.58
4	Alcohol	1.05 ± 0.12	4.50 ± 0.17	5.29 ± 0.12	7.43 ± 0.23
5	Color	14.33 ± 0.14	8.58 ± 0.14	18.44 ± 0.07	6.88 ± 0.05
6	Bitterness	6.40 ± 0.09	7.58 ± 0.15	5.99 ± 0.60	6.97 ± 0.06
7	Proteins	2.05 ± 0.07	1.98 ± 0.07	2.01 ± 0.00	2.14 ± 0.10
8	Specific gravity	1.04 ± 0.00	1.00 ± 0.00	1.00 ± 0.01	0.98 ± 0.01

Many beverages have been prepared after fermentation of millets and using microorganisms like yeasts, acetic acid, and lactic acid bacteria. Germinated millet served to enhance the amylolytic enzymes for starch degradation in many cases. Usually, yeast and lactic acid bacteria are involved in spontaneous fermentation (Okafor 1983) . Indigenous millet brewing technologies are nearly similar to modern brewing technology, except for the batch size and the scale of the level used. Millet-based fermented beverages include Bensaalga, Boza, Burukutu, Bushera, Doro, Fura, Jandh, Kodokojaanr, Koko, Koozh, Kunun-zaki, Madua Apong, Malwa/Ajon, Mangisi, Ogi, Oshikundu, Oti-oka, Shakparo Ale, Togwa, Uji, Zoom-koom, Xiao mijiao, etc. (Amadou 2019).

Rabadi is a popular natural cereal-based lactic fermented milk beverage of North-Western semiarid regions of India. Traditionally, Rabadi is made by admixing cereal flour with sour buttermilk in different ratios. This prepared mixture is fermented in sunlight for 3–4 h and boiled thereafter. The prepared product is either consumed directly or after dilution by adding milk or buttermilk. Rabadi, prepared by pearl millet fermentation (*Pennisetum typhoideum* (L.)) flour with buttermilk, is a traditional popular beverage of the North-Western states of India. A process for pearl millet-based Rabadi-like fermented milk beverage was attempted and skim milk and flour of 24 h germinated pearl millet grains (FGG-24 h) were used as sources of solids (Modha and Pal 2011).

Apong and *Madua Apong* are two widely used beverages of Arunachal Pradesh, which are produced traditionally from rice and millet, respectively, under uncontrolled fermentation; in low productivities. The present study was conducted with the main aim of improving traditional bioprocessing. The improved production technology for *Apong* and *Madua Apong* was developed by scientific interventions. As a result, more quantity of *Apong* (6–7 L/kg) and *Madua Apong* (5–6 L/kg) was extracted under laboratory conditions as compare to traditional preparations

(3–4 L/kg). Biochemical analysis shows higher alcohol contents (7.6%) with lesser carbohydrate (7.1 g/100 g), total reducing sugar (5.2%), and ash contents (1.02%) in laboratory prepared *Apong* as compared to traditionally prepared (8.5 g/100 g, 6% and 1.03–1.04, respectively) beverage. Similar results were also obtained in laboratory prepared versus traditionally prepared *Madua Apong* (alcohol—7.8%/5.5–6.2%; carbohydrate—5 g/100 g/5.5 g/100 g; total reducing sugar—3.2%/4%) (Karuna Shrivastava et al. 2015).

Skimmed milk was blended with three millets—finger millet, pearl millet, and sorghum—for beverage preparation. The different techniques such as soaking, sprouting, and extraction of milk were used for processing. The millet milk and skimmed milk ratio was finalized by Mixture Design based on physicochemical characters like viscosity, sediment, acidity, wheying off, and sensory responses. The overall acceptability of the optimized sample (7.1) was very close to the predicted value. The nutritional analysis was performed using an optimized combination of millet milk (Sudha et al. 2016).

Pearl millet is one of the important staple foods of nearly 60% of the Namibian population. It is consumed in the form of porridge, flour, and fermented acidic beverage, named as ontaku or oshikundu. This preparation comprises pearl millet meal, sorghum, or pearl millet malt with water. Ontaku has a lower shelf life, mainly of less than a day at ambient conditions. Its brewing processes are different but many still are not standardized. Freshly prepared ontaku is nonalcoholic however over time it can have an alcohol content of up to 1.6%. The quality and other malt phenolic compounds and fermentation metabolites like methanol and butyrate of ontaku have not been studied much. The production of ontaku with consumer-acceptable qualities and improved safety coupled with predictable ingredients and processing conditions need standardization (Embashu and Nantanga 2019).

15.5 Raw Materials Used in the Production of African Traditional Cereal Beverages

A wide range of cereal grains is used for the preparation of fermented foods and beverages. Cereal grains such as barley (*H. vulgare* L.), sorghum (*Sorghum bicolor* (L.) Moench), maize (*Zea mays* (L.)), millets (pearl and finger millets (*Pennisetum glaucum* (L.) and *Eleusine coracana*)) and are the most common substrates used in Africa to produce a different range of beverages (Sekwati-Monang 2011). Raw materials used in the production of African traditional cereal beverages are listed in Table 15.7 (Solange et al. 2014).

Sorghum and finger millet based materials were used to prepare three flavored ready-to-drink beverages. Desiccated coconut, cocoa powder, and fresh carrots were used for getting good color and flavor. Laboratory scale trials and benchtop sensory evaluation were carried out for optimization. The total suspended solids ranged from 43.8 to 44.4 _Brix. Consistency, firmness, cohesiveness, and index of viscosity did not differ significantly ($p > 0.05$) among the different treatments. Optimized beverages were acceptable by the consumers as depicted by the overall acceptability

Table 15.7 Raw material and location of African traditional cereal beverages

Raw materials	Name of beverages	Class	Location
Sorghum	Amba	Alcoholic beverage	Cameroon
	Bushera	Nonalcoholic beverage	Uganda
	Bili-bili	Alcoholic beverage	Tchad
	Burukutu	Alcoholic beverage	Northern part of Nigeria
	Dolo	Alcoholic beverage	Burkina Faso
	Gowé	Nonalcoholic beverage	Benin
	Ikigage	Alcoholic beverage	Rwanda
	Kaffir beer	Alcoholic beverage	South Africa
	Kunun-zaki	Nonalcoholic beverage	Northern part of Nigeria
	Oti-oka	Alcoholic beverage	South-Western region of Nigeria
	Pito	Alcoholic beverage	Nigeria, Ghana
	Chakpalo	Alcoholic beverage	Benin, Togo
	Tchapalo	Alcoholic beverage	Côte d'Ivoire
	Tchoukoutou	Alcoholic beverage	Benin, Togo
	Gowé	Nonalcoholic beverage	Benin
Maize	Busaa	Alcoholic beverage	Northern part of Nigeria
	Kaffir beer	Alcoholic beverage	Kenya
	Kunun-zaki	Alcoholic beverage	South African
	Mahewu	Nonalcoholic beverage	Northern part of Nigeria
	Pito	Nonalcoholic beverage	South Africa
	Tchapalo	Alcoholic beverage	Nigeria, Ghana
	Gowé	Nonalcoholic beverage	Benin

(continued)

Table 15.7 (continued)

Raw materials	Name of beverages	Class	Location
	Bushera	Nonalcoholic beverage	Uganda
	Gowé	Nonalcoholic beverage	Benin
	Mangisi	Nonalcoholic beverage	Zimbabwe
	Koko sour water	Nonalcoholic beverage	Northern part of Ghana
	Kunun-zaki	Nonalcoholic beverage	Northern part of Nigeria
	Malwa	Nonalcoholic beverage	Uganda
	Masvusvu	Nonalcoholic beverage	Zimbabwe
	Oti-oka'	Alcoholic beverage	South-Western region of Nigeria
	Tchapalo	Alcoholic beverage	Côte d'Ivoire
	Burukutu	Alcoholic beverage	Northern part of Nigeria
Sorghum and millet	Obiolor	Nonalcoholic beverage	Nigeria
	Pito	Alcoholic beverage	Nigeria, Ghana
	Burukutu	Alcoholic beverage	Northern part of Nigeria
Sorghum and maize	Kwete	Alcoholic beverage	Uganda
	Pito	Alcoholic beverage	Nigeria, Ghana
	Umqombothi	Alcoholic beverage	South African
Maize, sorghum, wheat, finger millet, and tef and barley	Borde	Nonalcoholic beverage	Ethiopia

score > 7 on a 9 point hedonic scale. Radical scavenging activity (RSA %) was significantly higher for coconut flavored beverage (76.6), whereas the total phenolic content was significantly ($p > 0.01$) higher in carrot-based beverage (142.2 mg/GAE) (Khwairakpam and Murugkar 2020).

15.6 Conclusion

Usually, millet beverages have been made by fermentation using microorganisms such as yeasts, lactic acid bacteria, and acetic acid bacteria. Dry malt and germinated millet are the main components of these beverages to enhance the amylolytic enzymes for starch degradation. Beverages are classified as alcoholic and nonalcoholic. The selection of a genotype for malt depends not only on kernel shape, size, color but also on the malt parameters such as malting yield, free amino nitrogen, hot water extract, Betaglucan, and diastatic power. Barley is preferred normally for malt preparation. Lager beer is a classical fermented beverage made from barley malt with blended unmalted cereal adjuncts. Pilot studies at IIMR revealed that sorghum grains can be used as a source of malt with high α-amylase activity at 96 h of germination and can also be used as a cereal adjunct. Among the nonalcoholic beverages, *Rabadi* is an indigenous natural cereal-based lactic fermented milk beverage popular in North-Western semiarid regions of India. Traditionally, Rabadi is prepared by fermenting pearl millet *(bajra)* flour with sour buttermilk in different proportions. The malt-based millet beverages have nutritional and health benefits.

References

Aisien AO (1989) Utilization of Sorghum in brewing lager beer in Nigeria. Proceedings of a symposium on the current status and potential of industrial uses of sorghum in Nigeria, 4-- 6 December, Kano, Nigeria. p 29

Amadou I (2019) Millet based fermented beverages processing. In: Fermented beverages. Woodhead Publishing, Cambridge, pp 433–472

Arri BK (1989) Problems associated with the use of Sorghum for lager beer production. Proceedings of a symposium on the current status and potential of industrial uses of sorghum in Nigeria, 4–6 December, Kano, Nigeria. p 29

Asiedu JJ (1992) Sorghum and millet. In: Processing tropical crops. Macmillan Press Ltd, London, p 189±223

Beta T, Rooney LW, Waniska RD (1995) Malting characteristics of sorghum cultivars. Cereal Chem (72):533–538

Embashu W, Nantanga KKM (2019) Pearl millet grain: a mini-review of the milling, fermentation and brewing of ontaku, a non-alcoholic traditional beverage in Namibia. Trans R Soc S Afr 74 (3):276–282

EtokAkpan OU (2004a) Preliminary study of the enzymolysis of sorghum and barley pentosans and their levels in worts. World J Microbiol Biotechnol 20(6):575–578

EtokAkpan OU (2004b) Changes in sorghum malt during storage. J Inst Brew 110(3):189–192

Faparusi SI, Olonboba MO, Ekundayo JA (1973) Microbiology of burukutubeer. Zeitschrift for Allgemeine Mikrobiologie 13:563–568

Goode DL, Arendt EK (2003) Pilot scale production of a larger beer from a grist containing 50% unmalted sorghum. J Inst Brew 109(3):208–217

Goode DL, Halbert C, Arendt EK et al (2002) Mashing studies with unmalted sorghum and malted barley. J Inst Brew 108(4):465–473

Graham GG, MacLean WC, Morales E (1986) Digestibility and utilization of protein and energy from 'nasha' a traditional Sudanese fermented sorghum weaning food. J Nutr 116:978–984

Hahn RR (1966) Sorghum as a brewing adjunct. Brewers Digest 41:70–76

Ikediobi CO (1989) Industrial production of Sorghum malt in Nigeria. In: Proceedings of a symposium on the current status and potential of industrial uses of sorghum in Nigeria, 4--6 December, Kano, Nigeria. p 32
Jaya K, Bala Ravi S, Ratnavathi CV, Suresh K, Bhalla JK (2001) Micro-scale malting studies in Sorghum. Poster presented at the International conference on new horizons in biotechnology, 18–21 April, Trivandrum, India
Khwairakpam B, Murugkar D (2020) Development of millet based ready-to-drink beverage for geriatric population. J Food Sci Technol 57(9):3278–3283
Lasekan OO, Idowu MA, Lasekan W (1995) Effect of germination and degree of grind (coarse/fine) on the extract and sugar content of sorghum malts. Food Chem 53:125–128
Loggerenberg M, Pretorius AJ (2004) Sorghum malting quality as affected by suitability of cultivars. South Afr J Plant Soil 21(3):192–195
Malleshi NG, Desikachar HSR (1986) Studies on comparative malting characteristics of some tropical cereals and millets. J Inst Brew 92:174–176
Modha H, Pal D (2011) Optimization of Rabadi-like fermented milk beverage using pearl millet. J Food Sci Technol 48(2):190–196
Muyanja CMBK, Kikafunda JK, Narvhus JA, Helgetun K, Langsrud T (2003) Production methods and composition of Bushera: a Ugandan traditional fermented cereal beverage. Afr J Food Agric Nutr Dev 3(1):10–19
Novellie L (1962b) Kaffir corn malting and brewing studies XII. Effect of malting conditions on malting losses and total amylases activity. J Sci Food Agric 13:121–123
Novellie L (1962c) Kaffir corn malting and brewing studies XIII. Variations of diastatic power with variety, season, maturity and age of grains. J Sci Food Agric 13:124–126
Novellie L (1968) Wallerstein laboratory communications (31):17–29
Ogundiwin JO, Ilori MO, Okelaye A (1989) Brewing of clear beer from sorghum grains of SK 5912 variety without addition of external enzymes to achieve Saccharification: a case study Proceedings of a symposium on the current status and potential of industrial uses of sorghum in Nigeria, 4–6 December Kano, Nigeria. p 31
Okafor N (1983) Processing of Nigerian indigenous fermented foods a chance for innovation. Niger Food J 1:32–37
Okafor N, Aniche GN (1980) Brewing a lager beer from Nigerian sorghum. Brew Distill Int 10:32–35
Okoh PN, Kubiczek RP, Njoku PC, Iyeghe GT (1989) Some compositional changes in malted sorghum (Sorghum vulgare) grain and its value in broiler chicken diet. J Sci Food Agric 49:271–280
Owuyama CI (1997) Sorghum cereal with larger beer brewing potential. J Microbiol Biotechnol 13:253–260
Owuyama CI, Okafor N (1990) Use of un-malted sorghum as a brewing adjunct. World J Microbiol Biotechnol 6:318–322
Palmer GH (1989) Cereals in malting and brewing. In: Palmer GH (ed) Cereal science and technology. University Press, Aberdeen, pp 161–242
Ratnavathi CV (2005) Annual Report, NATP – RNPS – 24. Developing Sorghum as an efficient biomass and bioenergy crop and providing value addition to the rain damaged kharif grain for creating industrial demand. NRCS, Hyderabad. p 30
Ratnavathi CV, Ravi SB (1991) Effect of different durations of steeping and malting on the production of alpha-amylase in sorghum. J Cereal Sci 14:287–296
Ratnavathi CV, Ravi SB, Subramanian V, Rao NS (2000) A study on the suitability of un-malted sorghum as a brewing adjunct. J Inst Brew 106(6):383–387
Ratnavathi CV, Boehr S, Patil JV, Nandini Nimbkar (2005) National agricultural technology project, RNPS-24, final report, pp 18–20
Sekwati-Monang B (2011) Microbiological and chemical characterisation of ting, a sorghum based gluten-free fermented cereal product from Botswana. PhD Thesis, University of Alberta, Edmonton, Canada. p 156

Shrivastava K, Greeshma AG, Srivastava B (2015) Improvement in traditional technology of rice and millet based fermented beverages of Arunachal Pradesh, North East, India through Scientific Approach. International Conference on Chemical, Environmental and Biological Sciences (CEBS-2015) (2018–19), 2015 Dubai (UAE)

Solange AKA, Georgette K, Gilbert F, Marcellin D, Bassirou B (2014) Review on African traditional cereal beverages. Am J Res Commun (2):5

Subramanian V, Rao NS, Jambunathan R, Murthy DS, Reddy BVS (1995) The effect of mashing on the extractability of proteins and its relationship to diastatic activity in sorghum. J Cereal Sci 21:283–289

Sudha A, Priyenka Devi KS, Sangeetha V, Sangeetha A (2016) Development of fermented millet sprout milk beverage based on physicochemical property studies and consumer acceptability data. J Sci Ind Res 75:239–243

Swanston JS, Rao NS, Subramanian V, Taylor K (1994) The influence of some aspects of grain quality on malting potential in sorghum. J Cereal Sci 19:91–97

Millet-Based Value-Added Food Products for Diabetics

16

Sarita Srivastava and Anju Bisht

Abstract

Diabetes mellitus is a metabolic disorder characterized by hyperglycemia owing to insufficient or inefficient insulin secretion. Globally half a billion people are living with diabetes and in India the rate is increasing at an alarming rate. This possesses a large and increasing burden and therefore immediate and collective actions are required to prevent diabetes. The four identified pillars of managing diabetes are drugs, exercise, education, and diet. Diet can play an important role in treatment and prevention of diabetes. Millets can be effectively explored to incorporate in the diet of diabetics. Millets contain certain bioactive components like non-starch polysaccharides, flavonoids, polyphenols, protein, certain vitamins, and minerals which attribute antioxidative, anti-inflammatory properties, and reduce insulin sensitivity. The nutraceutical potency of millets aids in lowering the glycemic index (GI) of food containing millets. The present chapter deals with the incorporation of millets in developing food products for diabetics. Various millet-based low GI value-added food products have been developed and proven to reduce blood glucose, lipid profile, glycosylated hemoglobin in diabetics through intervention studies. Thus, millets the underutilized but potential grains can be effectively used to formulate functional food products and can form an integral part of the diet for diabetics.

Keywords

Millet · Diet · Health · Disease · Glycemic index

S. Srivastava (✉)
Department of Foods and Nutrition, G.B. Pant University of Agriculture and Technology, Pantnagar, Uttarakhand, India

A. Bisht
Department of Home Science, MB Government PG College, Nainital, Uttarakhand, India

A. Kumar et al. (eds.), *Millets and Millet Technology*,
https://doi.org/10.1007/978-981-16-0676-2_16

16.1 Introduction

Diabetes Mellitus is a group of disease characterized by high glucose concentrations in blood owing to inefficient or insufficient secretion of insulin hormone. It is a chronic, progressive disease, recognized as one of the fastest growing threat to public health. The gravity of the threat can be better understood by the prevalence report published by WHO (2016). According to the report, the global prevalence of diabetes among adults over 18 years of age has risen from 4.7% in 1980 to 8.5% in 2014, which means the number of people with diabetes has ascended from 108 million to 422 million in two and half decade. It has been estimated that diabetes was the seventh leading cause of death (1.6 million deaths) in 2016. The prevalence is rising more rapidly in middle- and low-income countries. It has been estimated that globally, 463 million (9.3%) people were living with diabetes in 2019 and this number is expected to increase to 578 million in 2030 and 700 million in 2045, representing 10.2% and 10.9% of the adult population (Saeedi et al. 2019).

Unfortunately, India is also highly affected and is designated as the diabetic capital of the world. The numbers of diabetics are increasing at an alarming rate, from 51 million people in 2010 to an expected 87 million in 2030 (Snehalatha and Ramachandaran 2009). According to another estimate, the number of diabetic people living in India were 77 million in 2019 which is expected to rise to 134 million by 2045 (Saeedi et al. 2019). The Indians are more susceptible to diabetes partially because of their genetic constitution and partially because of environmental factors. Pancreatic β-cell dysfunction along with insulin resistance is the pathophysiological mechanism for the development of diabetes. The "thrifty gene hypotheses" have been postulated to understand the vulnerability of Indians for diabetes. The thrifty phenotype hypothesis proposes a disparity between intrauterine environment and adult life environment. The intrauterine malnutrition and low birth weight lead to changes in β-cells which help to thrive during early years but during later years responsible for high central obesity, insulin resistance, and higher plasma insulin levels leading to diabetes. The thrifty genotype hypothesis indicates a mismatch between ancestral genes and environment which has been disturbed in the modern era of feasting. Urbanization and socioeconomic transition have changed the environment and lifestyle, especially the wrong eating habits coupled with sedentary lifestyle which adversely affect insulin sensitivities and increase the risk of diabetes in Indians (Mohan 2004; Radha et al. 2011; Ramachandran et al. 2012).

Thus, there is a need to emphasize on modifiable risk factors like diet for the management of diabetes. Interventions are required to control the poor dietary choices and change in the lifestyle. Food plays a crucial role in control and prevention of diabetes. In recent years, functional foods are gaining a lot of importance in the management of diabetes. Millet is one such food which possesses nutraceutical properties and is being explored for the prevention of diabetes. Millets the small seeded grains have been used for human consumption for centuries, but owing to difficulties in its processing, the production and consumption have declined in the past few years. However, recently millets have again captured the attention of researchers because of their nutritional and therapeutic properties.

16.2 Bioactive Components in Millets

Researches amassed have shown that minor millets contain certain bioactive components which endow them with therapeutic properties. Millets are considered as a good alternative to commonly consumed cereal grains like rice, wheat, maize because they are superior and comparable in many nutritional attributes to their customary preferred counterparts. Millets are a rich source of gluten-free protein, essential amino acids, resistant starch, and dietary fiber. They are also a good source of micronutrients like iron, calcium, phosphorus, magnesium, zinc, niacin, riboflavin, thiamine, and folic acid. Therefore, incorporation of millets in conventionally eaten food improves the nutritional and therapeutic properties of the food. The high resistant starch, soluble fiber, insoluble dietary fiber, minerals, vitamins, and antioxidants in millets makes them an excellent ingredient for the formulation of diabetic food (Amadou et al. 2013; Kumar et al. 2018).

Finger millet is rich in non-starch polysaccharide arabinoxylan, flavonoids, and phenolic acids which are said to possess anti-oxidative, antiproliferative, and anti-inflammatory properties (Sood et al. 2019). The high fiber and polyphenols such as gallic, tannic, ferulic, caffeic, and coumaric acid affect carbohydrate digestibility and delay absorption of glucose thus making it a suitable food for diabetics (Kumar et al. 2016; Udeh et al. 2017; Sharma et al. 2018). Other small millets such as kodo, foxtail, little, and barnyard are also rich in fiber, protein, and phytochemicals like flavonoids and phenolic compounds which act as potent antioxidants (Prabha and Selvi 2016; Sharma and Saxena 2016). Foxtail millet and its bran-rich fraction have significant radical scavenging capacity and reducing power indicating good antioxidant potency (Suma and Urooj 2012). Consumption of millets increases serum leptin, decreases insulin resistance and inflammation because of the presence of these bioactive components (Ren et al. 2018).

In the past, animal studies have been conducted which are the evidences to support the positive impact of millets on diabetes. Millet protein increases insulin sensitivities and reduce blood glucose and triglyceride levels (Nishizawa et al. 2009). Kodo, little, and pearl millet were found effective in controlling blood glucose and improving lipid profile (Thilagavathi and Kanchana 2017). Administration of aqueous extract of foxtail millet seeds to diabetic rats showed a significant fall in blood glucose, glycosylated hemoglobin, LDL, VLDL, triglyceride and increase in HDL (Sireesha et al. 2011).

Intervention studies have further affirmed the role of millets in reducing blood glucose levels in diabetics. The low GI of dehulled barnyard grains (50) and dehulled and heat treated barnyard millet grains (41.7) has been reported. Administration of these, significantly lowered the blood glucose, VLDL, LDL, TC: HDL, LDL: HDL upon intervention. Similar reduction in blood glucose level on administration of foxtail millet was seen (Ugare et al. 2014; Ren et al. 2018). Millets are effective in maintaining blood glucose and blood lipids levels. Foxtail millet diabetic mix reduced fasting blood glucose, glycosylated hemoglobin, homocysteine concentration, blood lipid levels and increased insulin and VLDL levels in diabetic patients (Jali et al. 2012). Foxtail millet also showed lipid lowering effect in pre-diabetics

(Anusha et al. 2018). Similar results were seen by Itagi et al. (2012). They concluded that foxtail millet mix (80 g), with GI of 49.6 when consumed by diabetics in the form of rice, pancakes, *upma*, or *thalipattu* for a period of 4 weeks led to a fall in glucose, triglyceride, cholesterol, and increase in HDL levels. Finger millet based diets were also seen to reduce plasma glucose levels in NIDDM subjects probably because of the presence of high fiber and anti-nutritional factors which may reduce digestibility and absorption of starch (Kumari and Sumathi 2002).

It has been revealed that the fiber, especially from cereals cause the reduction in fasting blood glucose and insulin concentration, thereby significantly reducing the incidence of type 2 diabetes. Water-soluble fiber forms a viscous solution in the small intestine which reduces the concentration and mixing of macronutrients with digestive enzymes and thus delays the absorption of glucose which consequently decreases glucose and insulin levels. Insoluble fiber increase insulin sensitivity, gut transit time, and satiety by affecting gut hormones. The dietary fibers are inversely associated with inflammatory markers that lead to the initiation and progression of type 2 diabetes (McRae 2018; Weickert and Pfeiffer 2008).

16.3 Glycemic Index (GI)

The concept of Glycemic index (GI) was introduced in the early 1980s by Jenkins et al. (1981), who provided ranking of carbohydrates between 0 and 100, based on postprandial impact on blood sugar level. Since then GI has gained importance as a reliable and quick dietary epidemiological tool for the classification of food products according to their postprandial glycemic response.

GI is defined as the area under the blood glucose response curve after consumption of 50 g carbohydrate from a test food divided by the area under the curve after consumption of 50 g carbohydrate from a control food (either white bread or glucose). Generally speaking, it is a tool which measures how quickly a food may raise the blood glucose levels. On the basis of GI values, the foods are categorized into three groups: The high GI foods (>70), intermediate GI foods (>55–<70), and low GI foods (<55). The low GI diet facilitates reduction in blood glucose levels, glycosylated hemoglobin, fructosamine, and urinary c-peptide (Miller 1994). The millet-based food products have lower GI value because of the presence of bioactive components and therefore making them suitable for consumption by diabetics. Addition of millets to traditionally eaten food facilitates the reduction of GI. Various investigations have been conducted to develop functional food products for diabetics by incorporation of minor millets in the commonly eaten food items. Various low GI millet-based food products which may be used as a complete meal or breakfast items or snacks have been developed. Pearl millet, finger millet, and semolina based *uttapam* (a thick fermented pancake made traditionally from a mixture of rice and lentils or semolina) had a GI of 38.7 and pearl millet, finger millet, and *besan* (Bengal gram flour) made *cheela* (pancake traditionally made from Bengal gram flour) had a GI of 36.8 as compared to 38.9 and 44.07 for their respective controls (Kumari et al. 2014).

16.4 Millet-Based Food Products for Diabetics

Attempts have been made to modify the conventionally consumed food products into low GI diabetic food by the addition of millets. A list of various millet-based food products for diabetics is furnished in Table 16.1. Arora and Srivastava (2002) modified traditionally consumed food *khichadi* (a complete meal made from rice and pulses with other variations including addition of vegetables and seasoned with spices), *baati* (hard, unleavened bread), and *laddu* (sphere-shaped sweets which can be used as a substitute for dessert) into low GI products by addition of finger millet and barnyard millet. The GI of these products ranged from 25.53 to 36.7, classifying them as low GI products having hypoglycemic effect.

Baked products like buns and breads which are traditionally made from refined wheat flour have a high glycemic index. Addition of millets in these products enhances their nutritional properties, lowers GI without affecting their organoleptic characteristics. Buns made from finger millet were found to have a low GI of 36.57 and caused a significant reduction in fasting and postprandial blood glucose levels. They were also found to improve lipid profile in diabetics (Tiwari and Srivastava 2017). Pizza base prepared by incorporating finger millet and barnyard millet were found to have low GI of 36.2 and 36.6 compared to control (58.06) prepared from refined wheat flour (Shrestha 2017). Muffins made by replacing 50% of sweet potato flour by finger millet flour showed GI below 50, compared to sweet potato incorporated muffin having GI > 50 and control muffin made from refined wheat flour which had a GI of >70 (Ulfath et al. 2016). Bread made from finger millet and foxtail millet showed lower GI as compared to bread made from refined wheat flour. The composite bread formulated from two genotypes of finger millet had a low GI of 41.4 and 43.1, whereas foxtail millet bread had a GI of 49.5 compared to refined wheat bread which had a GI of 67.8 (Chhavi and Sarita 2012).

Foxtail millet has also been explored for developing low GI food products for diabetics. Foxtail millet was used to prepare *dosa* (fermented flat shallow fried pancake) and was compared with traditionally eaten rice-based *dosa* and it was found that foxtail millet *dosa* had a GI of 59.25 compared to rice-based *dosa*, which had a GI of 77.96. Subjects who consumed foxtail millet *dosa* had lower postprandial blood glucose levels than their counterparts who consumed rice-based *dosa* (Narayanan et al. 2016). Besides foxtail millet, kodo millet and barnyard millet were also explored as a better option to prepare *dosa* for diabetics as compared to traditional rice *dosa*. *Dosa* prepared using kodo millet and barnyard millet was observed to have good acceptability and low GI compared to traditionally eaten rice *dosa*. After consumption of kodo millet and barnyard millet *dosa*, the postprandial blood glucose levels were significantly lower than after ingestion of rice *dosa* by the diabetics, advocating the efficacy of kodo millet and barnyard millet for diabetics in comparison to commonly eaten rice (Nagalakshmi and Beatrice 2013). Like *dosa*, *Idli* is also a type of savory rice cake made by steaming fermented batter consisting of rice and black lentils. It was observed that on incorporating kodo millet at 60% level, kodo millet *idli* was found to be suitable for consumption by diabetics as the

Table 16.1 Millet-based food products suitable for diabetics

Food products	Millet type	GI of millet food product (test)	GI of traditionally consumed food (control)	Effect on biological parameters	Researchers
Khichadi	Finger millet	25.33	45.78	–	Arora and Srivastava (2002)
Laddu	Finger millet	34.62	44.36	↓ blood glucose ↓ blood lipids ↑ HDL	Arora and Srivastava (2002)
Batti	Finger millet	36.12	39.95	–	Arora and Srivastava (2002)
Buns	Finger millet	36.57	–	↓ blood glucose ↓ blood lipids ↑ HDL	Tiwari and Srivastava (2017)
Bread	Finger millet (VL146)	41.4	67.8	–	Chhavi and Sarita (2012)
Bread	Finger millet (PRM601)	43.1	67.8	–	Chhavi and Sarita (2012)
Muffins	Finger millet	<55	>70	–	Ulfath et al. (2016)
Pizza base	Finger millet	36.2	58		Shrestha (2017)
Pizza base	Barnyard millet	36.6	58		Shrestha (2017)
Khichadi	Barnyard millet	27.24	45.78	–	Arora and Srivastava (2002)
Laddu	Barnyard millet	34.6	44.36	–	Arora and Srivastava (2002)
Batti	Barnyard millet	36.71	39.95	–	Arora and Srivastava (2002)
Dosa	Barnyard millet	–	–	↓ postprandial glucose level	Nagalakshmi and Beatrice (2013)
Khichdi	Barnyard millet	34.96	62.5	–	Joshi and Srivastava (2016)
Noodles	Barnyard millet	42.07	–	–	Surekha et al. (2013)
Noodles	Barnyard millet and pulses	35.68	–	–	Surekha et al. (2013)
Noodles	Barnyard millet and vegetable	38.02	–	–	Surekha et al. (2013)

(continued)

Table 16.1 (continued)

Food products	Millet type	GI of millet food product (test)	GI of traditionally consumed food (control)	Effect on biological parameters	Researchers
Bread	Foxtail millet	49.53	67.82	–	Chhavi and Sarita (2012)
Burfi	Foxtail millet	37.5	43	↓ serum glucose ↓ lipid profile ↓ GHb ↑ HDL	Bisht and Srivastava (2013), Thathola et al. (2011)
Biscuit	Foxtail millet	50.8	68	↓ serum glucose ↓ lipid profile ↓ GHb ↑ HDL	Anju and Sarita (2010), Thathola et al. (2011)
Dosa	Foxtail millet	59.25	77.96	↓ postprandial glucose levels	Narayanan et al. (2016)
Bread	Foxtail millet	–	–	↓ fasting glucose level ↓ postprandial glucose level	Ren et al. (2018)
Dosa	Kodo millet	–	–	↓ postprandial glucose level	Nagalakshmi and Beatrice (2013)
Idli	Kodo millet	58.5	67.11	–	Yadav et al. (2013)
Sewaiupma	Kodo millet	65.49	69.14	–	Yadav et al. (2013)
Uttapam	Pearl millet and finger millet	38.7	38.9	–	Kumari et al. (2014)
Cheela	Pearl millet and finger millet	36.8	44.7	–	Kumari et al. (2014)
RTC flakes	Little millet	52.11	–	–	Patil et al. (2015)

GI of kodo millet *idli* was 58.5 in comparison to traditionally eaten rice-based *idli* which had a GI of 67.11 (Yadav et al. 2013).

Potential of kodo millet to be used as diabetic food was also investigated in form of *sewaiupma*. *Sewaiupma* is an extruded traditional food made from refined wheat flour. The kodo millet incorporated *sewaiupma* had a GI of 65.49 which was significantly lower than refined wheat flour control *sewaiupma* which had a GI of 69.14 (Yadav et al. 2013).

Foxtail millet was also investigated to prepare low GI biscuits (50.8) and *burfi* (37.5) (a sweet) and it was revealed that on intervention, the foxtail millet based products, biscuits and *burfi*, were able to reduce serum glucose, serum cholesterol, serum LDL, glycosylated hemoglobin, and serum VLDL. The products were able to increase serum HDL levels in diabetic subjects (Thathola et al. 2011). The GI of foxtail millet can further be reduced by application of additional coating of gum acacia (GI = 32), fenugreek seeds (GI = 28), curry leaves (GI = 30), and aloe vera (GI = 31). The uncoated foxtail millet had a GI of 33 and seasoned foxtail millet had a GI of 47 (Wahlang et al. 2018).

Effectiveness of barnyard millet was also examined by researchers and it was seen that addition of barnyard millet instead of rice in preparation of *khichadi* (usually a complete meal prepared by combination of rice with pulses and seasoned with spices) lowered the GI of product. The barnyard millet *khichadi* had a GI of 34.96 and rice-based *khichadi* had a GI of 62.5 thus making barnyard millet fit for consumption by diabetics (Joshi and Srivastava 2016). Barnyard millet was also utilized to prepare biscuits and *burfi* (a sweet) at 45% and 43% level of incorporation of millet, respectively. The GI of barnyard millet biscuit was found to be 68 which make it an intermediate GI food and GI of barnyard millet *burfi* was 45 which makes it fall in low GI food category (Anju and Sarita 2010; Bisht and Srivastava 2013). Noodles made from barnyard millet may also be utilized for diabetics. Three variation of barnyard millet noodles, the plain barnyard millet noodles, in combination with pulses, and in combination with vegetable, were found to be suitable for diabetics, as all the three types of noodles showed low GI of 42.07, 35.68, and 38.02, respectively (Surekha et al. 2013).

Little millet based food products contrary to rice are suitable for consumption by diabetics. Ready-to-cook flakes made from little millet were used to prepare *avalakki*—a common breakfast item used in North Karnataka, India. The mean GI value was 52.11 which was lower than conventional rice flakes (Patil et al. 2015). Likewise, pasta made by different combination of millets—kodo, little, and pearl millet and pulses—horse gram and soybean were found to be effective in controlling blood glucose and improving lipid profile (Thilagavathi and Kanchana 2017).

16.5 Conclusion

Millets, the small seeded grains are nutritious as they are rich in many nutrients. Nutrients and phytochemicals confer them with therapeutic attributes. Small millets like finger millet, barnyard millet, foxtail millet, little millet, and kodo millet have been found to be rich in fiber, phenolic compounds, and flavonoids which impart them with antioxidant potency and make them suitable for use by diabetics. Millets have been found to have low GI and therefore improve glucose and lipid profiles in diabetic subjects. The traditionally consumed food products can be modified into functional foods by the addition of millets, without altering their organoleptic properties. Various millet-based food products like *dosa*, *idli*, bread, buns, pancakes, *laddu*, etc., can be formulated by the addition of millets at different levels, for diabetics. Further researches may be conducted for designing complete meal based on millets for diabetics. Also, human clinical trials are required to standardize the quantity and duration of intervention in diabetics.

References

Amadou I, Gounga ME, Le G-W (2013) Millets: nutritional composition, some health benefits and processing - a review. Emir J Food Agric 25:501–508

Anju T, Sarita S (2010) Suitability of foxtail millet (*Setaria italica*) and barnyard millet (*Echinochloa frumentacea*) for development of low glycemic index biscuits. Mal J Nutr 16:361–368

Anusha B, Hymavathi TV, Vijayalakshmi V, Reddy P, Robert TP (2018) Lipid-lowering effect of foxtail millet (*Setaria italica*) and Quinoa (*Chenopodium quinoa wild*) in pre-diabetics. J Pharm Res Int 24:1–7

Arora S, Srivastava S (2002) Suitability of millet-based food products for diabetics. J Food Sci Technol 39:423–426

Bisht AT, Srivastava S (2013) Efficacy of millets in the development of low glycemic index sweets for diabetics. Mal J Nutr 19:215–222

Chhavi A, Sarita S (2012) Evaluation of composite millet breads for sensory and nutritional quality and glycemic response. Mal J Nutr 18:89–101

Itagi S, Naik R, Bharati P, Sharma P (2012) Readymade foxtail millet mix for diabetics. Int J Sci Nat 3:47–50

Jali MV, Kamatar MY, Jali SM, Hiremath MB, Naik RK (2012) Efficacy of value added foxtail millet therapeutic food in the management of diabetes and dyslipidemia in type2 diabetic patients. Recent Res Sci Technol 4:03–04

Jenkins DJ, Wolever TM, Taylor RH, Barker H, Fielden H, Baldwin JM, Bowling AC, Newman HC, Jenkins AL, Goff DV (1981) Glycemic index of foods: a physiological basis for carbohydrate exchange. Am J Clin Nutr 34(3):362–366

Joshi S, Srivastava S (2016) Barnyard millet as a substitute of rice in preparation of *khichdi* for diabetics. IJSR 5:1798–1802

Kumar A, Metwal M, Kaur S, Gupta AK, Puranik S, Singh S, Singh M, Gupta S, Babu BK, Sood S (2016) Nutraceutical value of finger millet [*Eleusine coracana* (L.) Gaertn.], and their improvement using omics approaches. Front Plant Sci 7. https://doi.org/10.3389/fpls.2016.00934

Kumar A, Tomer V, Kaur A, Kumar V, Gupta K (2018) Millets: a solution to agrarian and nutritional challenges. Agric Food Secur 7:31. https://doi.org/10.1186/s40066-018-0183-3

Kumari PL, Sumathi S (2002) Effect of consumption of finger millet on hyperglycemic in non-insulin dependent diabetes mellitus (NIDDM) subjects. Plant Food Hum Nutr 57:205–213

Kumari N, Gupta A, Sheikh S (2014) Development of low glycemic foods with the use of pearl millet and finger millet. Int J Sci Res 3:193–195

McRae MP (2018) Dietary fiber intake and type 2 diabetes mellitus: an umbrella review of meta-analyses. J Chiropr Med 17:44–53

Miller JC (1994) Importance of glycemic index in diabetes. Am J Clin Nutr 59:747–752

Mohan V (2004) Why are Indians more prone to diabetes? J Assoc Physicians India 52:468–474

Nagalakshmi K, Beatrice DA (2013) Comparative study of the glycemic response of kodo and barnyard millets in type 2 diabetic subjects. IJGHC 2:930–935

Narayanan J, Sanjeevi V, Rohini U, Trueman P, Viswanathan V (2016) Postprandial glycemic response of foxtail millet *dosa* in comparison to a rice *dosa* in patients with type 2 diabetes. Indian J Med Res 144:712–717

Nishizawa N, Togawa T, Park K-O, Sato D, Miakoshi Y, Inagaki K, Ohmori N, Ito Y, Nagasawa T (2009) Dietary Japanese millet protein ameliorates plasma levels of adinopectin, glucose, and lipids in type 2 diabetic mice. Biosci Biotech Bioch 73:351–360

Patil KB, Chimmad BV, Itagi S (2015) Glycemic index and quality evaluation of little millet (*Panicumm iliare*) flakes with enhanced shelf life. J Food Sci Technol 52:6078–6082

Prabha KC, Selvi S (2016) Nutrient and antioxidant evaluation of four underutilized minor millets. Int J Curr Microbiol App Sci 5:224–233

Radha V, Kanthimathi S, Mohan V (2011) Genetics of type 2 diabetes in Asian Indian. Diabetes Manage 1:309–324

Ramachandran A, Snehalatha C, Shetty AS, Nanditha A (2012) Trends in prevalence of diabetes in Asian countries. World J Diabetes 3:110–117

Ren X, Yin R, Hou D, Xue Y, Zhang M, Diao X, Zhang Y, Hu J, Wu J, Hu X, Shen Q (2018) The glucose-lowering effect of foxtail millet in subjects with impaired glucose tolerance: a self-controlled clinical trial. Nutrients 10(10):1509. https://doi.org/10.3390/nu1010159

Saeedi P, Petersohn I, Salpea P, Malanda B, Kururanga S, Unwin N, Colaguiri S, Guariguata L, Motala AA, Ogurtsuva K, Shaw JE, Bright D, Williams R (2019) Global and regional diabetes prevalence estimates for 2019 and projections for 2030 and 2045: results from International Diabetes Federation Diabetes Atlas, 9th Edition. Diabetes Res Clin Pract 157:1–10

Sharma S, Saxena J (2016) Phytochemical screening and quantitative estimation of total phenolic contents and total flavonoid content of grains of *Paspalum scrobiculatum*. Asian J Pharm Clin Res 9:73–76

Sharma A, Kumar RA, Sood S, Khulbe RK, Agarwal PK, Bhatt JC (2018) Evaluation of nutraceutical properties of finger millet genotypes from mid hills of northwestern Himalayan region of India. Indian J Exp Biol 56:39–47

Shrestha R (2017) Suitability of finger millet & barnyard millet incorporated pizza base in diabetes and cardiovascular disease. PhD (Human Nutrition) Thesis submitted to Department of Foods and Nutrition, G B Pant University of Agriculture and Technology, Pantnagar

Sireesha Y, Kasetti RB, Nabi SA, Swapna S, Apparao C (2011) Antihyperglycemic and hypolipidemic activities of Setaria italica seeds in STZ diabetic rats. Pathophysiology 18:159–164

Snehalatha S, Ramachandaran R (2009) Insight into the mechanism of primary prevention of type 2 diabetes: improvement in insulin sensitivity and beta cell function. "Genetic and epigenetic basis of complex diseases" conference in Centre for Cellular and Molecular Biology, December 2009

Sood S, Joshi DC, Chandra AK, Kumar A (2019) Phenomics and genomics of finger millet: current status and future prospects. Planta 9:1–21

Suma PF, Urooj A (2012) Antioxidant activity of extracts from foxtail millet (*Setaria italica*). J Food Sci Technol 49:500–504

Surekha N, Devi R, Naik R (2013) Development of value added low glycemic index barnyard millet (*Echinochloa frumentacea Link*) noodles. IJFANS 2:20–24

Thathola A, Srivastava S, Singh G (2011) Effect of foxtail millet (*Setaria italica*) supplementation on serum glucose, serum lipids and glycosylated hemoglobin in type 2 diabetics. Diabetol Croat 40-1:23–28

Thilagavathi T, Kanchana S (2017) A study on the effect of millet and pulse based pasta on blood glucose and lipid profile in alloxan-induced diabetics rats. IJPCBS 7:112–121

Tiwari N, Srivastava S (2017) Effect of finger millet (*Eleusine coracana*) buns supplementation on serum glucose and serum lipids level in type 2 diabetics. Asian J Dairy Food Res 36:337–340

Udeh HO, Duodu KG, Jideani AIO (2017) Finger millet bioactive compounds, bioaccessibility, and potential health effects - a review. Czech J Food Sci 35:7–17

Ugare R, Chimmad B, Naik R, Bharati P, Itagi S (2014) Glycemic index and significance of barnyard millet (Echinochloa frumentacea) in type II diabetic. J Food Sci Technol 51 (2):392–395

Ulfath TK, Nikhitha M, Praditi S, Kavitha GS, Myrene RD, Sangeetha P (2016) Development of value added product to reduce the risk of hypoglycemia in diabetic population. Eur J Pharm Med Res 3:514–518

Wahlang B, Joshi N, Ravindra U (2018) Glycemic index lowering effect of different edible coatings in foxtail millet. J Nutr Health Food Eng 8:404–408

Weickert MO, Pfeiffer AFH (2008) Metabolic effects of dietary fiber consumption and prevention of diabetes. J Nutr 138:439–442

WHO (2016) Global report on diabetes. WHO, Geneva

Yadav N, Chaudhary K, Singh A, Gupta A (2013) Evaluation of hypoglycemic properties of kodo millet based food products in healthy subjects. IOSR J Pharm 3:14–20

Genomics-Assisted Improvement of Grain Quality and Nutraceutical Properties in Millets

17

Anil Kumar, Ram Sewak Singh Tomar, Ajay Chandra, Dinesh Joshi, Sushma Tiwari, Prabha Singh, Rakesh Kumar Choudhary, and Vishnu Kumar

Abstract

Current projections suggest that the global population could reach nine billion by 2050. To meet the food demand of such a huge population, a 70% increase in global food production is projected by 2050. The conventional cereal crops might not be able to fulfill these requirements because the ultimate yield potential of most of these crops is currently at a plateau level. To replace the existing pressure on conventional crops, millets are the best alternatives in the near future considering the global food sustainability and nutritional reliability. By virtue of their

A. Kumar (✉)
Directorate of Education, Rani Lakshmi Bai Central Agricultural University, Jhansi, Uttar Pradesh, India

R. S. S. Tomar
Department of Biotechnology & Crop Improvement, Rani Lakshmi Bai Central Agricultural University, Jhansi, Uttar Pradesh, India

A. Chandra
G.B. Pant University of Agriculture and Technology, Pantnagar, Uttarakhand, India

D. Joshi
ICAR-Vivekananda Parvatiya Krishi Anusandhan Sansthan, Almora, Uttarakhand, India

S. Tiwari
Rajmata Vijayaraje Scindia Krsihi Vishva Vidyalaya, Gwalior, Madhya Pradesh, India

P. Singh
ICAR-Indian Grassland and Fodder Research Institute, Jhansi, Uttar Pradesh, India

R. K. Choudhary
Rani Lakshmi Bai Central Agricultural University, Jhansi, Uttar Pradesh, India

V. Kumar
Department of Genetics & Plant Breeding, Rani Lakshmi Bai Central Agricultural University, Jhansi, Uttar Pradesh, India

A. Kumar et al. (eds.), *Millets and Millet Technology*,
https://doi.org/10.1007/978-981-16-0676-2_17

exceptional nutritional profiles, they have a great potential to contribute to solve the malnourishment problems mainly in low-income food deficient developing countries. Also, the C4 photosynthetic pathway and its ability to combat the moisture stress make them a ultimate option for climate resilient agriculture. In spite of enormous perspective, attempt for the genetic advancement of millets is considerably lagged behind the conventional cereal crops. Research done so far in millets has decoded their exceptional nutritive value, broad adjustability, and phenotypic plasticity. Information on genetic mechanisms underlying important agronomic traits of millets including nutritional parameters is limited. To improve the crop very lesser efforts have been devoted by applying novel biotechnological and genomic tools. However, with the availability of draft genome sequences for some of the millets, there is wide scope to accelerate the progression of genetic improvement. This chapter is an attempt to gather the meager information available on the aspects of grain quality and nutraceutical development in millets by utilizing the potential of emerging genomics tools and techniques.

Keywords

Millets · Nutraceutical · Genomics · Undernourish · Micronutrient

17.1 Introduction

The population of world is primarily dependent on plant-based diets. While, the world is dedicated to achieve "Zero hidden hunger" by 2030, approximately more than 868 million people all over the world are presently "chronically undernourished." This alarming condition is even increasing due to the marginal supply of nutri-enriched food to eat. People have less access to essential micronutrients, viz., iron, zinc, calcium, etc., in their diet. People from resource-poor countries of Southeast Asia, Sub-Saharan Africa, and parts of Latin America, basically rely on nondiversified staple crops for their nutritional requirements. However, majority of these main crops including rice, wheat, and maize are frequently deficient in essential micronutrients and are unable to supply these nutrients to the people as per the recommended daily allowance (RDA) (Bouis and Welch 2010; Saltzman et al. 2013; Tiwari et al. 2016a, b). The nutritional security and health status of the growing population primarily depends on this plant diet. Nutrient-rich diet is the basic need for normal growth and development as more than half of the world population suffers from nutrition deficiency due to intake of insufficient and malnutrition diet (Zhao and McGrath 2009). The development of nutrient-rich food diet is called biofortification. This approach is very helpful to enrich nutrient sufficiency and delivering to the malnourished and starving populations (Bouis et al. 2011). In underdeveloped and developing countries like parts of Asia and Africa where resources are limited millets have a significant part by providing approximately 75% of total calorie intake. These "small-seeded grasses" include several species like pearl millet (*Pennisetum glaucum*), finger millet (*Eleusine coracana*), foxtail millet (*Setaria italica*), proso millet (*Panicum miliaceum*), barnyard millet (*Echinochloa* spp.), Kodo millet (*Paspalum scrobiculatum*), and little millet (*Panicum sumatrense*). These different species of millets are widely cultivated in parts of India, China, Japan, and Korea for various food and fodder purposes. Being nutritionally superior to many cereal crops like rice, wheat, maize, and soybean include a high amount of essential proteins, dietary fibers, iron, zinc, calcium, phosphorus, potassium, vitamin B, and essential amino acids (Saleh et al. 2013). Due to antinutrients in millets like phytates, polyphenols, and tannins the mineral bioavailability is reduced by chelating multivalent cations with Fe^{2+}, Zn^{2+}, Ca^{2+}, Mg^{2+}, and K^{+} (Abdel Rahman et al. 2005). These anti-nutritional factors had led millets to render the orphan status globally.

Several schemes are available which can be incorporated in biofortification of several crops with minerals. One of the promising approaches is to identify or design crop or genotypes effective against specific nutrients and triggering their molecular machinery for sequestrations to storage tissues. The relative receptivity of specific germplasm to particular nutrients largely relies on their genetic prospective for nutrient use efficiency (NuUE) of crops (Liu et al. 2017). NuUE of crop plants needs to be enhanced so that these plants or cultivars would be able to produce higher yields per unit of nutrient applied under specified conditions. Thus, the fertilizer formulation and fertigation regime have a significant impact on NuUEs of the crops. The vast genetic variability is found in millets compared to other cereal crops in key elements like iron, zinc, and calcium (Muthamilarasan and Prasad 2015). Also, millets are tolerant to both biotic as well as abiotic stresses which confer good insurance to the grower (Chopperla et al. 2018; Reddy et al. 2011). Minor millets have been given priority over other millets as the crop choice for quality improvement. However, the quality improvement can be achieved either by enriching the nutrients aggregation in milled grains or by minimizing the antinutrients to enhance the bioavailability of other minerals. Considering these points, this chapter highlights the major breakthrough in developing nutrient enriched, i.e., biofortified varieties using omics approaches to enhance grain quality and nutraceutical value. After observing similar work on other model crop plants we highlight the association of conventional breeding and genome-assisted tools and techniques in millets.

17.2 Genome Enabled Approaches for Improving Grain Quality and Nutraceutical Traits

17.2.1 Genetic Diversity Analysis

Availability of polygenic variability is a prerequisite for any crop improvement program. Phenotypic and nutritional characterization of small millets germplasm revealed immense genetic variation for economic traits (Upadhyaya et al. 2011; Yamunarani et al. 2016; Vetriventhan and Upadhyaya 2018, 2019). However, further confirmation with molecular markers is required for a better understanding of allelic diversity to ensure plant breeders to use the genetic variation more judiciously and precisely. Reports about the application of molecular markers in millets are very limited, which again were limited to implement conventional nonreproducible DNA markers such as random amplified polymorphic DNA (RAPD) for cultivars and germplasm characterization (Ramakrishnan et al. 2016; Rajendran et al. 2016). The millets grain quality improvement at genomic level has always been limited because unavailability of enough number and easily detectable codominant markers were not reported earlier. Further exhaustive efforts are required for the development of new, highly polymorphic microsatellite markers that would cover the locations in the genome which are currently meagerly available microsatellites (Babu et al. 2014; Ramakrishnan et al. 2016; Rajendran et al. 2016). Exhaustive germplasm characterization for economic traits including grain nutritional profile with next-generation sequencing (NGS)-based markers are required, using diversity panels and core collections extracted from global germplasm collections including wild relatives, for its use in various breeding programs. However, germplasm characterization studies utilizing next-generation sequencing (NGS)-based data leading to the identification of molecular markers, largely single nucleotide polymorphisms (SNPs) are yet to be implemented in small millets on large scale.

17.2.2 Next-Generation Sequencing (NGS)-Based Platforms

The availability of a reference genome is a major resource for trait characterization and crop improvement. The latest physical map of finger millet genome has 1897 total number of scaffolds with a N50 length of 2.6 Mb and 96% single-copy orthologs (Hatakeyama et al. 2017). The draft genome of finger millet was available (Hittalmani et al. 2017). Likewise, in the foxtail millet genome sequence, the overall length of total scaffolds is 423 Mb, with 28 Mb (6.6%) gaps (Zhang et al. 2012). The genome sequence of proso millet is known to possess both protein-coding genes (55,9300) and microRNA genes (339) (Zou et al. 2019). However, genome sequence information for many nutritionally important minor types of millet such as kodo millet, barnyard millet, and little millet is still awaited.

To determine the genome-wide nucleotide variations, re-sequencing of various crops using next-generation sequencing technology has geared up with good speed. The analysis of genes and transcripts using high-throughput techniques such as genotyping by sequencing, transcriptome, and epigenetic analysis played a significant role in identifying the genetic basis for various nutritional traits in different millets (Singh et al. 2014; Muthamilarasan and Prasad 2015). For instance, draft genome sequence of foxtail millet has accelerated the research for genetic gain in foxtail millet (Zhang et al. 2012). Foxtail millet was resequenced using Solexa sequencing technology and the Genome Analyzer II (GA II) to search the genetic variations responsible for yield-associated traits. Alignment of these nucleotides with the known reference genome resulted in detecting SNPs (single nucleotide polymorphisms), InDels, and SVs (structural variants). It leads to the development of novel and unique markers using re-sequencing which further advanced genome mapping.

17.2.3 Dissecting the Genetic Architecture of Quality Traits

To date, scanty information is available in small millets for uncovering the QTLs and genes governing grain quality and nutraceutical traits and their inheritance pattern. However, in some of the millet species, advances in molecular marker discovery and genotyping assays have undeciphered genomic regions governing grain quality traits. For instance, microsatellite markers based association mapping leads to identify the yield-associated traits including grain quality. The QTLs related to Ca content were identified in 113 finger millet genotypes using 23 SSR markers (Kumar et al. 2015); Two QTLs (OM5 and FM8) for tryptophan content and one QTL (FMO2EST1) for protein content using 120 SSR markers. The same QTLs were linked to opaque2 modifiers (Opm) gene (Babu et al. 2014). The identification of QTLs and putative candidate genes governing these traits in Ca accumulation are useful for the successful breeding program. In small millets, only the association mapping of diversity panels based on a limited set of markers was used for QTL identification. The identification of QTL in small millets will play a crucial role in detecting important traits and improvement.

While efforts are underway in generating genomic resources in small millets, biparental and multiparental mapping populations are lacking due to poor breeding efforts. Very beginning effort in the development of mapping resources generated F_2 mapping populations derived from the parents exhibiting sufficient phenotypic diversity for few traits (Anonymous 2018). However, repeatability of genetic measurements in heredity-based mapping resources like F_2 and backcross populations are restricted due to lack of heterozygous genetic constitution. Therefore, a current need in small millets genomic research is development of nearly homozygous recombinant inbred line (RIL) populations, which will be an ideal mapping resource for linkage map development. Till date, only one RIL population is reported in finger millet (Anonymous 2016). A major limitation associated with biparental mapping populations is low mapping resolution thereby limiting the precision of genetic maps and QTL mapping (Cavanagh et al. 2008). Intermating among multiple founder parents in advanced generations has been proposed as an effective strategy to enhance the mapping resolution and facilitate fine mapping of the genomic regions governing quantitative traits (Korte and Farlow 2013). Keeping this in view, novel mapping resources like multiple advance generation intercross (MAGIC) and nested associated mapping (NAM) populations have been introduced to accelerate genomics-oriented breeding in crop plants (Mackay and Powell 2007). At present, no efforts are directed towards the development of these mapping resources of immense value in small millets. Recent advancements in crossing techniques and availability of genetic male sterility in some of the millets (Gupta et al. 1997) will encourage the breeders to place emphasis towards the inclusion of multiple parents to generate MAGIC and NAM populations to accelerate genomic research in small millets.

17.2.4 Comparative Genomics for Quality Traits

Till today, several inputs and interventions have been incorporated in orphan crops, through comparative genomics approaches. It resulted in a better understanding of genetic and functional similarity with closely related species for detecting the region governing trait of economic importance. Recent availability of reference genome sequences in some of the small millets like foxtail millet (Zhang et al. 2012), finger millet (Hittalmani et al. 2017; Hatakeyama et al. 2017), and proso millet (Zou et al. 2019) will allow a detailed structural and functional comparison with model cereal genomes such as rice for genes regulating various essential micronutrients and health-promoting bioactive compounds. Such genomic comparisons have already demonstrated their utility in identifying the genomic regions in small millets. For example, using the genetic similarity in rice, maize, and sorghum genomes several regions and QTLs responsible for tryptophan and protein content in finger millet have been detected (Babu et al. 2014).

17.2.5 Next-Generation Molecular Markers

The advancement in genome sequencing and array based technologies using low-cost approaches have resulted in the identification of SNPs and facilitated high-throughput automated genotyping across the crop plants (Appels et al. 2018). The techniques have been particularly useful in assessing allelic diversity for nutritional traits in diversity panels and core collections and use this genetic variations to identify genomic regions (quantitative trait loci, QTLs) through genome-wide association studies (GWAS). Considering their immense utility in small millets genomics, Wallace et al. (2015) determined the genetic makeup of global barnyard millet collection through SNP genotyping. However, these approaches have not been reported in other minor millets where it could greatly help to decode the genetic control of nutritional traits with complex genetic architecture. The identification of SNPs in genotypes having variation for gluten texture (waxy locus) in barnyard millet, calcium density in grains of finger millet will result in improvement after successful breeding for nutraceutical development.

BAC libraries are important genomic resources for providing genetic applications such as molecular marker development, identifying genomic sequences, developing genetic and physical maps, and comparative genomics. These libraries may be utilized for sequencing the genes governing bioactive flavonoids in millets.

17.3 Nutri-Genomics

With the emerging advances in omics science, nutritional biology research provided insights that behavioral factors such as diet can also affect the modulation and expression pattern in genes as well as differential interaction of the nutrients. The integration of behavioral genetics and nutritional molecules with genomics gave birth to the new scientific area of research in nutritional biology popularly known as "nutri-genomics" (Kumar et al. 2016). Nutri-genomics comprised of quantitative techniques such as genome-wide identification of genes controlling nutrient biosynthesis pathways. Further, this information is integrated with functional genomics approaches (transcriptomics, proteomics, and metabolomics) and bioinformatics tools for quantifying the genetic structure of genes and their relation with human health (Mutch et al. 2005). The complete profiling of nutrient-responsive gene will be helpful in determining the variations affected by diet depending upon individual to individual behavior (Kumar et al. 2016).

17.4 Transcriptomics

Transcriptomics or gene expression profiling is one of the most important tools, widely used for determining nutrient-responsive genes and biological pathways involved in their accumulation and biosynthesis. For non-model cereal crops like small millets, with limited genomic information, RNA-sequencing can reveal underlying expression mechanism of nutritionally important genes (Kumar et al. 2016). Recently, a hypothetical model called tripartite model in which nutrient-responsive differential expression genes in finger millet has been studied (Kokane et al. 2018). For lesser explored minor millets such transcriptome-based models will provide crucial insights about expression patterns of genes regulating calcium transport and functional polymorphism. Comprehensive transcriptome analysis of developing spike has been carried out to decipher the underlying gene regularity mechanism of calcium accumulation in finger millet (Mirza et al. 2014). With the help of transcriptome profiling, 82 unique calcium sensor genes have been identified in the spike (Mirza et al. 2014). These unique calcium sensor genes were later categorized into Calcium sensors genes and calcium transporters genes (Sood et al. 2016). Further, the expression profiling of calcium transporters genes, CaM and CAX1 was found to be highly expressed in high grain calcium genotypes (GP-1 and GP-45) during the grain filling stage (Mirza et al. 2014). The functional characterization of calcium accumulation was provided by these candidate genes in finger millet. Since grains of other minor millets are also the reservoir of essential micronutrients, global transcriptome analysis is required for the exploration of the molecular mechanism underlying the micronutrient homeostasis pathway. This valuable information generated through global transcriptome analysis will be utilized for the development of functional markers such as EST-SSRs, SNPs, and intron-spanning region (ISR) associated with grain quality traits.

17.5 Proteomics

Proteomics refers to the structural and functional characterization and analysis of expression patterns of a set of proteins (Kussmann et al. 2006). The important area of plant proteomics utilized for the identification of highly specialized proteins are 2-D gel electrophoresis, chromatographic separation and quantification of the desired protein, mass spectrometry, and enzyme-linked immunosorbent assay (ELISA) (Jagadeesh et al. 2017). Similarly, with the emerging insights in next-generation proteomic tools such as co-immunoprecipitation (co-IP), protein microarray, *yeast-two hybrid* systems and gel-free shotgun proteomics approach are accelerating the protein–protein interaction studies in nutritional biology (Aslam et al. 2017). Proteomics is to be studied in understanding the grain quality traits in minor millets on a large scale. However, few reports have analyzed the significance of seed proteome in finger millet for the identification and characterization of proteins concerned with calcium accumulation. For instance, utilizing immune-detection technique Kumar et al. (2015) identified the calmodulin protein associated with the accumulation of calcium in finger millet genotypes. Furthermore, by applying peptide mass fingerprinting technology another calcium-binding protein in developing spikes, called calreticulin has been identified in finger millet (Singh et al. 2016). Identification of biomarkers associated with nutritional traits through cutting edge proteomic studies will explore the possible mechanisms involved in the nutrient homeostasis pathways in minor millets. The gluten-free nature of millet protein is the most important characteristic for providing it the status of health boosting and allergen free food supplement. Therefore, absolute quantitative proteomics of millet protein is recommended for getting deep insights into its nutraceutical and allergen free nature for declaring it as a nutraceutical dietary supplement.

17.6 Metabolomics

Metabolomics is one of the recent "omics" approaches that allow the high-throughput identification and characterization of nutritionally important low molecular weight biomolecules (metabolome). Utility of this emerging omics approach has been well demonstrated for the development of biomarkers, food quality assessment, and monitoring of diet-related diseases (Combs et al. 2013). Further, this technology has effectively been utilized for structural and functional assessment of secondary metabolites like polyphenols and flavonoids contributing towards nutraceutical properties in crop plants (Sumner et al. 2003). In minor millets, the complete profiling of the metabolites essential for uptake, transport, and accumulation of minerals and ions will give molecular insight into the discovery of possible metabolic pathways responsible for nutrient biosynthesis and accumulation in grains. Exhaustive metabolic profiling of minor millet grains is yet to be implemented. However, various studies have investigated the metabolome of finger millet grains (Chethan and Malleshi 2007). This study identified polyphenols (0.04–0.09%), phenolic acid (45.0 mg/100 g), benzoic acid derivatives (85%), and cinnamic acid derivatives as the major secondary metabolites of finger millet grains. In another instance, total bound phenolic acids, *p*-coumaric, and ferulic acid were identified as the major secondary metabolites and account for 50–99% and 64–96% in finger millet grains (Devi et al. 2014). While implementing these metabolomic studies, tools and techniques such as high-performance liquid chromatography (HPLC), gas chromatography–mass spectrometry (GC-MS), nuclear magnetic resonance (NMR), and electrospray ionization mass spectrometry (ESI-MS) were widely used (Devi et al. 2014). This valuable information on secondary metabolite profile of small millets may be added along with several omics data for deciphering the underlying genetic mechanism.

17.7 Conclusion

Minor millets are very nutritious on which the poor population of Asia and Africa rely heavily. However, the research work to utilize highly nutritious millet crops is inadequate and insufficient to combat malnutrition. The bioavailability of nutrients, essential amino acids, minerals, and vitamins need further study. It can be done by the extensive understanding of the pathways involved in nutrient uptake, translocation, and storage. The study leading to search of molecular markers such as SNPs and InDels linked to quality traits and candidate genes will help in dissecting complex traits. The encouraging results observed in finger millet suggested that studies pertaining to transporter pathways involved in transport and localization of macro and micronutrients in grains of other minor millets need exhaustive investigation through transcriptomics, proteomics, and metabolomics approaches. It will ultimately accelerate conventional and molecular breeding resulting in the characterization of germplasm for new sources of variation in nutritional traits.

With the emerging insights in genome sequences of some of the minor millets, it is peak time that genomic regions governing nutritional traits should be deployed in breeding programs. With the advent of next-generation sequencing platforms, rapid re-sequencing of thousands of small millet genotypes is possible soon. On the whole, the collective approach utilizing all the omics tools like genomics, transcriptomics, proteomics, and metabolomics with conventional and traditional breeding could promote millets as a model system for the advancement of quality traits and to be used as a staple crop in the world.

References

Abdel Rahman SM, Babiker EE, Tinay AH (2005) Effect of fermentation on antinutritional factors and HCl extractability of minerals of pearl millet cultivars. J Food Technol 3:516–522

Anonymous (2016) Annual report 2016–2017. ICAR-Vivekanada Parvatiya Krishi Anusandhan Sansthan, Almora, Uttarakhand, India. p 184

Anonymous (2018) Annual report 2018–2019. ICAR-Vivekanada Parvatiya Krishi Anusandhan Sansthan, Almora, Uttarakhand, India. p 184

Appels R, Eversole K, Stein N et al (2018) Shifting the limits in wheat research and breeding using a fully annotated reference genome. Science. https://doi.org/10.1126/scienceaar7191

Aslam B, Basit M, Nisar MA et al (2017) Proteomics: technologies and their applications. J Chromatogr Sci 55(2):182–196

Babu BK, Dinesh P, Agrawal PK et al (2014) Comparative genomics and association mapping approach for blast resistant genes in finger millet using SSRs. PLoS One 9(6):e99182

Bouis HE, Welch RM (2010) Biofortification-a sustainable agricultural strategy for reducing micronutrient malnutrition in the global south. Crop Sci 50:20–32

Bouis HE, Hotz C, McClafferty B et al (2011) Biofortification: a new tool to reduce micronutrient malnutrition. Food Nutr Bull 32:S31–S40

Cavanagh C, Morell M, Mackay I et al (2008) From mutations to MAGIC: resources for gene discovery, validation and delivery n crop plants. Curr Opin Plant Biol 11:215–221

Chethan S, Malleshi NG (2007) Finger millet polyphenols optimization of extraction and the effect of pH on their stability. Food Chem 105:862–870

Chopperla R, Singh S, Tomar RS et al (2018) Isolation and allelic characterization of finger millet (Eleusine coracana L) small heat shock protein EcHSP178 for stress tolerance. Indian J Genet 78 (1):95–103

Combs GF, Trumbo PR, McKinley MC et al (2013) Biomarkers in nutrition: new frontiers in research and application. Ann N Y Acad Sci 1278:1–10

Devi PB, Vijayabharathi R, Sathyabama S et al (2014) Health benefits of finger millet (*Eleusine coracana* L) polyphenols and dietary fiber: a review. J Food Sci Technol 51:1021–1040

Gupta SC, Muza FR, Andrews DJ (1997) Registration of INFM 95001 finger millet genetic male sterile line. Crop Sci 37:1409

Hatakeyama M, Aluri S, Balachadran MT et al (2017) Multiple hybrid de novo genome assembly of finger millet, an orphan allotetraploid crop. DNA Res. https://doi.org/10.1093/dnares/dsx036

Hittalmani S, Mahesh HB, Shirke MD et al (2017) Genome and transcriptome sequence of finger millet (*Eleusine coracana* (L) Gaertn) provides insights into drought tolerance and nutraceutical properties. BMC Genomics. https://doi.org/10.1186/s12864-017-3850-z

Jagadeesh DS, Kannegundla U, Reddy RK (2017) Application of proteomic tools in food quality and safety. Adv Anim Vet Sci 5(5):213–225

Kokane SB, Pathak RK, Singh M et al (2018) The role of tripartite interaction of calcium sensors and transporters in the accumulation of calcium in finger millet grain. Biol Plant. https://doi.org/10.1007/s10535-018-0776-5

Korte A, Farlow A (2013) The advantages and limitations of trait analysis with GWAS: a review. Plant Methods 9:29

Kumar A, Gaur VS, Goel A et al (2015) De novo assembly and characterization of developing spikes transcriptome of finger millet (*Eleusine coracana*): a minor crop having nutraceutical properties. Plant Mol Biol Rep 33:905–922

Kumar A, Sharma D, Tiwari A et al (2016) Genotyping-by-sequencing analysis for determining population structure of finger millet germplasm of diverse origins. Plant Genome 9(2):1–15

Kussmann M, Raymond F, Affolte M (2006) OMICS-driven biomarker discovery in nutrition and health. J Biotechnol 124:758–787

Liu D, Liu Y, Zhang W et al (2017) Agronomic approach of zinc biofortification can increase zinc bioavailability in wheat flour and thereby reduce zinc deficiency in humans. Nutrients 9:465

Mackay I, Powell W (2007) Methods for linkage disequilibrium mapping in crops. Trends Plant Sci 12:57–63

Mirza N, Taj G, Arora S et al (2014) Transcriptional expression analysis of genes involved in regulation of calcium translocation and storage in finger millet (*Eleusine coracana* L Gartn). Gene 550:171–179

Mutch DM, Wahli W, Williamson G (2005) Nutrigenomics and nutrigenetics: the emerging faces of nutrition. FASEB J 19:1602–1616

Muthamilarasan M, Prasad M (2015) Advances in Setaria genomics for genetic improvement of cereals and bioenergy grasses. Theor Appl Genet 128:1–14

Rajendran SRCK, Yau Y, Pandey D et al (2016) CRISPR-Cas9 based genome engineering: opportunities in agri-food-nutrition and healthcare. OMICS 19(5):1–16

Ramakrishnan M, Antony CS, Duraipandiyan V et al (2016) Assessment of genetic diversity, population structure and relationships in Indian and non-Indian genotypes of finger millet (*Eleusine coracana* (L) Gaertn) using genomic SSR markers. Springer Plus 5(120):1–11

Reddy INBL, Reddy DS, Narasu ML et al (2011) Characterization of disease resistance gene homologues isolated from finger millet (*Eleusine coracana* L Gaertn). Mol Breed 27:315–328

Saleh ASM, Zhang Q, Chen J et al (2013) Millet grains: nutritional quality, processing, and potential health benefits. Compr Rev Food Sci Food Saf 12:281–295

Saltzman A, Birol E, Bouis HE et al (2013) Biofortification: progress toward a more nourishing future. Glob Food Secur 2:9–17

Singh UM, Chandra M, Shankhdhar SC et al (2014) Transcriptome wide identification and validation of calcium sensor gene family in the developing spikes of finger millet genotypes for elucidating its role in grain calcium accumulation. PLoS One 9(8):e103963

Singh M, Metwal M, Kumar VA et al (2016) Identification and molecular characterization of 48 kDa calcium binding protein as calreticulin from finger millet (*Eleusine coracana*) using peptide mass finger printing and transcript profiling. J Sci Food Agric 96:672–679

Sood S, Kumar A, Kalyana Babu B, Gaur VS, Pandey D, Kant L, Pattnayak A (2016) Gene discovery and advances in finger millet [Eleusine coracana (L.) Gaertn.] genomics-an important nutri-cereal of future. Front Plant Sci 7:1634

Sumner LW, Mendes P, Dixon RA (2003) Plant metabolomics: large-scale phytochemistry in the functional genomics era. Phytochemistry 62:817–836

Tiwari S, Krishnamurthy SL, Kumar V et al (2016a) Mapping QTLs for salt tolerance in rice (Oryza sativa L.) by bulked segregant analysis of recombinant inbred lines using 50K SNP Chip. PLoS One. https://doi.org/10.1371/journalpone0153610

Tiwari C, Wallwork H, Arun B et al (2016b) Molecular mapping of quantitative trait loci for zinc, iron and protein content in the grains of hexaploid wheat. Euphytica 207:563–570

Upadhyaya HD, Ramesh S, Shivali S et al (2011) Genetic diversity for grain nutrients contents in a core collection of finger millet (*Eleusine coracana* (L) Gaertn) germplasm. Field Crop Res 121:42–52

Vetriventhan M, Upadhyaya HD (2018) Diversity and trait specific sources for productivity and nutritional traits in the global proso millet (Panicum miliaceum L) germplasm collection. Crop J 6:451–463

Vetriventhan M, Upadhyaya HD (2019) Variability for productivity and nutritional traits in germplasm of kodo millet (*Paspalum scrobiculatum* L), an under-utilized nutrients rich climate smart crop. Crop Sci 59:1095–1109

Wallace JG, Upadhyaya HD, Vetriventhan M et al (2015) The genetic makeup of a global barnyard millet germplasm collection. Plant Genome 8. https://doi.org/10.3835/plantgenome2014100067

Yamunarani R, Ramegowda GG, Thammegowda H et al (2016) Genetic diversity for grain Zn concentration in finger millet genotypes: potential for improving human Zn nutrition. Crop J 4:229–234

Zhang G, Liu X, Quan Z et al (2012) Genome sequence of foxtail millet (Setaria italica) provides insights into grass evolution and biofuel potential. Nat Biotechnol 30:549–554

Zhao FJ, McGrath SP (2009) Biofortification and phytoremediation. Curr Opin Plant Biol 12:373–380

Zou C, Li L, Miki D et al (2019) The genome of broomcorn millet. Nat Commun 10:436

Rural Entrepreneurship Development in Millet Processing

18

Subir Kumar Chakraborty and Shalini Chakraborty

Abstract

Millets are crops of antiquity. They have high nutritional value and can be used to develop a variety of value-added products. Along with these qualities the millets possess an additional benefit of no fuss agricultural practices, they do not require chemical fertilizers to grow, can coexist with weeds and pests, and can survive in rain-fed conditions. With all these positives these crops failed to stay in the list of the priority foods of twentieth century, part reasons can be that they were indigenous to the developing and underdeveloped countries and failed to stand up to the competition of the cereals from the developed world which came with the added advantage of being processing friendly. However, in the present time these crops are reviving and rising in prominence. The reasons can be the incidences of physiological disorders accompanying the daily consumption of wheat, rice and their processed foods. Also, there is a widespread realization of the extent of nutritional benefits associated with millets. Here lie the tremendous entrepreneurship opportunities for processing and value addition of millets which can be successfully pursued by the enterprising youth living in the production catchment. This chapter brings to the fore certain aspects of millet crops in terms of its processing possibilities, its role in the promotion of rural livelihood, the complete value chain of millets, and the scope for entrepreneurship in the area of millet processing and value addition.

S. K. Chakraborty (✉)
Agro Produce Processing Division, ICAR-Central Institute of Agricultural Engineering, Bhopal, Madhya Pradesh, India
e-mail: subir.kumar@icar.gov.in

S. Chakraborty
Fruit Research Station (Enthkhedi), Rajmata Vijayaraje Scindia Agricultural University, Gwalior, Madhya Pradesh, India

A. Kumar et al. (eds.), *Millets and Millet Technology*,
https://doi.org/10.1007/978-981-16-0676-2_18

Keywords

Cereals · Millets · Rural entrepreneurship · Value addition

18.1 Introduction

Millets are a group of crops, often referred to as coarse cereals. They are grown all over the world, common millets of India are *jowar*, *bajra*, *madua*, *kangni*, *sawan*, *kodo*, *kutki* or *samai*, and *barri*. Over the past 5000 years, historically millet cultivation is going on in drylands, hill, and tribal agricultural land of India. It is contributing to food, fodder, and nutritional security at the farm and domestic level (Dayakar et al. 2016).

Millet crops can survive at very little water; they can thrive in rain-fed agriculture. Skeletal soils are enough to support their root growth. Soil as less as 15 cm in depth is enough to support the growth of millets. Use of farmyard manure alone is enough for the growth of the millet crops. All these traits suggest that growing millet crops do not entail any carbon emissions in terms of irrigation and fertilizer requirement. The state does not have to carry the burden of a power bill or fertilizer subsidy for millet cultivation. All these outstanding qualities and immense benefits of millets farming are failing to arrest the steep fall in the gross cultivated area for these crops. In India, millets are cultivated in about 8.65% (2017–2018) of the gross cropped area of vast stretches of dryland and hilly area by small and marginal farmers (Anonymous 2018). A prominent characteristic of millets is their astonishing nutritional value (Sarita and Singh 2016); this trait can be boasted to increase its consumption for the benefit of the producer as well as the consumer.

There is a noticeable drive for low-calorie high fiber food products among the consumers. Food like millets can be fitted in any cuisine as part of a balanced diet. Many bodily disorders of the present times can be effectively managed by consuming foods with high fiber content (Sharma and Niranjan 2018). Millet, being a rich source of dietary fibers, nutraceuticals, micronutrients, phytochemicals, can be included in the daily diet to combat blood pressure, certain forms of cancer, obesity, cardiovascular diseases, and maintain a healthy gastrointestinal tract.

Across all types of millets, the general composition is 7–12% protein, 2–5% fat, 65–75% carbohydrates, and 15–20% dietary fiber. The millet cooking process and preparation are also similar to major staple food grains. Sadly, millets brimming with nutrition are neglected due to a lack of dissemination of technology for value addition and processing. Profitability in millet cultivation will increase by leaps and bounds once value addition potential in millets has been completely exploited. Till recent past, the growers and the potential consumers were not aware of the high nutritional value of the millets. Lack of knowledge did not provide enough drive for the local craftsman and scientific community to invent technologies and machines exclusive to millet processing and value addition. The growers of the millets were left with no choice but to adopt age-old practices and traditions for farming millet crops. This situation resulted in contributing to a decrease in the popularity of millets among the farming people and the consumers, alike.

Things are changing fast for the better, processed food industry and agrarian community has refocused their approaches in reviving this long neglected traditional crop of India by intervening in practices of production technology and through value addition operations customized for millets. Thus, these forgotten nutri-cereals having a rich store of carbohydrates and minerals (calcium, phosphorous, iron, etc.) are slowly being accepted in the processed food industry. One can find these nutri-cereals as raw, semi-processed, and finished product form in the shelves of the superstores and malls. At this juncture, it is ensuring that there lies an adequate quantum of scope for mass entrepreneurship opportunities in processing and value addition of millets. Entrepreneurs can take up the processing of millets in the production catchments itself. Such processing units will take away all the drudgery-prone postharvest operations that came in the way of popularization of these millets for domestic consumption, challenging the consumers' inclination for grains like rice, wheat, or sorghum. A typical entrepreneurship venture for millet processing shall include the following unit operations, viz., dehulling, destoning, cleaning, grading, milling, and sifting.

18.2 Health Benefits

Millets are crops of ancient times; they find mention in myths and tales. They have countless nutritional and physiological benefits. Nutritional features of millets override that of the major cereals, they are a source of good carbohydrates, micronutrients, and phytochemicals with nutraceutical properties (Table 18.1). Like cereals, millets also contain a large proportion of carbohydrates. Carbohydrates present in millets may be grouped into starch (65–70%) and (16–20%) of non-starchy polysaccharides (NSP). The features of the millet like lowering of blood cholesterol, prevention of constipation, and gradual release of glucose to the bloodstream during digestion can be attributed to NSP. The composition of the carbohydrates of millets is far superior to that of rice and wheat with respect to the

Table 18.1 Common names of millet and their nutrient content *vis-à-vis* major cereals

	Millet crop		Composition (per 100 g)				
S. No.	Hindi name	English name	Protein (g)	Fiber (g)	Mineral (g)	Iron (mg)	Calcium (mg)
1	*Chena, Barri*	Proso millet	12.5	2.2	1.9	0.8	14
2	*Kakum*	Foxtail millet	12.3	8.0	3.3	2.8	31
3	*Sawan*	Barnyard millet	11.2	10.1	4.4	15.2	11
4	*Bajra*	Pearl millet	10.6	1.3	2.3	16.9	38
5	*Kodon*	Kodo millet	8.3	9.0	2.6	0.5	27
6	*Kutki*	Little millet	7.7	7.6	1.5	9.3	17
7	*Ragi*	Finger millet	7.3	3.6	2.7	3.9	344
8	*Gehun*	Wheat	11.8	1.2	1.5	5.3	41
9	*Chawal*	Rice	6.8	0.2	0.6	0.7	10

glycemic index. Good fiber, magnesium, and vitamin-E content of millets make them suitable as diabetic food. The fiber present in the millet gives a sense of satiety to the consumer by decreasing the hunger which in turn is because of the slow digestion rate. Millet starch has high amylose content as compared with amylopectin, this contributes to slow absorbing and sustainable release of sugar to blood stream. Magnesium in millets helps in relaxing the blood vessels and cardiovascular system; it enhances nutrient delivery by improving the blood flow and maintains the blood pressure thus further protects the cardiovascular system. Millets play an avid role in the mitigation of diabetes; consumption of millets lowers the risk of diabetes. Not only do the millets help in reducing the blood glucose levels, but they also improve the insulin response. Magnesium is a very important mineral that is supplied in abundance by millets. Secretion of enough insulin is ensured by magnesium resulting in complete metabolism of glucose in the human body. Gut microflora present in human beings converts the millet lignans into mammalian lignans and entero-lactone. Millets are also a very good source of phosphorus which is an essential component of ATP, the storage unit of energy in the human body. Presence of phosphorus also ensures a healthy nervous system and cell structure.

The protein content of some millets is even better with respect to the average protein content of rice. The protein content ranges from 7% to 12% and fat content from 1.12% to 5%. The essential amino acid profile of millets is also far better as compared to wheat or maize. As compared to cereals, millets have a higher digestibility of the protein; this can be attributed to fewer cross-linked prolamins that are present in the millets. Important vitamins present in millet grains are riboflavin, thiamine, niacin, and folic acid. For example, among all the cereals pearl millet has the highest niacin content whereas, the amino acid of finger millet is unique for its richness of sulfur (Antony et al. 1996).

Ragi has high calcium content among all the cereal/millet grains and other millets are a good source of iron and phosphorus. Gluten is a structural protein that is found in wheat. Elastic characteristic of wheat dough is due to the presence of gluten. Gluten however is not tolerable to everybody, it causes a disease called "Celiac disease" which leads to inflammation of the small intestine causing unsatisfactory absorption of nutritionally important food components like iron, folic acid, calcium, and fat-soluble vitamins. The global occurrence of celiac disease is 1:3345. Adopting a diet free of gluten throughout the lifetime is the only option for the patients of this disease. Millets do not contain gluten and is suitable for gluten intolerant people. It can be envisioned that there will be a rise in gluten-free products in the future.

Along with all the health benefits that the millets promise, there are also some anti-nutritional factors that are present in the millets. Presence of the following antinutrients, tannin, phenolic, oxalate, phytate, protease, and amylase inhibitors, results in depleted assimilation of health promoting factors of the millets. The bioavailability of nutrients is adversely affected due to the presence of these compounds. Proteins of the millets are conjugated by phenols and tannins, while the iron and calcium are made unavailable by the phytate and oxalate. However, these anti-nutritional factors can be countered by appropriate processing measures.

18.3 Processing of Millets

Breakfast as a meal has acquired its due importance by the nutritional scientists. Urbanization has resulted in a changed lifestyle and a craving for convenience foods, here processed millets, can be one option that stands to be exploited by the health-conscious society (Chakraborty et al. 2019). Although millets can be compared nutritionally and in some aspects are far superior to the popular cereals, the food use of these crops is very much restricted to the traditional consumers and they have largely remained the food of the poorer and less privileged section of the population. The reasons for the unpopularity of millets are mainly (1) ignorance of common man to its health benefits, (2) paucity of technologies for appropriate primary processing as well as secondary and tertiary processing has resulted in absence of ready-to-use or ready-to-cook products from millets, and (3) millets cannot be easily cooked to soft grains like rice, as the sub-aleurone layers are hard and quite resistant to hydration. In addition, these grains have very strong characteristic flavor of their own. Further, these grains do not possess gluten, which is the characteristic of wheat, and hence do not lend themselves for the preparation of thin *roti*/chapatti (unleavened bread). The aforesaid factors account for the limited diversified food uses and status of economic prominence for the millets. Thus, these grains have remained the foods for the poor sections of society. However, studies towards the processing of millets have demonstrated their successful utilization for various traditional foods and also in newer convenience health-oriented products. Hence, efforts are needed for popularization and wider adoption of these technologies to promote them for diversified utilization among the nontraditional urban population.

Millets are grains with numerous opportunities they can find use in many diversified commodities and processes. Basic grain-based value addition operations like milling, malting, popping, and other technologies. Grains after undergoing these unit operations can blend with legumes and be used for the development of low-cost weaning and complementary foods.

18.3.1 Malting

Complementary food for weaning purposes can be prepared with barley malt and milk solids. Barley malt finds good market in the brewing industry; here malted millets can be a suitable option for making end-use specific food formulations. Finger millet and pearl millet malt can be successfully used for such purposes (Chethan and Malleshi 2007). Malting involves germinating grains and allowing it to sprout. Grains are steeped in water (four times) for a time period ranging between 16 and 24 h. The water is drained off and grains are allowed to germinate at 25–30 °C for 72 h. This time-temperature combination is just adequate for the germination and appearance of sprouts. Sprouted grains are dried by toasting it on a skillet at 80–110 °C for about 15 min. The toasted millets are ground into a fine powder and can be stored in airtight containers.

Germinated sorghum and millets are used for the preparation of alcoholic beverages. Malting causes a change in the composition of amino acids, leads to conversion of starch into sugars, and improves the availability of vitamins and minerals along with fat. Germinated grain flour has a diminished capacity to thicken when heated with water in comparison to ungerminated grain flour. Hence, three times more germinated flour is needed to make a paste of the same consistency and viscosity. Thus, a high-calorie density product can be obtained. The energy that young children can extract from the daily diet is often limited by the bulk of the amount of food that they can consume. Thus, with the use of germinated grains, calorie dense foods suitable for young children can be made. In India, the consumption of malted finger millet is a very common food. Perhaps, the nutritional superiority of malted finger millet is the reason behind it being popular over malted sorghum and malted maize.

18.3.2 Milling

Millet processing technologies are aimed at improving the grain quality in terms of marketability and end-use (Sahay and Singh 1996). There is enough production of millet in the country to support cottage level and small-scale processing units in the production catchments. Millets can be further processed to obtain flakes, RTE snacks, supplementary foods, extrudates, malted products, baby foods, geriatric foods, and health foods.

18.3.3 Popping

It is generally carried out with finger millets to develop a highly desirable aroma with crunchy texture. Popping is not very popular with other millets. In a standard popping process, the whole millet grains are made to achieve 10% moisture followed by an equilibration for 4 h in a closed container. This grain is then put in sand maintained at about 230 °C for about 15–20 s. Popped millets are good sources of dietary fiber and carbohydrates. There is an expansion of about 7–10 ml/g after popping. The extent of popping depends on the morphology of the grain, good popping takes place with a hard endosperm and medium thick pericarp. Destruction of lipolytic enzyme during popping results in the popped millet grain that has a very good shelf life. It can be consumed as a snack or can be ground into coarse grits for use as an ingredient to ready-to-eat foods. Popped millet could be consumed as snacks; it makes delicious and nutritionally balanced convenience foods for growing children and lactating mothers.

The product has the advantage of being low-cost food permits addition of other flavoring ingredients like groundnut, dried coconut, etc., for improving the taste and nutritional quality. Similar products are also available from rice and wheat but they will be devoid of the micronutrients and phytochemicals that millet-based products can offer.

18.3.4 Baking

Almost any type of flour can be used for baking. Dough made with flour and water is made into thin sheets of circular shape before being cooked in a hot griddle to make flatbread (Chakraborty et al. 2018). The batter can be based on sorghum, millet, or any other cereal and it may or may not be fermented. Common name for these flatbreads are *roti* and *chapatti*, might not leaven as wheat-based *roti*. However, there are reports that by local technical knowledge villagers in Uttarakhand are using an aqueous solution of a plant extract to make the dough for making *roti* that leavens to an acceptable extent.

18.3.5 Extruded Products

Extrusion cooking has been reported to lower the level of antinutrients while enhancing protein and starch digestibility. Inherent low-fat content makes millets a very healthy ingredient for extruded snack food. Extrusion, over time has become a very popular food processing technology of the present times, corn and rice are the most common raw material but the millets can also be used to prepare ready-to-eat (RTE) extruded products (Chakraborty et al. 2011). A sizeable portion of snack foods available in the market are extruded products. Extrusion process is a high-temperature short-time (HTST) process, the cooking temperature seals all the chances of survival of micro organisms. During extrusion, starch is gelatinized and protein is denatured, both these process results in improved digestibility (Asharani et al. 2010). Extrusion also leads to the inactivation of the anti-nutritional factor present in the millet. Also, extruded products can be fortified with vital nonheat-sensitive nutrients.

Since the seed coat or the bran affects the expansion ratio and also the eating quality of the extrudates, it is desirable to use refined grits and flour (Chakraborty et al. 2014). Equilibrating the flour to about 18% moisture content and extruding in a single- or twin-screw extruder at about 150 °C and 200 rpm, gives extrudates of highly desirable food qualities with an expansion of 1.5–2 times. The products will have a crunchy texture and the same can have a sweet or savory flavor upon a coating with traditional ingredients. Alternately, the raw material in the form of grits can be mixed with spices and condiments before extrusion to obtain tasty RTE snacks. Cold extruded products like noodles with 20% pearl millet flour can also be made and they are quite acceptable.

18.3.6 Storage of Millet Products

Although the millet grains are small seeds and fetch not very high price in the local markets, small value-added operations like grading and cleaning can improve the price of the product. It is difficult to store the millet grains for a long period of time, especially if they are dehusked or dehulled. The germ lies just beneath the hull, during dehulling the germ gets exposed, with high moisture content/high humidity in the environment there is a chance of insect and fungus infestation during storage. High moisture also leads to the release of mycotoxins/aflatoxins in the grains making it poisonous and dangerous for humans and livestock alike. All these facts can be summed up to be the reason behind the difficulty of storage of dehulled millets at domestic level or at any other level. This challenge of a shorter shelf life of millet grains and flour can be sorted out by adopting a suitable hydrothermal treatment with a set protocol.

18.3.7 Millet Flour

Very common utilization of millet has been as an ingredient to multigrain flour. Multigrain flour used domestically pan India is the widest adoption of millet flour. It is used rampantly with wheat or maize flour. Adding an optimum amount of millet flour in base flour can be used for pasta, biscuits, bread, noodles, etc., would certainly help the food industry to make use of millet flour on par with wheat and cornflour (Chakraborty et al. 2016). The resultant products shall possess all the nutritional advantage of millets.

18.3.8 Probiotic Millet Foods

This is a contemporary approach of improving the quality of micronutrients and dietary fibers in ready-to-eat products. The level of phytate in fermented millet products can be managed by the addition of selected lactic acid cultures belonging to *lactobacilli* family which exhibit phytase activity. This has lead to the introduction of innovative products which has nutritional benefits of millet and health-promoting virtues of lactic acid bacteria, one such product is Bajra *lassi*. Owing to their reported beneficial attributes, probiotic cultures are widely used in making various functional foods for young and old. Probiotics with millets also succeed in enhancing the bioavailability of minerals as well as acceptance by the consumers. The lactic acid beverages fermented with selected probiotic strains will have better acidification and cell count containing right amount of probiotic organisms. The serum cholesterol level is reduced by probiotic strains of *Lactobacillus acidophilus*. *Lactobacillus acidophilus* and *Bifidobacteria* synthesize niacin, thiamine, folic acid, riboflavin, vitamin K, and pyridoxine. Millets act as food substrate for probiotics and improve flavor, texture, and overall acceptability of the product.

18.4 Millet and Livelihood Promotion

Since millet is a mainstay of farming activity of the people living in the forest areas, mostly indigenous tribes, any technological intervention in changing the order of processing of the millet crops will have a direct or indirect effect on the livelihood of the tribal agrarian community (Ambrose et al. 2017).

18.4.1 Tribal Population in India

As has been mentioned earlier that millets could not keep pace with the growth of the popularity of wheat and rice, it was relegated to the fringe farmers or the tribal population in various parts of India. India has a widespread population of tribes. A generous estimate puts the tribal population of India to be four times the population of Australia. In India, the total geographical area occupied by forest, hills, inaccessible terrain is about 15% of the total landmass, the tribal population of the country resides mainly in these locations. Most of the communities have been living in isolation for centuries; their life has remained untouched by the modern times. As a result of this seclusion, the socioeconomic growth indicators are worrisome and alarming.

18.4.2 Tribal Women

The tribal population is stretched between life and livelihood; the women are no exception. In tribal communities, the role of women is substantially crucial. The role is not limited to cooking and household work alone, the women carry out the role of gatherers by visiting the forest for collection of minor forest produce, firewood, etc. Apart from this, women have to carry out drudgery-prone agricultural activities in the field and also participate in postharvest operations like cleaning, grading, milling, etc. The natural resources present in the local environment have hugely affected the lifestyle of tribal women and have kept them occupied. Millet cleaning and milling is carried out by women folks. A woman has to hand pound for more than 4 h to get enough dehulled millet grain for a family of four. Intervention in terms of mechanization of agriculture and postharvest operations shall provide a big relief to the farm women engaged in the agricultural work, especially in the area of millet cultivation, harvest, and value addition. Evaluation of the impact of developmental planning should be conclusive and decipherable. The process should be such that along with the socioeconomic system, the entire ecosystem should change inclusively for the tribal farm women.

Women should be engaged in activities of economic importance, involving self/wage-employment by means of the products that they are harnessing from the environment. The engagement should be such that it generated adequate resources to meet the household as well as personal requirements in a dignified way, and most importantly the process should be sustainably repeatable. Only a repeatable process will make the engagement meaningful and a way of life. The engagement should not be just money-making in nature but should add value and meaning to life.

18.4.3 Millet Processing as Livelihood Intervention

Forest-dwelling tribal farmers traditionally used to practice shifting cultivation. However, this practice gradually came to an end during the early twentieth century. Now, the tribal people took to settled agriculture mainly on the uplands. Here also they persisted with cultivating multiple types of crops. This ensures nutritional and social security at the cost of total yield in comparison with that of modern monocrop agricultural practices. Tribal people can comfortably be labeled as the earliest "social ecologists" which eulogized subsistence and conservation ethics.

Processing of the millets that the tribals are already growing in their land is a stand-out option for livelihood generation as well as limiting the drudgery of farm women to scalable levels. Nutritious millets have been ignored in all respects, less demand in the market resulted in no progress in the grain processing technology. India is sitting on a millet production of two million, yet there is absence of efficient technology for processing these grains even at village level. Dearth of up to the mark technology has compelled dependence on traditional methods of millet processing, these approaches are tedious, time-consuming, and drudgery prone for the women folk.

This has contributed to the decrease in popularity of the millets even among its traditional consumers. All the minor millet grains are small, but they vary in size within a particular type. This makes the removal of seed coat from their surface a tough task, the abrasive force requirement also varies. However, finger millet has a hard seed coat and they can be ground directly in a pulverizer and consumed as flour. All the other grains are to be first dehulled and then polished to make them fit for consumption. Women use a stone single running attrition type of mill to grind finger millet manually. This involves a lot of drudgeries. Milling of other minor millets is far tougher and require much more effort as well. Millets, other than finger millets are first beaten repeatedly in mortar with pestle to remove the husk. One has to work with a pestle for almost an hour to process 2 kg of grain. The magnitude of drudgery involved can be well understood and scaled. The onus of all these activities lies with the women of the tribal family. With changing time, the mindset of these women have changed, no one wants to take up this tortuous work. This has discouraged the use of these millets in household consumption, particularly when there is access to alternative grains like rice, wheat, or sorghum. Hence, it is imperative that some mechanized solution for milling and postharvest processing of these grains has to be commissioned at the earliest before these grains go further into oblivion. An important primary step is to promote and popularize household consumption and building a grain-based value chain.

18.5 Value Chain for Millets

Millets find mention in ancient Indian texts, folklore, and fables. As of now, millets are grown from Uttarakhand in the North to hill and hillocks of Tamil Nadu in the South, from semiarid Gujarat in the West to Arunachal Pradesh in the Northeast. The total production of millets was 10.28 million tons with average productivity of 1.16 t/ha.

In the times of jargons like, "doubling farmers' income" food processing is a "sun-rise industry," the connotations that these calls emanate indirectly point towards value addition of agricultural produce. Value addition involves a collection of activities aimed at increasing the monetary value of raw material, at times much more than ten times the cost. The value addition interventions can be introduced to a particular commodity only and only after understanding the value chain of the produce. Millet agriculture, especially minor millets, does not include major cost inputs, the seeds are broadcasted in the rainy season and the crop matures after 100–120 days. Minimum intervention is required in terms of agronomical practices and so on. Thus, any small amount of work done, right from threshing and obtaining the grains to cleaning, possible dehulling, and flour making and subsequent packaging; all of this results in adding value to the native raw material.

An author perceived value chain for millets is depicted in Fig. 18.1, where it can be seen that at every stage of unit operation there is are some valuable products that can be directly obtained and there can be a set of derived products as well. Overall, there is the creation or generation of enough value-added products all across the value chain.

18.5.1 Basic Unit Operations

All the operations that include harvesting the crops from the field to extracting the grains from the same and optional storage, all comes under this set of operations. The crops are traditionally harvested manually using a serrated sickle and are carried as a head load to the threshing units. Since the crops grow unattended for most part of the growing period and use the soil moisture alone for its growth, there is always a sizeable amount of weeds and trash that follow the harvested crops. These trash are removed before the crops are sent for threshing. Threshers (CAIE Multi Millet Thresher) are available and have good efficiency in handling almost all the types of minor millets crops. Threshed grains are either stored in village storage structures or as per tribal culture, 12 sacks are made for domestic consumption spread over 12 months of the year and the remaining grain is sold at the village mart or is bartered for other household goods.

Periodically, the grains are dehulled as per the domestic consumption needs. Dehulling is carried out by either hand pounding operation or can be carried out by millet mill (CIAE Millet Mill). The mechanical mill has a good efficiency for dehulling all the minor millets grains. Tribal farmers at times opt for dehulling the grains by mechanical mill and then getting the millet rice sold at the nearest city market through various marketing channels. Finally, in this series of unit operations, the end product is the dehulled millet grains. The biomass generated during these sets of operations can be utilized as animal feed or can also be further treated and processed to generate energy.

18.5.2 Secondary Unit Operations

These operations can be divided into two parts depending upon the raw material to be used. Threshed grain and dehulled grains both can be made to undergo a select set

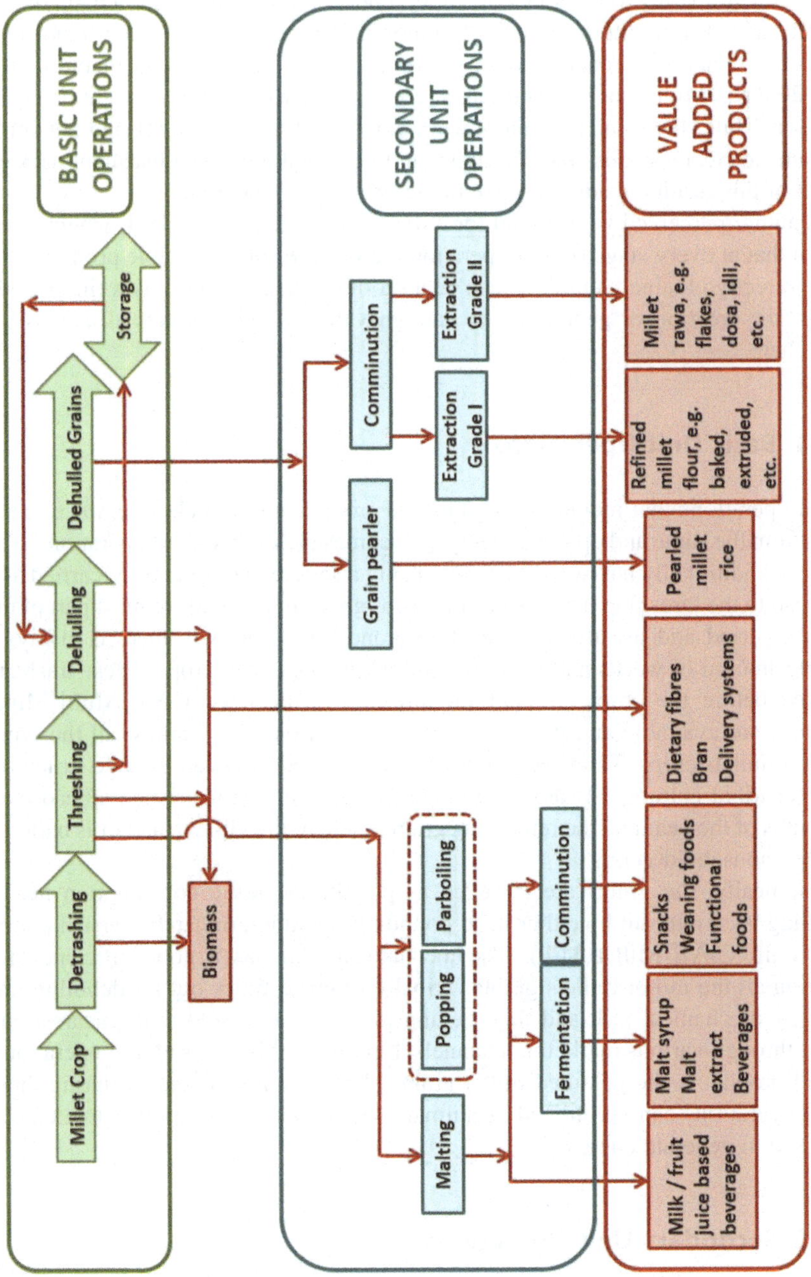

Fig. 18.1 Value chain for minor millets

of operations to get various end products. The threshed grains, i.e., the grains with the hull can be used for malting, popping, and parboiling. All these unit operations shall take care of the anti-nutritional factors of the grain while increasing the bioavailability of most of the nutritionally important elements present in the grains. Malted grains can be further processed for fermentation and comminution to get various value-added products. Dehulled millet grain can be either pearled to get a polished "Millet Rice" or the same rice can be further comminuted to get millet flour. The level of extraction can be changed to get flour of different consistencies, which may find varied end uses.

18.5.3 Value-Added Products

There is a wide scope for the value addition of millets to make a variety of products. Besides the millet rice and millet flour being the basic raw material, the products can be broadly classified as infant foods to geriatric foods and from functional foods to ready-to-eat snacks (Chakraborty et al. 2009). Also, the by-products can be utilized to make further products as standalone or can be complemented with some other ingestible item for making a product of multiplied value. These products invariably will be taken up in the other chapters of this book.

18.6 Entrepreneurship in Millet Processing

Rural enterprises can be set up with millets being the key raw material. There are a lot of promotional schemes from the governments of the developing countries to promote utilization of the indigenous crops by suitable intervention of technology and entrepreneurship. An enterprising rural youth can set up a "Millet Processing Center" at the production catchment of these crops to create employment and wealth for self in particular and for the community as a whole.

18.6.1 Millet Processing Center

A millet processing center shall provide an end-to-end solution to the farmers in terms of processing of the produce. A farmer shall have the option to take the threshed or the cleaned dehulled grain back home. An option should also be there to get the pearled millet rice or millet *rawa* (Hindi name for coarse ground cereal grain) or millet flour from the raw material. There will also be an arrangement for packaging of the product in desired quantity. The list of various machines that find their place in the millet processing center is given in Table 18.2.

Table 18.2 Machines installed at the Millet Processing Center

S. No.	Machinery	Utility
1	Millet thresher	Threshing of the harvested millet crop with more than 85% efficiency thus obtaining millet grains.
2	De-stoner cum grader	Cleaning all the trash and stone accompanying the millets from the field/threshing yard. There is an arrangement of replacement of sieves for different millets so as to increase the effectiveness of getting clean grains. The cleanliness of the grains shall decide the success of any further process with the grains.
3	Millet mill	Efficient dehusking of the millet grains to get millet rice as the end product. This machine works on the principle of attrition with two stone disks separated by an adjustable gap. A cyclone is also provided to separate the husk and broken from whole grains.
4	Grain polisher	Pearling of millet grains, i.e., to polish the dehusked grain. A sizeable part of the millet rice processed in the MPC is packaged and sold in the open market, polished grains always appeal to the customer more.
5	Pulverizer	This machine provides an option to the stakeholders to make flour out of the millets and maize brought to the MPC by local tribes. There is a "tribal bakery" within the campus of the MPC. It is a prevalent practice in this bakery of using millet and maize flour in making baked goods.
6	Flour sifter	Separation of coarse millet flour from fine millet flour is accomplished by this machine.
7	Packaging machinery	A bag sealer and a paddle-operated packet sealer for packing/packaging of the end product.

18.6.2 Performance Indices of Millet Processing

It is important to possess basic knowledge about the performance indices of the various machines used in the MPC. A tangible solution to monitor the performance of MPC can be accomplished with the help of this type of information. Some of the indices representing the performance of the MPC are:

Cleaning efficiency (η_c, %): Represented the efficiency of the destonner cum cleaner working at MPC. The ratio of cleaned grains of millet (W_c, kg) grain to the total weight (T_w, kg) of threshed millet grain and is expressed in percentage as below (Gbabo et al. 2015):

$$\eta_c = \frac{W_c}{T_w} \times 100$$

Millet milling efficiency (η_m, %): Milling is also called dehulling efficiency leads us to estimate the efficiency of the machine in removing the hull (husk + bran) from the grain surface. It can be expressed in percent, defined by the fraction of weight of milled grains (W_m, kg) to total weight (W, kg) of the raw grains (Bisen et al. 2014):

$$\eta_{m} = \frac{W_{m}}{W} \times 100$$

Polishing efficiency (η_p, %): Polishing efficiency of pearler can be explained as the ratio of weight of polished grains (W_p, kg) to total weight of milled grain (W_m, kg) of the grains:

$$\eta_{p} = \frac{W_{p}}{W_{m}} \times 100$$

Machine capacity (M_c, q/h): This machine parameter is represented as the quantity (Q, q) of millet processed per unit time (T, h). The mathematical expression is,

$$M_{c} = \frac{Q}{T}$$

Cost economics of MPC: The economics of running the MPC was also worked out for evaluating the profitability of this venture in terms of,

$$\text{Total cost, Rs} = \text{Fixed cost, Rs} + \text{Variable cost, Rs}$$

$$\text{Cost of operation, Rs} = \text{Total cost per kg}$$

$$\text{Total profit, Rs} = \text{Total revenue, Rs} - \text{Total cost of operation, Rs}$$

$$\text{Benefit Cost Ratio (BCR)} = \frac{\text{Total profit}}{\text{Total cost}}$$

18.7 Conclusion

It is well understood that the millets are highly nutritious, they have the potential to support the livelihood of the tribal population living in the fringes of the society, and that one can enjoy the charm of entrepreneurship by processing millet crops for various value-added products. Having said that, one needs to understand that overcoming the relegation of this crop can only be possible by a strong government hand, like the inclusion of millets in the well-oiled network of public distribution system. A subsequent rise in demand for millets shall surely promote the cultivation and prosperity of the farmers.

Value-added product should find a window of opportunity in terms of sale through all the government tourism offices, something like promoting millet tourism giving a glimpse of tribal life and indigenous agrarian practices can also go a long way in the popularization of millets. A key issue of concern in regard to value-added products is the absence of quality standards for millet-based products. Bureau of

Indian Standards should in all earnest create a sort of task force to formulate the standards for value-added millet-based products.

Sloganeering and advertisements with an endorsement from key scientific bodies like Indian Council of Medical Research (ICMR), New Delhi and National Institute of Nutrition (NIN), Hyderabad for the popularization of the millet-based products also needs to be taken up. The government should also strive to provide some additional business grants to entrepreneurs, apart from the one received by the state-of-the-art proposals, and allow them to enjoy government-assisted marketing. A well-planned concentrated effort is required by all the stakeholders to make a noticeable mark in the growth and prosperity of the millets and its inheritors in India and in the World.

References

Ambrose DCP, Annamalai SJK, Naik R, Dubey AK, Chakraborty SK (2017) Performance studies on millet processing machinery for tribal livelihood promotion. J Appl Nat Sci 9:1796–1800

Anonymous (2018) Ministry of Agriculture & Farmers Welfare, Government of India. http://www.nutricereals.dac.gov.in/

Antony U, Sripriya G, Chandra TS (1996) Effect of fermentation on the primary nutrients in finger millet (*Eleusine coracana*). J Agric Food Chem 44:2616–2618

Asharani VT, Jayadeep A, Malleshi NG (2010) Natural antioxidants in edible flours of selected small millets. Int J Food Prop 13:41–50

Bisen RM, Ramawat RB, Khope PB, Choudhary PS (2014) Design, development and performance evaluations of mini rice mill for domestic purpose. Int J Eng Res Technol 3:1291–1294

Chakraborty SK, Singh DS, Kumbhar BK, Singh D (2009) Process parameter optimization for textural properties of ready-to-eat extruded snack food from millet and pulse-brokens blends. J Texture Stud 40:710–726

Chakraborty SK, Singh DS, Kumbhar BK, Chakraborty S (2011) Process optimization with respect to the expansion ratios of millet and legume (pigeon pea) based extruded snacks. J Food Process Eng 34:777–791

Chakraborty SK, Singh DS, Kumbhar BK (2014) Influence of extrusion conditions on the colour of millet-legume extrudates using digital imagery. Irish J Agric Food Res 53:65–74

Chakraborty SK, Gupta S, Kotwaliwale N (2016) Quality characteristics of gluten free bread from barnyard millet - soy flour blends. J Food Sci Technol 53:4308–4315

Chakraborty SK, Kotwaliwale N, Navale SA (2018) Rheological characterization of gluten free millet flour dough. J Food Meas Charact 12:1195–1202

Chakraborty SK, Kotwaliwale N, Navale SA (2019) Selection and incorporation of hydrocolloid for gluten-free leavened millet breads and optimization of the baking process thereof. LWT-Food Sci Technol. https://doi.org/10.1016/j.lwt.2019.108878

Chethan S, Malleshi NG (2007) Finger millet polyphenols: characterization and their nutraceutical potential. Am J Food Technol 2:582–592

Dayakar Rao B, Vishala AD, Arlene Christina GD, Tonapi VA (2016) Technologies of millet value added products. Centre of Excellence on Sorghum, ICAR-IIMR, Hyderabad, India, p 48

Gbabo A, Ndagi B, Al Hassan Mohammed K, Lukman A (2015) Development and testing of a rice destoning machine. Int J Eng Res Sci Technol 4:134–141

Sahay KM, Singh KK (1996) Unit operations of agricultural processing. Vikas Publishing House, New Delhi

Sarita ES, Singh E (2016) Potential of millets: nutrients composition and health benefits. Int J Innov Res Sci Eng Technol 5:46–50

Sharma N, Niranjan K (2018) Foxtail millet: properties, processing, health benefits, and uses. Food Rev Int 34:329–363

Quality Management System in Millet and Sorghum

19

Manoj Kumar Tripathi, Rajpal S. Jadam, and Anil Kumar

Abstract

The quality of all types of food and their management are playing a major role in the agriculture and food processing sectors. The viewpoint of quality control and related parameters mainly affect and control quality promise. To enhance utilization, sorghum and millet grains need to be improved beyond the elimination of deterioration caused by diseases and insects. In previous years, numerous instances have arisen in the agricultural-food segments, for example, an incidence of foodborne illnesses and the manufacture of disease-risk food products. Sorghum and millets are mainly used in development of several food products and feed in a traditional manner. In current time these are regularly utilized in development of food products, exclusively healthy food and gluten-free food products. These extensive use directives preferred quality factors for quality evolution using selection of many biochemical approaches. Some common factors which are commonly concerned are physical characteristics, biochemical composition, and food safety. Systems for quality control in millets and sorghum from many authorized forums and their major aims are to simplify occupation in the grains. Many multidisciplinary methodologies are required presently for supporting the quality control and regulations in development of millets and sorghum-based food products and for value chain.

M. K. Tripathi (✉) · R. S. Jadam
ICAR-Central Institute of Agricultural Engineering, Bhopal, Madhya Pradesh, India
e-mail: manoj.tripathi1@icar.gov.in

A. Kumar
Directorate of Education, Rani Lakshmi Bai Central Agricultural University, Jhansi, Uttar Pradesh, India

A. Kumar et al. (eds.), *Millets and Millet Technology*,
https://doi.org/10.1007/978-981-16-0676-2_19

Keywords

Quality · Food · Safety · Fortificant · Mycotoxin · Agrifood chain

19.1 Introduction

An emergent need to care for millet research in the regions which are major sources of millet production for nutrition and health. In our country millet and sorghum farming is encouraged for diet, nutraceuticals, and health respects. The farming of millets and sorghum are favored by dry and high temperatures. These crops are very useful for poor farmers who are stressed with unsuitable soil conditions and limited farming contributions (Padulosi et al. 2015). The major issues which affect the decreasing movement of millet and sorghum production are little crop yield, extraordinary requirement of workers, challenging postharvest procedures, and absence of suitable rates of crop produce (King 2017). Due to favorable public delivery arrangement wheat and rice grains are easily available in the market which is the main causes of shifting of foodstuff intake in millet farming areas. Finger millet is an exceptional case among other millets where more technologies are available although drudgery of hulling is a major issue for farmers and producers. Some other issues related to millets production and consumption are investment during product formulation, marketing, and low financial status of societies for their consumption (Mondal et al. 2016). The reduced accessibility of millet and sorghum products in locally available marketplaces with extraordinary costs is mainly controlling millet-based product promotion. In the Current scenario, processed food products are one of the key factors which are playing an important role in promoting products and encouraging consumer requests for millets. For market value and product demand, the insurance of quality parameters of products is essential. Development of food standards and quality insurance is needed to fulfill the greater quality products and user demand. Suitable knowledge for enhancing product stability of millet and millet-based products is required which maintains the value and nutritional status of the developed food products. Different food standards are required immediately which mainly emphasize on product uniqueness, biochemical parameters status, nutritive value, and all food and related safety parameters which also differentiate the products from others.

In most of the developing nations, the process of urbanization is very fast and the lifestyle and food profile of the urban middle class is affected. Middle-class society appeals for value diets which are mainly developed from some traditional crops like millets and sorghum. The price of these adapted cereals is very high in the majority of rising nations. Therefore, agriculture businessmen are progressively replacing local available millet and sorghum grains for food producers (Taylor et al. 2006). There are several other factors than the price of food which play a major role in driving this change. Agricultural products produced by local growers are now highly acceptable because these products are very helpful for enhancing the financial status of developing countries.

This condition motivates private sector and government organization bonding and encourages the industrialized use of local agricultural products like millets and sorghums (Mackintosh and Higgins 2004). Food uses of sorghum and millet have been limited only to traditional consumers and areas of their cultivation. Processing of millet grains by means of traditional as well as contemporary methods for the development of value-added and suitability products would undoubtedly expand their food practices. Their utilization for the preparation of several millet ready-to-use (RTU) and ready-to-cook (RTC) products would be very useful in developing the consumption of millets in non-millet consumers and thus ensuring nutritional security. The grains of millet and sorghum have many health-related properties and have a high content of antioxidant compounds, phenolic compounds (Dykes and Rooney 2007) with gluten-free components (Ciacci et al. 2007). Demand for these quality-based products will be high in the future scenario. Therefore, the demands and intake of these nutrition and nutraceutical-rich foods are high in many urbanized areas and countries.

19.2 Brief of Sorghum and Millets

Millets and sorghum crops are tropical cereals. They are associated with the family *Paniceae*, finger millet and teff are associated with the family Eragrostideae. Table 19.1 indicates the brief of major farming varieties of sorghum and millets. Sorghum crop is an exclusively drought-tolerant crop which needs at least periodic rainfall of 400 mm (Dendy 1995). The rainfall requirement of millets mainly pearl

Table 19.1 Major millet and sorghum species (FAOSTAT 2008)

Commonly used name	Other public names
Finger millet *Eleucine coracana*	Ragi, Wimbi
Teff *Eragrostis tef*	Teff
White fonio *Digitaria exilis*	Fonio, Acha
Black fonio *Digitaria iburua*	Black acha
Barnyard millet *Echinochloa esculenta*	White and black millet
Brown-corn millet *Panicum miliaceum*	Common millet
Foxtail millet *Setaria italica*	German millet
Pearl millet *Pennisetum glaucum*	Bajra
Sorghum *Sorghum bicolor*	Jowar

millet is slightly lower than 300 mm. Sorghum production occupies the fifth most significant cereal next to wheat, maize, rice, and barley crops. The Farming performs for millets and sorghum is extremely inconstant. Sorghum is regularly farmed with some very high-yielding cultivars and programmed farming as in the case of some important nations, for example, the United State of America, Brazil, Mexico, and South Africa. In most developing countries millets and sorghum cropping is performed using non-mechanized and manual methods. Most of the millets are mainly farmed using non-mechanized at limited level (Food and Agriculture Organization 2008).

19.3 Grain Chemistry

Sorghum and most millets are kept in caryopses which have single-seeded fruit and fruit coats covering the seed. In contrast to millet, sorghum grains are fairly larger in grain size (Serna-Saldivar and Rooney 1995). In both cases, kernel morphology is comparable with major portions (Abdelrahman et al. 1984). Presence of some anti-nutritional compounds as tannins and other biochemical components in sorghum and finger millet are reported by Siwela et al. (2007). It was established that tannins in case of millets specifically in finger millet are placed as a thick layer in the testa (Earp and Rooney 1982). In seed, the aleurone layer contains several protein bodies, different enzymes, minerals, and some oil (Serna-Saldivar and Rooney 1995; McDonough et al. 1986). The endosperm layer contains some types of phenolic and due to the presence of these phenolic compounds they fluoresce. The germ layer is also rich in protein, oil, enzymes, and some micronutrients as minerals.

Sorghum, Pearl, and Proso millets are rich sources of protein and fat (Table 19.2). The major and most rich biochemical component is starch molecules (Serna-Saldivar and Rooney 1995). In the case of millets and sorghum, prolamin proteins are a key protein fraction which is located in the endosperm. Glutelins are the second key protein fractions and albumin, globulin, and glutelin are some minor protein fractions found in the protein matrix (Taylor and Schüssler 1986). Germ layer

Table 19.2 Major biochemical component of some millet and sorghum (Mohapatra et al. 2019)

Crops	Protein content (%)	Fiber content (crude) (%)	Ash content (%)
Millets			
Pearl	14.5	2.0	2.0
Finger	8.0	3.0	3.0
Proso	13.4	6.3	4.2
Fonio	8.7	8.0	3.8
Teff	10.9	2.4	2.2
Sorghum			
Sorghum	11.3	2.7	3.1

contains the majority of the lipid portion in case of millets and sorghum and this layer mainly goes outside during the decortication process.

The mineral components are mainly concentrated in pericarp, aleurone, and germ and therefore during refining, the probability of removal of minerals is high. Finger millet contains higher level of calcium compared to other millets and sorghum grains (Serna-Saldivar and Rooney 1995). Some important anti-nutritional compounds as phenolic and phytates are found in millets and sorghum and it is reported that they affect the enzymatic activity (Serna-Saldivar and Rooney 1995; Awika and Rooney 2004). Tannins (condensed) are rich in a majority of millets and sorghum seeds. It is also found in some research finding that condensed tannins are only concentrated in testa. Some reports also stated earlier that finger millet contains a good amount of condensed tannins which is located in the testa (Siwela et al. 2007). Phytate compounds are located in bran, aleurone, and germ therefore in the decortication process minimizes more than 50% phytic acid (Doherty et al. 1982).

19.4 Emerging Uses of Millets and Sorghum

The initial uses of sorghum are really limited currently but millets have the potential for many uses. The key evolving marketplace for sorghum is some drink industries. Sorghum drink industries are mainly growing very fast in some regions of Africa. Such types of developments are very helpful to enhance the economic conditions of a country. In the case of some developed countries there is a minor but rising business opportunity for development of gluten-free products such as bakery products, breads, and muffins specifically from sorghum, for example, in the USA and Australia (Taylor and Emmambux 2008). Several sorghum- and millet-based food options are present which include cookies, pasta, extruded snacks, and breakfast products (Awika and Rooney 2004). The fortification of sorghum and sorghum-based brans in the above food products can enhance the intensities of phyto ingredients which are reported for health benefits and ultimately development of healthy food (Rooney and Awika 2005). Biologically active phytochemical extraction from millets and sorghums is the best option from economic as well as food industries point of view (Awika and Rooney 2004).

Sorghums (pigmented) reported as noble anthocyanins source which is comparatively more stable than any plant fruit anthocyanins in terms of pH (Awika et al. 2004) and rich in several function properties for food uses (Awika and Rooney 2004). Millet and sorghum-derived brans are very important sources of phenolic and could be used for several food purposes as natural antioxidants, stabilizer during fat oxidation, and finally improve the quality of food (Sikwese and Duodu 2007). Some primary long-chain alcohol known as policosanols are rich in sorghum bran and have several health benefits (Hwang et al. 2004). In the current scenario, sorghum bran can be used as a good source of policosanols for industry use and also a promising cost-effective source for many therapeutic uses (Table 19.3).

Table 19.3 Recent application of millets and sorghum in food industries

Present uses	Emerging uses
Food stuffs	
Asian and African countries Flours, meals	*USA* Gluten-free baked goods
Asian and African countries Dumplings, porridges, and gruels	*USA* Ready-to-eat breakfast cereals
Asian and African countries Rice	*Japan* Noodles
Central America Flatbreads	*African countries* Instant porridges
African countries Couscous	*South Africa* Instant infant foods
Africa and India Malt	*African countries* Expanded snack foods
Beverages	
Asian, African, and European Nonalcoholic drinks (fermented)	*USA and Australia* Stouts, beers
African and Asian Cloudy and opaque beers	*China* Spirits
Industrial applications	
USA and Africa Starch	*USA* Bioethanol from starch

19.5 Limitations for Food Uses

Consumption of human food uses of millet and sorghum grain as human food is found to reduce from several years due to some important reasons. These crops are considered food for low-income families and avoided by middle and upper-class consumers. Many industrialized countries export low-priced wheat flour which is reported to be more cost-effective and economically suitable than millets and sorghum-based products. Majority of buyers choose those agriculture products which have favorable for supply convenience, color, sensory and with high shelf life. Millet and sorghum food items are fulfilling the above conditions. Some other problems associated with these grains are infection of fungus and head bugs in farming zones. In many reports, the suspected reduced nutritional value of sorghum and millets are due to the presence of tannins and poor protein digestibility properties. Some important nutritionists believe that all sorghum varieties contain tannins which limit them for food uses. Environmental circumstances during grain development critically mark the appearance of the grain. Bugs are a major problem for grain and the bugs injured parts of crops are very appropriate for fungi infection. Presence of warm and moist situations during the growth of crops adversely affects the value of crops and grains. Fungal infection causes discoloration of the grains which are responsible for major rejection of grains in market places.

19.6 Strategies for Overcoming Constraints

Development of noble food based on millets and sorghums must require remarkable and satisfactory characteristics. With the help of several food-based approaches we can achieve acceptable business and value-added products.

19.6.1 Nutritional Value

The nutritive value and functionality of the components (protein, oil, and starch) of sorghum grain should be improved, but not at the price of increased grain deterioration. A very few reports are available which indicate to improve the content or functionality of other components, e.g., lipids, fiber, minerals, and antioxidants. The composition of protein, starch, and other biochemical content are depending upon crop varieties and environmental conditions (Butler 1990).

The quality of protein plays a critical role in the diets of poor countries because their major diets are based on cereals (FAO and WHO 1991). Use of sorghum along with some legumes can help to minimize the problem of amino acid deficiencies. Several millet cultivars are rich in albumin, globulin, and glutelin with a high concentration of lysine and other essential amino acids. Nutritionally, tannin sorghum has lower protein quality than do their regular counterparts due to reduced protein utilization. In monogastric animals high-tannin sorghum consistently reduces protein retention and growth by 5–10% (Butler and Rogler 1992). Some varieties of sorghum with improved protein digestibility and their impact on food texture were reported by Chandrashekar and Mazhar (1999).

The sorghum kafirins have been the subject of much research and appear to be related to endosperm hardness and decreased protein digestibility (Chung et al. 1998). The amounts of α- and γ-kafirins and of protein bodies were greater in vitreous than in floury endosperm areas (Doherty et al. 1987) and in hard relative to soft endosperm sorghums. The floury endosperm contains more albumins and globulins, but glutelins were evenly distributed throughout the endosperm. Kafirin (El Nour et al. 1998) produced by the chemically mutated, high-lysine sorghum had increased digestibility; however, these traits appear to be recessive and the grain is susceptible to molds. Considerable research still is needed to produce sorghums with an improved amino acid content, protein digestibility, harder endosperm texture, and grain mold resistance, since increased digestibility almost always increases the susceptibility of the grain to post maturity molds and deterioration. Improved protein digestibility in sorghum, with easier-to-digest kafirins, is caused by more invaginations in anatomical regions of grains. Since more amino acids are hydrolyzed and absorbed from kafirins, the lysine deficiency is greatest when less kafirin is hydrolyzed. Increased digestibility of kafirins increases the change in sorghum amino acid (essential) as kafirin is deficient in essential amino acids (El Nour et al. 1998).

19.6.2 High-Value Food Products

Major approach in the development of suitable and stable food products from millet and sorghum is the formulation of value-added food products with low cost and easy processing methodology in comparison to foods products and supplements available in the markets. Formulation must use the mentioned general steps: (1) recognize expensive foodstuffs plus place markets (2) use of cost-effective and reproducible methodologies, and (3) instruct growers and manufacturers.

19.6.3 Grain Quality

The major purpose of plant breeders of millet and sorghum is to target production, product features, grain biochemical quality, and resistance to infection against fungus and bugs. The grain yield and grain quality is a major consideration for breeders. Therefore, the main priority in this regard is to develop and improve the locally available varieties and might be used for making valuable food products. Some millet varieties are available with excellent milling characteristics with light color food products.

19.6.4 Tannins in Millet and Sorghum

Millet and sorghum have significant amounts of tannins (condensed). Catechins are major tannins in sorghum which is responsible for the reduction of feed proficiency and also the processing of the grain. All sorghum contains many phenolic substances which have properties of common phenolic. For analysis of condensed tannin, Vanillin-HCL method can be preferred for best results.

19.6.5 Grain Quality: Molds, Insects, and Weathering

Cultivation of white, tan plant sorghum is useful in reduction of problems of molds, weathering, and bugs faced by sorghum. This approach is found very effective in some regions of Africa. Presence of biochemical components as condensed tannins and some special plant parts of sorghum also favors resistance against molds and weathering (Waniska et al. 1992). Presences of several proteins with antimicrobial characteristics are reported in sorghums which causes more resistance against molds. Sorghum never develops aflatoxins prior to harvest but in the presence of atmosphere avoids aflatoxin development.

19.7 Major Processing Parameters and Valuation

19.7.1 Milling Process

Millets and sorghum are mill dried before development in the final products. In the milling process the grain becomes small and in fine form and bran is also removed which is outer covering of the grains. Chandrashekar and Mazhar (1999) also reported the biochemical and physiological role of starch and protein in the hardness of endosperm in many cereals such as sorghum and millet. It is also reported in many studies that the grain kernel density plays a major role in the hardness of grains which is mainly measured by defining test weight. The method of test weight is used for grading millet and sorghum in many countries (Gomez et al. 1997). Reichert et al. (1986a, b) developed a common method for the evaluation of sorghum and millets. Presence of higher corneous endosperm fraction shows harder endosperm (Rooney and Miller 1982).

19.7.2 Value of Porridge Development

Millets and sorghum are universally used for the preparation of gruels and porridge. As per organoleptic and sensory point of view, proper texture of the final product is necessary. The texture properties of above-formulated food products are affected by many factors such as the milling process, biochemical composition of grain, and particle size (Kebakile et al. 2008). The content of starch fraction with amylase and amylopectin ratio have a great influence on the chemical composition of grains and formulation of a variety of food products. In case of millet texture of the food product such as porridge is influenced by the quality and finally product value (Kebakile et al. 2008).

19.7.3 Value of Baked Food Development

The quality required for the production of baked food from millets and sorghum is reported in several studies. Some studies have mentioned that better baked products are prepared with floury and soft endosperm because hard flour has high water absorption than soft flour (Munck 1995). Soft endosperm flour is very much related to softness and better rolling capability (Yetneberk et al. 2004). Gritty texture products are formed with corneous endosperm in sorghum and millet which is not suitable for baked product development due to limited starch expansion (Ezeogu et al. 2008).

19.7.4 Malting and Brewing Quality

Malting of millets and sorghum for preparation of beer at commercial level is preferred in many countries. In some regions of Africa traditional and cloudy types of bears are developed with malting of millets and sorghums. In some cases, stout beers are produced commercially using malted sorghum. High germination of grain during the malting process is always favored for the preparation of good quality malt and further development of products from malted grains (Kent and Evers 1994). The major problem associated with millet and sorghums is their poor storage ability in high humidity and temperature.

19.7.5 Phytochemical Quality

Phytomolecules are not essential nutrients and play a major role in health and disease with disease protective functions. Presence of these phytochemicals in food is not compulsory for the purpose of nutrition and normal metabolism. Presence of several phenolic compounds and phytochemicals in several foods has been found to be very effective in several chronic diseases and also protect human cells. In case of millets and sorghum several phenolic compounds are reported with strong antioxidant activity and they may be categorized mostly into flavonoids compounds, phenolic acids, and condensed tannins (Hahn et al. 1984; Robbins 2003).

Condensed tannins found in millets and sorghum are known as proanthocyanidins (Butler 1982). Dykes and Rooney (2006) reported various methods for quantitative analysis of various phenolic compounds. The occurrence of condensed tannins which mainly affect the millet quality is determined by Chlorox bleach method (ICC 2008) which is a very cost-effective and easy method. In this method, black color pigments appear on the surface which is not found in case of non-tannin samples (Waniska et al. 1992). In some cases, this test also produced some false results with infected sorghums (Dykes and Rooney 2006). Recently this test is used frequently for finger millet (Siwela et al. 2007). Some other biochemical methods are also applied for analysis purposes such as Folin-Ciocalteu (ISO 1988) methods and precipitation of proteins with tannin (Hagerman and Butler 1978). Several enzymes have been used for assay procedures like alkaline phosphatase, amylase, trypsin, etc. (Schofield et al. 2001; Awika et al. 2004).

19.7.6 Anti-nutritional Quality

Presence of several anti-nutritional compounds in millets and sorghums affects several biological functions. The major anti-nutritional compounds are lectin, tannin, phytic acid, protease inhibitors, saponin, and several phenolic compounds. These anti-nutritional compounds can reduce the rate of mineral absorption and digestibility in humans as well as animals. Their presence in grains makes them unsuitable for food product development and consumption. Presence of these factors in food may cause negative effects on several organ functions. Reichert et al. (1986a, b) reported that most of the polyphenolic compounds and tannins are rich in bran portions. Presence of high-tannin level is correlated with low in vitro protein digestibility. In some millet, formation of insoluble complexes with oxalic acids and calcium molecules were reported which affect the biological availability of minerals and can cause mineral deficiency.

19.8 Quality Management Tools

Sorghum and millets are extensively used for poor people as well as for animals using traditional methods. But in recent times people are aware of the value of their health and importance of diet and increasingly they are involved in the production of several health-based food products. In Majority of countries, demands for healthy food and beverages are increasing. Extensive uses of phytochemicals containing crops are increased with directive preferred features and they are evaluated using selection of several approaches. The major quality concern factors are biochemical, physical, and food safety parameters. Many legal forums are established for management in sorghum and millets and their major aims are to simplify the procedure of quality management in the grains. Many multidisciplinary methodologies are required presently for supporting the qualities in the value chain. In case of managing, many numerical approaches must be needed for data analysis. Agreeing to the several experts, the following seven numerical approaches are formulated for quality examining and understanding mathematical documents and these are the tools; use of similarity drawing, interlink drawing, tree drawing, matrix drawing, prioritizing conditions, method choice program plan, and activity network diagram.

19.8.1 Total Quality Management

Customer benefits and linking every trade with product manufacturing is the primary goal of total quality management (TQM).

These are the important principles of TQM functions;

1. Include societies.
2. Procedures handle.
3. Accountability.
4. Everybody has a role.
5. Avoid complications immediately.
6. Prepare and deliver properly to customers.
7. Regular quality improvements.
8. Manage quality regularly.
9. Planning for quality improvement.
10. Use fault free standard.

19.8.2 Ten Steps to Managing the Quality

Maintenance is built following ten steps;

1. Follow novel planned thought
2. Recognize clients
3. Fixed client necessities
4. Think for anticipation
5. Diminish prolonged surplus
6. Follow constant enhancement policy
7. Organized policy for procedure enhancement
8. Decrease deviation
9. Use a stable approach
10. Relate all

19.8.3 Assurance of Food Safety and Food Quality

Assurance of food safety and food quality is major issue faced during the development of a system for the management of grains which aims to create confidence at the local level. In this perspective, some needful and essential quality parameters must be achieved. The important and major issues related to these quality parameters are explained in Table 19.4. Education for quality and safety of raw materials as well as final products must be required in the agricultural field and agriculture-based industries. The regulatory bodies Codex Alimentarius only allowed those food products which have no harmful and safety issues. The major risks related to fresh foodstuffs are biological risk, chemical risk, and physical risk. In agriculture and food industries, all have accountability for maintaining food product quality to avoid food contamination and deterioration at production, dealers, processors, and traders. In many countries, the related persons are not very aware of food quality and safety issues which reduce the progress in this area. Food safety issues always effect different food processing practices such as production, processing and handling, packaging, transport, and storage.

Some important ideas needed for these issues are:

Table 19.4 Agricultural food products quality issues

Nutritional properties	Serviceable properties	Organoleptic properties	Asepsis properties
• Supply of nutrients • Nutritional value label • Ingredients • Energy • Additives	• Expense • Packing utility • Customer demonstration service	• Aroma • Flavor • Texture • Others	• Elementary atmospheres• For harmful effects cortication • Insurance for Human health

- Appropriate agriculture assets warranting food safety and food quality.
- Establish links for educating quality and safety aspects.
- Establish a business chain for liability of causative harmless and healthy foods.

Policies for agro-industrial sector for maintaining product quality and safety is very important to establish a system for quality management and they include the following;

1. Execution and maintenance in the production chain.
2. The application of HACCP system principles for quality and safety. The HACCP system is a plan for the identification and prevention of dangers for particular food types and process technology. In this process identify the required critical points and conditions for minimization of risk. HACCP is specifically planned for mainly one process for food or for processing technology.
3. The execution of a management system at a higher management level is for observing and assessment of the total scheme and submission. The aims of several private and public organizations for setting standards are to optimize the safety and quality parameters during preparation and processing.
4. Several public and private standards objective to settling optimal quality and safety at several stages of the production and processing chain.

19.8.4 Limitations in Food Chain

Tremendous agriculture and food are developing currently using sorghum and millets and everyone has a deficiency of continuous supply of suitable quality of grain for food development. Some important factors play a major role in supply, processing, and food developments as follows:

- Seed availability for framing.
- Crop production.
- Pre and postharvest system.
- Storing system for seeds and grains.
- Management during processing.
- Transport at different level.
- Products development and processing.
- Selling.

Availability of suitable quality grain and enough amounts for processing purposes are major limitations. Suitable and efficient methods of removal of impurities during several processing operations are also one of the challenges in this area. In case of millet and sorghum, variable size and quality grains are available in the markets which clearly mark the requirement of processors. Some millet and sorghum varieties are resistant and able to fight the infections of molds and bugs and ultimately have very good quality for processing and food development. Such varieties are mainly preferred for processing and food product development by food processors. It is an urgent requirement of methods to assess the quality parameters of these varieties to support food processors. For quick and accurate analysis some standard methods for quality determination are necessary. These facilities must be essential and approachable to processors and also for producers. The successful execution of this activity always requires good linkage and communication between all the players of this field.

19.9 Standards for Quality Management

Important standards are available for quality management of different food forms of millet and sorghums (Codex Alimentarius 1996). The major aims of these standards are to protect food quality and safety from several concern factors. The level of tannins in case of sorghum is a major issue and the content of tannin must be checked during product development (Table 19.5). The standards for proximate composition and crude fiber for millets and sorghum are also available.

Table 19.5 Standard of codex food safety for sorghum (CAC1996)

Some essential factors	Description
General	Grains • Harmless and appropriate for food purposes • Without any unusual aromas and alive pests • No hazard for consumption
Moistness	• Not more than 14.5–15.0%
Shortcomings	• Maximum defects up to 7–8.0% • Maximum up to 1.5–2.0% (inessential substance) and not more than 0.5% (inorganic matter)
Defects	• Maximum defects up to 8.0% • Not more than 2.0% extraneous matter
Contaminated kernels	• Absence of contaminated seeds • Avoid those seeds which are universally familiar for health problems
Heavy metals	• Absent of human hazardous heavy metals
Pesticide residues	• Not more than established maximum residue limits
Mycotoxins	• Not more than established mycotoxin residue limits
Hygiene	Good manufacturing practice • No intolerable substance • Follow the permissible limit of microbes • Follow the permissible limit of parasites • Follow the permissible limit of microbial origin toxic material

19.10 Summary

Increased grain yield must occur along with improved grain quality attributes. Some important attributes as traceability concern of food products are currently increasing a major necessity for food standards to justify food quality and food safety aspects. Traceability is a procedure to specify all the steps of the food chain from production to public distribution. Product traceability delivers about all the steps and responses in agricultural and agriculture-based industries and plays a major role in emergency situations for food industries. In case of all types of food supply chain information will be required for product tracing in management for quality and safety as also essential for development of HACCP system. The methodologies for traceability in the food system are similar and vary with cost and efficiency. The variation in the ease of pericarp removal and making flour with very fine texture need to be determined and exploited. Grains with a hard endosperm generally have less grain molding and yield finer flour. Increased lysine content and protein nutritional value in hard endosperm sorghums are needed to maintain parity with high-quality protein maize. The incorporation of the bioactivity of tannins and phenols should not diminish from food quality attributes while providing biochemical mechanisms for resistance against fungal and insect. Millets and Sorghum food chain and quality management systems must be developed in such a manner that increases their utilization at each level.

References

Abdelrahman A, Hoseney RC, Varriano-Marston E (1984) The proportions and chemical compositions of hand-dissected anatomical parts of pearl millet. J Cereal Sci 2:127–133

Awika JM, Rooney LW (2004) Sorghum phytochemicals and their impact on human health. Phytochemistry 65:1199–1221

Awika JM, Rooney LW, Waniska RD (2004) Properties of 3-deoxyanthocyanins from Sorghum. J Agric Food Chem 52:4388–4394

Butler LG (1982) Relative degree of polymerization of sorghum tannin during seed development and maturation. J Agric Food Chem 30:1090–1094

Butler LG (1990) The nature and amelioration of the antinutritional effects of tannins in sorghum grain. In: Mertz ET, Ejeta G, Rooney LW, Schaffert R, Yohe J (eds) Sorghum nutritional quality. INTSORMIL, Lincoln, NE, pp 191–205

Butler LG, Rogler JC (1992) Biochemical mechanisms of the antinutritional effects of tannins, in phenolic compounds in food and their effects on health. Am Chem Soc Symp Ser 506:289–304

Chandrashekar A, Mazhar H (1999) The biochemical basis and implications of grain strength in sorghum and maize. J Cereal Sci 30:193–207

Chung KT, Wong TY, Wei CI, Huang YW, Lin Y (1998) Tannins and human health: a review. Crit Rev Food Sci Nutr 6:421–464

Ciacci C, Maiuri L, Caporaso N, Bucci C, Del Giudice L, Massardo DR, Pontieri P, Di Fonzo N, Bean SR, Ioeger B, Londei M (2007) Celiac disease: in vitro and in vivo safety and palatability of wheat-free sorghum food products. Clin Nutr 26:799–805

Codex Alimentarius Commission (1996) Codex Alimentarius, vol 7. Cereals, pulses, legumes and derived products and vegetable proteins. Codex Standards: Whole and decorticated pearl millet grains; 170–1989 (Rev. 1–1995) Pearl Millet Flour; 172–1989 (Rev. 1–1995) Sorghum Grains;

173–1989 (Rev. 1–1995) SorghumFlour. Food and Agriculture Organization of the United Nations/World Health Organization, Rome
Dendy DAV (1995) Sorghum and millets: production and importance. In: Dendy DAV (ed) Sorghum and millets: chemistry and technology. St. Paul, MN, American Association of Cereal Chemists, pp 11–26
Doherty C, Faubion JM, Rooney LW (1982) Semiautomated determination of phytate in sorghum and sorghum products. Cereal Chem 59:373–377
Doherty CA, Waniska RD, Rooney LW, Earp CF, Poe JH (1987) Free phenolic compounds and tannins in Sorghum caryopsis and glumes during development. Cereal Chem 64:42–46
Dykes L, Rooney LW (2006) Sorghum and millet phenols and antioxidants. J Cereal Sci 44:236–251
Dykes L, Rooney LW (2007) Phenolic compounds in cereal grains and their health benefits. Cereal Foods World 52:105–111
Earp CF, Rooney LW (1982) Scanning electron microscopy of the pericarp and testa of several sorghum varieties. Food Microstruct 1:125–134
El Nour IN, Peruffo AD, Curioni A (1998) Characterisation of sorghum kafirins in relation to their cross-linking behaviour. J Cereal Sci 28:197–207
Ezeogu LI, Duodu KG, Emmambux MN, Taylor JRN (2008) Influence of cooking conditions on the protein matrix of sorghum and maize endosperm flours. Cereal Chem 85:397–402
FAO/WHO (1991) Protein quality: evaluation; FAO/WHO nutrition meetings, Report Series 51. Food Agriculture Organization/World Health Organization, Rome, Italy
Food and Agriculture Organization (FAO) (2008) FAOSTAT. http://faostat.fao.org. Accessed Dec 2008
Gomez MI, Obilana AB, Martin DF, Madzvamuse M, Monyo ES (1997) Manual of laboratory procedures for quality evaluation of sorghum and pearl millet. ICRISAT, Patancheru, India
Hagerman AE, Butler LG (1978) Protein precipitation method for the quantitative determination of tannins. J Agric Food Chem 26:809–812
Hahn DH, Rooney LW, Earp CF (1984) Tannins and phenols of sorghum. Cereal Foods World 29:776–779
Hwang KT, Weller CL, Cuppett SL, Hanna MA (2004) Policosanol contents and composition of grain sorghum kernels and dried distillers grains. Cereal Chem 81:345–349
International Association for Cereal Science and Technology (ICC) (2008) Draft standards: estimation of sorghum grain endosperm texture; Determination of germinative energy of sorghum grain; No. 177 Detection of tannin sorghum grain by the Bleach test. ICC, Vienna
International Organization for Standardization (ISO) (1988) ISO/DIS 9648 Sorghum– determination of tannin content. ISO, Paris
Kẹbakile MM, Rooney LW, de Kock HL, Taylor JR (2008) Effects of sorghum type and milling process on the sensory characteristics of sorghum porridge. Cereal Chem 85(3):307–313
Kent NL, Evers AD (1994) Kent's technology of cereals, 4th edn. Elsevier Science, Oxford, UK, p 219
King EDIO (2017) Impact of reduced drudgery of women in production and post-harvest processing of small millets, MSSRF working paper. No. 9. M.S. Swaminathan Research Foundation, Chennai
Mackintosh I, Higgins B (2004) The development of a sorghum-based lager beer in Uganda: a model of co-operation between industry and government in the development of local ingredients for the production of quality lager beer and consequential benefits for the parties involved. Asp Appl Biol 72:235–245
McDonough CM, Rooney LW, Earp CF (1986) Structural characteristics of *Eleusine coracana* (finger millet) using scanning electron and fluorescence microscopy. Food Microstruct 5:247–256
Mohapatra D, Patel AS, Kar A, Deshpande SS, Tripathi MK (2019) Effect of different processing conditions on proximate composition, anti-oxidants, anti-nutrients and amino acid profile of grain sorghum. Food Chem 271:129–135

Mondal A, King IO, Roy S, Priyam S, Meldrum G, Padulosi S, Mishra S (2016) Making millets matter in Madhya Pradesh. Farm Matters 06:10–13. http://www.agriculturesnetwork.org/farmingmatters

Munck L (1995) New milling technologies and products: whole plant utilization by milling and separation of the botanical and chemical components. In: Dendy DAV (ed) Sorghum and millets: chemistry and technology. St. Paul, MN, American Association of Cereal Chemists, pp 223–281

Padulosi S, Mal B, King OI, Gotor E (2015) Minor millets as a central element for sustainably enhanced incomes, empowerment and nutrition in rural India. Sustainability 7:8904–8933. https://doi.org/10.3390/su7078904. http://www.mdpi.com/20711050/7/7/8904

Reichert RD, Tyler RT, York AE, Schwab DJ, Tatarynovich JE, Mwasaru MA (1986a) Description of a production model of the tangential abrasive dehulling device and its application to breeders' samples. Cereal Chem 63:201–207

Reichert ED, Tatarynovich JE, Tyler RT (1986b) Abrasisive dehulling of quinoa (*Chenopodium quinoa*), effect on saponins content as determined by an adapted haemolytic assay. Cereal Chem 63:471

Robbins RJ (2003) Phenolic acids in foods: an overview of analytical methodology. J Agric Food Chem 51:2866–2887

Rooney LW, Awika JM (2005) Overview of products and health benefits of specialty sorghums. Cereal Foods World 50:109–115

Rooney LW, Miller FR (1982) Variation in the structure and kernel characteristics of sorghum. In: Mertin JV (ed) Proceedings of international symposium on Sorghum grain quality. ICRISAT, Patancheru, India, pp 143–162

Schofield P, Mbugua DM, Pell AN (2001) Analysis of condensed tannins – a review. Anim Feed Sci Technol 91:21–40

Serna-Saldivar S, Rooney LW (1995) Structure and chemistry of sorghum and millets. In: Dendy DAV (ed) Sorghum and millets: chemistry and technology. St Paul, MN, American Association of Cereal Chemists, pp 69–124

Sikwese FE, Duodu KG (2007) Antioxidant effect of a crude phenolic extract from Sorghum bran in sunflower oil in the presence of ferric ions. Food Chem 104:324–331

Siwela M, Taylor JRN, De Milliano WAJ, Duodu KG (2007) Occurrence and location of tannins in finger millet grain and antioxidant activity of different grain types. Cereal Chem 84:169–174

Taylor JRN, Emmambux MN (2008) Products containing other specialty grains: Sorghum, the millets and pseudocereals'. In: Hamaker BR (ed) Technology of functional cereal products. Woodhead Publishing, Abington, UK, pp 281–335

Taylor JRN, Schüssler L (1986) The protein compositions of the different anatomical parts of sorghum grain. J Cereal Sci 4:361–369

Taylor JRN, Schober TJ, Bean SR (2006) Novel food and non-food uses for sorghum and millets. J Cereal Sci 44:252–271

Waniska RD, Hugo LF, Rooney LW (1992) Practical methods to determine the presence of tannins in sorghum. J Appl Poultry Res 1:122–128

Yetneberk S, De Kock HL, Rooney LW, Taylor JRN (2004) Effects of sorghum cultivar on injera quality. Cereal Chem 81:314–327

Demand Creation Measures and Value Chain Model on Millets in India

20

B. Dayakar Rao, Venkatesh Bhat, T. Niranjan, M. Sujatha, and Vilas A. Tonapi

Abstract

ICAR—India Institute of Millets Research (IIMR) Hyderabad has been instrumental in developing value chain on Millets, resurrecting Millets from being becoming museum crops through demand creation measures. Various innovative interventions include, first being farm level backward integration converting domestic cultivation to commercial cultivation for meeting health conscious urban consumer needs. Second, novel intervention was diversification of processing technologies through developing and retrofitting more than 50 machineries suitability to millets from milling, to many semi-processed technologies such as baking, cold extrusion, hot extrusion, flaking, popping, puffing, fermentation, and malting was standardized. More than 200 recipes were developed to suit various regional pallet and circulated among users in local languages. The prototype of various machinery are now being upscaled. The shelf life of processed products were enhanced from less than a month to 8 months thus offering consumers wide and convenient food options to choose for enhancing millet consumption and paving way for commercialization of millet foods. Third, intervention is nutritional evaluation of millets and their value-added products and clinical trials were conducted to substantiate the health benefits through empirical data ably supported by NIN. Thus, the data generated helps in the labeling of the products for commercialization. The key to commercialize the millets lies in creating USP (unique selling proposition) for flagging it as healthy and convenient option. A Successful and sustainable and replicable value chain model has been furnished through innovations by bringing interventions at on-farm level, processing diversification, nutritional evaluation, market building,

B. Dayakar Rao (✉) · V. Bhat · T. Niranjan · M. Sujatha · V. A. Tonapi
ICAR-Indian Institute of Millets Research, Hyderabad, Andhra Pradesh, India
e-mail: dayakar@millets.res.in

A. Kumar et al. (eds.), *Millets and Millet Technology*,
https://doi.org/10.1007/978-981-16-0676-2_20

entrepreneurship development and creating awareness, assuring sustainable food and nutritional security which are beneficial to not only the poor dryland farmer but also health conscious urban consumers.

Keywords

Value chain · Millets · Demand · Consumer

20.1 Introduction

Millets are grown in semi-arid regions which are best known for low carbon and water footprints formed an ideal staple choice to millions offering household food security and their fodder to meet the own livestock requirements. Millet grains are nutritionally superior in protein, dietary fiber, micronutrients, and phytochemicals when compared to rice and wheat. Millets are classified as major and minor millets. Major millets are sorghum (*Sorghum bicolor*), pearl millet (*Pennisetum glaucum*), and finger millet (*Eleusine coracana*). Minor millets are foxtail millet (*Setaria italica*), little millet (*Panicum sumatrense*), kodo millet (*Paspalum scrobiculatum*), barnyard millet (*Echinochloa frumentacea*), proso millet (*Panicum miliaceum*), and browntop millet (*Brachiaria ramosa* (L.) Stapf). These crops form a vital component of the dry land ecosystems globally, supporting millions of poor and food-insecure people and contributing 10% of foodgrain basket (Saleh et al. 2013). Millets thrive well under climatic extremities and can survive in the hottest, driest climate with a low water footprint. Its ability to cope under climate change with huge nutritional benefits catches the attention of policymakers globally.

Millets are not only a staple food, but also serve as fodder for livestock and fuel purposes. They are grown in marginal lands; they provide risk aversion, to the poor dryland farmers. Millets are known to possess nutritional properties with varied health benefits. Millet grains contain higher protein, fiber, calcium, and minerals (Gopalan et al. 1989), which provides an added advantage, over wheat and rice. This nutritional superiority made millets termed as "Nutri-cereals" (notified in gazette by the Government of India) and "Miracle grains (*AdbhutAnaj*)." The present pandemic situation in the world made people think about immunity and people are more particular about their food habits, whose knowledge on per capita intake of nutrients is on the rise. Evidently, there is a great need to revive millets into this space as a staple diet, flagging their nutritional superiority compared with other cereals in terms of micronutrients, vitamins, dietary fiber, and antioxidants.

20.2 Definition of the Value Chain

A series of business activities involved, that accounts and presents the value generated in a product or service from a specific input to the final consumption is called a value chain. The value chain study from input to the final product through intermediates is called value chain analysis (McKay et al. 1997). Value chain analysis involves identifying and mapping the relationships of four types of features (1) the activities performed during each stage of processing/product flow; (2) the value of inputs, processing time, outputs, and eventual value addition; (3) the spatial relationships, such as distance and logistics of the activities; and (4) the structure of economic agents, such as suppliers, the producer, and the wholesaler (Rao 2019).

20.3 Gaps in the Value Chain

- The absence of dedicated supply chains from smallholders to market outlets, which requires the improvement of logistic infrastructure and supply chain management practices.
- Lack of backward and forward linkages across conservation and production to consumption in millets.
- Though traditionally millets are known for diverse end uses, efforts to identify end-use specific genotypes are lacking.
- Lack of availability of efficient machinery for the primary processing of millets, which is increasing the processing cost and low-quality end product.
- Lack of suitable value-added technologies of the millets. Due to the absence of gluten, there are inconveniences in millet processing, increasing drudgery, and time-consuming.
- Non-availability of acceptable value-added millet health foods in the domestic market and the international market for a quality life.
- Lack of millet processing protocols and machinery for enhancing nutraceuticals for health benefits.
- Absence of standard guidelines for grains, products, and processing technologies.
- Absence of more sustainable and profitable business ventures in millets.
- Lack of suitable market strategies to generate demand for value addition at the farm gate and other startups in the sphere.
- Lack of policy support for millets in the supply-side and demand-side.

20.3.1 Need for Developing Value Chain

Despite a rich inter/intra-species diversity and wider climatic adaptability, the cultivation of diverse millet species/varieties is gradually decreasing in the recent past due to lack of awareness, value-added processing machinery, availability of value-added products, profitability, marketing support, policy neglect, etc. The research on millets food processing was neglected during the past seven decades as our country's policies were oriented primarily toward the green revolution led food security crops, such as wheat and rice, and nutritional security was not our priority, where millets have to lose their traditional ground. In order to increase production and productivity, inducing demand and scaling up its supply is the pre-requisite. One avenue for increasing demand is to focus on primary processing mechanization, value addition, nutritional labeling, and entrepreneurship development. Inducing demand and scaling up its supply is a need of an hour to attain nutritional security in those production regions through the development of millets value chain, production to consumption model approach. Millets, which were largely meant for home consumption can be targeted for commercialization, meeting health and nutrition needs of both rural and urban. Thus, developing a holistic value chain strategy will entail farmers income security and the nutritional security of consumers.

20.4 Scope of Value Chain on Millets

Intensive efforts on the millets will generate demand through the value addition of processed foods, driven by diversification of processing technologies, nutritional evaluation, and creating awareness. Identification and formation of new farmer producer organization to mobilize farmers in all the states. Suitable market strategies to generate demand for value addition at the farm gate and other startups in the sphere. Thus, to augment production and consumption, a workable and long-lasting value chain model is needed. Therefore, there is a need to strengthen this area.

The creation of demand on millets through the value addition of processed foods through diversification of processing technologies, sustainable machinery, and value-added product technologies for different processing are being identified for commercialization of products. Upgradation of the existing machinery based on an entrepreneur's requirements to demonstrate the technologies beyond the pilot scale has become crucial. Although research and development along with the whole value chain, including at the production end, are still needed, there has not been a concerted effort to create a Demand-Pull.

Now state and central government focuses on nutritional security, which was missed out in the past. However, the efforts were mainly concentrated on health and wellness in "niche markets," now with government policy intervention, the supply-side efforts of the value chain are grossly addressed to enhance production while there are efforts to strengthen demand-side factors in creating captive government markets for millets for mainstreaming millets in public-funded schemes (ICDS and Mid-Day Meal programs). There is an urgent need to create widespread public awareness on the goodness in millets and change the image on millets and millet products by taking a marketing approach, understanding the target audience, segmenting the audience, and creating communications that are not just "informative" but emotionally engaging and exciting.

20.4.1 Components of the Value Chain

See Fig. 20.1.

Fig. 20.1 Various components of Millet value chain

20.4.2 Value Chain in Millets

Crop improvement programs at IIMR and centres of AICRPs on millet crops have been working on the enhancement of productivity of these crops in view of the decreasing area, high nutritional value, and the importance of sustenance of millet-based rural communities. Varieties for different needs such as short-duration varieties, biofortified varieties, dual-purpose types, varieties adapted to specific seasons and soil types, varieties tolerant to pests and diseases, and varieties suitable for mechanical harvesting have been developed.

Crop production technologies for efficient conservation and utilization of moisture, integrated pest and nutrient management, cultivation in rice fallows, summer production, etc., have been developed. The yield potential of the newer varieties has been increasing over decades. New cultivars have been developed in all eight millets 31 high-yielding varieties and hybrids developed by ICAR and SAUs have been notified during the last 2 years. Public sector cultivars notified at national level (recommended for multiple states) are available in all millet crops (Table 20.1).

- Crop production technologies for efficient conservation and utilization of moisture, integrated pest and nutrient management, cultivation in rice fallows, summer production, etc., have been developed.
- Adaptability of millets in nonconventional areas/seasons have been evaluated for the horizontal spread of these crops. Outcomes of sorghum in rice fallows (in coastal AP and Odisha) and summer season (Indo-Gangetic region) have been highly encouraging.
- Profiling of millet varieties for nutritional composition has been undertaken to promote the millets for nutritional health benefits. Composition of protein and minerals for combating malnutrition, dietary fiber, and antioxidants for the

Table 20.1 Numbers of cultivars of millet crops notified during 2010–2019

Millet crop	No. of cultivars	National releases	Private sector	Public sector	Public hybrids
Sorghum	37	15	6	31	5
Pearl millet	73	51	44	29	20
Finger millet	27	5	0	27	
Foxtail millet	6	3	0	6	
Barnyard millet	2	2	0	6	
Little millet	9	3	0	9	
Kodo millet	5	2	0	5	
Proso millet	5	2	0	5	
Total	164	83	50	118	25

management of lifestyle diseases and stress are important for advocating millets as healthy foods.

20.4.3 Millets for Fodder for Livestock Animals

Millets form the major supplier of green and dry fodder (including stover) in India and their role becomes important during the lean period of winter and summer months. Besides the use of stovers, green forage of sorghum and pearl millet are produced in an estimated area of 3.5–4 m ha. About 20–60% of dry fodder supply in semiarid regions is dependent on sorghum and pearl millet. Finger millet and barnyard millet fodders possess high palatability and are also used for making hay or silage. Proso millet green plants are good fodders for cattle and horses. A large number of millet varieties and hybrids of forage purpose have been developed by IIMR, other ICAR institutes, and AICRPs. The multi-cut forage sorghum hybrid CSH 24MF is now estimated to have spread over at least 20% of multi-cut sorghum area in the country, based on the seed production estimates, by replacing private hybrids. Dual purpose sorghum varieties with high stover yield without compromising grain yield have been developed at IIMR.

20.4.4 On-Farm Production and Farm Gate Processing

ICAR-IIMR has made specific interventions with end product specific cultivars making the farmers move from domestic cultivation to commercial cultivation. As a consortium under the National Agricultural Innovation Project (World Bank funded) on "Creation of demand for millet foods through Production to Consumption System value chain" and subsequently upgradation of Directorate of Sorghum Resaerch to ICAR-Indian Institute of Millets Research has made interventions in on-farm production (Rao et al. 2010).

20.4.5 Farm Level Backward Integration Through On-Farm Production Intervention

On-Farm Production Intervention is for transforming domestic cultivation to commercial cultivation for meeting health conscious urban consumer needs. On-farm intervention in PPP mode with ITC (ABD) Ltd, State Agriculture university (ANGRAU), and National Institute on Nutrition implemented interventions in the value chain. Later tie-up with Britannia Industries Ltd strengthened the science of processing for commercialization. Through piloting of intensification of Sorghum cultivation in Maharashtra and Telangana states with the help of ITC (ABD) impacted in shifting millet cultivation from marginal lands to better lands. ITC did the procurement through e-*choupal* and grading, bulking and aggregation providing procurement through their e-*choupal* outlets. Farmgate value addition is enabled by them in providing demo processing machines at farmgate leading to increased incomes through primary processing-based value addition by their conversion to commercial cultivation of millets. Thus, IIMR has gained expertise in linking farmers with markets in PPP mode. IIMR, identified end product specific cultivars thereby establishing lead in this segment with the financial support of Britannia. This work was strengthened across various product ranges especially bakery technologies.

Identifying cultivars with specific end products was attempted successfully by analyzing the physical, biochemical, digestibility of protein, and amino acid profile and correlating the results to get the best suitable cultivars. ICAR-IIMR has provided on-farm technical backstopping to the farmers with hybrids, which are suitable for specific endproduct and analyzed the impact of the interventions through the e-*choupal* model of ITC Ltd. Through this model, about 3000 farmers were growing improved cultivars in about 1000 acres for 4 consecutive years in Parbhani, Nanded (Maharashtra), and Adilabad districts (Erstwhile AP).

Under the NAIP project, it is estimated that through technical backstopping of improved cultivars there is an increased net income to the farmers by 69% rabi and 222% Kharif season in 4 years. Productivity enhanced in *rabi* sorghum through improved technologies. The farmers were given product-specific cultivars of sorghum and were extended PPP mode on-farm extension services. The produce was procured by ITC ABD Ltd through their choupal sagar network by paying market price for the FAQ grain. ITC have bulked/aggregated and supplied to Other actors in the chain processors and made them available all around the year through continuous supply chain. This has impacted the farmers who shifted the cultivation of sorghum from marginal to better lands as they realized better profits. Thus, the impact is evident that the change in their behavior from domestic cultivation to commercial mode. Farmgate value addition, through processing, was realized by these dryland farmers a significant enhancement in their incomes by 2–3 times thus by the word of mouth many villages were influenced to take up the model as their realized first-time profits from sorghum cultivation.

Developed cultivar-based products and identified suitable cultivars for specific end products; as a part of this study, ICAR-IIMR collaboration work has been done with the M/s Britannia Pvt. Ltd to identify specific sorghum cultivar for biscuits manufacturing for large production. A GAP good agricultural practices for obtaining high productivity under dryland was tested, developed, and popularized among Maharashtra State farmers. The technology package includes sowing on ridges and furrows for high water-use efficiency, sprinkler irrigation system, seed treatment with *Rhizobium* and phosphorus solubilizing bacteria, foliar application of nitrogen through the spray of 2% urea/DAP/KNO_3, and IPM for *Helicoverpa armigera.*

These efforts were being extended to other parts and the policy advocacy worked out and through launch of Intensification of Nutritional Security in Millet Promotion (INSIMP) of Government and subsequent National Sub Mission on Millets under NFSM and through Technological backstopping of IIMR the backward integration is renewed through the latter program and ICAR-IIMR has given a responsibility to strengthen 100 millet-based FPO's in the country as they are being operated through an arm of DAC & FW, i.e, Small Farmers Agribusiness Consortium (SFAC) of which Six FPO's are entrusted to ICAR-IIMR to model nutri-cereals based and backstop remaining FPO's launched by the SFAC.

The FPO's are empowered financially to engage in farm agribusiness with regard to not only farm production but also in processing and value addition with Government support. This may be a path-breaking in millets as they are smallholder families and such an effort will ensure storage of grain for the whole year that is important for establishing business around the value-added products of this commodity. Some of the recent efforts under National Mission on millets some interventions where IIMR is actively involved are given below:

20.5 Seed Hubs

A total of 25 breeder seed production centers and 19 certified seed production hubs are created across the country to provide better access to quality seeds to millet farmers. Apart from the 100 processing clusters that are being set up in FPOs/KVKs' for demonstration, additional 500 such processing technology clusters may be made available to the local entrepreneur/progressive farmers/SHGs/NGOs as an attachment to local processing units. Primary and Secondary Processing facilities will be provided at each center either to KVKs/FPOs. IIMR as a nodal agency would enable the State Departments in identifying the suitable machinery in consultation with CIPHET/CIAE/CFTRI and aid in the purchase of them through State Agriculture Departments.

20.6 Establishment of Demonstration Centers

Each of these FPOs will be connected to a KVK/ICAR institute/NGO/AICRP and will serve as demonstration cum training centers for the furtherance of post-harvest technologies. As a part of the marketing support, each of these FPOs is encouraged to start a Millet Kitchen. The Technology backstopping will be provided by

ICAR-IIMR/AICRP. ICAR-IIMR has established six model FPOS's in four states that are focused on nutri-cereals in addition to the formation of 50 FPO's across millet-growing states by SFAC.

20.7 Third Novel Intervention Was Diversification of Value-Added Processing Technologies

The prototype of various machinery are now being upscaled. The shelf life of processed products was enhanced from less than a month to 8 months thus offering consumers wide and convenient food options to choose from for enhancing millet consumption and paving way for the commercialization of millet foods. The Replication of successful value-added processing, promotional and marketing interventions in pearl millet and small millets. Value addition efforts conducted in Sorghum are being replicated in other millets by diversified processing and optimization of conditions for suiting sorghum processing such as flaking, extrusion (hot and cold), biscuits making, parboiling including primary processing, dehulling, and milling. The existing machines are also tested for millet's suitability and new machinery for mechanization is also envisaged to be developed. Thus, the consumer is provided with convenient options.

20.7.1 Primary Processing

The primary processing of major millets such as sorghum, bajra, and ragi is relatively easier than other minor millets as the former grains are naked while the latter have outer inedible seed coat which needs to be dehulled. The machinery for sorghum primary processing was retrofitted earlier through our effort but now market vailable dehullers is an issue which is being addressed, meanwhile we validated and assessed the currently available machineries for dehulling of minor millets to suggest customers suitable machinery with higher efficiency. There is a difference in the number of layers of epicarp of these small grains, in case of little millet it has a single outer layer while kodo millets have seven outer layers making the latter as most vulnerable among five minor millets for dehulling. Therefore, no single machine despite variable RPM cannot be prescribed for dehulling of all the minor millets.

20.7.1.1 Husk Weight Analysis of Small Millets

A study was carried out at ICAR-IIMR to find the husk weight in the total grain weight and observed foxtail millet, little millet, proso millet, barnyard millet, and kodo millet contain about 21.54%, 19.65%, 18.52%, 19.02%, and 29.8% in the total grain.

20.7.1.2 Improving the Dehulling Efficiency by Parboiling Pretreatment

To improve the dehulling efficiency and minimize the broken percentage during milling (dehulling) of millets. Pretreatment (Parboiling) has been done for small millets (Foxtail Millet, Barnyard Millet, Kodo millet, Proso Millet, and little millet) and calculated the head recovery and broken percentage. Dehulling was performed by using the small portable Impact Huller (devised by M/s Dhan foundation). The results showed increased head recovery and decreased broken percentage in parboiled millet grains. Parboiled Kodo millet grains showed the highest (27.99%) and parboiled little millet grains showed the lowest (5.41%) head recovery when compared to nontreated grains (Figs. 20.2 and 20.3).

20.7.1.3 Inclusion of an Efficient Mesh for Reciprocating Sieve

Most of the manufacturer supplies a square cut mesh for processing millets in destoner, which either leads to sieving down the grains or ends up being stuck in square pockets. The best way to overcome these hurdles is to include a dimple mesh capable of providing effective motion to the grains with minor spillage.

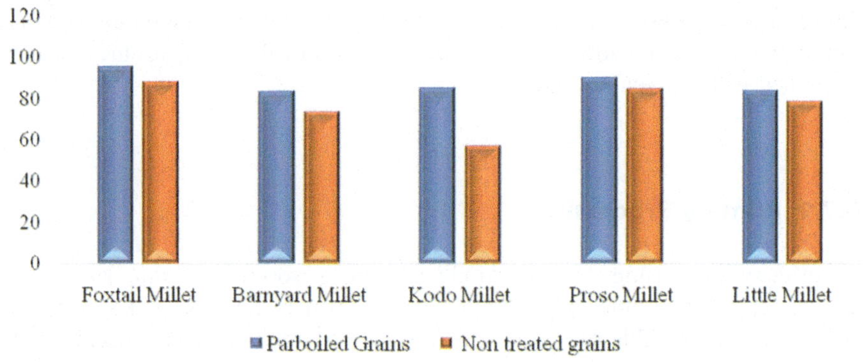

Fig. 20.2 Head recovery of parboiled and nontreated grains

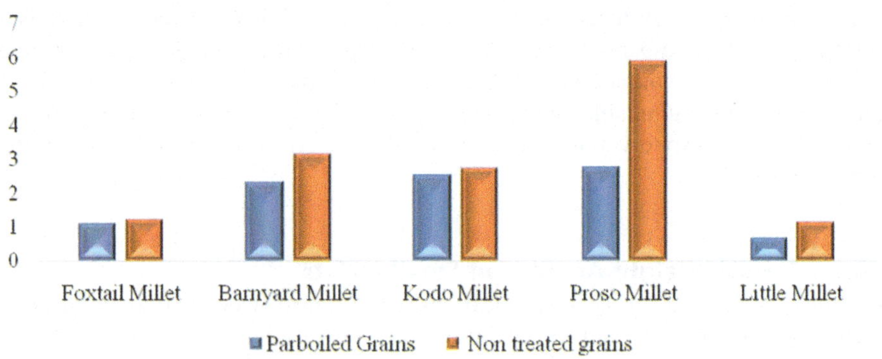

Fig. 20.3 Broken percentage of parboiled and nontreated grains

20.7.1.4 Roti Making Machine

To remove the drudgery involved in preparing rotis, IIMR, in collaboration with a Private company, has designed a device for preparing flattened dough suitable for cutting into circular shape and baking to prepare unleavened pancakes of desired thickness from sorghum/millets such as other non-glutinous cereals including multi-grain flour. The developed roti-making machine is used in large numbers inexpensively and hygienically without requiring fuel or electric power and enhancing the production efficiency of pancakes within a short period.

20.7.1.5 Retrofitting Machinery

Machinery is available for processing rice and wheat products. There is no specific machinery available for the processing of millets. The machinery available may not well suitable for millets as they lag in gluten protein and some inconveniences in processing. So there is a need to develop or retrofit the machinery available. IIMR has retrofitted more than 40 machineries, developed and optimized the machinery for millet value addition.

20.7.2 Secondary Processing

As part of millet processing interventions, more than 80 sorghum and millet-based value-added novel products (Ready-to-Eat, Ready-to-Cook) were developed. (Of which 30 products were commercialized under the *Eatrite*—an IIMR registered brand). More than 500 recipes were developed so far, including regional cuisines. The diversification of processing technologies attempted during Stage I of Value chain development are now translated for other millets such as milling, extrusion both hot and cold, flaking, puffing, popping, parboiling, and malting through Our Centre of Excellence. Many Instant breakfast products are developed with sensory and nutritional properties in place. Another 30 millet recipes were introduced targeting the children under ICDS and MDM anchored by WCD and MHRD Departments of Government of India.

20.8 Recently Developed Value-Added Products

20.8.1 Development of Sorghum Analog Rice

A study was designed at ICAR-IIMR to develop Rice Analogues from sorghum and studied all the factors and feasibility of formulations optimized. Sorghum rice analogs were developed from sorghum millet flour and broken rice flour through extrusion processing. Millet-based rice analog was developed, based on the DOE (data optimal mixture design) and was altered after a few trials based on the responses recorded after each trial. The optimum extruder condition was established at cutter speed, screw speed, and pressure of 7 Hz, 3 Hz, and 12 bar, respectively. Based on the most desirable qualities of the product.

20.8.2 Development of Millet-Based Express Food: Little Millet Veg Pulao Mix

Developed little millet instant ready-to-eat veg pulao mix (express food). Little millet veg Pulao mix was optimized with dehulled millets with different cooking times and different particle sizes with and without soaking analyzed the cooking quality parameters (gruel loss, cooked weight, etc.), nutrient parameters, rehydration for 3 min, and estimated sensory quality parameters by using Hedonic scale (Table 20.2). Instant millet veg pulao mix samples prepared from little millet rice were showed nutritionally superiority in terms of protein, fiber, ash, fat, and minerals, i.e., calcium and iron compared with control samples.

A study was conducted to standardize the Ragi bread by using different blends and analyzed physicochemical and sensory analysis of the products. The three different blends T_1 (40:60) (Ragi:Refined wheat); T_2 (50:50) (Ragi:Refined wheat); T_3 (30:70) (Ragi:Refined wheat) were selected (Table 20.3). It was found that blend T_1 was having a very hard texture and have low acceptability due to higher millet portion as it is devoid of gluten protein, which plays a crucial part and enhances the texture in the bread processing. Based on sensory, T_3 sample has a high overall acceptability score (7.5) on a 9-point hedonic scale due to the higher portions of gluten protein from wheat and having good protein content (8.40 ± 0.23).

20.8.3 Development of Foxtail Millet Choco Chip Cookies

Experiments were designed to develop low fat and crispy healthy cookies from Foxtail millet by using different types of emulsifiers. Different blends were formulated (Table 20.3) and analyzed for the spread ratio, hardness, and sensory attributes. Addition of 1.0% lecithin, the spread ratio (7.8613) and hardness (2651.7) value was close to the control sample (spread ratio 12.9524, harness 4788.9) than other blends. Blend with 1% lecithin with Choco chips scored the highest sensory score (8) on 9-point hedonic scale (Table 20.4).

20.8.4 Standardization of Foxtail Millet Vermicelli

A study has been conducted on the standardization of foxtail millet-based vermicelli process at different moisture levels. In this study, total solids loss, cooking time, cooked weight, cooked volume, bulk density, swelling index, and overall acceptability were estimated. Three blends were formed A (60:40), B (70:30), and C (80:20). Among A, B, and C total solids loss was very low in blend A. Cooking time was higher in sample C (6:35:30) and lower in sample A (5:35:00), and cooked weight and cooked volume was vice-versa. It was observed that the swelling index was higher in sample A lower in sample C. It was reported that overall acceptability was higher in sample A with 8.5 score based on a 9-point hedonic scale. Finally, the

Table 20.2 Cooking parameters of soaked grains

	Gruel loss (%)		Cooked weight (g)		Yield (%)	
Cooking time	Soaking	Without soaking	Soaking	Without soaking	Soaking	Without soaking
25 min	6.96 ± 0.5	6.07 ± 1.77	148.9 ± 1.0	171.3 ± 12	54.4 ± 1.2	57 ± 2.8
30 min	7.55 ± 1.7	10.94 ± 0.94	124.2 ± 37	160.5 ± 7.7	60.2 ± 2.0	57.7 ± 2.8

Table 20.3 Quality parameters for bread making

Trial code (Ragi:Refined wheat)	Moisture (%)	Water activity (%)	Protein (%)	Alcoholic acidity (%)	Baking losses (%)	Overall acceptability
T_1–40:60	32.26 ± 0.35	0.928 ± 0.00	6.05 ± 0.71	1.10 ± 0.00	8.83 ± 0.76	6.1
T_2–50:50	32.29 ± 0.15	0.926 ± 0.01	6.87 ± 1.26	1.00 ± 0.00	7.17 ± 0.58	6.7
T_3–30:70	31.09 ± 0.10	0.918 ± 0.00	8.40 ± 0.23	1.02 ± 0.04	12.00 ± 0.57	7.5

Table 20.4 Quality parameters of foxtail millet choco chip cookies

S. No.	Trial code	Spread ratio	Hardness	Overall acceptability
1.	Cookies (50% fat + 50% oil)	10.42 ± 0.53	1786.3 ± 0.15	7.5 ± 0.3
2.	Cookies (100% fat + no emulsifier)	4.22 ± 0.25	1654.2 ± 0.14	7.2 ± 0.9
3.	Cookies (50% fat + 50% oil + GMS (glycerol mono sterate))	4.24 ± 1.51	1898.2 ± 0.45	7.9 ± 0.5
4.	Cookies with (50% fat + 50% oil + lecithin (0.5%))	4.29 ± 0.55	2551.2 ± 0.97	8.1 ± 0.3
5.	Cookies with (50% fat + 50% oil + lecithin (1.0%))	7.75 ± 0.56	2651.7 ± 0.93	8.2 ± 0.2
6.	Cookies (50% fat + 50% oil + lecithin (1.0%) + Choco chips)	7.86 ± 0.57	2551.4 ± 0.76	8.2 ± 0.4
7.	Wheat based cookie (100% oil)-control	12.95 ± 0.99	4788.9 ± 0.5	8.9 ± 0.9

Table 20.5 Quality parameters of foxtail millet vermicelli

S. No.	Trial name	Total solids loss (%)	Cooking time (min)	Cooked weight (g)	Cooked volume (ml)	Swelling index	Overall acceptability
1	A	12.62	05:35	86.55	92.5	245.98	8.50
2	B	15.08	06:30:00	83.46	92.5	233.71	7.00
3	C	18.05	06:35:30	74.66	84.5	198.42	6.00

study has concluded that foxtail millet-based vermicelli with A-60:40 (Foxtail millet:Wheat) 26% moisture (on dry basis) has good quality values and higher acceptability nature (Table 20.5).

20.8.5 Development of Sorghum Yogurt

A study was carried out at ICAR-IIMR to develop sorghum yogurt, a nondairy yogurt, purely from plant-based source and lactose-free was developed. In the experiment, milk was extracted from sorghum. The milk extracted was cooked to carry out the complete gelatinization of starch. Different blends of sorghum milk and soya milk were prepared with different concentrations. Yogurt culture was added at 15% concentration. Analyzed for nutritional parameters, microbial parameter, and lactic acid. It was found that Sorghum soya blend containing 60% sorghum milk and 40% soya milk gave the best-set yogurt. Obtained values for the optimized blend were total solids, 5.305, titratable acidity as 0.07425% as lactic acid, and 6.31 g/ 100 g of protein. The product so developed was found to be superior in nutritional value, microbiological stability, and acceptability over the control sample. The shelf life of the product so developed was up to 7 days under refrigeration.

20.8.6 Optimization and Development of All Millet Puffs

Puffs from millets were developed using a puff gun machine with varied moisture conditions ranging from 21% to 23% for Sorghum, pearl millet, Kodo millet, proso millet, barnyard millet, and little millet and 0.8 Psi pressure, which was found optimum. For Sorghum, the best suitable cultivar was found to be M 35-1, which yields the highest percentage of puffs (70%). Process conditions are being optimized for other millets for increasing its recovery.

20.8.7 Sorghum Flaking Using Edge Runner Technology

A process was developed to produce sorghum thick flakes. Edge runner is retrofitted for sorghum flakes as the machine designed for rice flaking machines and optimized the processing conditions. Using paddy flaking equipment, that is, edge runner (flaking machine) and roaster, made it possible to produce flakes from Sorghum. In this method, sorghum grain was hydrated to equilibrium moisture content, incubated to remove the surface moisture, and subjected to a high-temperature short time (HTST) treatment. During HTST treatment, starch granules of grain get gelatinized, and these grains were fed to the edge runner, where the gelatinized endosperm gets flattened, and bran is eventually separated, pulverized, and collected from another outlet. Sorghum flakes obtained from this technology resemble rice flakes and is texturally more appealing due to quick hydration characteristics and less chewy properties. The yield recovery fractions from this technology depend on cultivar type, nature of endosperm, and the optimized conditions, and the resultant output is 50–65%. The broken flake powder is a by-product that can be used for making value-added sweets for increasing the overall profitability of flaking for commercial purposes.

20.8.8 Standardization of Pearl Millet Flakes Processing

A study has been piloted to standardize the pearl millet flakes by differentiating soaking timings (T_1-13, T_2-16, and T_3-19 h). Among the three treatments, T_1-13 h (soaking) has got the highest yield—75.36 ± 0.5% with 14.40 ± 0.4% husk yield and T_1 got the highest sensory score as 8.0/9.0, T1 is good among all three treatments (Table 20.6).

20.8.9 Standardization of Pearl Millet Puffs Processing

Experiments were piloted to standardize Pearl millet puffs for different moisture levels and tempering time. Tempering the grains with 21% moisture for 30 min (T_2) was resulted with good puff yield of about 90% and considered the standardized optimum conditions. This study has concluded that over-tempering and less

Table 20.6 Pearl millet Roller flakes recovery

S. No.	Sample code	Flakes recovery (%)	Husk recovery (%)	Flakes broken (%)	Overall acceptability
1.	T_1-13 h	75.36 ± 0.5	14.40 ± 0.4	7.54 ± 0.7	8.0 ± 0.1
2.	T_2-16 h	70.23 ± 0.4	20.41 ± 0.6	8.56 ± 0.3	7.6 ± 0.4
3.	T_3-19 h	68.89 ± 0.8	21.46 ± 0.2	9.09 ± 0.2	6.5 ± 0.7
4.	Control (Sorghum-16 h standardized process)	80.20 ± 0.8	16.96 ± 0.5	3.20 ± 0.4	8.5 ± 0.1

Table 20.7 Recovery of pearl millet puffs

S. No.	Sample code	Moisture levels	Tempering time (min)	Puffs yield (%)	Wastage (%)	Overall acceptability
1	T_1	17	30	72 ± 0.57	28 ± 0.45	8.0 ± 0.18
2	T_2	21	30	90 ± 0.12	10 ± 0.89	8.1 ± 0.65
3	T_3	24	30	76 ± 0.59	24 ± 0.88	8.0 ± 0.56
4	T_4	17	60	78 ± 0.42	22 ± 0.93	7.6 ± 0.45
5	T_5	21	60	80 ± 0.91	20 ± 0.94	7.7 ± 0.36
6	T_6	24	60	70 ± 0.42	30 ± 0.45	7.4 ± 0.25

moisture conditions result in less puffing yield as all the moisture levels go down. T_2 has been selected for puffs manufacturing and has an appealing taste with good texture (Table 20.7).

20.8.10 Millet Flaking Using Roller Flaking Technology

The roller flaking machine has been retrofitted and millet pretreatment conditions were optimized for making roller flakes. Millet grain was hydrated to equilibrium moisture content, incubated to remove the surface moisture, and subjected to high-temperature short-time (HTST) treatment. During HTST treatment, starch granules of grain get gelatinized, and these grains were fed to a roller flaker, where the gelatinized endosperm gets rolled, and bran is eventually separated, pulverized, and collected from another outlet. All millet grains can be made to flakes from this technology. The flakes are then to be roasted or blistered and they resemble corn flakes texturally. These have quick hydration characteristics and less chewy properties and can be used as breakfast cereals. Further studies are to be conducted for recovery fractions and shelf life analysis from this technology.

20.9 Developed and Standardized Ready-to-Eat Snacks (RTE) Using Extrusion Technology

Millets can also be extrusion cooked to prepare ready-to-eat products successfully. The products will have a crunchy texture and can be coated with traditional spice and condiments. These products being ready-to-eat nature will have greater scope for use as weaning and supplementary foods. The studies were carried out with millet, corn, wheat, and rice flour in different blends. Different blends are formulated using Sorghum, foxtail, kodo, finger, little, barnyard, and proso with maize grits (50:50, 60:40, 70:30) and found 50:50 as the best composition blend, analyzed for physical properties like expansion ratio, water absorption index, water solubility index, true density, bulk density, color, texture, and nutritional properties. Preconditioning was optimized to prepare the flours for extrusion cooking and moisture content was adjusted for 21–23% for all the formulations. Extrusion cooking was carried out using a twin-screw extruder at optimized extrusion parameters, viz., temperature: 90 and 130 °C for two different heating zones, die diameter: 3 mm, screw speed: 35 Hz, and cutter speed: 15 Hz.

20.9.1 Instant Mixes (*Idli, Upma, Kichadi, Pongal*)

Instant and Ready to Constitute (RTC) foods have become well-established products in western countries. Millet-based instant mixes such as upma, pongal, khichdi, payasam, bisibellibath, idli, and dosa using different millets and multi-millet were developed at ICAR-IIMR. Since millet offers several health benefits, there exist greater scope for its use in the production of traditional and contemporary convenience mixes.

20.9.2 Ragi Instant Vegetable Soup Mix

Ragi instant vegetable soup mix is developed with ragi flour, dried vegetables, and seasonings using milling, sieving, and drying technologies. Different combinations were formulated and the most acceptable blend was optimized through the sensory analysis. Further shelf life studies were carried out and found that the instant ragi soup mix's shelf life is 6 months.

20.9.3 Processing Effect on the Micronutrients Fe and Zn in Value-Added Sorghum Products

In addressing iron deficiency (anemia) and zinc deficiency (associated with poor growth and impaired immune), there is a huge need to develop measures to increase the micronutrient dietary intake with fortification or dietary supplementation. Studies were conducted to estimate the iron and zinc values in different processed products of the Sorghum and were observed that there was an increase in iron content in dehulled *idli rawa* (3.09 mg) compared to normal (1.39 mg) and parboiled (1.76 mg). Similarly, iron content increased in parboiled *khichdi rawa* (2.645 mg) compared to normal (1.655 mg) and dehulled (0.86 mg) after processing. The zinc values also increased in dehulled and parboiled grains compared to whole sorghum grain, and the highest value was observed in parboiled grain (5.0 mg).

20.9.4 Study on Fortification of the Value-Added Sorghum Products by Using Natural Foods

Garden cress processed seed (10%) was used for natural iron fortification and gingelly processed seed (20%) for zinc fortification, and value-added products like biscuits, pasta, and vermicelli were developed and standardized. The iron values of sorghum pasta increased from 1.2 to 1.9 mg and zinc value from 1.56 to 1.81 mg in sorghum pasta by incorporated 10% garden cress. In Sorghum, vermicelli iron increased from 1.2 to 1.5 mg, zinc 1.56 to 1.72 mg, and incorporated 5% garden cress.

20.9.5 Identification of Cultivar Suitability of Sorghum for Biscuits and Bakery Products

Pure sorghum cookies were developed and commercialized under the joint project between Britannia Industries Ltd and IIMR (2011–2014). The cookies were developed using 12 sorghum cultivars (flour particle size 60 mesh) with the standard recipe and were analyzed for starch (%) and dietary fiber. The mineral composition of sorghum cookies was determined using an atomic absorption spectrometer. The B vitamins content was determined using the HPLC method. A ten member semi-trained panel carried out the sensory evaluation of prepared sorghum cookies through rating products on a 9-point Hedonic scale. The starch percentage was highest in C-43 (47.06%) and lowest in CSV 15 (42.15%) sorghum varieties. Dietary Fiber (%) was highest in CSH 14 (9.27%) and lowest in CSV 20 (5.45%). Phosphorus content was highest in C43 and lowest in CSH 14, and potassium content was highest in C43 and lowest in CSH 14. It was found that Sorghum had trace amounts of niacin, and the amount ranged from 0.11 to 1.06 mg/kg. The presence of thiamine and riboflavin was <0.5 (mg/100 g), and also the level of folic acid was <5.0 (μg/100 g). Among all the different cultivars, it was found that the scores were highest in CSH 23, CSH 13R, and CSV 18R and lowest in CSV 15. Hence, cookies that have high nutritional value can be made from Sorghum.

20.9.6 Innovation in Value Addition of By-Products from Sorghum Foods

To enhance the profitability of value-added sorghum products, the value addition of by-products was also sought. Developed roasted flakes peda from by-product after roasted flakes preparation, bran peda—by-product after milling of sorghum Bran Soup by-product after dehulling Sorghum. There was a 30–50% increase in returns due to the value addition of by-products concerned.

20.9.7 Preprocessing Treatment of Sorghum for Enhancement of Shelf Life

Preprocessing of sorghum grain was developed, which is the process of soaking sorghum grain, drying till required moisture content, after which the grain is roasted in a grain roaster and dried until 7% moisture content. As the soaked grain gets saturated, dried, and roasted, it leads to gelatinization of starch and thus reduces the polyphenol content, which enhances the shelf life. The sorghum grain was soaked for 16 h at room temperature, drained and dried until 21% moisture content for 1 h, and roasted in a grain roaster at 290 °C for 45 s and allowed to cool, dried at 60 °C for 4 h in a tray drier or air-dried till 7% moisture in room temperature and milled to flour. This processing of Sorghum helps to preserve all the nutritive substances and vitamins and also decreases the cooking time of the product and enhances the shelf life from 2 to 4 months.

20.9.8 The Enhanced Shelf Life of Sorghum Flour and Multigrain Flour

Short shelf life of sorghum grain has been regarded as the bottleneck to sorghum marketing and ultimately resulted in decline in consumption. The shelf life of sorghum flour was assessed using infrared heating. Sorghum flour treated under infrared heating for 8.5 min at 120 °C enhanced its storability from its normal 45 days to 90 days.

20.9.9 Sensory Evaluation

At CoE, ICAR-IIMR Sorghum-based foods were evaluated for their sensory properties and compared with traditional rice and wheat-based foods. Standard procedures were followed in preparing the Sorghum and rice/wheat-based foods. These products were consumed daily at IIMR and subjected to sensory evaluation.

20.10 Fourth Intervention Is Nutritional Evaluation of Millets and Their Value-Added Products

The key to commercialize the millets lies in creating USP (unique selling proposition) for flagging it as a healthy and convenient option.

20.10.1 Nutritional Evaluation

IIMR in collaboration with NIN under the NAIP project estimated nutritional composition and organoleptic studies of 30 sorghum-based products earlier. Now ICAR-IIMR has performed nutritional profiling for all the newly developed/refined millet value-added technologies (flours, rawa, puffs, extruded products, flakes, cookies, muffins, cakes, bread, cookies, muruku, etc.) developed. Standard methods are followed in estimating macro and micronutrients like protein, fat, ash, fiber, energy carbohydrate, iron, calcium, magnesium, manganese, phosphorus, etc. (Rao et al. 2018).

20.10.2 Clinical Trials to Generate Evidence for Health Benefits

20.10.2.1 Diabetic Population

Earlier clinical studies were conducted on 150 type-2 diabetic patients for 60 days, jointly NIN and IIMR collaboration. Evaluated the effect of sorghum diet (50% sorghum and 50% rice) on the type-2 diabetic patients glycosylated hemoglobin and lipid profile. Anthropometric and biochemical indices before and after diet were analyzed. It was found that there was an increase in hemoglobin with a significant reduction of fasting glucose and glycosylated hemoglobin. Thus, evidence-based studies conducted and the findings which were in favor of sorghum were published for wider acceptability and labeling information to new business enterprises for commercilaization of sorghum and millets.

20.10.2.2 Clinical Study Among School Children

A study was conducted on 50 school children under age group of 9–12 years. This study was carried out in collaboration with NIN under the NAIP project. The children were divided into two groups experimental and control. The experimental group was given the sorghum diet (50% sorghum and 50% rice), and the control group was given the regular rice diet for complete 8 months. The children were tested for anthropometric and biochemical indices before and after diet. The results showed a significant increase in serum, hemoglobin, albumin, total protein, folic acid, vitamin-B12, ferritin, calcium, and iron levels in the experimental group.

20.10.2.3 Glycemic Index and Glycemic Load

Evaluated sorghum foods effect and compared with rice, wheat foods on the Evaluated Glycaemic index (GI) and Glycaemic Load (GL). The study was conducted on ten healthy volunteers (20–40 years) who were diabetic-free. The GI and GL of Sorghum-based foods were significantly lower ($p < 0.05$) than wheat/rice-based foods. Sorghum-based foods coarse *rawa* (semolina) *upma*, flakes *poha*, pasta, and biscuits were having low GI, and *roti* and fine rawa upma were having medium GI, whereas all the sorghum products and wheat/rice-based were having high GL. The GI and GL of Sorghum-based foods were lower than wheat or based foods. A low GL sorghum diet can be achieved by choosing small servings of foods, and intake of low GL foods would help to prevent and control diabetes. Recently efforts are being made to study the evidence for health claims with the help of NIN and other medical Institutions. Further, ascertain the bioavailability of major nutrients in millet-based foods.

20.11 Entrepreneurship Development

By disseminating technologies through signing MoU's and licensing to close to 75 Startups and backstopping them. Further, Training and capacity building of various entrepreneurs belonging to ten states for about 8–9 years continuously with the funding of DAC & FW.

A Centre of Excellence on sorghum and Millets is established to cater research on food processing, quality control facility, and technology showcasing and establishing linkages through EDP programs. Now the CoE is upgraded as Centre of Excellence on Nutricereals catering to R&D on value addition, showcasing technologies, establishing linkages, capacity building of stakeholders, and linking farmers with markets.

These value-added technologies generated are being demonstrated to various interest groups and stakeholders across the country involved in millet/sorghum processing. A well-developed quality control lab facility is provided through the Center of Excellence (CoE), through which different R&D trials are done for entrepreneurs. This facility stands at the center for all the innovations in value-added technologies and different millet recipes. Thus, the Centre of Excellence on Millets has the necessary facilities and expertise, serves as the lead R&D center on food processing catering various states in the country for addressing various issues on the value chain, primarily food processing and nutrition and integrating with agriculture and EDP funded by National Food Security Mission (NFSM).

As a part of Entrepreneurship development two programs are being organized every month, "Start-up Ignition—Entrepreneurial Opportunities in Millets Production, Processing & Value-Addition" for aspiring entrepreneurs willing to start a business in the Millets domain and "Cooking with millets" one-day event mainly to create awareness about recipes of millets developed and how to prepare recipes with millets especially in the absence of gluten. Since 2011, more than 30,000 participants from various backgrounds hailing from various millet and non-millet-growing states. They have been trained under these capacity-building programs to provide hands-on experience of processing, machinery operation, and shelf life testing, sensory evaluation, marketing, labeling, branding, packaging, business plan preparations. To support and handhold the entrepreneurs in millets space, a Technology Business Incubator (NUTRIHUB) on Nutri-cereals was established in the year 2016 with the support of the Department of Science and Technology (DST), GOI. This incubator serves as a hub for those startups that are working on Millets.

20.12 Intervention on Commercialization

With products in place, IIMR launched first-ever brand on millets "Eatrite" which was an instant success as consumers were on the lookout for healthy alternatives in midst of lifestyle diseases. This was backed with Aggressive campaigning undertaken by both ATL and BTL communication strategies which was outsources in the form of broadcasting in electronic, social media effectively to build brand millet and create awareness in malls, exhibitions, and wet sampling in malls and public parks and hospitals involving doctors, nutritionists, chefs, and celebrities. Products were tested in Food bazaar and Heritage Retails stores.

20.13 Promotion of Millets and Creation of Awareness

ICAR-IIMR has aggressively campaigned to increase awareness through untiring roadshows, health and nutrition campaigns, media coverage, and institutional visits. The popularization of Sorghum and millets as health and convenient food and its suitability in tackling lifestyle diseases is loudly disseminated through roadshows, exhibitions, wet sampling in public parks, and commercial malls in a specially designed Mobile van "millet rath." Several media, including social media, electronic media, and print media, are aggressively utilized to raise the message "eat millets stay healthy." Organized melas and exhibitions on sorghum to showcase sorghum processing technologies by participating and put up an Eatrite stall in events organized all over India, for the promotion of sorghum products to touch every corner of India.

An exhibition van was designed as JowarRath and started a grand show organizing *JowarRath* more than 3000 km participation in Krishi Parivartan Yatra (14th to 26th May 2014) from Hyderabad to New Delhi, covering prominent ICAR institutes enroute. Farmers and other stakeholders found the JowarRath very interesting, and it had created a huge impact on all the places visited. Promotional materials such as Jowar jingle, Millets Public Service Message (PSM), EATRITE Hindi commercial ads were shot and were telecasted in National Electronic media. Every year IIMR organizes a National event to create a platform for all the millet stakeholders for discussing millets.

IIMR hosted national and international events involving various stakeholders. The awareness is created through publications depicting the value-added technologies, recipes, and nutritional health benefits of millet books in different languages. Commercial advertisements were telecasted in prominent print, electronic and social media.

Close to 70 MOU's with public and private institutes were made to undertake inclusive millets promotion with ICRISAT, CFTRI, NIN, and many private Players and state Governments.

20.14 Policy Advocacy

The country's policies are oriented primarily towards green revolution-led food security crops such as rice and wheat, and nutrition security was not our priority where millets have to lose their traditional ground. After decades of neglect and the recent Indian Government's focus shift to nutritional security, ICAR-IIMR has diversified its mandate from crop improvement to value-added processing to revive millet consumption in the country, since 2008. ICAR-IIMR became technology-ready for the processing of value-added and convenient foods in sorghum, to begin with, and now strengthened their R&D even on other millets too.

Significant interventions on sensitizing policymakers and line departments of Government and forging linkages with stakeholders IIMR-led consortium has organized several workshops regional, National, and International and first time successful with DAC & FW Millet promotion Rs 100 Crore program launched INSIMP in the parliament 2012 in millet-growing states under "Rashtriya Krishi Vikas Yojana" (RKVY), for the first time. This scheme was introduced exclusively for the promotion of millets. This scheme combines extraordinary policy factors such as demonstration, inputs, seed, post-harvest technology, cognizance raising, capability construction, and research. IIMR was successful in rolling out 200 processing clusters through IIMR technologies all around the country.

Three National Centres of Excellence (CoEs) were established in the year 2010–2011 under this scheme; Centres of Excellence on Sorghum at Indian Institue of Millets Research (Formerly Directorate of Sorghum Research), Hyderabad, Centres of Excellence on pearl millet at CCSHAU, Hisar, and Centres of Excellence on small millets at University of Agriculture Sciences, Bengaluru. This scheme encouraged the progressive farmers, SHG's, FPO's for the setting of the millet processing facilities through the technical backstopping from these COE's.

Under this scheme, IIMR has developed 60+ technologies, promoted, popularized, and trained different stakeholders on millet technologies and recipe making. The outcome of the scheme was to increase the production, area, productivity, and processing of millets. The scheme was confined only to the selected districts has been the main constraint. Later in 2014–2015, the scheme was merged in National Food Security Mission (NFSM) (12th 5-year plan) and named as National Food Security Mission on coarse cereals. Now, the program topped up with more features in a holistic manner in the form of National Mission on Millets under NFSM for 5 years in 14 states launched in 2018. In both, the programs IIMR was entrusted as a nodal institute for implementation especially in providing Technical backstopping.

We are also in liaison with NItiAyog, MOFPI, APEDA, MHRD, and GoI departments as well as state Governments Missions. The technological handholding of various state governments who have set up the Missions (Odisha, Karnataka, Tamil Nadu, and Telangana State Governments), Millet Boards (Proposed by Government of Andhra Pradesh).

Millets lost their importance in favor of coarse cereals. So it is important for policymakers to realize the tremendous potential of millet production and processing for retaining National nutritional and health security. Policy intervention in the value chain's supply-side efforts is grossly addressed to enhance production while there are no efforts to strengthen demand-side factors in creating government markets for millets by mainstreaming millets in public-funded programs.

To attain nutritional and health security, there is a great need to revive millets as a staple diet. Only major millets (jowar, ragi, bajra) are supported with Minimum Support Price (MSP), leaving behind all the minor millets; governments should extend their support to small millets so that the farmer gets benefited and, in turn, production increase. The policymakers also need to address the issues to achieve the target of doubling the farmer's income by 2021–2022. Indian Government is giving for the development of the food processing industry, in India announced about 10,000 crores for assisting 2,00,000 micro food processing units. This will help the new budding entrepreneurs who are willing to step into millet processing.

20.15 Mainstreaming of Millets in Public Funded Programs Include

Introducing them in PDS, Piloting in Mid-Day Meal programs of Various millet-growing states, making them as a part of nutrition in ICDS, Encouraging millet-based startups and SMEs, providing seed support to the diversified products, promoting nutri-cereal and health benefits of millets, scaling up of technologies would increase price affordability and accessibility, research studies to strengthen the database on health benefits, bioavailability would increase the uptake of millets consumption and will provide evidence. Encouraging startups in the nutri-cereals domain and connecting them with rural and urban markets through policy push can help develop the nation's nutritional and health security. That can only happen when a policy push can happen.

20.16 Strategies to Promote Production of Millets in India

20.16.1 Increase Production and Productivity of Millets

- Provision of quality seed—Breeder seed and certified seed production of the latest high-yielding varieties to enhance the seed replacement ratio in millets—Breeder seed production and Seed hubs have been initiated. Additional 700q breeder seed produced during 2018–2019 in millet crops. This can provide quality seeds for an additional 10% area. The quality seed to farmers is provided at subsidy to farmers, reducing the cost of production and increasing productivity.

- DAC & FW has initiated the provision of minikits of improved production technologies across major millet-growing states. GoI has declared higher MSP for major millet crops. This is expected to motivate farmers to expand the area under millet crops where the area under these crops is reducing otherwise.
- Cultivation of millets in the north-east has received a fillip under NEH and TSP programs. Millets processing machines have been made available in six clusters in tribal areas across five states.
- Promotion of Millet-based FPOs in major millet-growing states have been undertaken. This will help increase area under millets and increase millet-based farmers' income.

20.16.2 Incentivizing Cultivation

- Share of producers in the consumer rupee should be higher.
- Something similar to Karnataka offering 10K/ha sowing.

20.16.3 Farm Gate Processing

- Primary processing machinery is required for value addition.
- Proper training on machinery handling should be given.
- Branding and Labeling for marketing in Urban areas.

20.16.4 Farmer Producer Organizations

- Strengthening FPOs to focus on entire value chain and to sustain as a group.
- Support for value addition in terms of secondary processing.

20.16.5 Procurement

- Local procurement at designated spots.
- Declaration of MSP, in case of small millets.
- Incentivizing farmers.

20.16.6 Mainstreaming Millets in Public Funded Programs

- Introducing them in PDS.
- Piloting in Mid-day Meal Programs of Various millet-growing states.
- Making them as a part of nutrition in ICDS.
- Identifying Recipes and training staff at Anganwadis.

20.16.7 Enhanced Role of Private Sector

- Private sector catering to urban and health-conscious consumers.
- Encouraging millet-based startups and SMEs.
- Providing seed support to the diversified products.
- Licensed products from IIMR, CFTRI, etc., can be made available to consumer market thus offering solution to lifestyle diseases.
- More involvement of big players like Britannia, ITC in promotion of millets.

20.16.8 Awareness Creation

- Promoting nutri-cereal and health benefits of millets.

20.16.9 Technology Scaling

- Scaling up of technologies would increase price affordability and accessibility.
- Research studies to strengthen the database on health benefits.
- Bioavailability would increase the uptake of millets consumption and will provide evidences.
- Which will help in fulfillment of regulatory gaps.

20.17 R&D on Nutri-Cereals for Enhancing Nutritional Security

- Profiling of nutri-cereals cultivars, clinical validation and bioavailability studies.
- Identification of specific cultivars suitable for processing of specific products.
- Shelf life enhancement.
- Development of National database and establishing grades and standards in small millets.
- Mechanization of primary processing machinery for developing efficient dehulling machinery in small millets.

20.18 Handholding State Governments in State's Millets Missions

- ICAR-IIMR is handholding technological and entire value chain aspects of various state governments in setting up the Millet Missions (Odisha, Karnataka, Tamil Nadu and Telangana State Governments).
- Millet Boards (Proposed by Government of Andhra Pradesh) and developing strategies for millets promotion in MP, Chattisgarh, UP, and Kerala states.
- Entrepreneurship Development programs in Uttarakhand, Jharkhand, Rajasthan, Gujarat, and Punjab.

- Technology backstopping on entire value chain aspects to the NE States Assam, Tripura, and Sikkim.

20.19 Developing Linkages with Private and Public Sectors and NGO's

- Working in close tandem with NITI Aayog as Technical advisor on framing policies on Millets promotion in the country.
- Strong linkages are established not only with state Governments but Other Departments of GoI such as DAC & FW, MOFPI, APEDA, WCD, MHRD, Ministry of Health, MSME, and DST to strengthen the Value chain in scaling up efforts on Mainstreaming Millets for inclusion in their programs such as in MDM, ICDS, PoshanAbhiyan, tribal welfare.
- Working with private bodies such as ITC, Britannia, Parle, MTR, Patanjali, and HLL.
- NGO's such as DHAN foundation, DDS, Sahaja group, WASSAN, in linking with grassroot workers.
- Establishment of FPO's through SFAC and NABARD.
- Strong linkages with ICMR, CSIR, and Other Public organizations with their arms such as NIN, CFTRI, IIFPT, IIT Delhi, IIM Bengaluru, ISB, IIT Kharagpur, IIT Roorkee, and IIT Assam in addition to strong linkages and joint programs with ICAR institutes.

20.20 Scaling Up of Efforts on Various Components of the Value Chain on Millets

Interventions through the millets' value chain project have been successful in an overall context; through an integrated approach and active participation by all stakeholders in the value chain, they have all benefited, from farmers to consumers. The recent efforts of the IIMR-led intervention in a PPP under NAIP have led to the commercialization of some millet products, though upscaling is a challenge.

20.20.1 Upscaling and Replication

Upscaling challenges are numerous, although post-NAIP, the issues are being addressed in the country in an assertive manner; by R&D, private sectors, NGOs, and policymakers, especially regarding the development of machinery suitable for various processing technologies, primary processing, product development, marketing, and entrepreneurship development. Pilot-scale studies are being attempted in some states to assess the feasibility of the mainstreaming of millets in the Mid-day Meal Scheme. The private sector has created multigrain formulations with millets as major ingredients, and Britannia Industries' collaborative work has been done with

IIMR and developed sorghum biscuits and bakery products on a pilot scale and identified suitable sorghum cultivars for cookie making to enhance millet utilization. IIMR upscaled some millet processing equipments from lab to pilot, pilot to commercial production. Established millet processing lines for baking, flaking, puffing, primary processing, milling, muruku, chikki, and packaging and made it available for new startups and providing facilities for incubates.

The replication of our Value chain could be in African countries where millets like sorghum and ragi are quite popular, though they have a value chain it is disjointed in sorghum, of late there have been some efforts to strengthen this. In Asia, specifically Middle Eastern countries, millets value chain can be replicated which has good potential, having the advantage of being an integral part of the Indian continent. The USA is not averse to such value chains that are nutri-centered and millet has strong potential in a nation where gluten-free market is being established more conspicuously. The USA and Europe have many migrants settled from India and, indeed, replication is possible wherever Indian communities are settled across the continents. And while the pilot-scale value chain intervention may not have proved entirely successful in India, a number of the components are worth emulating to enable the success of millet promotion for the nutritional security of consumers. The IIMR could only disseminate the technologies to the participating farmers and partner entrepreneurs; through INSIMP (Initiative for Nutritional Security through Intensive Millets Promotion), training programs are being organized for farmers' groups, NGOs, Krishi Vigyan Kendras (a grassroot level institution designed and devoted to impart need-based and skill-oriented vocational training to the practicing farmers, in-service extension personnel and to those who wish to go in for self-employment through "learning by doing"), entrepreneurs and government servants in all major millet-producing states. This training will enhance capacity building on the various aspects of value addition in millets.

20.20.2 Horizontal Expansion Within India

Millets being a nutri-cereal, there is scope to expand its market in the food sector if it is available in a convenient form. Once again, reviving their production and demand will contribute enormously to the nutritional and health security of the country. With successful commercialization of sorghum products at the pilot scale in the twin cities of Hyderabad and Secunderabad through Heritage retail stores, IIMR has further spread the marketing operations of their "eatrite" products in Mumbai through Big Bazaar's Future Retail stores. Being a government agency, the institute's main objective is to promote millets across the country by persuading more and more potential firms to include or exclusively take up sorghum and millets in their business.

The aggressive promotional and health awareness programs distributed through roadshows, exhibitions, seminars and workshops, and advertisements over the past 10 years have created much awareness, especially in Hyderabad. Taking advantage of the growing awareness among consumers and policymakers, more than 50 firms have signed memoranda of understanding (MoUs) with IIMR, and Future Retail Ltd. group (Big Bazaar) have signed an MoU to market millet products not only in Mumbai but also to expand horizontally across the country, especially to NCR Delhi.

Mainstreaming millets even in nontraditional states to begin with pilots will augur well for Millets in ICDS, MDM, and tribal and Defence programs.

20.20.3 Export Avenues

Enhancing infrastructural facilities and logistic support for handling the grain and other value-added products for exports will make sorghum export competitive. Of late there is a renewed interest in projecting millets as a value-added export product which will not only enhance export competition but also make it profitable through enhanced utilization. The Agricultural and Processed Food Products Export Development Authority (APEDA) was established by the government of India, entrusted with responsibility for export promotion and development of various agro products. APEDA has approved its support to finance a project proposal by IIMR to develop domestic markets for export purposes. This might give a new approach for millet exports, with processed and RTE millet products for the first time potentially exported from India to the Middle East, China, the USA, and African countries.

20.21 Conclusion

Thus, globally the constraints and opportunities of value chain development of millet are similar, although some factors are particular to a country. As the model of the millet value chain is successful under NAIP project, it can be replicated in other parts of the world where millets are the main crop. For future sustainability, policymakers and concern department should take initiative effort to inspire and encourage the millets farmers to go for commercialization through various approaches such as inclusion of millets in PDS (public distribution systems), subsidize machinery for processing of millets and national-wide campaign of millet flagging as health benefit food. Most research proposes the development of processed food (ready-to-eat/ready-to-cook) using millet as the way forward, so efforts should be directed towards the development of new product and processing technologies to overcoming the constraints and enhances the shelf life of millets and their value-added products. Sustainability and reliability of the model for millet promotion, policy changes on the part of the responsible central/state governments are required. These should include the inclusion of millets in the PDS, procurement of millets, subsidies for millet cultivation, special funds for R&D, and promotional and awareness programs on millet promotion and their nutritional importance, encouraging startup entrepreneurs for value addition in millets and subsidies for processing machinery development. Of course, the role of the private sector is equally important, especially with increased health awareness among the urban population.

The value chain project, which focused on sorghum, is replicable with other millets. With the various components of the value chain replicable beyond India, millet interventions are possible in other countries; especially relevant as lifestyle diseases have become an issue irrespective of country, leading to consumers searching for alternative cereals in their diet. Since millet, especially sorghum, products are new to consumers in nontraditional areas, firstly commercialization needs to go side-by-side with their promotion as health foods. Secondly, the products need to be available on the shelves of as many stores as possible in the targeted areas. Thirdly, they need to be sold by a reputable firm/s for positive perception and word-of-mouth promotion. In view of the above, IIMR has developed a protocol to shortlist the potential entrepreneurs based on their credibility. Technology incubation will allow them to produce the food products using the facility provided by the lead center for market testing so that once it is successful, the firm itself could beneficially start their own production and market it under its own brand, while IIMR retains the intellectual property rights to the technologies. The product standards need to develop to export millet products through the APEDA.

References

Gopalan C, Rama SBV, Balasubramanian SC (1989) Nutritive value of Indian foods. National Institute of Nutrition, Indian Council of Medical Research, Hyderabad, pp 50–59

McKay A, Morrissey O, Vaillant C (1997) Trade liberalisation and agricultural supply response: issues and some lessons. Eur J Dev Res 9:129–147

Rao BD (2019) Sorghum value chain for food and fodder security. In: Breeding sorghum for diverse end uses. Woodhead Publishing, Cambridge, pp 409–419

Rao BD, Patil JV, Rajendraprasad MP, Reddy KN, Devi K, Sriharsha B, Kachui N (2010) Impact of innovations in value chain on sorghum farmers. Agric Econ Res Rev 23:419–426

Rao BD, Kulkarni DB, Kavitha C (2018) Study on evaluation of starch, dietary fiber and mineral composition of cookies developed from 12 sorghum cultivars. Food Chem 238:82–86

Saleh ASM, Zhang Q, Chen J, She Q (2013) Millet grains: nutritional quality, processing, and potential health benefits. Compr Rev Food Sci Food Saf 12:281–295

Role of Nutrihub Incubation for the Development of Business Opportunities in Millets: An Indian Scenario

21

B. Dayakar Rao and Sri Devi Nune

Abstract

Innovation has become a buzzword in these recent times and it has become important to have innovative, scalable, and viable startups and in fact for attracting investments into the firm too. In this ever-changing business world and especially in a situation like COVID, robust and scalable business ideas and business models have occupied a crucial role. So, in order to bring in this innovation and for making scalable business models, the support of incubation is very huge. Needless to say that incubation facility provides the required cushion for the startups to save on fixed expenses like office space, manufacturing setup, and so on. Hence, they only have to spend on operational expenses. Also, through the experienced business incubation team with the incubator, they can have a focused approach to their businesses. This chapter deals on the concept of generic incubation and different business and incubation opportunities that have opened up during the development of value chain on millets by ICAR-Indian Institute of Millets Research. Incubator is branded as Nutrihub engaged in handholding the startups, through enabling various services maturing them to be graduated from the Technology Business Incubator and establishing a self-sustaining business entity. Hence, incubation forms a major role in driving this. Further, in this chapter investment ventures for startups and in a detailed way talks about the Journey of Nutrihub.

Keywords

Nutrihub · Millet start-ups · Business and opportunity

B. Dayakar Rao (✉) · S. D. Nune
ICAR-Indian Institute of Millets Research, Hyderabad, Andhra Pradesh, India
e-mail: dayakar@millets.res.in

A. Kumar et al. (eds.), *Millets and Millet Technology*,
https://doi.org/10.1007/978-981-16-0676-2_21

21.1 Introduction

Millets are termed as nutri-cereals and were staples, meeting food security to millions till the Green Revolution. The negative effects of the green revolution led to decline of millets as the policies were pro-rice and wheat, which as food security was the main focus at that point of history. The ill-effects of lack of food diversity led to nutritional insecurity. Due to gluten in rice and wheat many products, which are convenient to prepare and are tastier, are made available. This led to an even further decline of millets. In modern India, we are observing more lifestyle diseases and people suffering from stress. Thus, the revival of millets has come to the forefront. ICAR-IIMR led consortium efforts to establish a value chain on millets in order to create demand for almost a decade, initiated this revival by coming up with value-added technologies and converting them for commercialization (Dayakar Rao et al. 2011). In order to create a spiral effect of their efforts created a plethora of business opportunities in the millets with USP in place (Dayakar et al. 2014). Technology licensing to entrepreneurship has evolved into an ecosystem of a pipeline of startups. These startups needed handholding mechanism to further their journey. Thus, IIMR ventured into incubation, starting with Agri-Business Incubation (ABI) of ICAR and then moving into Technology Business Incubation (TBI) funded by the Department of Science and Technology, Government of India. The first part of this chapter deals with Incubation, different types of incubation, and investment path for a startup. The second part deals with *Nutrihub's* success story.

21.2 Efforts of IIMR in Reviving and Developing of Millets Value Chain

ICAR-Indian Institute of Millets Research (IIMR) efforts led to developing the value chain of millets. They engaged multiple stakeholders like Britannia and ITC to take this initiative forward. Nutritional labeling and sensitization on health benefits of millets is one more facet wherein they tried to increase awareness among consumers. They felt until and unless they provide millets in the form suitable for consumption to the consumers both in the ways of texture and taste and also in the form which can be easily cooked, millets cannot be brought into the limelight again. Hence, they developed value-added processing technologies of millet foods for various consumer segments (Ali et al. 2003). Diversification of processing technologies through innovative technologies (as they are devoid of gluten and makes processing a challenging proposition) led to the development of value-added products such as millet-based pasta, vermicelli, flakes, extruded snacks, puffing, etc., which are either ready-to-eat or ready-to-cook products (Dayakar et al. 2015). Breaking the jinx of inconvenience which was the primary cause of decline in consumption of the millets in the past. Through their continuous efforts, they have come up with 60+ value-added technologies and these are also available for commercialization. Machinery development and their scaling up for encouraging the commercial production both at farmgate and meeting the health and convenience food markets.

Not only this, in order to boost further demand and to take millets closer to the consumers, they have come up with their own brand known as "*Eatrite.*" This brand was the first of its kind to make the value-added products in millets available to the consumer. Taking cues from this, now there are more than 400+ millet brands available in the market. In order to further this movement of millets, they have come up with Nutrihub, their hub for Millet entrepreneurship development. Through Nutrihub, they are handholding more than 125+ startups. They also provide them access to funding through RKVY-RAFTAAR. Through this, the startups can avail grant of up to 5.0 lakhs and 25.0 lakh rupees. They also have mentors from known organizations like NITI AAYOG. Also, they have a well-developed R&D lab through which startups can do their trials and testing, along with their new product development.

Not only this they are also working on the policy front to make millets a part of different public-funded programs like PDS, Mid-Day Meal Scheme, and so on (Hallgren et al. 1992). Through the efforts of IIMR, 2018 has been declared as the National year of millets; 2023 is going to be declared as the International year of millets very soon by UN. Thus, they are working on interventions at each stage of the value chain of millets and thus working towards a collective goal of doubling the farmer's income by 2022. The following are the key highlights of their efforts.

- A decade of experience of value chain development on millets.
- Multiple stakeholders engagement for promotions.
- Nutritional labeling and sensitization on health benefits of millets.
- Value-added processing technologies to make a variety of millet foods for various consumer segments.
- Technology licensing.
- Machinery development and their scaling up for encouraging the commercial production both at farmgate and meeting the health and convenience foods markets.
- Working for mainstreaming of millets.

21.3 Business Opportunities Emerged Out of This Value Chain Development

See Fig. 21.1.

21.4 Types of Business Model Drivers

Before we dwell into details of each of the above opportunities, let us have a look at the different business model drivers and their objectives.

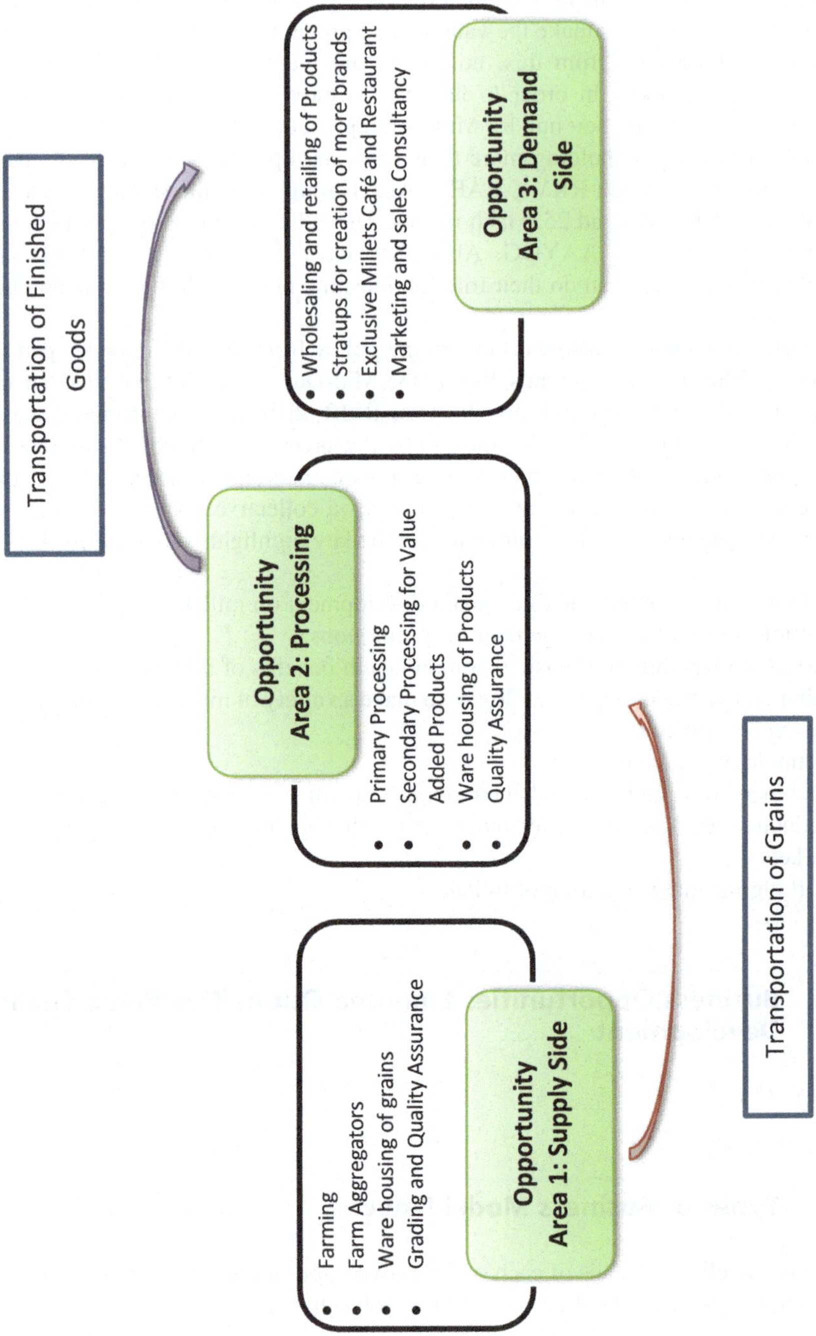

Fig. 21.1 Business opportunities in millets value chain

Driver	Type	Key objectives
If the driver is a smallholder	Then the type is producer-driven	New markets and higher market Price
If the driver is (a) Large farmers (or) (b) Processors (or) (c) Exporters (or) (d) Retailers (or)	Then the types is buyer-driven	Stabilize market position, increase supply, and assure that supply
If the driver is (a) Traders (or) (b) Wholesalers (or) (c) NGOs (or) (d) Other support agencies	Then the type is intermediary-driven	Supply more discerning customers

Source: Vorley et al. (2009)

Basically in millets we are more of producer driven and needs a lot of efforts to further into other business models.

21.5 Opportunity Areas Explained

21.5.1 Farming

Cultivars which are suitable for specific value-added products which can be made out of millets are developed through the extensive work of IIMR. This gives farmers develop partnerships with those firms who are into manufacturing of these products, in the lines of contract farming. This is something similar to the Tomato farmers making partnership with Pepsi for their tomato flavored chips. Hence, this opened up an avenue for those who wanted to come into agriculture.

21.5.2 Farm Aggregators

Other than direct-to-consumer approaches and institutional relationships, farm aggregation is one way which helps farmer's access local markets and markets outside their areas. Farmers product will fetch much higher bargaining power and price than when they sell their products individually.

21.5.3 Warehouses for Storage of Grains

This is a very key link in the marketing and ensures that the grains are in proper form for further processes. Storage also forms a major role in the continuous supply of millets to the market and for further value addition processes. Not only this storage also helps in the stabilization of prices by adjusting demand and supply.

21.5.4 Grading and Quality Assurance

The grades and standards in millets are still in a nascent stage and this opens up a very promising area for those who want to venture into this.

21.5.5 Primary Processing

The initial stages of cleaning, grading, dehulling, and dehusking come under this stage. Farmers and farmer groups also can establish primary processing thus bringing farm gate level processing. This fetches more value to the products of the farmers.

21.5.6 Secondary Processing

One of the major reasons for the lowest consumption of millets is their ease of cooking and taste. In order to address these issues, there are a lot of technologies which are developed in the market to produce different value-added products out of millets like pasta, vermicelli, atta, noodles, cookies, puffs, namkeen, and so on.

21.5.7 Marketing and Sales Consultancy

In order to increase the demand for these products, marketing plays a major role and majority of the startups keep struggling to get this support. Hence, this is also one more upcoming area for setting up a business.

21.5.8 Transportation

Similar to warehousing, transportation also plays a crucial role in connecting the farmers to the customers. As we see many of the millet farmers are still producing at subsistence level and in small quantities. Hence, transportation plays a major role in the millets value chain. Aggregation of transport is also one more avenue at this stage.

21.6 Future Business Prospects

Few of the future business prospects are

21.6.1 Biofortification

With nutrition taking the top seat during this COVID era, fortification of millets would be an interesting area to explore. Fortification with other minerals and making products out of them would be an upcoming field in this area.

21.6.2 Beverages

Making beer and dairy products out of millets is still an area to catch up with the Indian market. Millet Milk, Millet Curd, Millet Icecreams, Millet Cool Drinks are a few to name in the plethora of options available to explore here.

21.6.3 Biofuels

India is the largest producer of millets. If we can make fuel out of millets, then our dependency on other countries will drastically come down. Making biofuel out of millets especially Sorghum is another untouched area. Also, making jaggery and other products from Sorghum can be explored further.

21.6.4 Plant-Based Protein Meat

India being a vegetarian prominent country, this is a new area which is opening up for coming up with vegetarian protein alternatives for meat. Millets, due to their goodness of nutrition can offer a promising opportunity in this.

21.7 Startup's Life cycle

As can be seen above, there are a lot of opportunities which emerged out of millets value chain. For any startup, before we examine the importance of Incubation, let us see what are the different stages in a startup lifecycle (Fig. 21.2). As it can be seen throughout this lifecycle, any startup passes through the following important phases which decided its make or break:

1. Validating the idea of Product or Service. This happens majorly at two stages of the startup via Idea stage and Prototype/POC stage.
2. Once the product reaches POC stage, next is to concentrate on its product development and come up with minimum viable product. This is followed by test marketing the product, which is the Go-to-Market stage of the startup.
3. Post the product is ready and is test-marketed, it is ready for commercialization. Startups which are in growth or acceleration phase are usually in this phase.

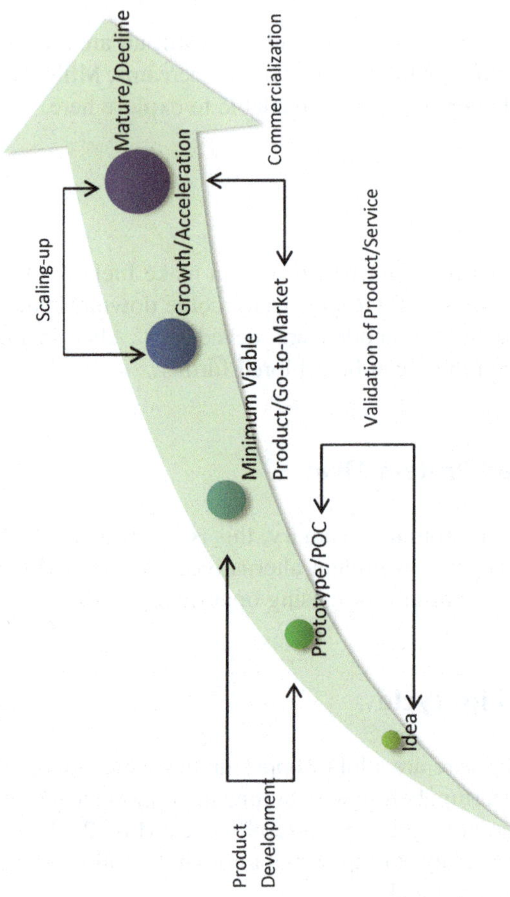

Fig. 21.2 Stages a startup go through

4. After this phase, the startups usually go for scaling up their production and reach a saturation phase where they start to decline.

21.8 Fundraising Stages of a Startup

As can be observed in the evolution of a startup, the startup will have different funding requirements. Any startup typically would go through the following funding rounds:

1. *Boot Strapping*: This is not an actual fundraising as the funds are put into the system by the startup founders only. This is the prevalent round in the prototype/ POC stages of the startup before they come out with the actual product and test their market.
2. *Funding from Friends and Family*: This is more informal route which founders will take to raise funds, after investing their owe money into the startup. Difficulty at this stage is the valuation of the company, as no professional advisory is taken in arriving at the valuation of the company.
3. *Angel/Angel Syndicates Investment*: This is the first stage of asking for a professional investment round of funds. Angel investors may be a single person or a group of persons who are known as Angel Syndicates. These are early-stage investors and therefore expect a fair amount of equity for an early bet on the Company, higher risk-return ratio, and continuous dilution at each round of fundraising.
4. *Seed Funding*: This funding is for those who have proved their idea and potential viability in the market. This is spearheaded by professional investment firms or early-stage venture capitalists. This stage is also sometimes referred to as Pre-Series A funding.
5. *Series A Round*: Venture capitalists usually participate in this stage. The company may not be having profits at this stage but by seeing its metrics and due diligence is done to arrive at the valuation of the company. This money is pumped into the startup to negate the cashflow and accelerate the growth of the startup. Usually, money or stock is put in this stage as a return.
   Here we should follow 15:5:1 ratio in getting in touch with the investors. If we pitch for 15 Investors, there will be 5 Investors who might show interest in the proposal and 1 Investor will finally put in the money into the venture
6. *Series B/C and Later Rounds*: Series A is followed by Series B and further rounds. Lot of VCs pump in their money for the acceleration of the company.
7. *Mezzanine and Bridge Financing*: Majority of the startups' journey would be towards IPO or get acquired by any large corporate or they look for some management buyouts. To get this, Company tap into Mezzanine or bridge financing, which is a short-term debt to support these business opportunities while preparing for IPO, acquisitions, management buyouts, or leveraged buyouts. Such debt is often for 6–12 months. Sometimes bridge funding is raised between seed funding and Series A funding for equity exchange to offset the

negative cash flow or finance the working capital requirement till series A funding.

Finally, when Company hits the IPO or goes for Acquisition/Buyouts, investors actually earn a return on their investment by redeeming their equity. This is an exit point for founders and investors but before that startup has to hustle for various funding rounds.

21.9 Renowned Investors in India

Majorly the following are the ones who invest in seed funding or before stages of a startup:

1. Friends/Family
2. Crowdfunding
3. Early Stage Venture Funds (VCs) like 500 Startups, Y Combinator
4. Angel Investors like AngelPad

The following are the potential investors for Series A funding stage:

1. Accelerators
2. Super Angel Investors, and
3. Venture Capitalists

Potential Investors for series B and beyond rounds are:

1. Venture Capitalists
2. Late Stage VCs
3. Private Equity Firms
4. Hedge Funds, and
5. Banks

The most popular investors in India are:

(a) 50k Ventures—They back technology startups at the idea and early stage of their lifecycle.
(b) Aarin Capital—They invest in early and mid-stage startups in MedTech, FinTech, BioTech, and so on. They are specialized in direct and fund of funds investment.
(c) Aavishkar Venture Capital—It mainly invest in those startups which provide solutions for the rural or semi-urban community. This specializes in early-stage and expansion capital.
(d) Calcutta Angels—This platform has been created for providing angel investments, mentorship, and guidance to the eligible startups in East India.

(e) Carpediem Advisors—This is a private equity firm specializing in buyouts, investments in small-sized and medium-sized businesses.
(f) Endiya Partners—This investor specializes in investing in idea-stage startups and those businesses in technology, consumer, and healthcare sectors. This venture capital firm partners with entrepreneurs in building startups.
(g) Hyderabad Angels—This venture capital invests in agriculture, retails, cleantech, healthcare, telecom, and education sectors in India.
(h) Indian Angel Network—This is a Delhi based venture capital firm that brings together highly successful CEOs, Entrepreneurs, and Investors in India and across the world for investing in startups or early-stage ventures.
(i) Kalaari Capital—This venture capital firm is in existence since 2006 and has invested in more than 127 companies till date. It majorly invests in technology-related startups.

There are many more investors which are spread across sectors and who are actively investing in startups.

21.10 Incubation Support to Startups

If we observe, throughout the lifecycle of startups, there needs to vary drastically from one stage to another stage. In order to handle these variable needs of startups came the concept of Incubation and Incubators.

Incubators are on rise in recent days due to the facilities they offer to startups and businesses with growth mindset. Incubators typically offer all the facilities that are required by a startup, which helps them lower their costs than figuring out and spending from their own pocket. Incubators basically eliminate the inherent risks that are associated with any venture.

Startups when given right direction and proper orientation reach success on a massive scale very easily. Many unicorns and successful startups have spurred their growth by getting incubated at incubators for spurring their growth. The following are the main reasons, why any startup would go for incubation.

21.10.1 Networking

During the initial phase of a startup, a good product or service is very crucial, for it to become sustainable. But in order to spur its growth, networking is very essential. Networking is nothing but building linkages with other startup founders, leaders, and professionals within your industry. If you want to build this network, Incubators provides you the right platform. Usually, any incubator houses dozens of startups and companies that are looking to build their businesses and grow them. This makes it easier for the startup to exchange information and build relationships with the ecosystem in your industry. Some of these networking opportunities will result in Joint ventures, Asset sales, Strategic Partnerships, and so on.

21.10.2 Accessibility of Resources

Every startup would need some administrative support and internet access. Being associated with an incubator will help the startup to get access to these resources easily including software required for the startup and business tools, which otherwise were inaccessible to the startups.

Majority of these also host regular workshops for the numerous startup partners at the incubator. These workshops often focus on topics such as future forecasting, business fundamentals, how to obtain funding, legal structures, and quick prototyping. All of these workshops are typically provided in one location, which means that you will not need to travel every time you think a workshop will be beneficial for your company or team.

21.10.3 Partnerships

When you are associated with an incubator, there are high chances of you getting partnerships with other startups that might offer you a competitive advantage over others in the industry. Sometimes incubators offer services to a single industry or a group of industries in a similar field thus making it easy for the startups to develop partnerships.

21.10.4 Expensive and Sophisticated Equipment at Affordable Price

With the limited funds what a startup will be having, it will be quite difficult for them to meet the capital expenses for their production/machinery. Incubators offer these facilities at quite affordable and competitive prices. In addition to this, you will also receive training on these during the startup's association with the incubator.

21.10.5 Low-Cost Space and Flexibility

Renting an office space of their own to run a startup's business will quite huge on their pocket. Incubator provides coworking spaces at a much cheaper and affordable price. Some incubators also provide options to accommodate the needs of startups and to adjust their budget. For example, at Nutrihub, ICAR-IIMR, they offer incubator space in three different options via Physical, Physical Shared, and Virtual Incubation. This gives plenty of choices for the startups to choose from.

21.10.6 Mentorship by Industry Leaders and Eminent Personalities

Every incubator will have a mentor network which will have eminent personalities from industry and renowned institutes as members of this group. Being provided with advice from individuals who have already built successful businesses can prove to be invaluable when you are looking to grow your own company. Learning from the experiences that other successful entrepreneurs have had will help you avoid some of the more common mistakes that a startup can make.

21.10.7 Access to Funding

Majority of the incubators build up investor relationships and relationships with funding agencies like banks thus facilitating the funding requirement of the startups. Also, some of the incubators will be running the Government grant schemes thus taking the fruits of these grants to reach the right people. For example, NIDHI PRAYAS, NIDHI-SSS from DST, RAFTAAR from RKVY are few such programs which are facilitated through competent incubators and grants are disbursed to the qualified startups post pitching.

21.10.8 Business Facilitation

Some incubators will have business teams to facilitate the business plan preparation of the startups, offer advice in their business model and connect them with the relevant personnel for their mentoring advice. They basically help the startups to have focus on their business thus achieving their business goals and objectives that they have set for their growth.

21.10.9 Accelerates the Growth of the Startup

The resources at the incubator help in the exponential growth of the startup. When you are attempting to expand your horizons to an industry which has pin throat competition, the networking, funding, and mentorship opportunities that you can gain access to through an incubator can pay dividends.

In nutshell, an incubator provides 360° support to the startup which is very much essential for them to stay afloat in the business and to start their journey towards their growth phase (Fig. 21.3).

21.11 Different Types of Incubators

There are four different types of incubators which we find majorly in India. They are:

1. *Government Backed Incubators*: These incubators are funded by the state or central Government. One such incubator is We-Hub, which is an incubator for handholding women entrepreneurs and is funded by the Government of Telangana.

Fig. 21.3 Advantages of an incubator

2. *University Backed Incubators*: These incubators are housed by renowned universities of India. IIM Ahmedabad, IIM Bangalore, IIT Bombay, IIT Kharagpur, and so on such universities are housing these incubators and are providing world-class facilities to the startups.
3. *Corporate Backed Incubators*: There are certain incubators which are backed by corporate giants. Mahindra E-Cole is one such incubator in India.
4. *Private Incubators*: These are backed by private persons. Wadhwani centre for Entrepreneurship Development is one such incubator facility.

Also, incubators are present based on the sector to which they cater to:

(a) *AgriTech/FoodTech Incubators*—Those incubators which support the startups which are into agriculture and innovations in this space comes under this. AIP of ICRISAT is an example of this incubator type.

(b) *MedTech Incubators*—These are the incubators which provide handholding support to those startups which are offering solutions in the healthcare domain. AIC-CCMB in Hyderabad is one such type.
(c) *DeepTech Incubators*—These are the incubators which houses those startups which work on AI, Machine Learning, Robotics, and so on. CIIE of IIM Ahmedabad is one such incubator.
(d) *FinTech Incubators*—India is becoming a hub for FinTech in recent smart interventions like PhonePe, Paytm. Start Tank by Paypal is one of the incubators which is concentrating on FintechStartups.

Nutrihub belongs to the category of AgriTech/FoodTech Incubators.

21.12 Nutrihub's Journey of Success

ICAR-IIMR has the mandate of R&D on eight different millets. It has a vision of doubling farmer's income by 2022. The consequences of the Green revolution led to a decline in Millets, our ancient grains, from the consumer's diet and made consumers shift to more convenient forms of products made out of Rice and Wheat. It was not a cakewalk after this for ICAR-IIMR to revive millets into consumer's meal. IIMR in order to create demand for millets, healthy and nutritious technologies should be developed, in convenient forms for consumers to cook/eat can be made. The journey started with the development of machinery suitable for millets using existing machinery of Rice and Wheat.

Through a decade long work, now IIMR is able to develop more than 60+ value-added technologies, which are available for commercialization. Till now more than 75 companies took more than 85+ technologies from IIMR (Table 21.1).

Table 21.1 Commercialization details of IIMR

	Startups turnover in Cr		Startups revenue in Cr		IIMR revenue in Cr
Year wise	Physical incubates	Virtual incubates	Physical incubates	Virtual incubates	Commercialization[a]
2014–2015	–	0.8	–	0.24	0.3
2015–2016	–	1.5	–	0.45	0.45
2016–2017	–	3.0	–	0.90	0.62
2017–2018	7.0	15.0	2.1	4.50	0.97
2018–2019	9.2	18.00	3.3	5.30	0.99
2019–2020	15.3	22.4	7.20	9.70	1.20

[a]Through technology licensing and supporting Agri-business incubation

However, IIMR could not get much-expected results out of this commercialization, in taking things closer to consumers. Hence, it came up with its own brand in millets known as "*Eatrite.*" It translates to "Eating it Right" thus prompting the consumers that they are eating better products in terms of nutrition and complex carbohydrates that have health benefits to reduce diabetes, Cardiovascular diseases, Cancer, etc. (FAO 1995; Ratnavathi and Patil 2013). This is the first-ever brand among all the ICAR institutes.

This step led to taking millet value-added products closer to the consumers. Taking cues from this, IIMR ventured into Entrepreneurship development, to drive demand for these products at a faster pace (Ogidi and Abah 2012). Thus, came into existence their incubator which is known as "Nutrihub," in the year 2016. The name has been taken from the word "Nutri-Cereals," which is another name for Millets. Hence, this incubator is a hub for those startups which are working on Millets.

Nutrihub, started its journey as Agri-Business Incubator, under ICAR. In 2017, it bagged the prestigious NIDHI-TBI project from DST thus making it one of the 152 TBIs in India. Through this project, infrastructural space and physical space were developed, which the startups can avail and make use of, instead of spending money on their own premises. It has developed physical space, for housing 25 startups. Fifteen startups got graduated in their first cohort and a new batch is onboarded in the incumbent cohort. As a part of their incubation support, it provides the following facilities to the incubates:

1. *Technology Consultancy*: The startup companies incubated avails the technologies developed at CoE, IIMR, for commercialization under their own brands.
2. *Infrastructural Support*: 60+ technologies what we have are spread across nine different processing lines, including the packaging line, which has a capacity of more than 1 tonne/h capacity.
3. *R&D Lab Support*: Any startup needs nutritional analysis, shelf life study, and R&D support for their new product development, attached to the "Centre of Excellence" lab which the startups can avail of.
4. *Domain Consultancy*: Nutrihub has backstopping of more than 50+ scientists from IIMR. Hence, the required domain expertise can be availed by startups from them being an Incubate.
5. *Mentoring Support*: Nutrihub has eminent personalities from prestigious institutes like NITI AAYOG, IIT Kharagpur, CFTRI, NIN, and so on. Incubates can avail their mentoring support, if need maybe.
6. *Business Consultancy*: Nutrihub has a dedicated support of business team for supporting startups, for taking care of any business-related support to the startups.
7. *Access to Funding*: They have Hyderabad Angels, one of the VC funding networks on their board. They are also running a program under RKVY-RAFTAAR for providing grant-in-aid support to the startups.
8. *Marketing Support*: Nutrihub helps in facilitating startups for the government programs like ICDS, Mid-Day Meal Scheme, and so on. Also, other online options of Flipkart and GeM portals are explored for onboarding startup products onto their platforms.

Nutrihub got strengthened with RKVY-RAFTAAR in the year 2018. Under this program, Nutrihub is providing handholding support to more than 125+ startups and thus creating more than 250 direct employment through these startups, till now. They cater to these startups' needs through two handholding support programs.

The first one being Agriprenuership Orientation program, known as "NEST." This is a program, wherein the startups come up with an idea and want acceleration in the product. Through this program, 2 months of rigorous subject (millets) and business knowledge training is imparted to the participants, along with an opportunity to intern with their incubates. After the completion of this training program, the participants get an opportunity to pitch for a grant-in-aid of upto 5 lakh rupees. The second program is more of market acceleration for those startups, which are already having their viable product ready. This program is known as "*NGrain*" and it is mostly a "One on One" customized training program, as per the needs of the startup. After the completion of this training program, the participants get an opportunity to pitch for a grant-in-aid of upto 25 lakh rupees.

In order to provide more networking opportunities to the startups in the ecosystem, Nutrihub actively participates in ecosystem events like "Rejig Startups" which was spearheaded by all Telangana Incubators/Accelerators. Also, it hosted some other cross-border incubation exchange programs like "Maitri" which is between India and Brazil.

Nutrihub so far has provided R&D support to more than 100 startups in bringing their products into light and conducted Nutritional studies and shelf life studies for majority of the products of the startups. Not only this, Nutrihub plays a major role in connecting its incubates to different Governmental bodies. Through the courtesy of ICAR, parent body of IIMR, all the secretaries of GoI are being supplied with millet-based products of IIMR's own brand and of its startups. Also, two of its startups M/s Boinpally's Agro Food Products Pvt. Ltd. and M/s M for Millets Pvt. Ltd. are supplying millet products in Mid-Day Meal Scheme.

A brief snapshot of the startups associated with them is given below:

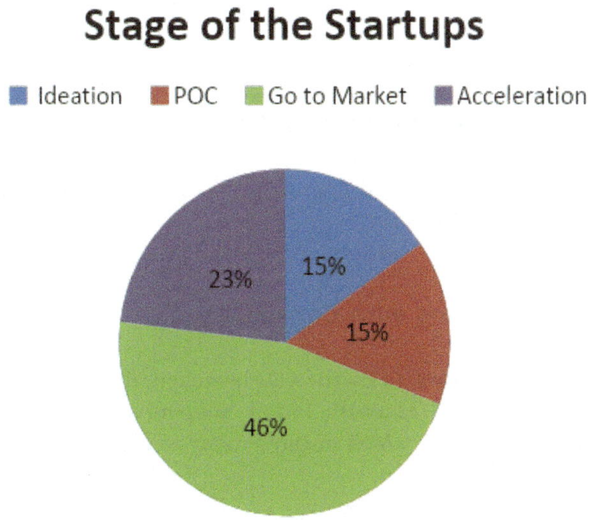

As shown above, there are some startups which have passed through their break even stage but COVID 19 is posing serious challenges for them to proceed further. All the startups across the Startup cycle are having tough times due to this pandemic. Majority of them faced working capital-related issues and policy-level interventions are very much required for making these startups progress.

21.13 Achievements of Startups So Far

Few of our startups already bagged Venture C funding from external funding agencies. M/s Innerbeing Wellness Pvt. Ltd., M/s Fountainhead Foods Pvt. Ltd., M/s M for Millets Pvt. Ltd. have received VC funding.

Also, few have carved a distinction for themselves in India by bagging awards like

(a) National Siridhanya Award for Best Industrious Startup-2019 was won by M/s Innerbeing Wellness Pvt. Ltd.
(b) Millenova bagged "Best Indian Social Enterprise" award at ISB.
(c) Eco Jivan, won the Karnataka Startup Challenge and bagged 50 lakh rupees from the Government of Karnataka.
(d) Nutrimagic won "Best Millet Product of the Year" award at Time2Leap Awards.
(e) M for Millets bagged the investment led by Omnivore and Bayer group.

In nutshell, Nutrihub is truly becoming a "one stop hub" for all Nutri-Cereals related startups by providing different facilities and linkages with the government and most importantly by providing funding support to the needy and promising startups. Thus, it is making every effort in spreading the goodness of millets to every nook and corner of the society.

Finally, we would like to conclude this article with the tagline "Eat Millets and Stay Healthy" for consumers and "Reach Nutrihub-A One Stop Hub for all Your Nutri-Cereals Related Startup needs" for Startups.

21.14 Popularization Efforts of IIMR

IIMR has advocated three pronged approaches to take millets closer to its target groups. The three pillars of this are;

21.14.1 For Popularizing with Consumers

Eatrite Brand: A brand known as Eatrite was launched, through which around 30+ products like Idly Rawas, Vermicelli, Pastas, Soup mixes, and so on are sold. This is the first brand in Millets, which inspired many to come into millets.

Branding: Post the inception of Eatrite brand, an agency is roped in to do the brand promotion through various commercials, the snippets of which are shown below.

Sales Counter: Efforts in branding will be in vain if we do not take proper steps for distribution. One such effort is the institutionalization of the sales counter at IIMR campus in Rajendranagar, Hyderabad.

Millets Exhibition Van: A vehicle was customized to keep these value-added products and it was branded as "Millet Rath" to roam through the city and popularize millets.

Packaging: The labeling of the Eatrite branded products gives nutritional information provided by NIN. NIN conducted a nutritional study of the developed products on diabetic patients and school children.

Efforts with Corporates: IIMR has signed MOUs with big corporate giants, like Britannia, 24 Mantra, and so on, to make millets as a part of their product basket. Horizontal expansion was also planned through established market players.

Millets Marathon: Millets Marathon is also one such event, which is conducted every year to sensitize fitness enthusiasts about millets' nutritional benefits. Every year around 1000 runners participate in the event in 5, 10, 21, and 32 km marathons.

21.14.2 For Popularizing with Farmers

Choupal Haats: In order to create awareness on an improved package of practices on millets cultivation, an initiative known as choupal haats was started. These haats are conducted in villages on a normal haat day or engineered specifically to attract the farmers and rural women/consumers.

Initiative with ITC: We have partnered with ITC, to encourage more farmers cultivate millets. It was initially a gigantic task to get them to sow millets but once they were shown monetary realization of their yield, slowly farmers started cultivating millets in their cultivable land.

21.14.3 For Popularizing with Entrepreneurs

Nutri-Cereals Conclave: In order to popularize millets with all the stakeholders in the ecosystem and especially with the startup community, Nutri-Cereals Conclave is being conducted every year at National level. In 2019, this conclave was a 2-day event, which was themed as "Invest in Millets for Future Nutritional Security."

Handholding Support to Mainstream Startups: In order to bring in more number of startups into millets, we are running two programs under RKVY-RAFTAAR scheme. The first program is for product acceleration, which entails the participant for a grant-in-aid of upto 5 lakh rupees post the completion of 2 months of training. The second program is for market acceleration, which entails the participant for a grant-in-aid of upto 25 lakh rupees.

21.14.4 For Popularizing with Other Stakeholders

Chefs Connect: Chefs connect is very important for grain like millets to be taking their place in the restaurant's menu and to take them closer to customers fast. As a part of this, we do make sure to take part in all chefs related events and "Reclaiming Traditions and Building the Future" is one such event of chefs of the western India culinary association.

21.15 Nutrihub Startup's Success Story

TROO GOOD

M for Millets

Brand Name: Troo Good

Founder & CEO: Mr. RAJU BHUPATI

Millets although have never been new to Indians but in the last few decades lot of alternative grains snaffled the share which once was dominated by millets. This ensued the fall of consumption of millets in the daily meal therefore contributed to a rapid declination of production. Knowing the versatile nature of millet grains and the significant rewards that it offers to the environment Troo Good, a company that was founded by Raju Bhupati and Padma in the year 2018, brought a tangible plan to bring out affordably nutritious products using millets. Taste largely drives the buying decision by most of the population, so it is a paramount factor to be addressed while making the products. Troo Good introduced the snacks by infusing a combination of millets which are very affordable and tasty. Millet Chikki with a unique formulation is the poster product of Troo Good and the company has sold 85 Million millet chikkis in the last 2 years therefore shaping the company as the front runner disseminating awareness of Millets.

Millet benefits can best be made available to the society by tying up with the government and supplying them through different diet schemes and PDS. This will eventually spiral up the consumption of millets among the consumers. Troo Good certainly understood this and developed meaningful affiliations with various Govts across India and created large-scale infrastructure to accommodate the progressive trajectory of growth.

With an annual turnover of over INR 25 crores and ARR INR 35+ crores Troo Good established two state-of-the-art factories and maintains 200+ workforce with dedicated R&D divisions driven by world-class nutritionists whose mission is to bring novel millet products at an affordable price. Troo Good while procures the raw material from the market yards but also developed stronger links with farmers and FPOs to cater better economy of scale.

21.15.1 Products Range

21.15.2 Bouquet of Few of the Startups with Nutrihub

MeGo TM
Millet Snack on the Go
Lippia TM
universal taste
DMR
FOOD PRODUCTS
Eat Millets for Healthy Life
Hapup TM
NATURALLY MADE ♥ NATURALLY LOVED
Foods
Nutrania
Magical Millets Mania
Nutri Magic
Millets & More...
Millet Break
Elite Nutri
sign of great nutrition
Nandini
FOODS
Millet Basket
impeccable innovations
Explore us. Explore success.
pkp
pidata kinda pappu

21.15.3 Products of Few of the Startups with Nutrihub

Solace
Foxtail
Cookies
MilletChef
MILLET FLOUR
Siri
MULTI MILLET
FLAKES
Ragi
COOKIES
MILLET
MUESLI
HOPE
MANGO
PORRIDGE
Ragi Cookies
Good Life's
Natural Products
ENERGY PROTEIN MALT
Natural RAAGI Protein Malt
Organic Ayush
Eco Jivan
Bisi Bele Bath
Hapup

References

Ali SZ, Meera MS, Malleshi NG (2003) Processing of sorghum and pearl millet for promoting wider utilization for food. In: Proceedings of an expert meeting. ICRISAT, Patancheru, India, pp 169–187

Dayakar Rao B, Patil JV, Nirmal Reddy K, Sriharsha B, Kachui N, Kiranmai E (2011) Millets value chain to pilot innovative agri-business approaches & farm entrepreneurial development – a success story. In: International conference on innovative approaches for agricultural knowledge management, NASC complex, New Delhi, India, 9–12 November

Dayakar RB, Patil JV, Nirmal Reddy K, Soni VK, Srivastava G (2014) Sorghum: an emerging cash crop, 1st edn. Cambridge University Press, New Delhi, India

Dayakar RB, Bhargavi G, Kalpana K, Vishala AD, Ganapathy KN, Patil JV (2015) Development and standardization of sorghum pasta using extrusion technology. J Food Sci Technol 52:6828–6833

FAO (1995) Sorghum and millets in human nutrition. Food and Nutrition Series 27. FAO, Rome, pp 23–26

Hallgren L, Rexen F, Petersen PB, Munck L (1992) Industrial utilization of whole crop sorghum for food and industry. In: Proceedings of the international workshop on policy, practice, and potential relation to uses of sorghum and millets, 8–12 February 1988. ICRISAT, Bulawayo, Zimbabwe, pp 121–130

Ogidi AE, Abah DA (2012) The impact of sorghum value chain on enterprise development: a holistic diagnosis of some actors in Benue State, Nigeria. Int J Agric 4:79–92

Ratnavathi CV, Patil JV (2013) Sorghum utilization as food. J Nutr Food Sci 4:42–47

Vorley B, Lundy M, MacGregor J (2009) Business models that are inclusive of small farmers. In: da Silva CA, Baker D, Shepherd AW, Jenane C (eds) Agroindustries for development. FAO, UNIDO, CABI

The manufacturer's authorised representative in the EU is Springer Nature Customer Service Centre GmbH, Europaplatz 3, 69115 Heidelberg, Germany. If you have any concerns regarding our products, please contact ProductSafety@springernature.com

Printed and bound by CPI Group (UK) Ltd, Croydon, CR0 4YY
30/09/2026
02225330-0015